Morphological Evolution, Aptations, Homoplasies, Constraints and Evolutionary Trends

Catfishes as a Case Study on General Phylogeny and Macroevolution

Morphological Evolution, Aptations, Homoplasies, Constraints and Evolutionary Trends

Catfishes as a Case Study on General Phylogeny and Macroevolution

Rui Diogo
University of Liege
Liege, Belgium

CRC Press
Taylor & Francis Group
Boca Raton London New York

CRC Press is an imprint of the
Taylor & Francis Group, an **informa** business
A SCIENCE PUBLISHERS BOOK

CRC Press
Taylor & Francis Group
6000 Broken Sound Parkway NW, Suite 300
Boca Raton, FL 33487-2742

First issued in hardback 2019

ISBN-13: 978-1-57808-291-9 (hbk)

Library of Congress Cataloging-in-Publication Data

Diogo, Rui.
Morphological evolution, aptations, homoplasies, constraints and evolutionary trends: catfishes as a case study on general phylogeny and macroevolution/Rui Diogo.
p. cm.
Includes bibliographical references and index.
ISBN 1-57808-291-9
1. Catfishes--Phylogeny. 2. Macroevolution. I. Title
QL637.9.S5D56 2004
597'.49--dc22 2004056834

Visit the Taylor & Francis Web site at
http://www.taylorandfrancis.com

and the CRC Press Web site at
http://www.crcpress.com

Preface

The catfishes, or Siluriformes, are included in the superorder Ostariophysi, a major group of teleosts including, apart the Siluriformes, the Gonorynchiformes, the Cypriniformes, the Characiformes and the Gymnotiformes. With their 34 families, about 437 genera and more than 2700 species, catfishes represent about one third of all freshwater fishes and are one of the most diverse and economically important groups of fresh and brackish fishes in the world. They have a particularly wide and complex geographical distribution, being found in North, Central and South America, Africa, Eurasia, South-East Asia, Japan and Australia, with fossil catfishes having even being reported in Antarctica.

The amazing diversity and complexity of order Siluriformes renders very difficult the study of higher level phylogeny and thus general evolution of this group as a whole. As highlighted by De Pinna (1998), in such a diverse and complex group, historical processes such as phylogenetic convergences, parallel adaptations to similar habitats or evolutionary reversions (e.g. in subterranean catfishes) may expectedly be quite frequent. Such frequency of alterations in historical processes raise serious problems in correctly inferring the relationships among catfish groups from all continents, however, and consequently from similar habitats within different regions of the globe. Moreover, such broad general studies are further complicated by another serious drawback, related to the logistics of obtaining and including an appropriate number of representatives of all major groups of catfishes in the phylogenetic analysis. And not to be overlooked is the challenging and time-consuming task of analysing and controlling, for each of these numerous terminal taxa, the even more numerous characters necessarily obligatory to undertake such a general phylogenetic analysis in a satisfactory manner.

The logistical difficulties probably explain why in spite of long-standing scientific interest in the phylogeny of Siluriformes the only cladistic analyses to date dealing with the interfamilial relationships of the order as a whole are Mo's published work (1991) and De Pinna's unpublished thesis (1993).

Logistical difficulties might well account for neglect of muscular characters in traditional studies of catfish interrelationships as these are often not readily detectable nor readily documented and, as just mentioned, analysing and controlling a large number of such features in each taxon constitutes a rather difficult task. Moreover, museums are much more reluctant to loan the specimens necessary for complete muscular dissection than specimens for osteological observations.

Given all the obstacles mentioned above and the associated fact that the only general cladistic studies on Siluriformes are those of Mo (1991) and De Pinna (1993), it is not surprising that the vast majority of catfish researchers continue to view siluriform higher level phylogeny as a largely unresolved issue (see Alves-Gomes, 2001; Chardon et al., 2003; Gayet and Meunier, 2003; Ng, 2003; Teugels, 2003). One of the major purpose of this book is to analyse the puzzling, intriguing, higher-level interrelationships of catfishes, and to discuss the general evolution of these fishes. To achieve this purpose meant devoting a large part of the work to a cladistic analysis of the higher level phylogeny of the group. Some terminal taxa not included in previous analyses, and particularly a large number of characters traditionally excluded from those analyses, such as those concerning catfish myology, have been incorporated in this analysis, which lends particular importance to complex structures. The cladistic analysis of catfish general phylogeny and the focus on their complex structures thus pave the way for a discussion on the evolution of these complex structures within the whole order Siluriformes and on catfish general evolution. Lastly, the focus on catfish higher level phylogeny and evolution allows a broader, theoretical discussion on general phylogeny and macroevolution.

Liege, Belgium
February 2nd, 2004

RUI DIOGO

Acknowledgement

This project greatly benefited from the precious help of P. Vandewalle and M. Chardon, to whom I owe a deep debt of gratitude. Since they accepted me in the Laboratory of Functional and Evolutionary Morphology, in 1998, they have introduced me to the anatomy, functional morphology, phylogeny and systematics of Vertebrates in general and of Teleostei in particular. They have shown me how interesting are the Siluriformes, not only in what concerns the study of many different issues concerning the order itself, but also in what refers to the ample and diverse implications that the analysis of these issues has for a general discussion on theoretical Biology. They have not only spared their time continuously to discuss various points concerning the project and biological sciences in general, but also introduced me to Belgium and common life in this beautiful country. I also want to thank very, very much my laboratory's colleague, E. Parmentier. He is really one of the brightest young scientists I have ever met, and his persistence, the remarkably ability that he has to solve all the different type of problems, and the courage he has to enter and to get deep involved in all type of scientific areas were really inspiring for me.

A very special thanks to the late G. G. Teugels (Musée Royal de l'Afrique Centrale), for kindly providing several specimens studied in this work, for participating in so many projects and for discussing catfish phylogeny and systematics with me, and for allowing me to undertake bibliographical research on his Museum, which revealed to be fundamental for this work. I am also particularly thankful to R. Vari, as well as their colleagues S. Weitzman, J. Williams and S. Jewett (National Museum of Natural History), for accepting me in his amazing Museum during two academic years, for providing a large part of the specimens analyzed in this work, and for reading and commenting several papers included in this work. I am thankful to I. Doadrio (Museu Nacional de Ciencias Naturales) and F. Poyato-Ariza (Universidade Autonoma de Madrid) for also receiving me in their laboratories and for valuable scientific discussions.

I acknowledge J. Cambray (Albany Museum of Grahamstown), P. Laleyé (Université Nationale du Bénin), P. Duhamel (Muséum National D'Histoire Naturelle), W. L. Fink, D. W. Nelson and H. H. Ng (Museum of Zoology, University of Michigan), and M. Stiassny (American Museum of Natural History) for providing numerous specimens studied in this work. Thanks to R. Bills and P. Skelton (South African Institute for Aquatic Biodiversity) for the kind donation of three austroglanidid specimens to the Laboratory of Functional and Evolutionary Morphology, as well as to B. Hall, F. Galis, T. Grande, T. Abreu, C. Ferraris, J. Lundberg, M. Brito, M.M. de Pinna, P. Skelton, R. Reis, L. Soares-Porto, P. Bockmann, A. Zanata, E. Trajano, B.G. Kapoor, F. Meunier, C. Oliveira, P. Peng, M. Hardman, S. He, D. Adriaens, F. Wagemans, H. Gebhardt, M. Ebach, A. Wyss, J. Waters, B. Perez-Moreno, G. Cuny, A. Choudhury, M. Vences, S.H. Weitzman, L. Cavin, F. Santini, J.C. Briggs, L.M. Gahagan, M. Philiphe, J.G. Maisey, J. Alves-Gomes, T. Roberts, M.J.L. Stiassny, R. Winterbottom and M. Gayet, P. Lecointre and L. Taverne for their helpful criticism, advice and assistance.

I would like to thank very, very much G. Arratia. Being closely in contact with her work while editing the book "Catfishes", seeing the seriousness and the remarkable quality of her research, and hearing from her many and many comments and, sometimes, I must say, hard critiques, has been extremely important for me and for my personal formation as a scientist. Thank you so much.

My special thanks to all my friends, particularly to Pedro Brito, Gregory Piskula, Joao Malcato, Pedro Castro, Henry Evrard, Denoel Mathieu, Neo Toumbos, Thomas Stokart and particularly Claudia Oliveira. Thank you very much to you, Alejandrita Pelito Lindo. A very, very, very special thanks to my parents, Valter and Fatima, to my brothers, Hugo and Luis, and to my grandfather, Raul.

I really would also like to thank R. Primlani, who kindly invited me to publish this work as a book of the prestigious company Science Publishers Co. Thank you very much for the confidence in my work and for the several projects we have together. I hope I will not disappoint you.

A particularly thorough review of the whole book was done by Margaret Majithia, from Oxford & IBH Publishing Co., for which I am especially grateful.

Finally, thanks to all those who were involved in administering my PhD scholarship (PRAXIS XXI/BD/19533/99, Fundação para a Ciência e a Tecnologia, Portuguese Government) and other grants and/or awards received during the last years, without whom this work would really not had been possible.

Thanks to ALL of you!!!

Contents

1

Catfishes: Introduction

1.1 PHYLOGENETIC POSITION WITHIN TELEOSTEI

The catfishes, or Siluriformes, a distinctive order comprising 34 families, about 437 genera and more than 2700 species (de Pinna, 1996, 1998; Ferraris and de Pinna, 1999; Teugels, 2003), represent about one-third of all freshwater fishes known worldwide. They are one of the economically important groups of fresh and brackish fishes in the world (Teugels, 1996), as well underscored by the 'Asian-American economical catfish war' very much in the business newspapers these days (see, e.g., Roy, 2003). Catfish are characterised, as their name indicates, by the presence of barbels surrounding the snout region (see Fig. 1.1). They have a wide geographic distribution and are found in

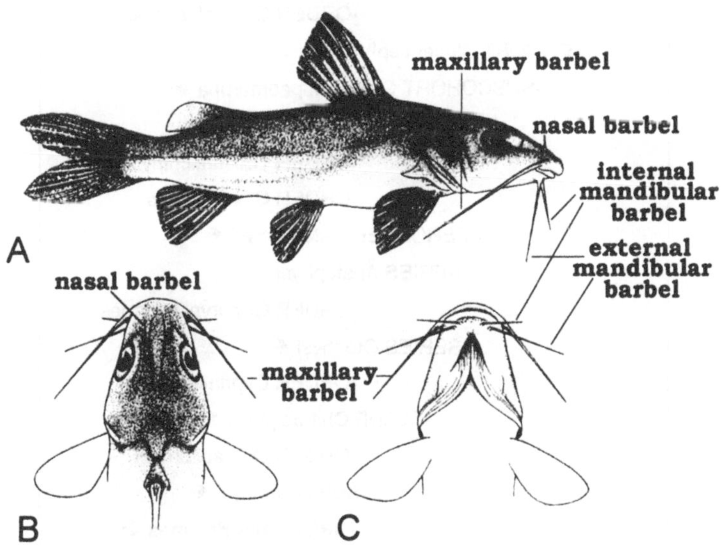

Fig. 1.1 Maxillary, mandibular and nasal barbels of a generalised catfish, *Chrysichthys polli*. A) Lateral view of the body. B) Dorsal view of the head. C) Ventral view of the head (modified from Risch, 1987).

North, Central and South America, Africa, Eurasia, South-East Asia, Japan and Australia, with fossil catfishes having been recorded even in Antarctica (Grande and Eastman, 1986).

The Siluriformes have long been included in the Ostariophysi. This superorder contains slightly more than 25% of the teleost species and about 80% of all freshwater fishes, including, apart from the Siluriformes, the Gonorynchiformes, Cypriniformes, Characiformes and Gymnotiformes (Fig. 1.2) (Nelson, 1994; Teugels, 1996) [Note: as a precautionary measure, the extraordinary and remarkable tentative new ostariophysan order Sorbininardiformes of Taverne (1999) has not been included here because more data and a greater general consensus concerning its status are needed.] The ostariophysans are recognised, as their name indicates (osteon= bone;

TELEOSTEI
-†some basal teleostan fossil
  -SUPERCOHORT Elopomopha
        -ORDER Elopiformes
        -ORDER Albuliformes
        -ORDER Anguilliformes
  -SUPERCOHORT Osteoglossocephala ←
    -COHORT Osteoglossomorpha
        -ORDER Osteoglossoidei
    -COHORT Clupeocephala ←
      -SUBCOHORT Ostarioclupeomorpha ←
        -SUPERORDER Clupeomorpha
          -ORDER †Ellimmichthyiformes
          -ORDER Clupeiformes
        -SUPERORDER Ostariophysi ←
          -SERIES Anatophysi
            -ORDER Gonorynchiformes
          -SERIES Otophysi ←
            -ORDER Cypriniformes
            -CLADE Characiphysi ←
              -ORDER Characiformes
              -CLADE Siluriphysi ←
                -ORDER Siluriformes ←
                -ORDER Gymnotiformes
      -SUBCOHORT Euteleostei

Fig. 1.2 The phylogenetic position of the Siluriformes within the Teleostei, based on Arratia's 1997 classification of the basal Teleostei.

fysis= bladder), by a specialised set of anterior vertebrae associated with the swim bladder. This specialisation is significantly less developed in members of the series Anatophysi (Gonorynchiformes), however, than in members of the other ostariophysan series, the Otophysi (Cypriniformes, Characiformes, Gymnotiformes and Siluriformes), where a true chain of Weberian ossicles interconnects the swim bladder and the labyrinth organ (see, Fink and Fink, 1981, 1996; Gayet and Chardon, 1987; Poyato-Ariza, 1996; Grande and Poyato-Ariza, 1999).

The phylogenetic position of the catfishes within the Ostariophysi was the subject of extended discussion in the past century. For much of the century, and in a large number of studies concerning the interrelationships of ostariophysans, the Siluriformes were seen as the most plesiomorphic otophysans, the Characiformes and the Gymnotiformes often considered sister-groups (see Regan, 1911a, b; Greenwood et al., 1966; Rosen and Greenwood, 1970; Roberts, 1973). Such a view was sternly called into question in 1981, however, when Fink and Fink promoted the first explicit cladistic analysis of ostariophysan phylogeny. According to their study, later reinforced by another study by these authors (Fink and Fink, 1996), the Siluriformes are, in fact, a specialised group of Otophysi constituting, together with the Gymnotiformes, the Siluriphysi (Fig. 1.2). The characiforms, often viewed as related to the gymnotiforms, were placed by Fink and Fink (1981, 1996) as the sister-group of the Siluriphysi. The cypriniforms are placed at the very base of the otophysans (Fig. 1.2).

Fink and Fink (1981: Fig. 1.1) presented several non-homoplasic synapomorphies to support the sister-group relationship between catfishes and gymnotiforms, namely: 1) dorsal portion of mesethmoid compressed and appears slender in dorsal aspect; 2) absence of intercalar; 3) reduction of eye size; 4) absence of sclerotic bones; 5) infraorbital series consisting largely or entirely of canal-bearing portions of bones; 6) supraorbital bone absent; 7) ectopterygoid greatly reduced posteriorly or absent; 8) endochondral portion of the metapterygoid triangular and appearing to be equivalent to the anterior half of the metapterygoid in primitive otophysans; 9) Opercle with triangular shape; 10) primordial ligament attaching on the anterodorsal, and not posterior tip of anguloarticular; 11) presence of only one pharyngobranchial tooth-plate; 12) articular process of intercalarium absent; 13) third neural arch with anteroventral process articulating with or fused to a dorsal prominence on the second centrum; 14) transverse process of fourth centrum with ovoid anterolateral face; 15) os suspensorium with elongate anterior horizontal process; 16) horizontal projection of pleural ribs; 17) well-developed Baudelot's ligament attaching to skull; 18) posterior fin rays offset posteriorly from anterior ray; 19) flanges for muscle attaching proximally on ventral ray halves about equal in size to those on dorsal ray halves; 20) medial radial ossification absent along entire length of both dorsal anal fin pterygophores; 21) when caudal fin present, anal fin rays articulate directly with proximal radials, and distal radials reduced; 22) presence of electroreception; 23) anterior lateral line nerve with recurrent branch innervating electroreceptors of trunk (for more details concerning these characters, as well as their phylogenetic implications, see Fink and Fink, 1981).

Some of Fink and Fink's (1981) characters have, however, been sharply criticised, in particular by Mireille Gayet, a palaeontologist who has elaborated several criticisms, some of which very pertinent and incisive (Gayet, 1986a, b; Gayet and Chardon, 1987). Other characters presented by Fink and Fink (1981) but not mentioned in Gayet's critique, such as those related with catfish suspensorium, for example, appear to be based on a seemingly erroneous interpretation of the skeletal components of this complex structure in the Siluriformes (see Diogo et al., 2001a; Diogo and Chardon, 2003). It is also noteworthy that neither of the two major independent studies which purportedly supported Fink and Fink's work (Dimmick and Larson, 1996; Arratia, 1992) in actuality has directly supported their phylogenetic hypothesis on ostariophysan interrelationships. The independent analyses performed in these two studies resulted, as a matter of fact, in a phylogenetic scenario quite different from that proposed by Fink and Fink (1981). Only a posteriori recombination of the results of these independent analyses with the latter's characters allowed a consensus with the latter's scenario. The phylogenetic analysis based on genetic data performed by Dimmick and Larson (1996) pointed out a clade formed by the Characiformes and the Gymnotiformes, with the catfishes a sister-group of this clade and the Cypriniformes the most basal Otophysi. The phylogenetic analysis based on morphological data performed by Arratia (1992) pointed out a clade formed by the Gymnotiformes and the Siluriformes, as shown by Fink and Fink, but in Arratia's study the clade was the sister-group of a clade formed by both the Characiformes and the Cypriniformes (for more details, see Arratia, 1992; Dimmick and Larson, 1996).

However, all the contra points listed above notwithstanding, it has to be acknowledged that concerning the sister-group relationship between siluriforms and gymnotiforms, the numerous synapormorphies presented by Fink and Fink (1981), the additional characters presented by these authors in 1996, and the lack of studies substantively challenging this sister-group relationship, renders Fink and Fink's scenario the most acceptable one available at the present time (see Fig. 1.2).

With respect to the phylogenetic position of the Ostariophysi within the Teleostei, the case is somewhat similar to that referring to the phylogenetic position of the catfishes within the otophysans. During much of the last century the Ostariophysi were seen as somewhat derived teleosts belonging to the Euteleostei and, hence, not closely related to the Clupeiformes (see, e.g., Lecointre, 1995, for an historical overview of this subject). However, in the last decades this view was severely disputed through independent studies undertaken by various researchers analysing different types of data. These strongly support a sister-group relationship between ostaryophysans and Clupeiformes and thus the exclusion of Ostariophysi from Euteleostei (see, e.g., Lecointre, 1995, and references therefrom; Arratia, 1997; Inoue et al., 2003; Saitoh et al., 2003).

The phylogenetic position of catfishes within ostariophysans and of the latter fishes within the teleosts can thus be resumed by the basal teleostean classification presented by Arratia (1997), and given here in Figure 1.2 as the hierarchical framework for the present work.

1.2 CATFISH FAMILIES

As mentioned above, 34 catfish families are commonly recognised today, of which two, †Hypsidoridae and †Andinichthyidae, comprise fossil remains exclusively. A brief presentation of each of these 34 families, based to a great extent in, but, of course, less detailed than, the excellent up-to-date systematic overview recently published by Teugels (2003), is given below. The presentation is complemented by Table 1.1 which overviews the principal published cladistic studies providing relevant information on the autapomorphies of each of these families, as well as on the relationships among their genera (for a more detailed presentation, see Teugels, 2003). It should be noted that this presentation does not focus on the interrelationships among the different families since this issue is taken up in the next Section.

Family Akysidae

This family includes 4 genera and about 27 species. Its monophyly was supported by the studies of Mo (1991) and de Pinna (1996) and its phylogeny studied by de Pinna (1996) (see Table 1.1). Two subfamilies are presently recognised, the Akysinae and the Parakysinae. With respect to external morphology, the akysids, small-sized (about 10 cm) stream catfishes known from southern China and South-East Asia, are recognised by the presence of four pairs of barbels, strong dorsal and pectoral spines, and a long adipose fin (absent or represented by a ridge in *Parakysis*). The head, body and fins are covered with unculiferous plaques, with some of those situated on the body greatly enlarged and arranged in longitudinal rows.

Family Amblycipitidae

The monophyly of amblycipitids, which include 3 genera and about 22 species, is well supported (see Table 1.1). Their interrelationships were studied by Chen and Lundberg (1994). These torrent catfishes occur in fresh water in southern Asia, from India to southern Japan, and are recognised externally by a small-size, robust body, depressed head, four pairs of barbels, short dorsal and pectoral spines, smooth body covered with thick skin, and an adipose fin more or less confluent with the caudal fin.

Table 1.1 List of principal cladistic studies published to date providing relevant information on the phylogenetic relationships among the genera and/or on the autapomorphies of the various catfish families. Note: a (p) after the reference of a certain study indicates that the respective study only provides information about the relationships among part of the family; the three subfamilies of the Pimelodidae, i.e., Pimelodinae, Pseudopimelodinae and Heptapterinae, are separately presented here (for explanations, see text).

Family	*Relationships among the different genera of the family*	*Autapomorphies to support monophyly of the family*
Akysidae	De Pinna, 1996	Mo, 1991; De Pinna, 1996
Amblycipitidae	Chen and Lundberg, 1994	Mo, 1991; Chen and Lundberg, 1994; De Pinna, 1996; Diogo et al., 2003b
Amphiliidae	He et al., 1999; Diogo, 2003b	Diogo, 2003b
Andinichthyidae	Family with only a single genus	Gayet, 1988
Ariidae	NA*	Mo, 1991; Oliveira et al., 2002 (but see comments in text)
Aspredinidae	De Pinna, 1998	De Pinna, 1996; Diogo et al., 2001b
Astroblepidae	Family with only a single genus	Schaefer and Lauder, 1986; Schaefer, 1990; Howes, 1983a; De Pinna, 1998
Auchenipteridae	Curran, 1989; De Pinna, 1998; Soares-Porto, 1998 (p)	Curran, 1989; De Pinna, 1998; Diogo et al., in press-a
Austroglanididae	Family with only a single genus	Mo, 1991
Bagridae	Mo, 1991; Maeda et al., 1994 (p); Ng, 2003	Mo, 1991; Diogo et al., 1999
Callichthyidae	Reis, 1998a	Schaefer, 1990; Reis, 1998a
Cetopsidae	NA	De Pinna and Vari, 1995
Chacidae	Family with only a single genus	Brown and Ferraris, 1988
Clariidae	NA	Diogo and Chardon, in press
Claroteidae	Mo, 1991	Mo, 1991
Cranoglanididae	Family with only a single genus	Diogo et al., 2002a
Diplomystidae	Arratia, 1987, 1992	Arratia, 1987, 1992
Doradidae	De Pinna, 1998	De Pinna, 1998
Erethistidae	De Pinna, 1996	De Pinna, 1996; Diogo et al., in press-b
Heptapterinae	Ferraris, 1988a (p); Lundberg et al., 1991a; Bockmann, 1994 (p)	Lundberg and McDade, 1986; Ferraris, 1988a; Lundberg et al., 1988, 1991a; De Pinna, 1998
Heteropneustidae	Family with only a single genus	Diogo and Chardon, in press
Hypsidoridae	Family with only a single genus	Grande, 1987; Arratia, 1992 (but see Mo, 1991: 195; Grande and De Pinna, 1998: 471)
Ictaluridae	Lundberg, 1975a, 1982, 1992	Grande and Lundberg, 1988; Lundberg, 1992
Loricariidae	Howes, 1983a (p); Schaefer, 1987, 1991 (p), 1998 (p); Armbruster, 1998 (p); Montoya-Burgos et al., 1997 (p), 1998	Howes, 1983a; Schaefer and Lauder, 1986, 1996; Schaefer, 1987, 1990; De Pinna, 1998
Malapteruridae	Family with only a single genus	Howes, 1985a
Mochokidae	NA*	Mo, 1991

(*Contd.*)

(Contd.)

Nematogenyidae	Family with only a single genus	Arratdia, 1992; De Pinna, 1998
Pangasiidae	Pouyaud et al., 2000 (p)	Diogo et al., in press-c
Pimelodinae	Lundberg et al., 1991b; De Pinna, 1998	Lundberg et al., 1988, 1991b; De Pinna, 1998
Plotosidae	NA*	Oliveira et al., 2001
Pseudopimelodinae	NA*	Lundberg et al., 1991a; De Pinna, 1998
Schilbidae	NA*	NA*
Scoloplacidae	Family with only a single genus	Schaefer et al., 1989; Schaefer, 1990
Siluridae	Bornbusch and Lundberg, 1989 (p); Bornbusch, 1991a (p), 1995; Howes and Fumihito, 1991 (p)	Bornbusch, 1991b; Howes and Fumihito, 1991
Sisoridae	De Pinna, 1996 (p); He, 1996 (p)	De Pinna, 1996; Diogo et al., 2002b
Trichomycteridae	De Pinna, 1988 (p), 1989ab (p), 1992, 1998; De Pinna and Starnes, 1990 (p); Costa, 1994 (p); Costa and Bockmann, 1994 (p)	De Pinna, 1989b, 1992, 1998

*NA: not available.

Family Amphiliidae

The monophyly of this family including 9 genera and about 60 species was supported (*contra* He et al., 1999: see below) by a phylogenetic study of Diogo (2003b) in which three monophyletic subfamilies were recognised: Amphiliinae, Doumeinae and Leptoglanidinae. The amphiliids, endemic to tropical African fresh waters and with a maximum total length of 195 mm but usually much smaller, have three pairs of barbels, the nasal pair being absent. Dorsal and pectoral fins are absent (except in *Leptoglanis, Trachyglanis* and *Zaireichthys*) but adipose fin present (in *Trachyglanis* it is preceded by a spine). Some specialised genera present a series of imbricate bony scutes on the body.

Family †Andinichthyidae

Family †Andinichthyidae comprises a single genus, †*Andinichthys*, established for a single fossil species, †*Andinichthys bolivianensis*, from the Maastrichtian of Tiupampa, Bolivia. The most remarkable feature of this species is the presence of a well-developed, deep 'supratemporal commissure' on the posterior region of the cranial roof (for more details, see Gayet, 1988; Arratia and Gayet, 1995; Gayet and Meunier, 2003).

Family Ariidae

This family includes 21 genera (including the genus *Ancharius:* see below) and about 153 species. The monophyly of Ariidae was supported by studies of Mo (1991) and Oliveira et al. (2002), but its interrelationships remain largely unknown (see Table 1.1). Moreover, it should be noted that the characters listed by these authors for the monophyly of Ariidae refer to all members of the group except the puzzling genus *Ancharius* from Madagascar, which is traditionally considered an ariid but has been placed by some authors in its own family (Anchariidae) or in the family Mochokidae (see Glaw and Vences, 1994; Ferraris and de Pinna, 1999; Teugels, 2003, for more details on this subject). Most ariids are marine, but some are confined to fresh waters, as, for example, precisely the members of genus *Ancharius*. They are found worldwide in tropical and subtropical regions and are externally recognised by a robust body compressed posteriorly, three pairs of barbels (nasal pair missing), dorsal and pectoral fins with a strong spine and a forked caudal fin.

Family Aspredinidae

Includes 13 genera and about 35 species. Its monophyly was supported by De Pinna (1996) and Diogo et al. (2001b) and a cladogram of its interrelationships (based on an unpublished Ph.D. thesis, by Friel) is presented in De Pinna (1998: Fig. 1.17). Three subfamilies are recognised: Bunocephalinae, Aspredininae and Hoplomyzontinae. The aspredinids are distributed throughout most of South America, with some members (the aspredinins) occurring in the sea, in brackish water and in estuaries and tidal portions of rivers. Externally, these 'ugly', peculiar catfishes are recognised by broadening of the head and anterior part of the body and a slender compressed tail. Their body usually bears knobs and sometimes a series of small plates is present along the lateral line and base of the dorsal and anal fins. There are three pairs of barbels (nasal pair missing). In addition to the three pairs of barbels, some species have numerous small barbels on the anterior part of the body. The dorsal fin is small and often spineless, adipose fin absent and the leading pectoral ray spiny.

Family Astroblepidae

The monophyly of this family comprising a single genus, *Astroblepus*, of about 54 species, is well supported (see Table 1.1). Astroblepid catfishes, known from montane regions in Panama and western South America up to Peru, have an elongated body presenting a dorsal fin with a strong spine and an adipose fin that may be present or not. Their mouth is inferior, forming a sucker disc. They possess two pairs of barbels, the maxillary and nasal ones.

Family Auchenipteridae

This large family includes 21 genera and about 107 species. Members are small to medium in size and confined to fresh waters in South America and Panama, although some are tolerant to brackish and salt water. The monophyly of the family is well supported and its interrelationships likewise relatively well studied (see Table 1.1), with two subfamilies being recognised, the Centromochlinae and the Auchenipterinae. The auchenipterids have an elongated, laterally compressed body. They usually have three pairs of barbels (nasal pair missing), except in *Ageneiosus, Tetranematichthys* and one species of *Entomocorus,* which have only a maxillary pair (sometimes even rudimentary). The dorsal fin is small, adipose fin may be present (small) or absent, anal fin may be very long and pectoral fin exhibits a strong spine.

Family Austroglanididae

This small family includes a single genus, *Austroglanis,* whose three species are known only from the Orange-Vaal and the Olifants river systems in southern Africa. Its monophyly was well supported by the study of Mo (1991). Austroglanidids are small catfishes externally recognised by the presence of three pairs of barbels (nasal pair missing), strong dorsal and pectoral spines, a rather small adipose fin positioned posteriorly on the body and some rheophilic adaptations.

Family Bagridae

The Bagridae is a large family, including 18 genera and about 144 species, which occurs in fresh waters in Central, Southern and South-East Asia except for species of genus *Bagrus,* endemic to Africa. Mo (1991), Maeda et al. (1994) and Ng (2003) have provided a cladistic account of the interrelationships of its members, and Mo's 1991 and Diogo et al.'s 1999 studies supported its monophyly (see Table 1.1). Two subfamilies are recognised, Ritinae and Bagrinae. With respect to external morphology, bagrids are recognised by a moderately elongated body, compressed posteriorly and depressed in the head region. Four pairs of barbels are present (two in *Rita*), the dorsal and pectoral fins have spines, and an adipose fin is present.

Family Callichthyidae

The callichthyids are mostly known from forest streams in a large part of South America and from Panama and Trinidad. The monophyly of this family including 8 genera and about 172 species, was supported by Schaefer (1990) and Reis (1998a), and its phylogeny studied in detail by Reis (1998a). Two subfamilies are recognised, Callichthyinae and Corydoradinae. Callichthyidae are characterised by a relatively short body covered with two rows of bony plates, up to two pairs of maxillary barbels and one pair of mental barbels, and the eventual presence of fleshy flaps. The snout is blunt, the mouth inferior, the dorsal and adipose fins present a strong spine, and in some genera the pectoral fin also has a strong spine. Some callichthyid species can practice aerial respiration and are able to move on land.

Family Cetopsidae

The monophyly of this family including 2 subfamilies, Helogenidae and Cetopsinae, 6 genera and about 22 species, was well supported in the work of De Pinna and Vari (1995), but the interrelationships of its members are not known (see Table 1.1). The cetopsids, confined to fresh waters in South America, are externally recognised by an elongated, naked body, three pairs of barbels (nasal pair missing), and dorsal and pectoral fins lacking pungent spines. Anal fin base long and fin with numerous rays; adipose fin minute or absent in adults.

Family Chacidae

Chacidae includes the single genus *Chaca* with its three species. Its monophyly was supported by Brown and Ferraris (1988). Chacids occur in fresh waters from the Ganges in India to Borneo in South-East Asia, and are characterised by a unique, long, broad and flattened head, posteriorly compressed body, three pairs of barbels (minute nasal barbel may be present), numerous cutaneous flaps or cirri on head and body, and dorsal and pectoral fins preceded by a spine.

Family Clariidae

Clariidae includes 15 genera and about 89 species. However, its monophyly seems only to be corroborated when the genus *Heteropneustes* of family Heteropneustidae is also included in it as some of its members are seemingly more closely related with *Heteropneustes* than with other clariids (for more details see Chardon, 1968; De Pinna, 1998; Diogo and Chardon, in press). Although clariids are the subject of extensive study, no cladograms of their interrelationships based on explicit cladistic analyses have been published thus far (see Table 1.1). Clariids are known as air-breathing or walking catfishes (most, but not all, present a well-developed suprabranchial organ, formed by extensions of the second and fourth epibranchials). They occur in fresh waters in Africa, extending to Syria and southern Turkey, the Indian subcontinent and in South-East Asia, with only *Clarias* being common to both continents. Concerning their external morphology, clariids are characterised by an elongated body with a long, spineless dorsal fin, long anal fin, adipose fin of moderate to large size in some genera (e.g. *Heterobranchus, Dinotopterus*), pectoral fin with a leading spine, and presence of four pairs of barbels. In some extremely elongated genera (e.g. *Channallabes, Gymnallabes*) the paired fins are reduced or absent and the dorsal and anal fins are confluent with the caudal fin.

Family Claroteidae

Mo (1991) described this family, which includes 13 genera and about 78 species, for part of the genera previously included in Bagridae. In the same work, the author provided support for its monophyly, as well as an account of its interrelationships, with two subfamilies recognised, Claroteinae and Auchenoglaninae. The external morphology of claroteids is similar to that in bagrids, with the body moderately elongated and compressed posteriorly, and the head depressed and usually presenting four pairs of barbels (three in *Auchenoglanis*). Dorsal and pectoral fins with strong spines and an adipose fin present.

Family Cranoglanididae

Diogo et al. (2002a) listed some autapomorphies to define this poorly studied small family including the single genus *Cranoglanis* and three species known from fresh waters in Yunnan Province in China and in North Vietnam (see Table 1.1). Externally, cranoglanidids are recognised by the large, inferiorly placed eyes, four pairs of barbels, strong dorsal and pectoral spines, a small posteriorly placed adipose fin and a high number of anal fin rays (35-41).

Family Diplomystidae

According to Teugels' 2003 overview the family includes a single genus, *Diplomystes*, with about six species endemic to the Austral subregion of South America and occurring in fresh waters in central and southern Chile, and from San Juan to Patagonia in Argentina, most of which are highly threatened. The phylogenetic relationships among these species were studied by Arratia (1987, 1992). Arratia also provided good support for the monophyly of the family as a whole. Diplomystids are easily recognised by the presence of more than one row of functional teeth along most of the ventral margin of the maxilla (see below the different phylogenetic interpretations of this character by different authors), exclusive presence of maxillary barbels and whole body covered with large papillae and tubercles. Dorsal and pectoral fins with strong leading spine and a relatively long adipose fin present.

Family Doradidae

This large family of South American catfishes includes 30 genera and about 71 species. Its monophyly was supported by De Pinna (1998) (based on an unpublished Ph.D. thesis by Higuchi). A cladogram of its interrelationships (also based on Higuchi's unpublished thesis) is presented in De Pinna (1998: Fig. 1.14), with three subfamilies being recognised, Platydoradinae, Astrodoradinae and Doradinae. Doradids are externally recognised by a thick-set body, generally covered laterally with a row or series of bony plates which may bear strong, spiny scutes, three pairs of barbels (nasal pair missing), dorsal and pectoral fins presenting a strong spine and usually an adipose fin.

Family Erethistidae

The monophyly of this small Asian family, which includes 6 genera and about 13 species that were previously assigned to the Sisoridae, was supported by De Pinna (1996) and Diogo et al. (in press-b). Its interrelationships were studied by De Pinna (1996), with two subfamilies being recognised, Erethistinae and Continae. Externally, erethistids are somewhat similar with sisorids, presenting four pairs of barbels, a small dorsal fin that sometimes bears a spine, and an adipose fin.

Family Heteropneustidae

The Heteropneustidae, as presently defined (see commentaries above, in the presentation of the family Clariidae), includes a single genus, *Heteropneustes*, and two species occurring in fresh waters in southern Asia from Pakistan to Thailand. Some autapomorphic features exclusive of *Heteropneustes* were described by Diogo and Chardon (in press). Externally, members of this genus are characterised by an elongated body, four pairs of barbels, short spineless dorsal fin, very long anal fin that may be confluent with the caudal fin, and a pectoral spine connected with a venom gland.

Family †Hypsidoridae

Family †Hypsidoridae includes a single genus, †*Hypsidoris*, which includes two fossil species, †*Hypsidoris farsonensis* from the Eocene Green River Formation of Wyoming and †*Hypsidoris oregonensis* from the Eocene Clarno Formation of central Oregon. The most remarkable features of these two fossil species are the presence of maxillary teeth and the highly developed coronoid process of the lower jaw (for more details and commentaries concerning these characters, see Grande and De Pinna, 1998).

Family Ictaluridae

The monophyly of this North American family including 7 genera and about 46 species was supported by studies of Grande and Lundberg (1988) and Lundberg (1992). Its phylogeny was also the subject of cladistic analyses by the latter author (Lundberg, 1975a; 1982; 1992). Ictalurids have a moderately elongated body, four pairs of barbels and an adipose fin. The dorsal (except in *Prietella*) and pectoral fins have a strong spine and, in some species of *Noturus*, the pectoral spine has a poison gland at its base. Some ictalurid species are troglobitic and share morphological features that seemingly correlate with their subterranean habitats, such as a depigmented body, absence of eyes and reduced lateral line.

Family Loricariidae

This is the largest and one of the most studied catfish families; it includes six subfamilies—Lithogeninae, Neoplecostominae, Hypoptopomatinae,

Loricariinae, Ancistrinae and Hypostominae, about 95 genera and more than 650 species. Its monophyly is well supported and the phylogenetic relationships among its members have been the subject of numerous cladistic studies (see Table 1.1). Loricariids are freshwater catfishes from Central and South America, with some genera reported from brackish water, however. Externally, loricariid catfishes are recognised by an elongated body compressed ventrally and generally covered with bony plates, a ventrally placed mouth often modified into a sucking disc and a pair of maxillary barbels (sometimes extremely reduced) connecting the upper and lower lip. Nasal and mandibular barbels absent; dorsal and pectoral fins have a flexible spine with numerous odontodes on these spines as well as in the snout region.

Family Malapteruridae

The monophyly of this African family comprising a single genus, *Malapterurus*, with 3 species was supported by Howes (1985a). Morphologically, the malapterurids or electric catfishes (they possess an electric organ of muscular origin that produces violent electric discharges up to 450 volts) are recognised by a more or less elongated cylindrical body, three pairs of barbels (nasal pair missing), absence of dorsal fin and presence of adipose fin, and spineless pectoral fin.

Family Mochokidae

The monophyly of Mochokidae, a large African family including 10 genera and about 177 species, was supported by Mo (1991). However, the interrelationships within the family are not known (see Table 1.1). The mochokids present a robust body, slightly compressed posteriorly and three pairs of barbels (nasal pair missing), with mandibulary barbels and sometimes maxillary ones also, being branched in some genera. Dorsal fin with strong spine, followed by an adipose fin; caudal fin generally forked and pectoral fins presenting a well-developed pectoral spine. In *Chiloglanis* the upper and lower lip expand and unite to form a sucker.

Family Nematogenyidae

Some unique autapomorphies have been described by Arratia (1992) and De Pinna (1998) to diagnose the single species included in this family, *Nematogenys inermis* from central and southern Chile. The members of this species, highly endangered, are externally recognised by an elongated body, three pairs of barbels (only one mandibular pair), dorsal fin situated midbody, pectoral fin with leading spine, and absence of a dorsal spine and an adipose fin.

Family Pangasiidae

Pangasiidae includes 4 genera and about 28 species occurring in southern and South-East Asia. Some autapomorphies of this family have been described by Diogo et al. (in press-c), and a cladistic analysis of part of this family was published by Pouyaud et al. (2000) (see Table 1.1). The external morphology of pangasiids consists of a laterally compressed body, short dorsal fin with one or two spines, small adipose fin, long anal fin with numerous fin rays, strong pectoral spine and two pairs of barbels (maxillary and mandibular).

Family Pimelodidae

Pimelodidae, a family of freshwater catfishes from South America, Central America, southern Mexico and the Caribbean Islands comprising 54 genera and about 312 species, is one of the largest and most diverse Neotropical groups. The external morphology of members of this family shows a high degree of diversity. The body is naked and somewhat elongated; there are three pairs of barbels (nasal pair absent) and in some genera, maxillary barbels are longer than the body. The dorsal spine is sometimes absent, adipose fin always present, and pectoral spine may/may not be present. Actually, nowadays most authors attribute the notable diversity of Pimelodidae to the fact that the family is a heterogeneous assemblage comprising 'three major well-defined monophyletic groups, currently ranked as subfamilies, namely, Pimelodinae, Heptapterinae and Pseudopimelodinae' that do not form a monophyletic 'Pimelodidae' clade (De Pinna, 1993 : 313). That is why I decided in a recent detailed overview of the phylogeny and systematics of this clade to treat these three subfamilies as separate families but retain their subfamilial names to avoid unnecessary nomenclatural complication (Table 1.1). It can be seen from this Table that all three pimelodid subfamilies have been the subject of several studies with reference to their monophyly but only Pimelodinae and Heptapterinae are relatively well studied with respect to their phylogenetic intrarelationships.

Family Plotosidae

Plotosidae includes 10 genera and about 32 species occurring in the western Pacific and the Indian Ocean from the east coast of Africa to Australia. About half of the species are confined to fresh water and occur in Australia and New Guinea. Some autapomorphies of this family have been described by Oliveira et al. (2001). The phylogenetic relationships among the plotosid genera have not been studied so far (see Table 1.1) The external morphology of plotosids consists essentially of an elongated body with a pointed tail, four pairs of barbels, two dorsal fins, and a pectoral fin with a strong spine.

Family Schilbidae

The monophyly of this large and diverse family, containing 15 genera and about 55 species of pelagic fishes from the fresh waters of Africa and southern Asia was questioned by Mo (1991). Regan (1911b) recognised three schilbid subfamilies, namely Schilbinae, Siluranodontinae and Ailiinae. Schilbids are externally recognised by a laterally compressed body, two to four pairs of barbels, dorsal and adipose fins that may/may not be present, a very long anal fin with numerous fin rays and a pectoral fin presenting a strong leading spine.

Family Scoloplacidae

Schaefer et al. (1989) and Schaefer (1990) provided strong evidence to support the monophyly of this small family including the single genus *Scoloplax* and 4 species endemic from fresh waters of Brazil, Peru and Bolivia. Scoloplacids are small catfish with a somewhat compact body, covered with two bilateral series of odontode-bearing plates. Maxillary barbels are well developed and mandibular barbels may/may not be present. Dorsal fin, as well as pectoral one, with spine; adipose fin absent.

Family Siluridae

The phylogeny of this Eurasian family including 10 genera and about 79 species is relatively well studied, and its monophyly well supported (see Table 1.1). As mentioned by Teugels (2003), the external morphology of silurids differs somewhat from that found in other catfishes. The head and body are compressed, nasal barbels absent, one and in some genera two pairs of mandibular barbels present, dorsal spine flexible, pectoral spines usually weak, anal fin very long, and adipose fin absent.

Family Sisoridae

Sisoridae is a large family including 16 genera and about 97 species. Its members are bottom-dwelling catfishes ranging in size from 20 mm to 2 m and occur in mountain rapids but also in large rivers in southern and eastern Asia, with one genus (*Glyptothorax*) also known from the Tigris-Euphrates basin in Turkey, Syria, Iraq and Iran and from the Black Sea drainage of Turkey. Monophyly of the family was supported by De Pinna (1996) and Diogo et al. (2002b) and its phylogeny was the subject of cladistic analyses by He (1996) and de Pinna (1996), with the latter author recognising two subfamilies, Sisorinae and Glyptosterninae. Sisorids present a more or less thickened leathery skin with unculiferous tubercles or plaques, four pairs of barbels, small dorsal fin, adipose fin and, in genera inhabiting fast-flowing mountain streams, a mouth developed into a sucker and a belly with special adhesive modifications.

Family Trichomycteridae

The last family to be presented, Trichomycteridae, is a very large and diverse family including 41 genera and about 183 species occurring in Costa Rica, Panama and South America, including southern Patagonia. Eight subfamilies are presently recognised—Copionodontinae, Trichogeninae, Trichomycterinae, Vandelliinae, Stegophilinae, Tridentinae, Glanapteryginae and Sarcoglanidinae. Although trichomycterids are usually cited as an example of parasitic catfishes, only a few are truly parasitic, with the vast majority being free-living predators (see de Pinna, 1998, for more details on this subject). The phylogeny of the Trichomycteridae was the subject of some cladistic analyses and the monophyly of the family is well supported (see Table 1.1). With respect to their external aspect, trichomycterids usually have an elongated naked body, two pairs of maxillary barbels and a pair of nasal barbels (mandibular barbels may/may not be present), a spineless dorsal fin, and usually also spineless pectoral fins.

1.3 HISTORICAL OVERVIEW OF HIGHER LEVEL PHYLOGENY OF CATFISHES

Several studies on the relationships between the various siluriform families have been undertaken since the catfishes were designated a formal group per se for the first time by Rafinesque in 1815. Important classic studies are, for example, Cuvier (1817), Bleeker (1852, 1862), Günther (1864), Gill (1872), Eigenmann and Eigenmann (1890), Bridge and Haddon (1893), Boulenger (1904), Goodrich (1909), Regan (1911b) and Chardon (1968).

However, the number of studies devoted to this subject has considerably increased in the last three decades due to the renewed impetus provided by the advent of cladistics in the second half of the last century (De Pinna, 1998; Diogo, 2003a). This Section provides an overview of those cladistic studies published in these last decades dealing with the higher level phylogeny of the Siluriformes, based mainly on an extensive Chapter published by the author (Diogo, 2003a) in the two-volume book "Catfishes". As mentioned in that Chapter, when the author refers to phylogenetic studies dealing with the relationships between various catfish families, he refers only to published cladistic studies. Among the several reasons for following this procedure, detailed in that chapter, one of the most important is to minimise the confusion associated with such a puzzling issue as the phylogeny of Siluriformes. In fact, despite their evident importance, many of the 'precladistic' (see De Pinna, 1998) studies dealing with catfish relationships are highly confusing, grouping certain taxa due to the presence of both plesiomorphic and highly homoplasic characters, or simply (several times) without giving a clear explanation for doing so. Therefore, to compare clearly, objectively and coherently the conclusions of all these complex, highly confusing studies using such 'precladistic' methodologies with the phylogenetic results of those studies following a cladistic methodology is extremely demanding. Furthermore, an excellent and detailed extensive historical overview of the most important 'precladistic' works on the taxonomy and classification of siluriforms (see above) was just recently published by De Pinna (1998).

Of the explicit cladistic works published to date on catfish phylogeny, the vast majority concerned the intrarelationships within a particular catfish family (see Table 1.1). The only published cladistic studies presenting original, explicit cladograms on the interfamilial relationships of either a part or the whole of the order Siluriformes are those of Howes (1983a), Grande (1987), Schaefer (1990), Mo (1991), Arratia (1992), De Pinna (1992, 1996, 1998), Lundberg (1993) and He et al. (1999). A précis of each is given below:

Howes, 1983a (Fig. 1.3)

In his Figure 1.22, Howes presents a hypothesis on loricarioid relationships (shown in Fig. 1.3 here), based on both his own observations and an unpublished thesis by Baskin, 1972 (Howes, 1983a: 341-342). According to this hypothesis, Astroblepidae and Loricariidae form a clade that is the sister-group of Scoloplacidae, with these three families constituting in turn the sister-group of Callichthyidae. Also, according to this hypothesis, the clade formed by these four families is the sister-group of Trichomycteridae, with Nematogenyidae the next sister-group and hence, the most basal of the six loricarioid families. In order to support the phylogenetic hypothesis illustrated in his Figure 1.1, Howes (1983a: 342) advanced several derived morphological characters (involving mainly the osteology, myology and external morphology of the cephalic region, the Weberian apparatus and the caudal fin) to support its different nodes. The clade including all six loricarioid families is diagnosed as 'swim bladder encapsulated, divided into separate vesicles; some part of the cranium contributing to encapsulation'. The clade including all loricaroid families excluding the Nematogenyidae is defined by 'claustrum and intercalarium lacking from Weberian vertebral series'. The characters uniting Callichthyidae, Scoloplacidae, Loricariidae and Astroblepidae are: 'posttemporal contributing to distal portion of swim-bladder capsule; derived hypural fusion pattern; low number of principal caudal fin rays'. Finally, the Scoloplacidae, Loricariidae and Astroblepidae are united due to the presence of a 'connecting bone between the 1st rib and the 2nd pterygophore'. The Loricariidae and Astroblepidae are grouped into a monophyletic clade by the presence of a retractor premaxillae muscle, the medial division of the protractor hyoideus, as well as the 'reverted lip', 'lateropterygium' and the '6 fused anterior centra'.

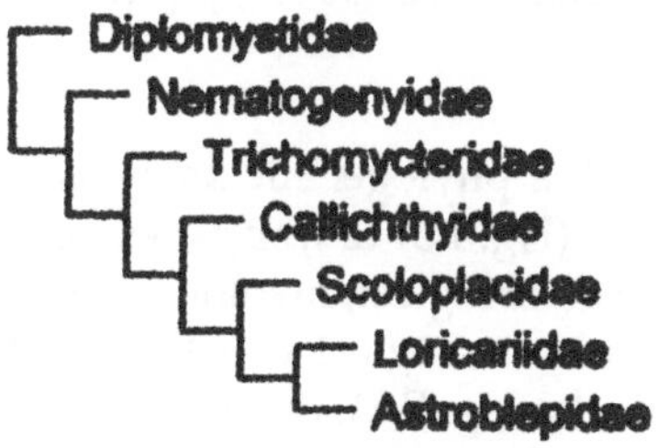

Fig. 1.3 Hypothetical relationships among the loricarioid families (the family Diplomystidae is used as the outgroup) according to Howes' 1983a paper.

However, Howes called attention to some derived characters that conflict with his phylogenetic hypothesis. The single derived character defining the clade including all non-nematogenyid loricaroids (see above), for example, conflicts with the 'lateral insertion of the dilatator operculi muscle', a derived character shared by both Nematogenyidae and Trichomycteridae (Howes, 1983a: 341-342).

Grande, 1987 (Fig. 1.4)

In 1987, Grande, based on his own osteological observations of a fossil catfish from the Eocene Green River Formation of Wyoming originally described by Lundberg and Case (1970), †*Hypsidoris farsonensis*, and comparisons with other catfishes, proposed an original hypothesis on the higher level phylogeny of Siluriformes (Fig. 1.4). According to this hypothesis the fossil catfish family †Hypsidoridae is the sister-group of all other non-diplomystid catfish families (= Siluroidea), with the clade formed by the latter plus †Hypsidoridae (= Siluroidei) being the sister-group of Diplomystidae (Fig. 1.4). Grande (1987: 48) listed three characters to diagnose the suborder Siluroidei, namely: 1) '17 or fewer principal caudal rays'; 2) 'an extension of lamellar bone over the ventral surface of the fifth centrum'; 3) 'fifth centrum joined closely to the complex centrum'. He (1987: 48) listed five characters to diagnose the superfamily Siluroidea: 1) loss 'of maxillary teeth'; 2) loss or reduction of 'distal expansion of the maxilla'; 3) loss of 'elongate mesial process of the maxilla'; 4) reduction of 'palatine either to an extremely small bone, or to a rod-shaped bone'; 5) 'long, interdigitating sutural contacts between the ceratohyal and epihyal'. Grande's 1987 hypothesis contradicts the phylogenetic hypothesis formulated in the original description of †*Hypsidoris farsonensis*, in which this fossil species was described as a member of the family Ictaluridae (Lundberg and Case, 1970: 451-456).

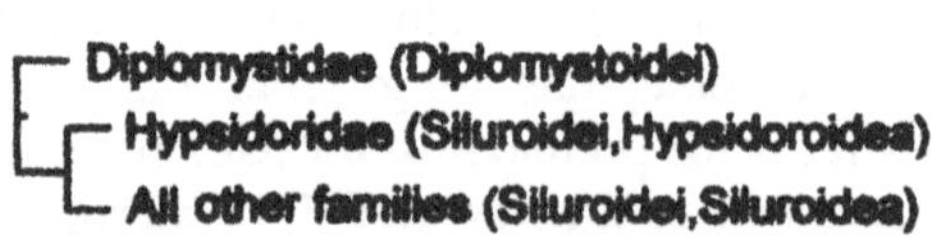

Fig. 1.4 Hypothetical relationships among the Diplomystidae, †Hypsidoridae and the other catfish families according to Grande's 1987 paper.

Schaefer, 1990 (Fig. 1.5)

Schaefer (1990) undertook a phylogenetic analysis to infer the relationships among loricarioid families (Fig. 1.5) as well as between the scoloplacid species. This study was based on osteological, myological and arthrological structures of the cephalic region and osteological structures of the axial and caudal fin. The numerical analysis (using PAUP) of a data matrix of 72 characters × 9 terminal taxa, which did not include autapomorphic characters, resulted in a single, most parsimonious cladogram with 77 steps and a Consistency Index (CI) = 0.842. Concerning the interrelationships between Loricariidae,

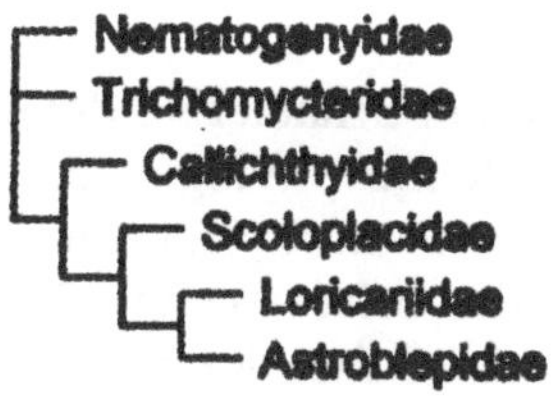

Fig. 1.5 Hypothetical relationships among the loricarioid families according to Schaefer's 1990 paper.

Astroblepidae, Scoloplacidae and Callichthyidae, this cladogram (Fig. 1.5) is similar to that of Howes (1983a). The synapomorphies listed by Schaefer (1990: 204) to diagnose the clade including these four families were: 1) 'loss of the mesethmoid lateral cornua'; 2) 'fusion of pterotic and supracleithrum'; 3) 'loss of canal in the lachrimal-antorbital'; 4) 'loss of tight attachment of premaxillae with the neurocranium'; 5) 'four or fewer branchiostegal rays'; 6) 'dorsal hypurals fused with the compound caudal centrum'; 7) 'presence of mesethmoid-maxillary ligament'; 8) 'presence of mesethmoid-premaxillary ligament'; 9) 'presence of retractor tentaculi muscle'. Synapomorphies listed by Schaefer (1990: 204) to define the clade formed by astroblepids, loricariids and scoloplacids included: 1) 'loss of open cranial fontanels'; 2) 'loss of cranial aperture which receives the cleithral dorsal process'; 3) 'presence of bifid jaw teeth'; 4) 'loss of interoperculum'; 5) 'ventrolateral shift in articulation of rib on sixth centrum'; 6) 'presence of lateral bone'; 7) 'loss of pterygoethmoid ligament'; 8) 'loss of cranial attachment'; 9) 'loss of interoperculomandibular ligament'; 10) 'shift in origin of retractor tentaculi muscle'; 11) 'bifurcation of hyohyoideus muscles'. Finally, the synapomorphies listed by Schaefer (1990: 204) to support the sister-group relationship between loricariids and astroblepids included: 1) 'loss of contact of mesethmoid posterior process with the frontals'; 2) 'presence of a hyomandibula-metapterygoid suture'; 3) 'ventromedial rotation of the mandibles'; 4) 'ankylosis or suture between sixth centrum and Weberian complex centra'; 5) 'loss of vertebral parapophyses'; 6) 'presence of expanded transverse shelf on first anal fin'; 7) 'geniohyoideus bilaterally subdivided'; 8) 'presence of an expanded oral disk'; 9) 'right and left sides of lower jaw not tightly associated at midline'; 10) 'presence of intermandibular cartilage'; 11) 'presence of juxtaposed nostrils'.

The most significant difference between Schaefer's (1990) and Howes' (1983a) studies is that Schaefer (1990) did not place the Trichomycteridae as the sister-group of the clade formed by these four families, but instead in an unresolved trichotomy including that clade, the Trichomycteridae and the Nematogenyidae (Fig. 1.5). However, it should be noted that, as stressed by Schaefer (1990: 174), such an unresolved trichotomy was not the consequence of his own phylogenetic results, but of considering the relationships between the nematogenyids, trichomycterids and the remaining loricarioids as 'unresolved *a priori*'.

Mo, 1991 (Fig. 1.6)

One year after the publication of Schaefer's 1990 paper, Mo (1991) published a study dealing mainly with the phylogenetic relationships of Bagridae, including a generic-level revision and phylogeny of the family. In addition, Mo (1991) included a somewhat brief (46 pages in a total of 216 pages plus 63 unnumbered figures) analysis of the higher level phylogeny of siluriforms. This analysis, based on osteological characters of the cephalic region, the Weberian apparatus, pectoral girdle and the various fins, but also on a few soft and/or myological characters, resulted in two considerably different cladograms. One (Fig. 1.6A), based on an 'unweighted' numerical analysis

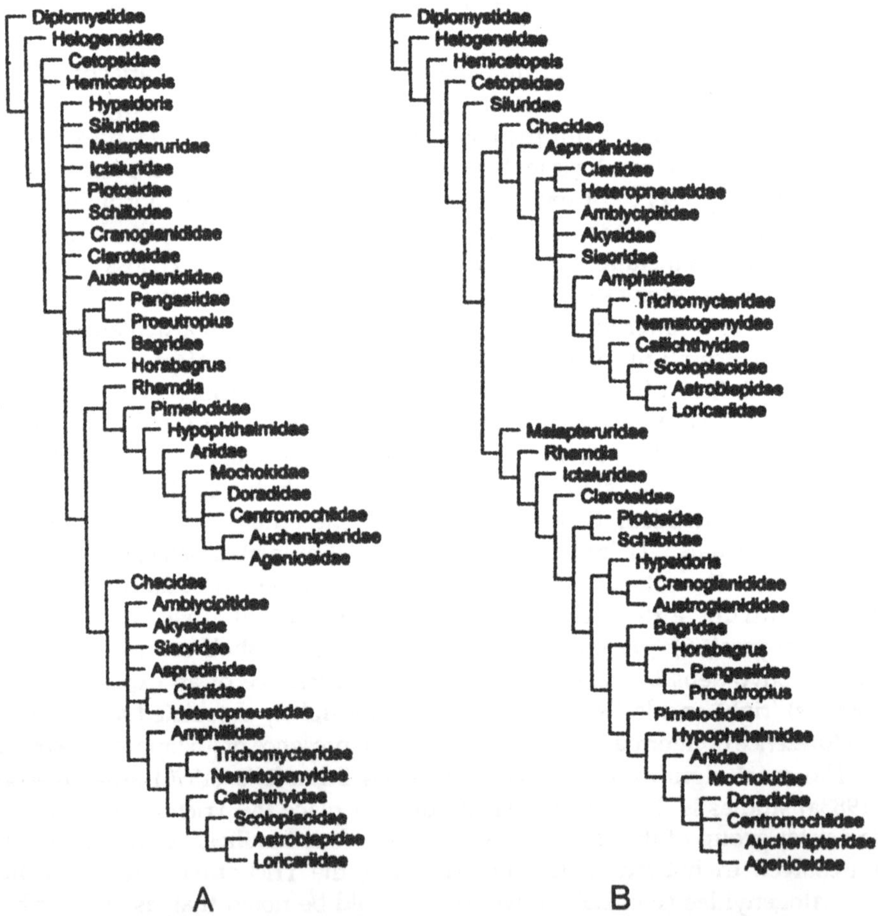

Fig. 1.6 Hypothetical relationships among the major groups of the Siluriformes according to Mo's 1991 paper. A) Cladogram produced from the numerical analysis of 126 unweighted characters. B) Cladogram produced from the numerical analysis of 126 characters with a weighting (4) on one of them, namely the "number of vertebrae united to the complex vertebra" (Mo, 1991: 193).

(using Hennig 1986) of a data matrix of 126 characters × 40 terminal taxa (which did not include autapomorphic characters), had 602 steps, CI = 0.31, RI (Retention Index) = 0.64 and was only poorly resolved. The other (Fig. 1.6B) was based on a 'weighted' numerical analysis (also using Hennig 1986) of the same data matrix in which to one of the 126 characters ('number of vertebrae united to complex vertebra') was given a phylogenetic weight 4 times that of the other characters. It had 549 steps, CI = 0.36, RI = 0.72 and, apart from two trichotomies, was completely resolved. According to Mo (1991: 193), this 'weighting' was due to the "morphological stability and consistent distribution of this character in comparisons with those conflicting features".

One of the main conclusions of Mo's (1991) work was the separation of 'Bagridae' into three monophyletic units, namely Bagridae (sensu Mo, 1991), Claroteidae, and Astroglanididae; according to Mo, these are more closely related to other catfish families than to each other (Fig. 1.6).

Another important conclusion of this work is the suggestion that Cetopsidae occupies a markedly basal position in the order Siluriformes (although the 'Helogeneidae', 'Cetopsidae' and '*Hemicetopsis*' of Mo were not, as commonly accepted today, grouped in a single monophyletic taxon, they were placed in a rather basal position among Siluriformes in both Mo's cladograms). In Mo's cladogram I (which he clearly seems to prefer), the cetopsids and diplomystids are distinguished from all other siluriforms by the presence of two characters: 1) 'interdigital union of the two coracoids'; and 2) 'ramus mandibularis nerve runs inside hyomandibular for a distance' (Mo, 1991: 204). Of these two characters, only the latter is listed in Mo's cladogram II to diagnose the clade composed by all non-diplomystid, non-cetopsid catfishes.

Another important, but confusing, conclusion drawn by Mo (1991) is the phylogenetic position of the fossil catfish †*Hypsidoris*. In his cladogram II (Fig. 1.6B), †*Hypsidoris* is placed in a far more derived position than in Grande's 1987 cladogram (Fig. 1.4). However, in his cladogram I (Fig. 1.6A), †*Hypsidoris* is placed in an unresolved polychotomy, with the phylogenetic position of this genus thus uncertain.

With respect to the other catfish groups, their phylogenetic position is also, with just a few exceptions, quite uncertain not only as a consequence of the poor resolution of Mo's cladogram I (Fig. 1.6A), but more importantly the significant differences between it and his cladogram II (Fig. 1.6B). These few exceptions are discussed below.

One exception concerns the relationships among the loricarioid families, which are essentially similar to those proposed by Howes (1983a) and Schaefer (1990); the only difference is that in Mo's cladograms Trichomycteridae and Nematogenyidae are considered sister-groups (Fig. 1.6). This sister-group relationship is supported, according to Mo (1991: 204, 208), by the fact that trichomycterids and nematogenyids, contrary to other loricaroids, have 'nasal barbels situated at anterior nostrils'.

In both of Mo's cladograms (1991) the loricarioids and amphiliids are grouped in a clade closely related to the Sisoridae, Akysidae, Amblycipitidae,

Clariidae, Heteropneustidae, Aspredinidae and Chacidae (Fig. 1.6). In cladogram I the clade including Amphiliidae and Loricaroidei is diagnosed by the 'posterior portion of the palatine reduced into a bony lamina or short spinelike process without distal cartilage' (Mo, 1991: 204), while in cladogram II no character defines this clade. In Mo's cladogram I the clade including the loricaroids, amphiliids, sisorids, akysids, amblycipitids, clariids, heteropneustids, aspredinids and chacids is justified by a 'computer generated node' (Mo, 1991: 204), while in cladogram II the clade including all these groups is diagnosed by the 'absence of extrascapular' (Mo, 1991: 207).

Both Mo's cladograms suggest a monophyletic clade consisting of the Auchenipteridae, Doradidae, Mochokidae, Ariidae, Hypophthalmidae and Pimelodidae (although the 'Auchenipteridae', 'Ageneiosidae' and 'Centromochlidae' of Mo were not, as is now commonly accepted, grouped in the family Auchenipteridae, all these three groups were placed in this clade in both of Mo's cladograms). In Mo's cladogram I this clade is defined by a single character, namely the 'anteriorly thickened and rounded or convex mesethmoid' (Mo, 1991: 204). In his cladogram II this clade is defined not only by this character, but also by two other features, namely the presence of 'four infraorbitals' and the 'enclosed aortic canal in the complex vertebra' (Mo, 1991: 208).

De Pinna, 1992 (Fig. 1.7)

One year after the publication of Mo's work, De Pinna (1992) described a new subfamily of the Neotropical catfish family Trichomycteridae, the Copionodontinae. In the same work, he provided a phylogenetic analysis of the interrelationships among trichomycterids, as well as among those catfishes and other loricarioids. The 27 characters included in that analysis consisted mainly of osteological characters of the cephalic region, Weberian apparatus, dorsal fin, pelvic fin and pectoral girdle, but also included a few soft and/or myological characters. The handmade comparison of these characters resulted on a fully resolved cladogram with a CI (autapomorphic characters not included) = 0.78 that, like both cladograms of Mo (1991), suggested a sister-group relationship between Trichomycteridae and Nematogenyidae. However, it should be noted that De Pinna's 1992 phylogenetic analysis was completely independent of Mo's (1991) (when writing his paper, De Pinna was unaware of Mo's results). In fact, the main reason that led De Pinna to propose the

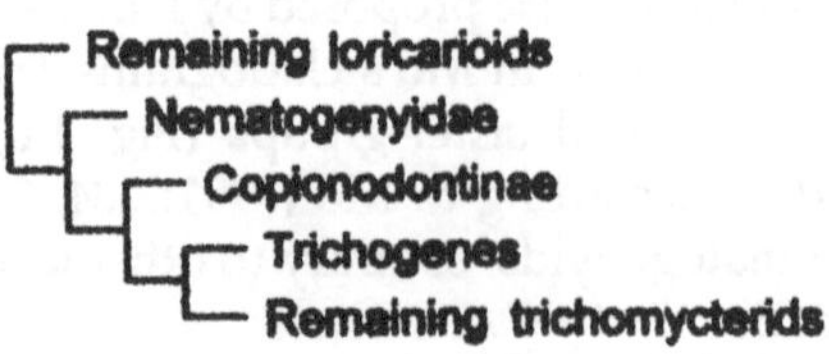

Fig. 1.7 Hypothetical relationships among the trichomycterids, as well as among these fishes and other loricarioids according to de Pinna 's 1992 paper.

sister-group relationship between Trichomycteridae and Nematogenyidae was 'to a major extent induced by the inclusion of copionodontines and Trichogenes in the analysis of lower loricarioid relationships' (De Pinna, 1992: 175). According to De Pinna (1992: Fig. 1.23), this inclusion indicated that, of the four derived characters traditionally used to support a sister-group relationship between trichomycterids and the other non-nematogenyid loricarioids, only one ('transformator process of tripus absent') indeed represented the plesiomorphic situation for trichomycterids. The other three ('intercalarium absent', 'ductus pneumaticus absent' and 'superficial ossification covering ventral surface of articulation between complex vertebrae and basioccipital') represented instead, an apomorphic configuration exclusively present in a restricted group of derived trichomycterids (Fig. 1.7: 'Remaining trichomycterids'). Consequently, the grouping of all non-nematogenyid loricaroid families, that could no longer be supported by more than a single derived character, was parsimoniously discarded by De Pinna (1992) in favour of the sister-group relationship between Trichomycteridae and Nematogenyidae, supported by three derived characters. These three characters are: 1) 'mesial juncture between scapulo-coracoids without interdigitations'; 2) 'first dorsal fin pterygophore inserted posterior to neural spine of ninth free vertebra'; 3) 'absence of dorsal fin spine and locking mechanism' (De Pinna, 1992: Fig. 1.23).

Arratia, 1992 (Fig. 1.8)

Arratia's 1992 work is a detailed, extensive study dedicated to the development, morphological variation and homologies of the suspensorium of certain siluriform and non-siluriform ostariophysans. It also provides an analysis,

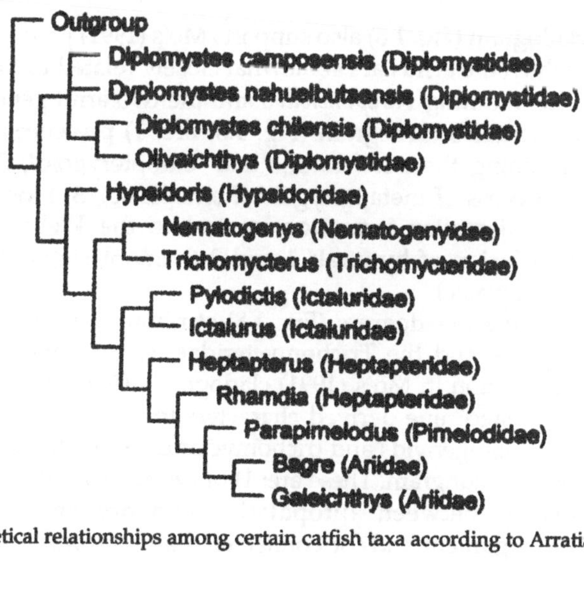

Fig. 1.8 Hypothetical relationships among certain catfish taxa according to Arratia's 1992 paper.

based on the suspensorial features examined, as well as some other characters described previously in other studies (Fink and Fink, 1981; Arratia, 1987; Grande, 1987), of the phylogenetic relationships among the different ostariophysan orders, and also among certain catfish groups. With respect to the relationships among certain siluriforms, Arratia's (1992) analysis resulted in four practically identical cladograms, the only difference concerning the relationships among diplomystids. As this Section essentially concerns the interfamilial relationships of the Siluriformes, I shall only refer to the cladogram illustrated in Arratia's (1992) Figure 46A (for a detailed explanation of the methodology followed to produce the other three cladograms, as well as a discussion of the differences between them, see Arratia, 1992: 126-129). This cladogram (Fig. 1.8), based on a numerical analysis (using PAUP) of a data matrix of 75 characters × 15 terminal taxa, in which 69 characters were ordered and 6 unordered, corresponds to the consensus of two equally parsimonious trees (CI = 0.672) with 137 steps. It supports Grande's (1987) hypothesis, according to which the fossil catfish †*Hypsidoris* occupies a rather basal position within the Siluriformes. Arratia (1992: Fig. 46A) listed eight uniquely derived characters and two homoplasic ones to diagnose the clade constituted by all non-diplomystid, non-hypsidorid catfishes examined in her study. These characters are: 1) maxilla without long anterior process; 2) maxilla rudimentary; 3) articulation between autopalatine and maxilla double, lateroventrally oriented; 4) absence of autopalatine extension dorsal to dermal entopterygoid; 5) absence of dermal ectopterygoid; 6) absence of dermal entopterygoid; 7) presence of link between 'entopterygoid' and prevomer (homoplasic); 8) loss of notch separating processus basalis and posterodorsal part of metapterygoid (homoplasic); 9) presence of three or four pairs of barbels; 10) absence of a supraneural bone above the Weberian apparatus in adults.

Arratia's cladogram (Fig. 1.8) also supports Mo's (1991) phylogenetic results, according to which the ariids are somewhat closely related to the pimelodids. The six characters uniting *Parapimelodus* and the two ariid genera, *Bagre* and *Galeichthys*, in Arratia's cladogram (Fig. 1.8) are: 1) presence of a sesamoid ectopterygoid joining the autopalatine and 'entopterygoid'; 2) presence of ectopterygoid process of metapterygoid (homoplasic); 3) blood vessels running in tubelike lamellar formation ventral to the Weberian apparatus (homoplasic); 5) fusion of hypurals 1 and 2 (homoplasic); 6) branched sensory canals (homoplasic).

However, Arratia's cladogram (Fig. 1.8) attributes a rather basal position to Nematogenyidae and the Trichomycteridae, two families that occupy a rather derived position in Mo's (1991) cladograms (Fig. 1.6). Indeed, Arratia (1992: Fig. 46A) listed nine derived characters to separate the diplomystids, †hypsidorids, nematogenyids and trichomycterids from all the other catfishes represented in her cladogram. These are: 1) presence of a rod like autopalatine; 2) no articulation between autopalatine and prevomer (homoplasic); 3) presence of a ligament and/or connective tissue between 'entopterygoid' and lateral ethmoid (homoplasic); 4) presence of a metapterygoid-'entopterygoid' ligament (homoplasic); 5) hyomandibula articulating with autosphenotic; 6) absence of prootic in the hyomandibular fossa; 7) presence of bony extension over the ventral surface of the fifth centrum (homoplasic); 8) presence of suture between pterosphenoid and parasphenoid (homoplasic); 9) blood vessels in a groove partially surrounded by lamellar walls in the ventral part of the Weberian apparatus (homoplasic).

Another significant aspect of Arratia's cladogram is the fact that the heptapterin genus *Rhamdia* appears as more closely related to the clade formed by *Parapimelodus* (Pimelodinae), *Bagre* (Ariidae) and *Galeichthys* (Ariidae) than to the heptapterin genus *Heptapterus*. The features listed by Arratia (1992: Fig. 46) to unite the genera *Galeichthys, Bagre, Parapimelodus* and *Rhamdia*, and thus to separate these genera from Hepapterus are: 1) fusion of abdominal centra 2-6 or more; 2) presence of a small, elongate pharyngobranchial attached to the epibranchial and medial aspect of the hyomandibula (homoplasic).

<u>Lundberg, 1993</u> (Fig. 1.9)

In an overview of certain clades formed by African and South American freshwater fishes and their respective implications for the continental drift theory, Lundberg (1993) provided a phylogenetic hypothesis (Fig. 1.9) concerning the relationships among some catfish taxa. This hypothesis was based on a handmade analysis of 12 osteological and myological characters of the cephalic region, dorsal fin and Weberian apparatus previously described by other authors and/or personally observed by Lundberg (Lundberg, 1993: 180). Lundberg's hypothesis (Fig. 1.9) is practically identical to Mo's (Fig. 1.6), with the addition of the Eocene fossil catfish "†*Titanoglanis*" as the sister-group of the clade constituted by Mochokidae, Auchenipteridae and Doradidae (the 'Auchenipteridae' and the 'Ageneiosidae' of Lundberg correspond to the Auchenipteridae of this work). It should be noted, however, that Lundberg (1993) was seemingly unaware of Mo's study since he makes no mention of it. The two characters listed by Lundberg (1993: 180) to support the sister-group relationship between "†*Titanoglanis*" and the clade including Mochokidae, Auchenipteridae and Doradidae were: 1) 'posterior edge of supraoccipital truncated, not draw out to form a process'; 2) 'middle nuchal plate with anterolateral processes contacting posttemporal-epioccipital region of skull'.

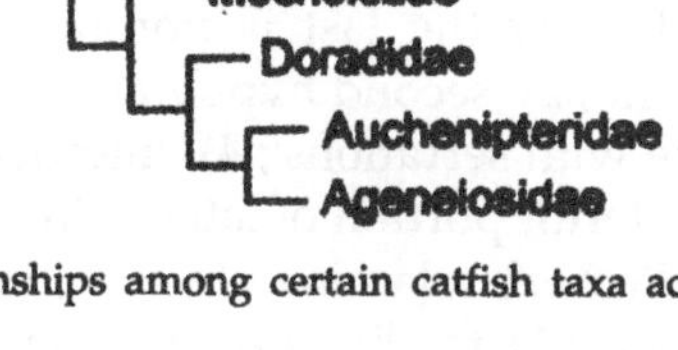

Fig. 1.9 Hypothetical relationships among certain catfish taxa according to Lundberg's 1993 paper.

De Pinna, 1996 (Fig. 1.10)

Based on a phylogenetic comparison of a considerable number of characters of the cephalic region, Weberian apparatus, pectoral fins and girdle, vertebrae, dorsal fin, pelvic fins and girdle and caudal fin, De Pinna (1996) developed a hypothesis on the relationships among the Asiatic Amblycipitidae, Akysidae, Sisoridae and Erethistidae and the South American Aspredinidae. He also examined the relationships among some genera of these families. His numerical analysis (using Hennig 1986) of a data matrix that had 112 characters × 21 terminal taxa resulted in a single, completely resolved, most parsimonious cladogram with 167 steps, CI = 0.70 (autapomorphic characters included) and RI = 0.79 (Fig. 1.10).

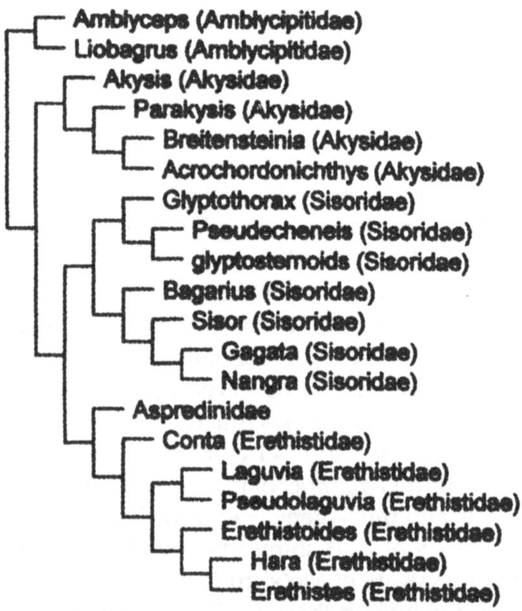

Fig. 1.10 Hypothetical relationships among the Sisoroidea according to de Pinna's 1996 paper.

One of the most significant conclusions of De Pinna's work was that the Sisoridae of previous authors was a paraphyletic assemblage, with a subunit of it (subsequently named Erethistidae by De Pinna) being more closely related to the Neotropical Aspredinidae than to the remaining taxa previously assigned to the Sisoridae (Fig. 1.10). Five synapomorphies were listed by De Pinna (1996: 64) to diagnose the clade constituted by Erethistidae and Aspredinidae, of which only the last is non-homoplasic: 1) 'mandibular laterosensory canal absent'; 2) 'second hypobranchial unossified'; 3) 'anterior margin of pectoral spine with serrations'; 4) 'internal support for pectoral fin rays small in size'; 5) 'anterior portion of lateral line running closely in parallel to lateral margin of Weberian lamina'. In turn, ten synapomorphies were listed by de Pinna (1996: 61) to diagnose the clade formed by these two families plus the Sisoridae sensu stricto, eight of which are homoplasic: 1) 'posterior portion of supracleithrum ankylosed to margin of Weberian lamina' (homoplasic); 2) 'parapophysis of fifth vertebra strongly flattened and expanded' (homoplasic); 3) 'parapophysis of fifth vertebra long, almost or quite reaching lateral surface of body wall'; 4) 'humeral process or region around it connected to anterior portion of vertebral column by well-defined ligament-state 3' (homoplasic); 5) 'posterior part of Weberian lamina extensively contacting parapophysis of fifth vertebra'; 6) '(reversal of) anterior half of segments of pectoral-fin spine elongate, almost parallel to axis of spine' (homoplasic); 7) 'coracoid with ventral anterior process' (homoplasic); 8) '(reversal of) second dorsal fin spine with medial ridge along its anterior surface, forming bilateral longitudinal pouches' (homoplasic); 9) 'ventral arms of first dorsal-fin spine with posterior subprocesses attached dorsal to their tip' (homoplasic); 10) 'basipterygium with ventral longitudinal keel, anteriorly extending alongside internal arm' (homoplasic).

In addition, De Pinna's (1996) work suggested the existence of a monophyletic clade formed by the Sisoridae, Erethistidae, Aspredinidae and Akysidae which, in turn, together with family Amblycipitidae formed superfamily Sisoroidea (Fig. 1.10). Three synapomorphies were listed to define the clade including Sisoridae, Erethistidae, Aspredinidae and Akysidae, namely: 1) 'supratemporal fossae present' (homoplasic); 2) 'supracleithrum strongly attached to skull'; 3) 'posterior nuchal plate with anterior process forming facet for articulation with anterior nuchal plate' (De Pinna, 1996: 60). With respect to superfamily Sisoroidea, De Pinna (1996: 59-60) listed seven synapomorphies, namely: 1) 'posterior center of ossification of palatine compressed and expanded vertically' (homoplasic); 2) 'articular region of lateral ethmoid elongated as a process, with articular facet for palatine at tip'; 3) 'parapophysis of fifth vertebra strong and attached to ventral side of centrum, directed directly transversely to centrum'; 4) 'humeral process or soft tissue around it connected to anterior portion of vertebral column by well-defined ligament'; 5) 'segments of pectoral fin spine very oblique, almost parallel to axis of spine, not evident' (homoplasic); 6) 'dorsal spine with medial ridges along anterior surface, forming bilateral longitudinal pouches' (homoplasic); 7) 'ventral tip of first dorsal fin pterygophore and corresponding neural spines with contacting facets'.

<u>De Pinna, 1998</u> (Fig. 1.11)

Two years after the publication of his 1996 work, De Pinna published an overview of the phylogenetic relationships of the Neotropical Siluriformes, which included a not completely resolved cladogram expressing the relationships among the major groups of the whole order (Fig. 1.11). As explained by him (1998: 289-290), this cladogram was mainly derived from his 1993 unpublished thesis, with: 1) 'some resolution added on the sisoroid-aspredinidid part of the tree based on the results of De Pinna (1996)';

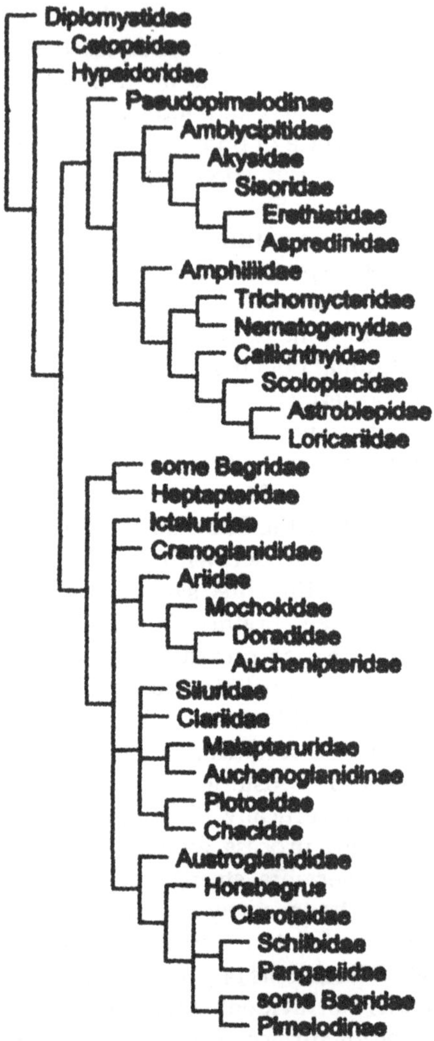

Fig. 1.11 Hypothetical relationships among the major groups of the Siluriformes according to de Pinna's 1998 paper.

2) 'position of the Ariidae from Lundberg (1993)'; and 3) 'position of †Hypsidoridae left unresolved'.

One of the most remarkable aspects of de Pinna's 1998 cladogram is the fact that although mainly derived from his unpublished thesis (de Pinna, 1993), of which the greater part was realised parallelly and thus independently, of Mo's (1991), several points of this cladogram coincide with Mo's phylogenetic results. In fact, small differences excepted, both De Pinna (1998) and Mo (1991) agree: 1) in a rather basal position of Cetopsidae within the siluriforms; 2) in the relationships among the different loricarioid families and a close relationship between these families and Sisoridae (sensu lato), Akysidae, Aspredinidae and Amblycipitidae; and 3) in the relationships among Mochokidae, Auchenipteridae, Doradidae and Ariidae (however the position of Ariidae in De Pinna's 1998 cladogram is based on Lundberg's 1993 paper).

But there are also some significant differences between the cladogram of De Pinna (1998) and the phylogenetic results of Mo (1991), of which one of the most notable is De Pinna's suggestion that both the Bagridae and Claroteidae sensu Mo (1991) are polyphyletic groups.

Another important aspect of De Pinna's 1998 cladogram (Fig. 1.11) is that it constitutes the first published cladogram providing an explicit hypothesis concerning the phylogenetic position of the three Pimelodidae groups, i.e., Pseudopimelodinae, Pimelodinae and Heptapterinae. In this cladogram the pseudopimelodins form, together with the loricarioids and sisoroids, a monophyletic unit that is the sister-group of a clade with the heptapterins and some bagrids as its more basal taxa (Fig. 1.11). With respect to the pimelodins, the cladogram suggests a sister-group relationship between these catfishes and some bagrids, with the clade formed by these two groups being included with the claroteins, schilbids, pangasiids, *Horabagrus* and austroglanidids in a clade included in a large, unresolved pentatomy (Fig. 1.11).

Unfortunately, except for the interrelationships among the loricaroid families, as well as some other specific cases, De Pinna's 1998 paper does not directly provide the phylogenetic characters that support the interfamilial relationships illustrated in that cladogram (these characters are given only in de Pinna's 1993 unpublished thesis). Consequently, neither the characters concerning the polyphyly of the Bagridae and Claroteidae sensu Mo (1991) nor the characters concerning the phylogenetic position of Pimelodinae, Pseudopimelodinae and Hepapterinae within the Siluriformes are included in De Pinna's 1998 paper.

<u>He et al., 1999</u> (Fig. 1.12)

The He et al. (1999) study is dedicated mainly to the phylogeny of the African family Amphiliidae but also includes an analysis of the relationships between this family and some other taxa, namely Diplomystidae, Amblycipitidae, †Hypsidoridae, Bagridae, Sisoridae and *Leptoglanis* (the phylogenetic position of this genus, transferred from the Bagridae to Amphiliidae in Bailey and Stewart's 1984 paper, was considered uncertain a *priori* by He et al.). The characters used in this study were osteological features of the cephalic region, Weberian apparatus, vertebrae, pectoral girdle, dorsal fin, caudal skeleton and the pelvic girdle. The phylogenetic study of He et al. (1999), based on a numerical analysis (using PAUP) of a data matrix with 73 characters including × 14 terminal taxa, resulted in a singlemost parsimonious cladogram with 190 steps and CI = 0.616 (0.603 excluding autapomorphic characters).

According with this cladogram (Fig. 1.12), neither the Amphiliidae or the Sisoridae are monophyletic groups. The doumein amphiliids are more closely

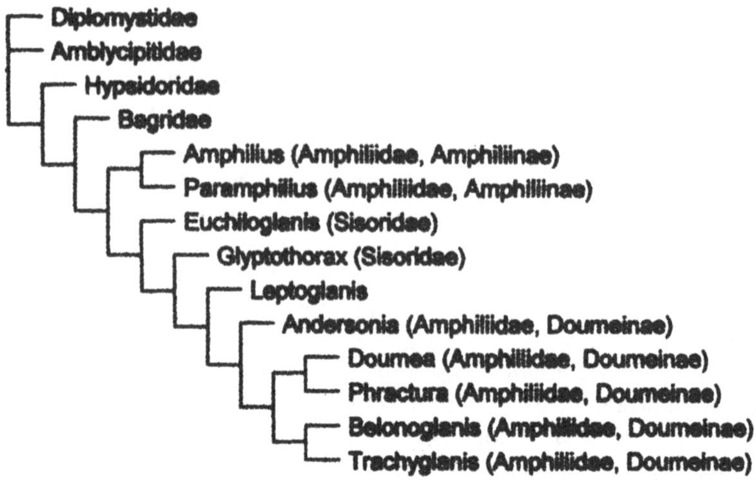

Fig. 1.12 Hypothetical relationships among the amphiliid genera, as well as among these genera and some non-amphiliid taxa, according to He et al.'s 1999 paper.

related to *Leptoglanis* and the sisorid *Glyptothorax* than to either the sisorid *Euchiloglanis* or the amphiliin amphiliids (Fig. 1.12). The characters listed by He et al. (indirectly given in their table 1) to support the clade composed by the doumein amphiliids, *Leptoglanis* and the sisorid *Glyptothorax* were: 1) no posterior fontanel (homoplasic); 2) posterodorsal process of supraoccipital short, slightly forked at its posterior end (homoplasic); 3) short maxillary without enlarged fanlike or forked posterior part; 4) fourth and fifth parapophyses of Weberian apparatus partly fused, thin and long (homoplasic); 5) proximal 1 and 2 of dorsal fin with independent nuchal plates; 6) all units of second dorsal spine fused.

Another interesting aspect of He et al. (1999) is placement of the Amblycipitidae in an unresolved trichotomy leading to this family, the Diplomystidae, and a clade constituted by the remaining catfishes examined, including the fossil catfish family †Hypsidoridae (Fig. 1.12).

1.4 CATFISH, AN EXCEPTIONAL BIOLOGICAL GROUP

As can be seen and as emphasised in the above Sections, catfishes are a remarkably diverse and widely distributed group, being 'the most diverse and widely distributed of ostariophysan groups' (De Pinna, 1998: 280). But catfishes are clearly not only remarkable in what concerns their amazing taxonomic diversity and biogeographic distribution. The diversity of these fishes is, in fact, remarkable in several other aspects, such as their anatomy, ethology, ecology or functional morphology.

For example, catfishes can attain lengths of up to 5 metres and weights of 330 kg, as does the European Wells *Silurus glanis* (see Fig. 1.13), rivalling even the famous, non-catfish osteoglossid *Arapaima gigas* in dimensions

Fig. 1.13 An example of giant catfish, the European Well Silurus glanis (for more similar photos on giant, as well as miniature, catfish, see the excellent survey given by Burgess, 1989).

(Burgess, 1989). The giant South American catfish *Paulicea lutkeni* has been said to approach and even surpass these sizes (Burgess, 1989). In contrast, there are some fully mature catfishes at lengths of no more than 35 mm, such as *Corydoras pygmaeus* (Burgess, 1989).

Several siluriforms present no structures related to sound production, while several others, e.g. the 'croaking' catfishes of family Doradidae, are famous for their sound production (for more details on this subject, see the up-to-date overview on catfish sound production given by Fine and Ladich, 2003). Also, although most catfishes are not poisonous, some, such as the plotosids, are particularly famous for the strong and painful poison released from the sharpened extremity of their pectoral spines (for more details on poisonous catfishes, see the recent overview on this subject by Perrière and Perrière, 2003). Some siluriforms are also renowned for their trophic performance. This is the case of species of the trichomycterid subfamily Vandeliinae which, by being exclusively hematophagous suckers of the blood from the gills of other fishes, are the only exclusive hematophagous jawed vertebrates other than some bats (De Pinna, 1998). Some of the vandellin catfishes are even popularly known in the Brazilian Amazon for their 'accidental' penetration of the urethras of humans and other mammals (see Burgess, 1989, for more details and some photos of these remarkable catfishes). Within the Siluriformes, some other groups are also famous for their 'aberrant' ecological preferences, e.g. members of the loricariid genus *Cochlidon* renowned for eating wood (see

Baras and Laleyé, 2003, for a recent overview on the ecology and behaviour of catfishes). Some catfishes are also ecologically famous for their subterranean habits, as well documented by Trajano (2003).

Biogeographically, catfishes are also an amazingly interesting, and highly puzzling, biological group. As already mentioned, they are primarily freshwater fishes, but families Plotosidae and Ariidae have considerable representation in marine environments. Catfishes are found in all continents, with catfish fossil remains having even been reported in Antarctica (see Grande and Eastman, 1986). One of the most striking and exemplary cases of puzzling biogeographic issues within the Siluriformes is the seemingly phylogenetic position of the South American aspredinids in the very middle of the Asian sisoroids hypothesised by De Pinna (1996, 1998) (see above). This, as stated by De Pinna himself, clearly would constitute an 'intriguing biogeographical phenomenon' (De Pinna, 1996: 77). Some palaeontological discoveries also provide such intriguing issues, as the finding of a seemingly ictalurid fossil form in Mongolia (see Stucky, 1982), when actually extant ictalurid catfishes are exclusively restricted to North American fresh waters. Or the finding of seemingly clariid and bagrid fossils forms in Europe when these two latter groups are actually exclusively present in Africa and Asia (for more details, and other similar, puzzling palaeontological discoveries, see the detailed up-to-date overview on catfish palaeontology and palaeobiogeography of Gayet and Meunier, 2003).

Representative examples of the high diversity and complexity of the Siluriformes could fill a whole book. In fact, the examples given in this chapter are only a very brief attestation of the indisputably amazing variation and complexity of this remarkable group of fishes, either from a taxonomic, anatomical, morphofunctional, ethological, biogeographic, or ecological point of view. That is why the Siluriformes constitute a very appropriate case study for a broader discussion on general macroevolution and phylogeny.

2

Methodology and Material

2.1 PHYLOGENETIC METHODOLOGY

As already mentioned, much of this book deals with an analysis of the higher level phylogeny of catfishes. The systematic procedure employed in this analysis is cladistic methodology, as extensively presented and discussed in the excellent book of Kitching et al. (1998). Parsimony was employed to find the hypothesis best supported by the analysed data, using both the Hennig86 (Farris, 1988) and the Nona Winclada (Nixon, 2002) computer programs. The Implicit Enumeration algorithm (ie*) of Hennig86 was employed in the search for the most parsimonious cladograms, with Nona Winclada used to check the most parsimonious results found with this algorithm. Hennig86 was chosen mainly because of its efficiency in calculation time (more specific information on the options and capacities of the program are provided in Fitzhugh, 1989). Tree manipulations and diagnostics were done using the ingenious and updated program Nona Winclada. Autapomorphies for the different terminal taxa examined were actively searched for and included in the analysis.

Multistate characters were ordered whenever possible, following the basic principles presented in Kitching et al. (1998). The same phylogenetic weight and hence the same phylogenetic importance was given to all characters used in the analysis, following Wiens (2001). In fact, as pointed out by him, most phylogenetic studies neglected the fact that by using ordered and unordered characters without corrective weighting, a different weight and thus different phylogenetic importance is, in reality, given to the different characters. For example, if species A and B share state 2 of a certain ordered character X and if species A and C share state 2 of a certain unordered character Y, this will favour a grouping of A and B, since they share two evolutionary steps on the cladogram (from state 0 to state 1 of character X, and then from state 1 to state 2 of this character, since this is an ordered character), while A and C share only one evolutionary step on the cladogram (directly from state 0 to state 2 of character Y, since this is an unordered character). Thus, a different weight and hence a different phylogenetic importance is, in reality, attributed to the characters X and Y since, after all, A and C share, like A and B, state 2 of a phylogenetic character. Therefore, in this simple example, in order to render the phylogenetic importance of the character Y equal to that of the character X, the phylogenetic weight of the character X should be multiplied by $\frac{1}{2}$, that is, by 1/(n-1), where n is the total number of states of the character. In this way, the sharing of state 2 of the unordered character Y by A and C is ranked as equally important as the sharing of state 2 of the ordered character X: both characters have a total weight of 1. Consequently, in order to attribute the same phylogenetic importance to all characters used in the present analysis, i.e., in order to give an equal total weight of 1 to all these characters, the following procedure was undertaken: a) the weight of unordered characters, as well as of ordered characters with two states, was not changed, i.e., it remained 1; b) the weight of ordered characters with three, four and five states (there are no ordered characters with more than five states in the analysis) was multiplied, as described above, by a corrective factor of respectively, $\frac{1}{2}$, $\frac{1}{3}$ and $\frac{1}{4}$. However, it is important to note that the phylogenetic outcomes resulting from the two other alternative methodologies concerning the ordering/unordering of characters, i.e., the exclusive use of unordered characters and the use of ordered characters without normalisation of their total weights, will be carefully examined and discussed in Chapter 5.

Special attention was also given to the process of determining the polarity of the different character states used in the phylogenetic analysis. In order to facilitate and optimise this process, it was decided to use family Diplomystidae as the outgroup of all other siluriforms. In fact, the sister-group relationship between Diplomystidae and all the other siluriforms has long been recognised (see Eigenmann and Eigenmann, 1890; Bridge and Haddon, 1893; Regan, 1911b; Alexander, 1965; Chardon, 1968; Gosline, 1975) and subsequently fully supported by all cladistic analyses on catfish higher level phylogeny (see Ch.1, Sec. 1.3). Therefore, defining all non-diplomystids as the ingroup and the Diplomystidae as the basal outgroup clearly seems to constitute the wisest, appropriate, unambiguous and efficient option in the present work. The configuration found in the non-siluriform ostariophysans was, of course, also seriously taken into account for decisions on the polarity of the characters. In fact, as an important framework for decisions on character polarity, a detailed and extensive analysis on the homologies, anatomical variation and plesiomorphic configuration of the different morphological features examined, taking into account the situation found in both the Diplomystidae and the other ostariophysans, as well as data from different fields such as functional morphology, ontogeny and palaeontology, was carefully undertaken. Much of this analysis was presented in a series of papers recently published by the author and colleagues, such as: Diogo and Chardon (2000a) (adductor mandibulae complex), Diogo and Chardon (2000b) (structures associated with the mandibular barbels), Diogo and Chardon (2000c) (structures associated with the feeding mechanisms), Diogo et al. (2000a) and Diogo and Chardon (2001) (palatine-maxillary system), Diogo et al. (2001a) and Diogo and Chardon (2003) (suspensorium), Diogo and Vandewalle (2003) (cranial musculature), Diogo et al. (2003a) (barbels), or Diogo et al. (2001c) (pectoral girdle).

As noted above, one of the major differences between the present phylogenetic analysis and the two other cladistic analyses done so far on catfish general phylogeny (Mo, 1991 and De Pinna, 1993) is the importance given in the present work to catfish myology. Both Mo and De Pinna, as well as the vast majority of authors dealing with catfish phylogeny, included almost exclusively data based on osteology and/or external anatomy, with only a few characters being eventually related with myological features. The phylogenetic comparison of the present work attributes, however, a high importance to the configuration of both the osteological and muscular features. The inclusion of myological characters in the phylogenetic analysis is justified by, among others, three main reasons.

Firstly, inclusion of an important amount of myological characters on the phylogenetic analysis has the major purpose of including a set of characters that have traditionally been neglected in previous analyses (see Ch. 1). This allows an increase in total number of characters included in the analysis. What is more important, it promotes the use of previously unexplored data to create hypotheses regarding the general phylogeny of catfishes, which could then be compared with other hypotheses previously elaborated by other authors.

The use of myological characters associated mainly with complex structures is also explained by the purpose to decrease the theoretical interference of homoplasy in the search of phylogenetic relationships. This is clearly a delicate and complex issue. Homoplasy, including convergences, parallelisms and phylogenetic reversions, is commonly viewed as the major problematic issue in phylogenetic systematics. This issue seems even more delicate on attending to the theoretical evolutionary framework provided in Gould's 2002 excellent, well-documented, philosophical and sagacious book. Most authors, like Gould himself, argue that the use of complex structures theoretically decreases the probability of homoplasy in a phylogenetic analysis, since such characters are less susceptible to homoplasy (e.g. Chardon, 1968; Stevens, 1980; Szalay, 1981; Winterbottom, 1993; Galis, 1996; Borden, 1998, 1999; Nielsen, 1998; Gould, 2002). Some authors have explicitly stated, based on empirical data, that in this context muscular characters are particularly 'highly corroborative, i.e., non-homoplasic, and thus reliable characters reflecting the true phylogenetic pattern of relationships' (Borden, 1999 : 224; see also Winterbottom, 1993). Inclusion of a high number of muscular characters in the analysis thus allows testing of statements made by authors such as Borden and Winterbottom, as it permits a broad direct comparison of the use of myological *versus* osteological characters for inferring phylogenetic relationships.

Lastly, and as also emphasised by authors such as Borden (1998, 1999), Galis (1996) and Winterbottom (1993), the study of muscular characters associated with complex structures paves the way for a more detailed, integrated reflection on the functional morphology and evolution of these kinds of structures. Complex structures, such as catfish palatine-maxillary system, or catfish pectoral girdle, are constituted by not just a set of integrated bones, but also muscles, cartilages and ligaments. Analysing all these different types of morphological components allows a more adequate and satisfactory insight into the functional morphology and general evolution of structural complexes. For this reason, as well as for the other two main reasons explained above, it was decided to analyse all the bones, muscles, cartilages and ligaments of each siluriform structural complex included in the phylogenetic analysis of the present work.

The morphological features analysed in the present work embrace, thus, all the bones, cartilages, muscles and ligaments of the cephalic region (branchial apparatus excluded), Weberian apparatus and pectoral girdle. As is obvious, unfortunately such an analysis could not be extended to all the structural complexes of the whole catfish body, principally due to restrictions of time and/or economic possibilities. Thus, selectivity with careful and detailed reflection was necessarily paramount. Therefore, within the number of structural complexes that could be studied within the time and possibilities available, it was decided to include in the analysis those complex structures constituting the cephalic region in the broad sense (i.e, including the Weberian apparatus and pectoral girdle).

In catfishes, as in teleosts in general, the cephalic region, as defined here—encompassing also the anterior vertebra and the pectoral girdle—is, in reality, the major and most important region of the body. The study of all muscles, bones, ligaments and cartilages of the various catfish cephalic structural complexes thus allows a broader integrative insight into the whole catfish cephalic region, its functional morphology and consequently its general evolution. Also, the cephalic complex structures analysed in the present work are those for which there are more insights available in the literature, either concerning their comparative anatomy in different catfish groups or their evolution and/or functional morphology. This facilitates not only the anatomical comparisons necessary for the phylogenetic analysis, but also decisions such as those related with the polarity of the characters. This was precisely one of the main reasons for including in the analysis the structures of the pectoral girdle or Weberian apparatus for example, rather than those of the branchial apparatus (catfish branchial apparatus was never the subject of a detailed comparative or evolutionary study, and the catfish branchial musculature is particularly poorly studied, while the comparative anatomy and evolution of catfish pectoral girdle and Weberian apparatus have been subjected to well-documented, detailed studies, such as Bridge and Haddon, 1894; Dubale and Rao, 1961; Tilak, 1963; Alexander, 1964; Chardon, 1968; Lundberg, 1975b; Gosline, 1977; Brosseau, 1978; Arratia, 1987; Diogo et al. 2001c; Chardon et al. 2003).

Let it be noted, however, that in spite of this necessary selectivity of complex structures, the phylogenetic analysis of the present work constitutes, nevertheless, among cladistic analyses of catfish general phylogeny, the one including the largest number of morphological characters (440 morphological characters versus Mo's 1991 and de Pinna's 1993 analyses including 126 and 239 characters, respectively).

2.2 DELIMITATION OF TERMINAL TAXA

With respect to the terminal taxa included in the phylogenetic analysis presented here, great effort was expended to include representatives of all the 32 extant catfish families (see Ch. 1). This was a difficult task since many catfish families have few species or even a single species which, in some cases, are moreover seriously threatened. Members of these catfish families are, thus, usually scarce in museums. Moreover, the difficulty of finding available specimens in museum collections was even more toilsome due to the fact that the analysis of complex muscular characters obliged, as mentioned earlier, their almost complete dissection.

Here I really must thank once more Dr. Richard Vari, the late Dr. Guy Teugels and most particularly Dr. Michel Chardon whose incredible generosity, together with contributions by others (see Preface), made available representatives of 87 key-genera covering every extant catfish family examined in this phylogenetic analysis.

The choice of these 87 key-genera took into account two main aspects: 1) the evidence for monophyly of the family and 2) the phylogenetic relationships, when known, inside the family. The first of these two aspects had, of course, more preponderance for the selection of the genera representing each family, since this phylogenetic analysis is primarily directed to the higher-level phylogeny of the whole order Siluriformes. Thus, for example, families with more representative genera in the analysis are polygeneric families with a seemingly more problematic monophyletic status, e.g., Pimelodidae (9 genera included in the analysis), Amphiliidae (9 genera included in the analysis) and Schilbidae (5 genera included in the analysis) (see Ch. 1). For large families for which there is strong monophyletic evidence and detailed phylogenetic studies available, such as Callichthyidae or Sisoridae, the more important aspect was not the inclusion of a proportionally large number of genera but instead of key-genera representing the major subgroups within the family (e.g., a genus representing each of the four tribes within the Sisoridae—see De Pinna, 1996; and a genus representing each of the two subfamilies of the Callichthyidae—see Reis, 1998a).

It is also important, before passing to the list of genera included in the phylogenetic analysis of this work, to note a significant difference between the methodology of this analysis and that of the analyses promoted by Mo (1991) and De Pinna (1993). Most terminal taxa of the analyses of these authors were catfish families, while in the present work all terminal taxa refer to individual genera. There are thus 87 terminal taxa in this phylogenetic analysis. This procedure avoids the use of generalisations at the familial level. In other words, it precludes such generalisations as 'plotosids have a processus on bone A' or 'amphiliids have a well-developed muscle B', which in point of fact imply a loss of data due to the great morphological variation within genera of diverse and large groups such as, in this specific example, the Plotosidae and Amphiliidae. Of course one might retort that using genera as terminal taxa also implies a generalisation. True enough. But this time at a generic level, which is precisely the strength of the procedure applied in this study. Morphological variation within a genus is necessarily less than within the family in which this genus is included, unless, of course, the family is monogeneric.

The genera representing each family in the phylogenetic analysis are briefly listed below. (Notes: a complete list of all the specimens examined, their serial numbers and institutions, and their trypsin-cleared and alizarine-stained or alcohol fixed condition are provided in Section 2.3); for information regarding subfamilies, refer to sections 1.2 and 1.3).

Family Akysidae. Monophyly well supported, two subfamilies recognised. One genus from each subfamily included in the analysis: the type genus *Akysis* (Akysinae) and *Parakysis* (Parakysinae).

Family Amblycipitidae. Monophyly well supported, no subgroups recognised. Two genera included in the analysis: the type genus *Amblyceps* and *Liobagrus*.

Family Amphiliidae. It is well known that the monophyly of this family has been highly disputed; the family was moreover the subject of a major previous work by the author. Hence, all the nine recognised genera of the family, i.e., *Amphilius, Paramphilius, Leptoglanis, Zaireichthys, Andersonia, Belonoglanis, Trachyglanis, Doumea* and *Phractura*, included in the analysis.

Family Ariidae. No major ariid subfamilies are recognised. But the monophyly of the group of ariid genera other than the enigmatic genus *Ancharius* from Madagascar is relatively well supported and concensually accepted by catfish specialists. However, there is no consensus among these specialists as to whether that monophyletic group and the genus *Ancharius* do indeed constitute a natural clade. Therefore, the genus *Ancharius* was included in the analysis of the present work, together with two genera representing the traditional monophyletic group constituted by the remaining ariid genera, namely the type genus *Arius* and the genus *Genidens*.

Family Aspredinidae. Monophyly well supported, three subfamilies recognised. One genus from each subfamily included in the analysis: the type genus

Aspredo (Aspredininae) and genera *Bunocephalus* (Bunocephalinae) and *Xyliphius* (Hoplomyzontinae).

Family Astroblepidae. Family comprises a single genus, *Astroblepus*, included in the analysis.

Family Auchenipteridae. Monophyly well supported, two subfamilies recognised. One genus from each subfamily included in the analysis, namely the type genus *Auchenipterus* (Auchenipterinae) and *Centromochlus* (Centromochlinae). It was also possible to additionally include in the analysis genus *Ageneiosus*, representing the *Ageneiosus*-like group nowadays included in the subfamily Auchenipterinae but previously assigned to a separate, independent family, the Ageneiosidae (for more details on this subject, see De Pinna, 1998).

Family Austroglanididae. Family with a single genus, *Austroglanis*, included in the analysis.

Family Bagridae. Although the monophyly of this family was strongly supported by Mo (1991) and previously corroborated by Diogo et al. (1999), De Pinna (1998), based on his unpublished thesis of 1993, suggested a polyphyletic status for the Bagridae. De Pinna's 1998 suggestion was seriously taken into account, with each of the two main bagrid subgroups referred by this author represented in the present phylogenetic analysis. Thus, this analysis includes the type genus *Bagrus* and genus *Hemibagrus*, both included in one of De Pinna's subgroups, and genera *Rita* and *Bagrichthys*, included in De Pinna's other subgroup. This selection also respects the most orthodox, consensual view of the relationships within the family since both the bagrid subfamilies recognised by Mo (1991) are represented in the analysis (the smaller subfamily Ritinae by genus *Rita* and the larger subfamily Bagrinae by genera *Bagrus*, *Hemibagrus* and *Bagrichthys*).

Family Callichthyidae. Monophyly well supported, two subfamilies recognised. One genus from each subfamily included in the analysis: the type genus *Callichthys* (Callichthyinae) and genus *Corydoras* (Corydoradinae).

Family Cetopsidae. Monophyly well supported, two subfamilies recognised. Both subfamilies represented in the analysis: the type genus *Cetopsis* and *Hemicetopsis* from Cetopsinae and the only genus, *Helogenes*, of subfamily Helogeninae.

Family Chacidae. Family established for a single genus, *Chaca*, included in the analysis.

Family Clariidae. As referred in Chapter 1, the monophyly of Clariidae, a family with no major subgroups recognised, only seems to be corroborated when *Heteropneustes* from Heteropneustidae is also included in it. This is because some of the clariid taxa are seemingly more closely related to *Heteropneustes* than to other clariids. Therefore, three main groups are represented in the analysis: genus *Heteropneustes* from family Heteropneustidae (see below) and the two clariid groups apparently divided by this genus, with *Uegitglanis* representing one group and genera *Clarias* and *Heterobranchus* representing the other (see Chardon, 1968; De Pinna, 1998).

Family Claroteidae: As with family Bagridae (see above), the monophyly of Claroteidae was also supported by Mo (1991) but contravened in De Pinna's 1998 paper, based on his 1993 unpublished thesis. Therefore, as with Bagridae, the present analysis includes representatives of the two claroteid subgroups that seemingly do not form a monophyletic group for De Pinna, which correspond to the two claroteid subfamilies recognised by Mo (1991): the Auchenoglanidinae group is represented by its type genus *Auchenoglanis* and the Claroteinae group represented by its type genus *Clarotes* as well as genus *Chrysichthys*.

Family Cranoglanididae: Family with a single genus, *Cranoglanis*, included in the analysis.

Family Diplomystidae: Family with a single genus, *Diplomystes*, included in the analysis.

Family Doradidae. Monophyly well supported, three subfamilies recognised. One genus from each subfamily included in the analysis: the type genus *Doras* (Doradinae), genus *Anadoras* (Astrodoradinae) and genus *Acanthodoras* (Platydoradinae). It was also possible to additionally include in the analysis genus *Franciscodoras*, a highly plesiomorphic doradid genus not assigned to any of the three main doradid subfamilies (for more details on this subject, see de Pinna, 1998).

Family Erethistidae. Monophyly well supported, two subfamilies recognised. It was possible to include in the analysis the type genus of this small family, *Erethistes*, from the subfamily Erethistinae, as well as genus *Hara*, also from this subfamily. However, it was not possible to include in the analysis the single genus *Conta* of the subfamily Continae, due to the absence of available specimens for dissection in all the museums contacted. However, as the major aim of this analysis mainly concerns the interfamilial relationships of the order Siluriformes, the non-inclusion of a representative of this latter subfamily is somewhat minimised by the strong evidence supporting the monophyly of the whole family.

Family Heteropneustidae. A single genus, *Heteropneustes*, included in the analysis.

Family Ictaluridae. Monophyly well supported, no major subfamilies recognised. Two genera included in the analysis: the type genus of the family, *Ictalurus*, and genus *Amiurus*.

Family Loricariidae. The monophyly of this large family is well supported, with six subfamilies recognised. However, as explained above, for a family for which there is a so large amount of evidence for monophyly and so many phylogenetic studies, there is no need, in an analysis of higher level phylogeny of the whole order Siluriformes, to include representatives of each of these six subfamilies. Therefore, only representatives of three of these six subfamilies were included in the analysis, namely the type genus of the family, *Loricaria*, from subfamily Loricariinae, genus *Lithoxus* from subfamily Ancistrinae, and genus *Hypoptopoma* from subfamily Hypoptopomatinae.

Family Malapteruridae: Family with a single genus, *Malapterurus*, included in the analysis.

Family Mochokidae. Monophyly well supported, no major subfamilies recognised. Two genera included in the analysis: the type genus of the family, *Mochokus*, and genus *Synodontis*.

Family Nematogenyidae. Family with a single genus, *Nematogenys*, included in the analysis.

Family Pangasiidae. Monophyly well supported, no major subfamilies recognised. Two genera included in the analysis: the type genus *Pangasius* and genus *Helicophagus*.

Family Pimelodidae. As noted above, this is a traditionally problematic catfish family, with several authors considering its three subfamilies, Pimelodinae, Heptapterinae and Pseudopimelodinae, not constituting a monophyletic unit. Therefore, as mentioned in Chapter I, each of these groups is carefully taken into account in the present work. The monophyly of subfamily Pimelodinae is well supported, with three major groups usually recognised (for more details, see the overview of De Pinna, 1998): '*Calophysus* group', '*Pimelodus* group' and '*Sorubim* group'. Genera representing each of these three major pimelodin groups, viz. the type genus of the subfamily, *Pimelodus* ('*Pimelodus* group'), genus *Pseudoplatystoma* ('*Sorubim* group') and genus *Calophysus* ('*Calophysus* group'), were hence included in this analysis. It was possible to additionally include genus *Hypophthalmus*, a genus previously assigned to its own family Hypophthalmidae, but now included in the '*Pimelodus* group' (see De Pinna's 1998 overview for more details on this subject). With respect to subfamily Heptapterinae, its monophyly is also well supported, with two major subgroups recognised, namely the '*Nemuroglanis* clade' and 'basal clade' (see Lundberg et al. 1991a and De Pinna, 1998, for more details on this subject). Three genera were included in the present analysis: the type genus of the subfamily, *Hepapterus*, from the '*Nemuroglanis* clade', and two genera of other, more plesiomorphic heptapterin 'basal clade', *Rhamdia* and *Goeldiella*. With respect to the last pimelodid subfamily, i.e., Pseudopimelodinae, this is a small subfamily with only three genera and no major subgroups recognised (see Lundberg et al. 1991a; Teugels, 2003). Its monophyly, as in the case of the other two pimelodid subfamilies, is well supported. Two of the three genera of this small subfamily are included in the present phylogenetic analysis: the type genus of the subfamily, *Pseudopimelodus*, and genus *Microglanis*.

Family Plotosidae. Consensually considered a monophyletic group; no major subgroups recognised. Four genera included in the analysis: the type genus *Plotosus* and genera *Paraplotosus, Neosilurus* and *Cnidoglanis.*

Family Schilbidae. As already mentioned, the monophyly of this family was seriously questioned by Mo (1991). According to him (1991), Schilbidae is a non-monophyletic assemblage, with a '*Schilbe* group' representing the 'real schilbids', one phylogenetically distinct '*Ailia* group' being closer to Clariidae and Heteropneustidae, and the third, also phylogenetically distinct "*Pseudeutropius* group" closer to Bagridae or Pangasiidae. Mo's (1991) view was contested two years later in De Pinna's 1993 unpublished thesis. Although, unfortunately, De Pinna's (1993) results were not published, it is noteworthy that this author included three different groups of schilbids in his analysis of the higher level phylogeny of the Siluriformes, namely the '*Laides*', 'remaining *Schilbinae*' and '*Ailiinae*' groups, supporting Schilbidae monophyly. However, as these results remain unpublished, the well-structured and up-to-date overview on the state of the art of catfish taxonomy recently provided by Teugels (2003) continues to state that there is no available evidence to support the monophyly of Schilbidae, stressing the problematic nature of this family. Therefore, the five schilbid genera included in the present analysis were carefully chosen to represent five major groups that embrace simultaneously the various schilbid divisions considered by Regan (1911b) (who divided the family into subfamilies Schilbinae, Siluranodontinae, and Ailiinae, see earlier), Mo (1991) (*Schilbe* group, *Pseudeutropius* group, *Ailia* group) and De Pinna (1993) (*Laides, Ailia* group and *Schilbe* group). So, the genus *Schilbe* represents both the Schilbinae of Regan (1911b) and the *Schilbe* groups of Mo (1991) and De Pinna (1993). *Siluranodon* represents the subfamily Siluranodontinae of Regan (1911b). *Ailia* represents both the Ailiinae of Regan (1911b) and the *Ailia* groups of Mo (1991) and De Pinna (1993). *Pseudeutropius* represents the *Pseudeutropius* group of Mo (1991). Lastly, *Laides* represents the *Laides* group of De Pinna (1993).

Family Scoloplacidae. Family with a single genus, *Scoloplax,* included in the analysis.

Family Siluridae. Monophyly well supported, no subfamilies recognised. Two genera included in the analysis: the type genus of the family, *Silurus,* and genus *Wallago.*

Family Sisoridae. The monophyly of this large family is well supported, with two subfamilies recognised. It was possible to include in the analysis representatives of all the four sisorid tribes (two tribes in each subfamily) recognised by De Pinna (1996), namely *Sisor* (Tribe Sisorini, subfamily Sisorinae), *Bagarius* (Tribe Bagariini, subfamily Sisorinae), *Glyptosternon* (Tribe Glyptosternini, subfamily Glyptosterninae) and *Glyptothorax* (Tribe Glyptothoracini, subfamily Glyptosterninae).

Family Trichomycteridae. The monophyly of this family is well supported, with eight subfamilies recognised. This family is poorly represented in the phylogenetic analysis of the present work, with the two genera included, i.e., the type genus of the family, *Trichomycterus*, and the genus *Hatcheria*, both being from subfamily Trichomycterinae. The fact that these 2 genera belong to the largest trichomycterid subfamily, that these genera seem to occupy a somewhat basal position within this subfamily, and especially the strongly supported monophyly of the family somewhat attenuate, nevertheless, the importance of this specific issue in view of the more global aim of promoting a general analysis on the higher level phylogeny of the whole order Siluriformes.

2.3 MATERIAL, TECHNIQUES AND NOMENCLATURE

The cladistic analysis of catfish higher level phylogeny presented in this work is based on a phylogenetic comparison of morphological characters concerning the configuration of the bones, muscles, ligaments and cartilages of the cephalic region, Weberian apparatus and pectoral girdle of the catfishes examined. In the anatomical descriptions provided in this comparison, the osteological nomenclature concerning the cephalic region basically follows that of Arratia (1997). However, for the several reasons detailed in recent papers (Diogo et al. 2001a; Diogo and Chardon, 2003), with respect to the skeletal components of the suspensorium, the author follows the nomenclature of Diogo et al. (2001a). The myological nomenclature is based mainly on Winterbottom (1974a), but for the various adductor mandibulae sections, Diogo and Chardon (2000a) is followed. With respect to the muscles associated with the mandibular barbels, which were not descrived by Winterbottom (1974a), Diogo and Chardon (2000b) is followed. Concerning the nomenclature of the pectoral girdle bones and muscles, Diogo et al. (2001c) is followed. A complete list of the anatomical structures figuring in this work and their abbreviations is provided at the end of this book.

With respect to the biological material studied, the fishes are from the collection of the Laboratory of Functional and Evolutionary Morphology, Liege (LFEM), Musée Royal de l'Afrique Centrale, Tervuren (MRAC), Université Nationale du Bénin, Cotonou (UNB), Muséum National D'Histoire Naturelle, Paris (MNHN), National Museum of Natural History, Washington DC (USNM), American Museum of Natural History, New York (AMNH), and the South African Institute for Aquatic Biodiversity, Grahamstown (SAIAB). Anatomical descriptions were done after dissection of alcohol-fixed or trypsin-cleared and alizarine-stained specimens (following Taylor and Van Dyke's 1985 method). Dissections and morphological drawings were made using a Wild M5 dissecting microscope equipped with a camera lucida. The trypsine-cleared and alizarine-stained (c&s) or alcohol fixed (alc) condition of the studied fishes is given in parentheses following the number of specimens dissected. A complete list of the specimens dissected is given below:

Akysidae: *Akysis baramensis* LFEM, 2 (alc). *Akysis leucorhynchus* USNM 109636, 2 (alc). *Parakysis anomalopteryx* USNM 230307, 2 (alc); LFEM, 1 (alc).

Amblycipitidae: *Amblyceps caecutiens* LFEM, 2 (alc). *Amblyceps mangois* USNM 109634, 2 (alc). *Liobagrus reini* USNM 089370, 2 (alc).

Amphiliidae: *Amphilius brevis* MRAC 89-043-P-403, 3 (alc); MRAC 89-043-P-2333, 1 (c&s). *Andersonia leptura* MNHN 1961-0600, 2 (alc). *Belonoglanis tenuis* MRAC P.60494, 2 (alc). *Doumea typica* MRAC 93-041-P-1335, 1 (alc). *Leptoglanis rotundiceps* MRAC P.186591-93, 3 (alc). *Paramphilius trichomycteroides* LFEM, 2 (alc). *Phractura brevicauda* MRAC 90-057-P-5145, 2 (alc); MRAC 92-125-P-386, 1 (c&s). *Phractura intermedia* MRAC 73-016-P-5888, 1 (alc). *Trachyglanis ineac* MRAC P.125552-125553, 2 (alc). *Zaireichthys zonatus* MRAC 89-043-P-2243-2245, 3 (alc).

Ariidae: *Ancharius fuscus* AMNH 93702, 1 (alc); LFEM, 1 (alc). *Arius heudelotii* LFEM, 4 (alc). *Genidens genidens* LFEM, 2 (alc).

Aspredinidae: *Aspredo aspredo* USNM 226072, 1 (alc). *Aspredo sicuephorus* LFEM, 1 (alc). *Bunocephalus knerii* USNM 177206, 2 (alc). *Xyliphius magdalenae* USNM 120224, 1 (alc).

Astroblepidae: *Astroblepus phelpis* LFEM, 1 (alc); USNM 121127, 2 (alc).

Auchenipteridae: *Ageneiosus vittatus* USNM 257562, 1 (alc). *Auchenipterus dentatus* USNM 339222, 1 (alc). *Centromochlus hechelii* USNM 261397, 1 (alc).

Austroglanididae: *Austroglanis gilli* LFEM, 3 (alc); SAIAB 58416, 1 (c&s).

Bagridae: *Bagrichthys macropterus* USNM 230275, 1 (alc). *Bagrus bayad* LFEM, 1 (alc); LFEM, 1 (c&s). *Bagrus docmak* MRAC 86-07-P-512, 1 (alc); MRAC 86-07-P-516, 1 (c&s). *Hemibagrus nemurus* USNM 317590, 1 (alc). *Rita chrysea* USNM 114948, 1 (alc).

Callichthyidae: *Callichthys callichthys* USNM 226210, 2 (alc). *Corydoras guianensis* LFEM, 2 (alc).

Cetopsidae: *Cetopsis coecutiens* USNM 265628, 2 (alc). *Helogenes marmuratus* USNM 264030, 1 (alc). *Hemicetopsis candiru* USNM 167854, 1 (alc).

Chacidae: *Chaca bankanensis* LFEM, 3 (alc). *Chaca burmensis* LFEM, 2 (alc). *Chaca chaca* LFEM, 2 (alc).

Clariidae: *Clarias anguillaris* LFEM, 2 (alc). *Clarias batrachus* LFEM, 2 (alc). *Clarias ebriensis* LFEM, 2 (alc). *Clarias gariepinus* MRAC 93-152-P-1356, 1 (alc); LFEM, 2 (alc). *Heterobranchus bidorsalis* LFEM, 2 (alc). *Heterobranchus longifilis* LFEM, 2 (alc). *Uegitglanis zammaronoi* MRAC P-15361, 1 (alc).

Claroteidae: *Auchenoglanis biscutatus* MRAC 73-015-P-999, 2 (alc). *Auchenoglanis occidentalis* LFEM, 2 (alc). *Chrysichthys auratus* UNB, 2 (alc); UNB, 2 (c&s). *Chrysichthys nigrodigitatus* UNB, 2 (alc); UNB, 2 (c&s). *Clarotes laticeps* MRAC 73-13-P-980, 2 (alc).

Cranoglanididae: *Cranoglanis bouderius* LFEM, 2 (alc).

Diplomystidae: *Diplomystes chilensis* LFEM, 3 (alc).

Doradidae: *Acanthodoras cataphractus* USNM 034433, 2 (alc). *Anadoras weddellii* USNM 317965, 2 (alc). *Doras brevis* LFEM, 2 (alc). *Doras punctatus* USNM 284575, 2 (alc). *Franciscodoras marmoratus* USNM 196712, 2 (alc).

Erethistidae: *Erethistes pusillus* USNM 044759, 2 (alc). *Hara filamentosa* USNM 288437, 1 (alc).

Heteropneustidae: *Heteropneustes fossilis* USNM 343564, 2 (alc); USNM 274063, 1 (alc); LFEM, 2 (alc).

Ictaluridae: *Amiurus nebolosus* USNM 246143, 1 (alc); USNM 73712, 1 (alc). *Ictalurus furcatus* LFEM, 2 (alc). *Ictalurus punctatus* USNM 244950, 2 (alc).

Loricariidae: *Hypoptopoma bilobatum* LFEM, 2 (alc). *Hypoptopoma inexspectata* LFEM, 2 (alc). *Lithoxus lithoides* LFEM, 2 (alc). *Loricaria cataphracta* LFEM, 1 (alc). *Loricaria loricaria* USNM 305366, 2 (alc); USNM 314311, 1 (alc).

Malapteruridae: *Malapterurus electricus* LFEM, 5 (alc).

Mochokidae: *Mochokus niloticus* MRAC P.119413, 1 (alc); MRAC P.119415, 1 (alc). *Synodontis clarias* USNM 229790, 1 (alc). *Synodontis schall* LFEM, 2 (alc). *Synodontis sorex* LFEM, 2 (alc).

Nematogenyidae: *Nematogenys inermis* USNM 084346, 2 (alc); LFEM, 2 (alc).

Pangasiidae: *Helicophagus leptorhynchus* USNM 355238, 1 (alc). *Pangasius larnaudii* USNM 288673, 1 (alc). *Pangasius sianensis* USNM 316837, 2 (alc).

Pimelodidae: *Calophysus macropterus* USNM 306962, 1 (alc). *Goeldiella eques* USNM 066180, 1 (alc). *Hepapterus mustelinus* USNM 287058, 2 (alc). *Hypophthalmus edentatus* USNM 226140, 1 (alc). *Microglanis cottoides* USNM 285838, 1 (alc). *Pimelodus blochii* LFEM, 2 (alc). *Pimelodus clarias* LFEM, 2 (alc); USNM 076925, 1 (alc). *Pseudopimelodus raninus* USNM 226136, 2 (alc). *Pseudoplatystoma fasciatum* USNM 284814, 1 (alc). *Rhamdia guatemalensis* USNM 114494, 1 (alc).

Plotosidae: *Cnidoglanis macrocephalus* USNM 219580, 2 (alc). *Neosilurus rendahli* USNM 173554, 2 (alc). *Paraplotosus albilabris* USNM 173554, 2 (alc). *Plotosus anguillaris* LFEM, 2(alc). *Plotosus lineatus* USNM 200226), 2 (alc).

Schilbidae: *Ailia colia* USNM 165080, 1 (alc). *Laides hexanema* USNM 316734, 1 (alc). *Pseudeutropius brachypopterus* USNM 230301, 1 (alc). *Schilbe intermedius* MRAC P.58661, 1 (alc). *Schilbe mystus* LFEM, 3 (alc). *Siluranodon auritus* USNM 061302, 2 (alc).

Scoloplacidae: *Scoloplax distolothrix* LFEM, 1 (alc); USNM 232408, 1 (alc).

Siluridae: *Silurus aristotelis* LFEM, 2 (alc). *Silurus glanis* LFEM, 2 (alc). *Silurus asotus* USNM 130504, 2 (alc). *Wallago attu* USNM 304884, 1 (alc).

Sisoridae: *Bagarius yarreli* USNM 348830, 2 (alc); LFEM, 1 (c&s). *Gagata cenia* USNM 109610, 2 (alc). *Glyptosternon reticulatum* USNM 165114, 1 (alc). *Glyptothorax fukiensis* USNM 087613, 2 (alc).

Trichomycteridae: *Hatcheria macraei* LFEM, 2 (alc). *Trichomycterus areolatus* LFEM, 2 (alc). *Trichomycterus banneaui* LFEM, 2 (alc). *Trichomycterus immaculatus* USNM 301015, 2 (alc).

3

Phylogenetic Analysis

3.1 CHARACTER DESCRIPTION AND COMPARISON

A total of 440 morphological characters concerning essentially the configuration of the bones, muscles, cartilages and ligaments of the cephalic region, the pectoral girdle and fins and the Weberian apparatus of 87 genera representing all extant catfish families are compared. As explained in Ch. 2, sect. 2.1, the basis for this phylogenetic comparison was a meticulously detailed analysis of the homologies, anatomical variation and plesiomorphic configuration of the various morphological features examined. A great part of this analysis has appeared in a series of papers published recently by the author and colleagues, such as: Diogo and Chardon (2000a) (adductor mandibulae complex), Diogo and Chardon (2000b) and Diogo et al. (2003a) (structures associated with the mandibular barbels), Diogo and Chardon (2000c) (structures associated with the feeding mechanisms), Diogo et al (2000a), Diogo and Chardon (2001) (palatine-maxillary system), Diogo et al. (2001a) and Diogo and Chardon (2003) (suspensorium), Diogo and Vandewalle (2003) (cranial musculature), Diogo et al. (2003a) (barbels) and Diogo et al. (2001c) (pectoral girdle and fins).

In some cases, phylogenetic characters from other authors were also included in the phylogenetic comparison of the present work, which are respectively discriminated with the sentence "*character inspired from . . .*". However, it should be noted that these characters inspired from other authors refer to cases in which they were used in *an explicit phylogenetic sense* by the authors (e.g., when a certain author explicitly stated that a certain character Y defined a certain catfish family or a certain catfish monophyletic group), and not to those cases in which the characters refer to morphological features that were simply described in a purely morphological context (e.g., when a certain author simply described the presence of a particular morphological feature Z in a certain catfish A, with no indication of whether this feature is present in other catfishes or discussion of the phylogenetic signification of that feature).

The textual comparison presented here is complemented by a large number of anatomical figures. Due to evident limitations of time and space, it is not possible, of course, to provide a detailed series of illustrations for each of the 32 extant catfish families analysed. Therefore, a selection was obligatorily made, with greater emphasis usually given to those groups for which there are fewer anatomical descriptions available in the literature. So, for example, groups for which numerous anatomical descriptions, including a large number of anatomical illustrations, are already available in the literature, such as, e.g., non-nematogenyid loricarioids, silurids or ictalurids, are only textually described in the morphological comparison. However, for families for which there are almost no anatomical descriptions available in the literature, such as, e.g., the Nematogenyidae, Amphiliidae, Austroglanididae or Cranoglanididae, anatomical illustrations are provided. The vast majority of the families are nevertheless illustrated, with anatomical illustrations, both osteological and myological, provided for representatives of 22 of the 32 extant catfish families, namely: Amblycipitidae (Figs. 3.1. to 3.5), Amphiliidae (Figs. 3.6 to 3.19), Ariidae (Figs. 3.20 to 3.24), Aspredinidae (Figs. 3.25 to 3.29), Auchenipteridae (Figs. 3.30 to 3.32), Austroglanididae (Figs. 3.33 to 3.38), Bagridae (Figs. 3.39 to 3.42), Cetopsidae (Figs. 3.43 to 3.48), Chacidae (Figs. 3.49 to 3.51), Clariidae (Figs. 3.52 to 3.54), Claroteidae (Figs. 3.55 to 3.56), Cranoglanididae (Figs. 3.57 to 3.62), Diplomystidae (Figs. 3.63 to 3.73), Doradidae (Figs. 3.74 to 3.76), Erethistidae (Figs. 3.77 to 3.81), Heteropneustidae (Figs. 3.82 to 3.87), Nematogenyidae (Figs. 3.88 to 3.95), Pangasiidae (Figs. 3.96 to 3.99), Pimelodidae (Figs. 3.100 to 3.111), Plotosidae (Figs. 3.112 to 3.115), Schilbidae (Figs. 3.116 to 3.119) and Sisoridae (Figs. 3.120 to 3.122). With the data combined from the anatomical illustrations and the textual comparison of the 440 characters below concerning the configuration of both osteological and myological structures in 87 genera representing the 32 extant siluriform families, it is hoped that this work will provide a relatively extensive insight into catfish morphology.

Ventral Cephalic Musculature and Structures Associated with Mandibular Barbels

1. *Presence of mandibular barbels (unordered multistate character)*. Plesiomorphically catfish lack mandibular barbels [State 0: e.g. Fig. 3.69]. Among the catfish genera studied presenting mandibular barbels, one [State 1: e.g. Fig. 3.98] or two [State 2: e.g. Fig. 3.5] pairs may be present.
   - CS–0: No mandibular barbels **(*Diplomystes, Ageneiosus, Trichomycterus, Callichthys, Corydoras, Loricaria, Hypoptopoma, Lithoxus, Scoloplax, Astroblepus, Hatcheria*)**
   - CS–1: One pair of mandibular barbels **(*Nematogenys, Pangasius, Helicophagus, Silurus, Wallago, Rita*)**
   - CS–2: Two pairs of mandibular barbels **(all genera not mentioned in other CS)**

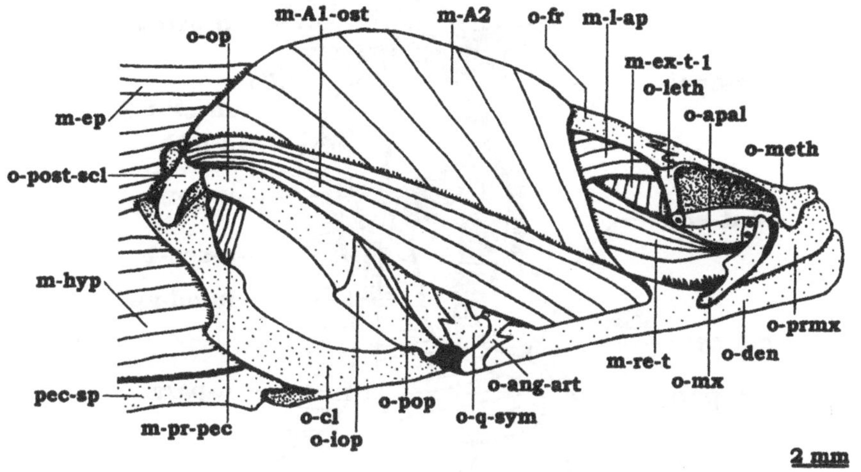

Fig. 3-1 Amblycipitidae: lateral view of the cephalic musculature of *Liobagrus reini*, all muscles exposed.

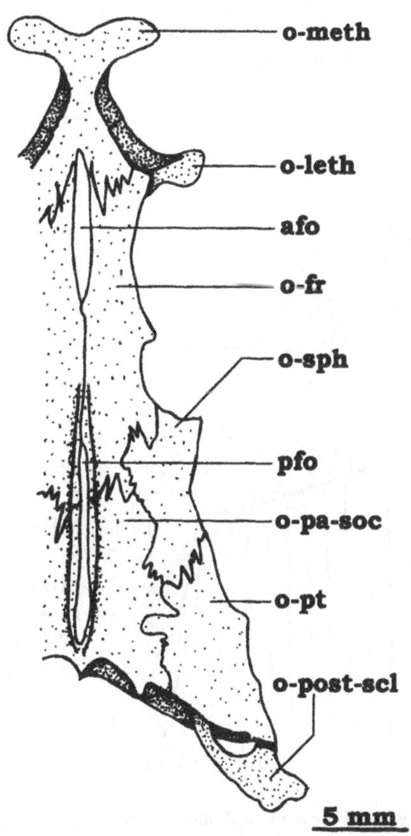

Fig. 3-2 Amblycipitidae: dorsal view of the neurocranium of *Liobagrus reini*.

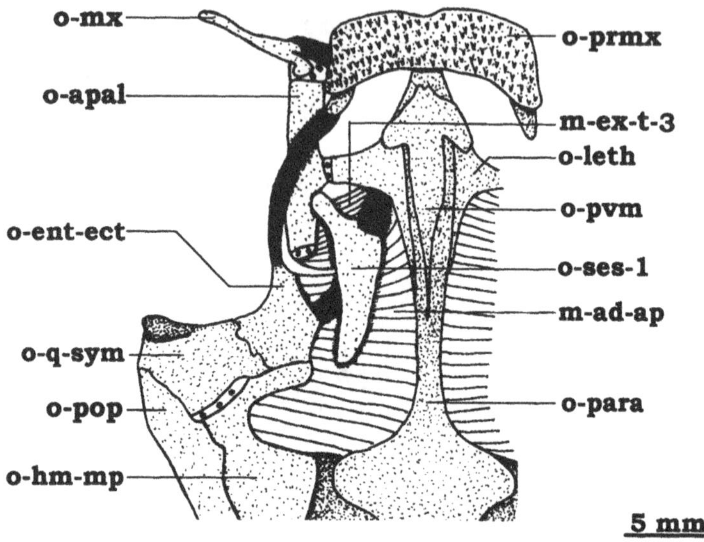

Fig. 3-3 Amblycipitidae: ventral view of the anterior region of the neurocranium of *Liobagrus reini*. Suspensorium, palatine-maxillary system and muscles and ligaments associated with these structures also illustrated.

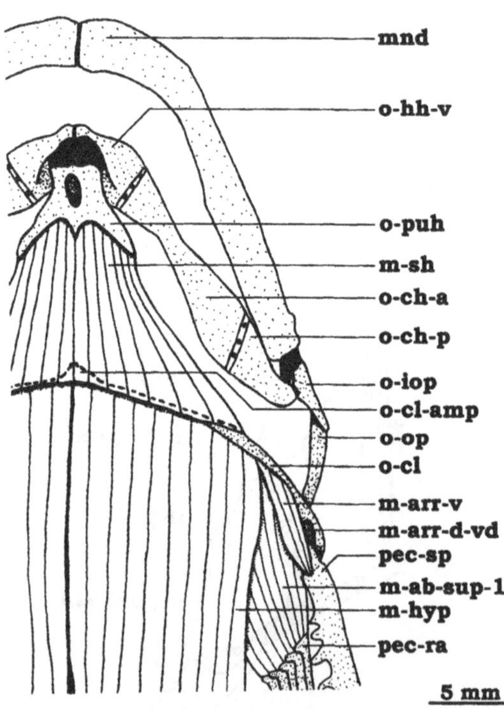

Fig. 3-4 Amblycipitidae: ventral view of cephalic region and pectoral girdle of *Liobagrus reini*. Muscles associated with pectoral girdle also illustrated.

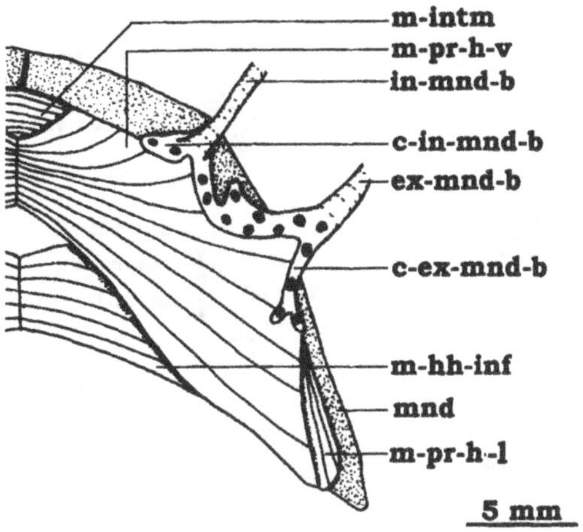

Fig. 3-5 Amblycipitidae: ventral view of cephalic musculature of *Liobagrus reini*, all muscles exposed.

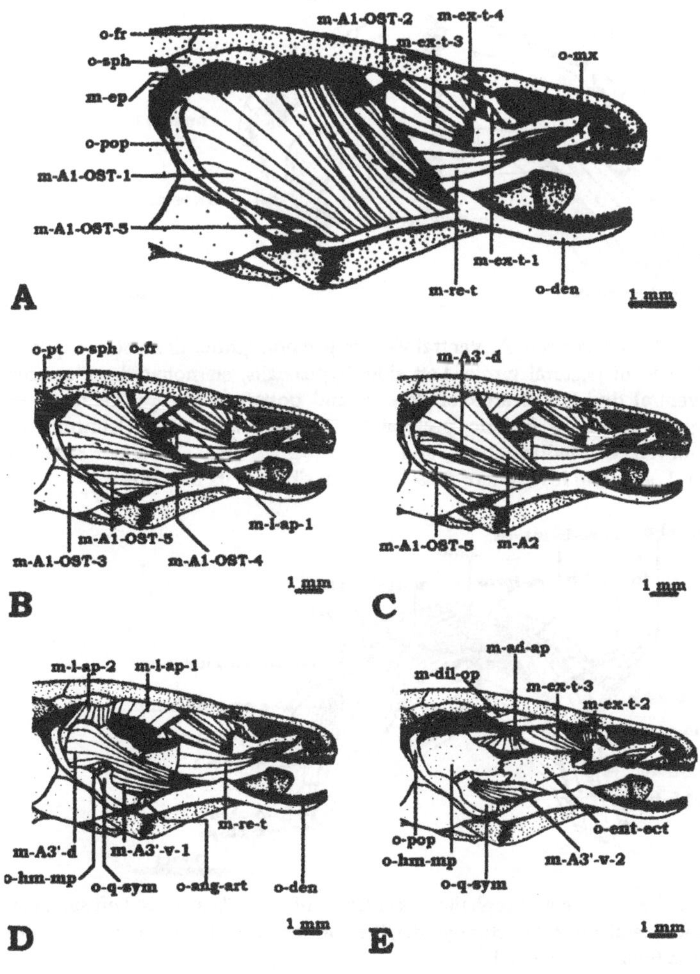

Fig. 3-6 Amphiliidae: lateral view of right cheek musculature of *Amphilius brevis*. **A)** All muscles exposed. **B)** A1-OST-1, A1-OST-2 and musculus epaxialis removed. **C)** A1 OST-3 and A1-OST-4 removed. **D)** A1-OST-5 and A2 removed. **E)** A3'-d, A3'-v-1, retractor tentaculi, levator arcus palatini and sections 1 and 4 of extensor tentaculi removed.

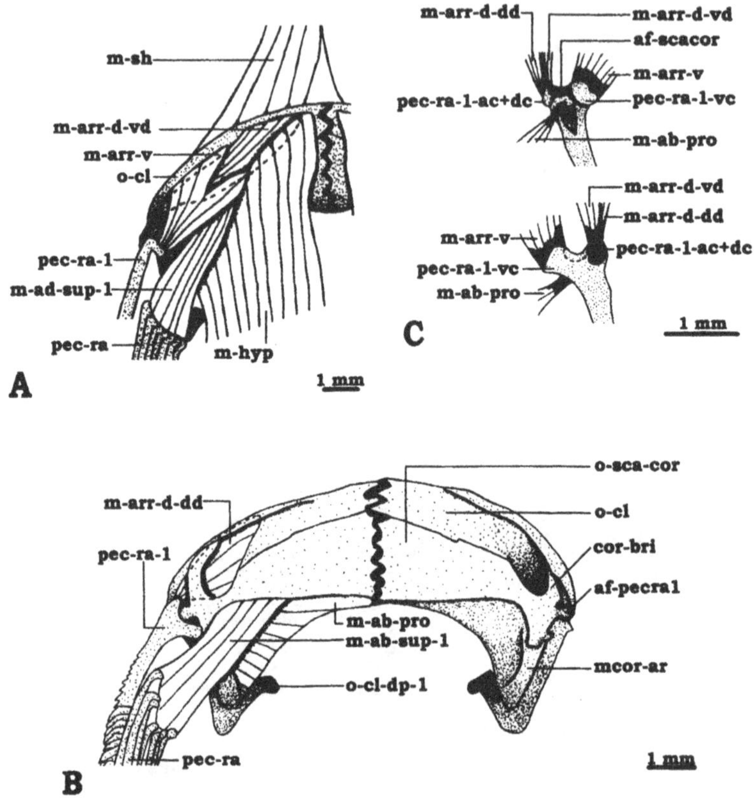

Fig. 3-7 Amphiliidae: *Amphilius brevis*. **A)** Ventral view of pectoral girdle, all muscles exposed. **B)** Ventral view of pectoral girdle. Left side: hypoaxialis, sternohyoideus, arrector ventralis, ventral division of arrector dorsalis and posttemporo-supracleithrum removed; right side: the pectoral fin, pectoral ray 1, posttemporo-supracleithrum and muscles removed. **C)** Median (above) and ventral (below) view of anterior region of pectoral ray 1, showing insertion of muscles responsible for its movement.

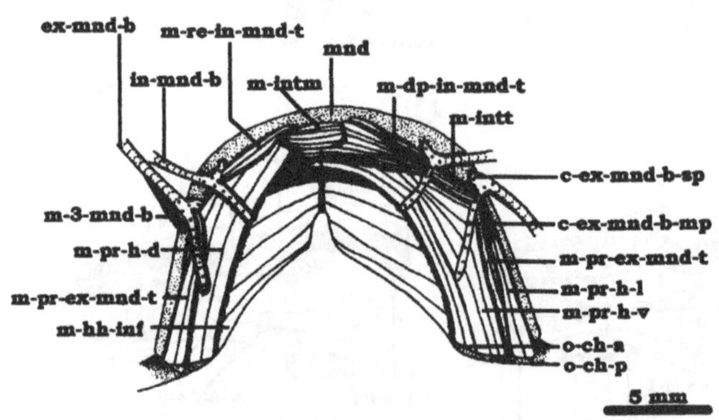

Fig. 3-8 Amphiliidae: Ventral view of cephalic musculature of *Amphilius brevis*. Left side: pars ventralis and lateralis of protractor hyoidei, the intertentacularis and depressor interni mandibularis tentaculi removed.

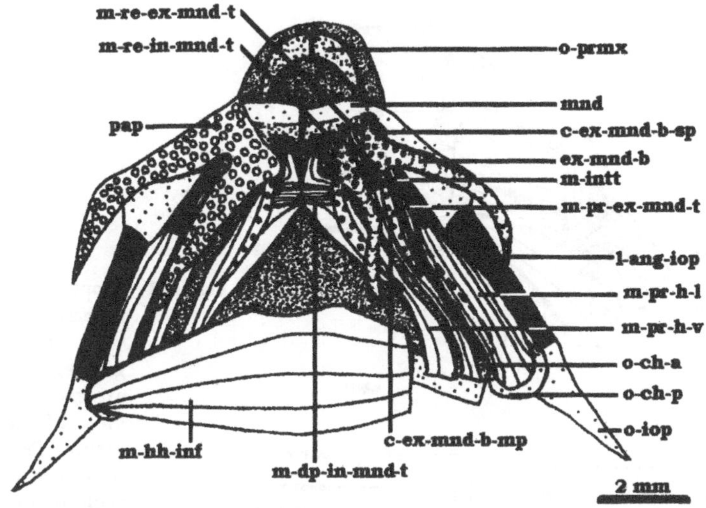

Fig. 3-9 Amphiliidae: ventral view of cephalic musculature of *Phractura brevicauda*. Right side: papillae associated with mandibular barbels removed.

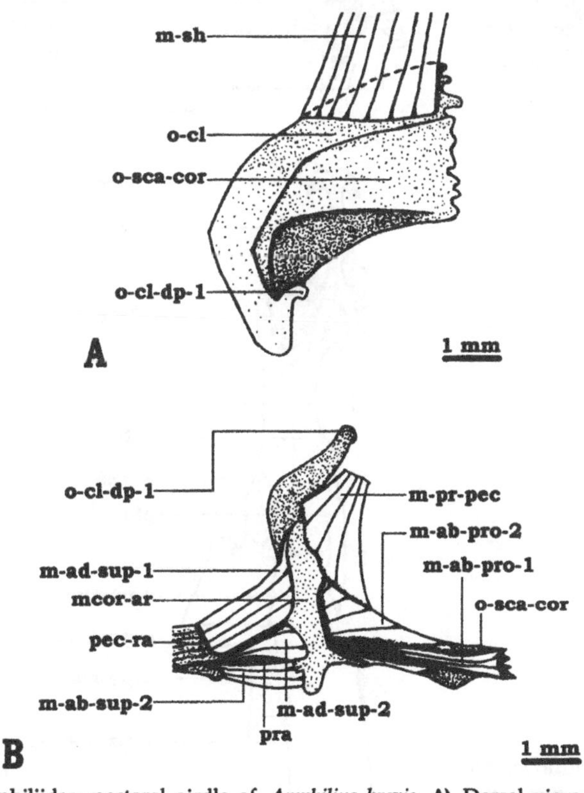

Fig. 3-10 Amphiliidae: pectoral girdle of *Amphilius brevis*. **A)** Dorsal view, pectoral fin and posttemporo-supracleithrum removed. **B)** Posterior view, posttemporo-supracleithrum removed.

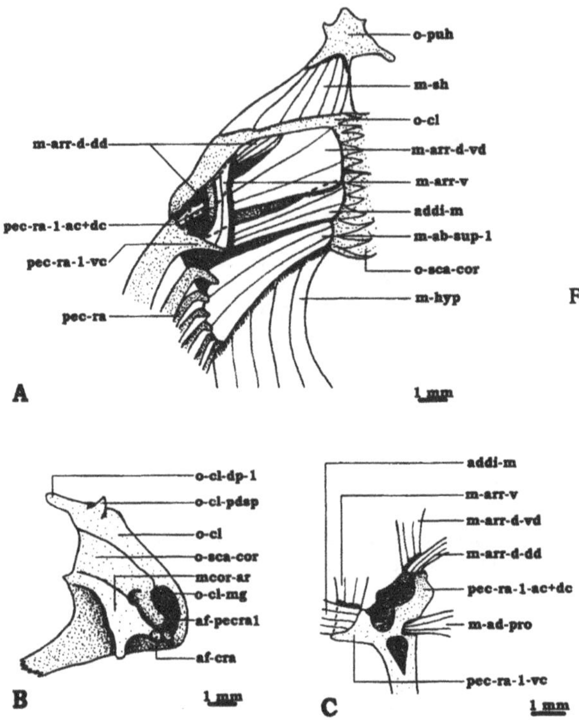

Fig. 3-11 Amphiliidae: *Phractura brevicauda*. **A)** Ventral view of the pectoral girdle, all muscles exposed. **B)** Posterior view of the pectoral girdle. Pectoral fin, posttemporo-supracleithrum and all muscles removed. **C)** Median view of anterior region of first pectoral ray, showing insertions of muscles responsible for its movement.

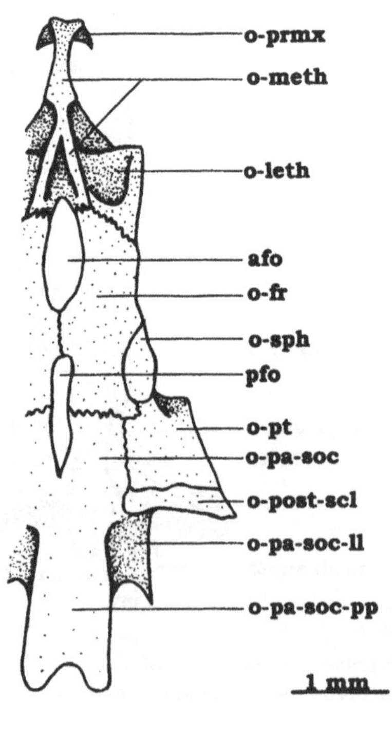

Fig. 3-12 Amphiliidae: dorsal view of neurocranium of *Trachyglanis ineac*.

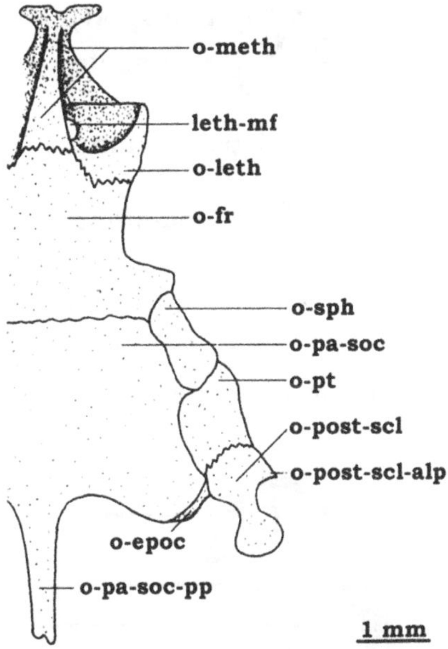

Fig. 3-13 Amphiliidae: dorsal view of neurocranium of *Zaireichthys zonatus*.

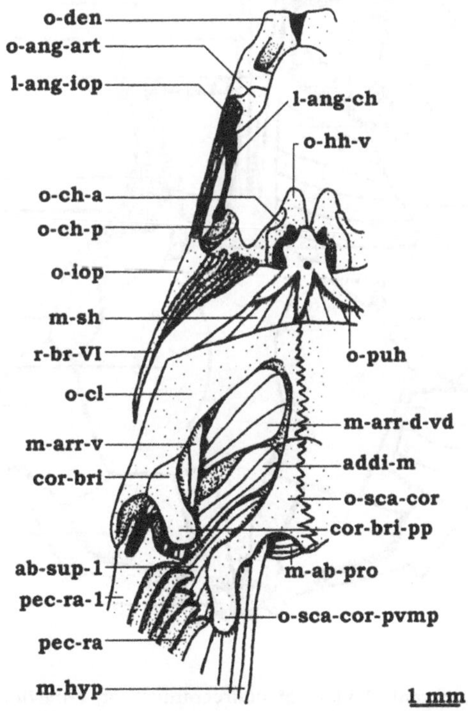

Fig. 3-14 Amphiliidae: ventral view of splanchnocranium, pectoral girdle and pectoral girdle muscles of *Belonoglanis tenuis*.

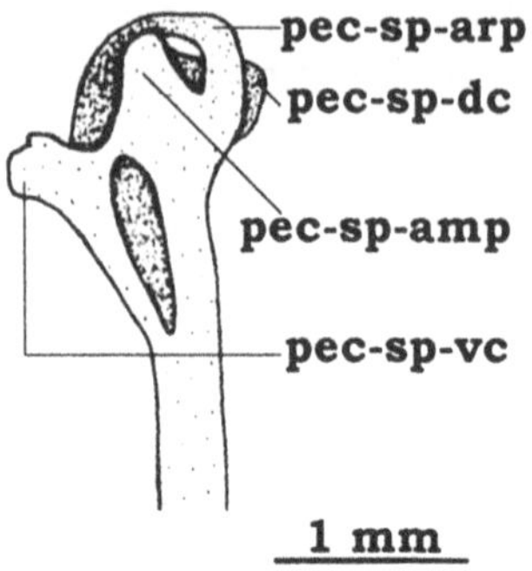

Fig. 3-15 Amphiliidae: mesial view of anterior portion of first pectoral ray of *Belonoglanis tenuis*.

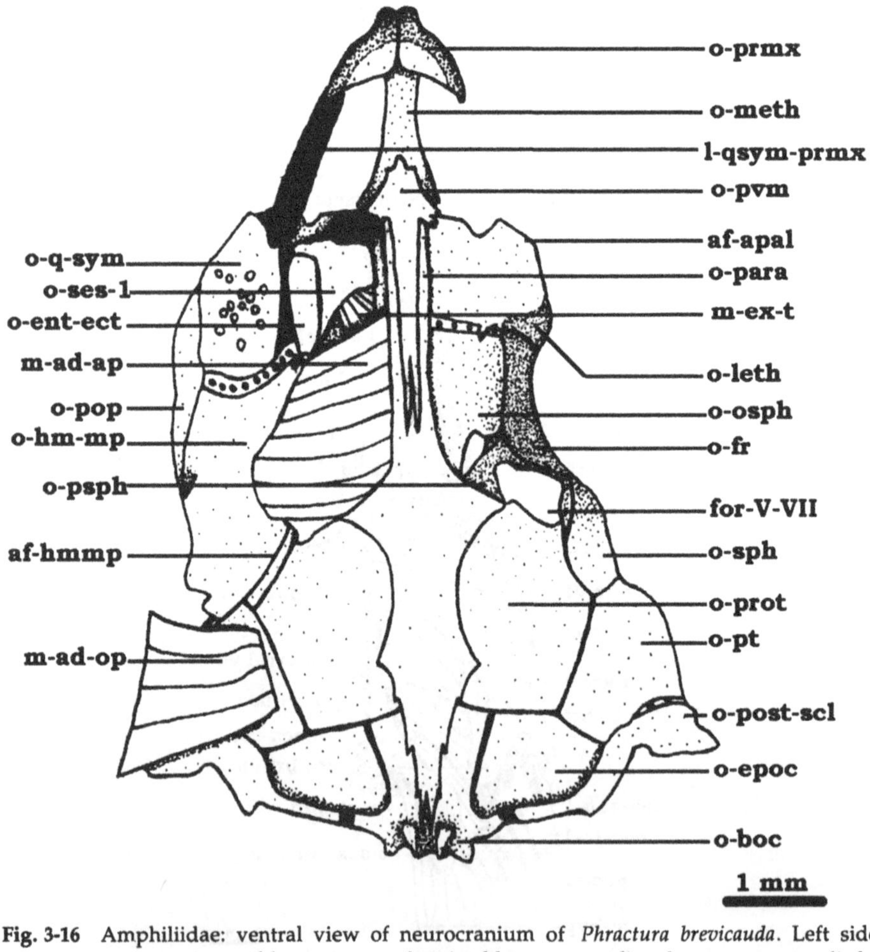

Fig. 3-16 Amphiliidae: ventral view of neurocranium of *Phractura brevicauda*. Left side: suspensorium, adductor arcus palatini, adductor operculi and extensor tentaculi also illustrated. Premaxillary teeth removed.

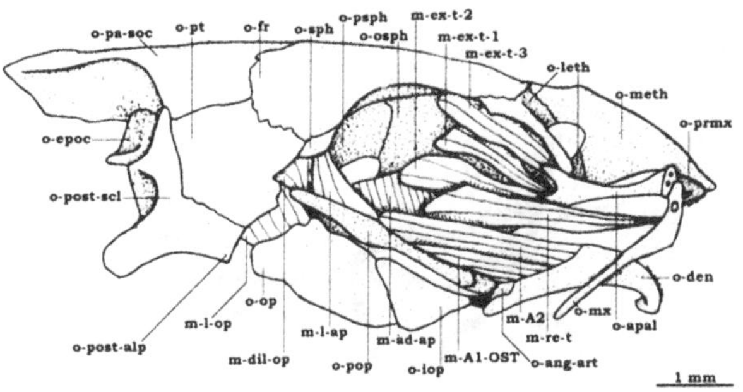

Fig. 3-17 Amphiliidae: lateral view of cephalic musculature of *Leptoglanis rotundiceps*.

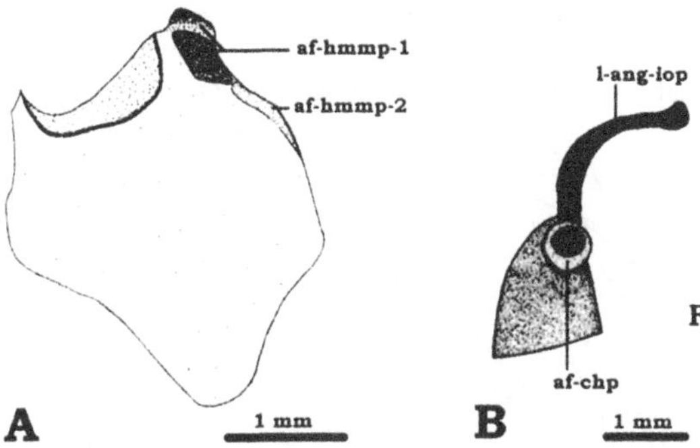

Fig. 3-18 Amphiliidae: *Phractura brevicauda*. **A)** Mesial view of opercular bone. **B)** Mesial view of interopercular bone.

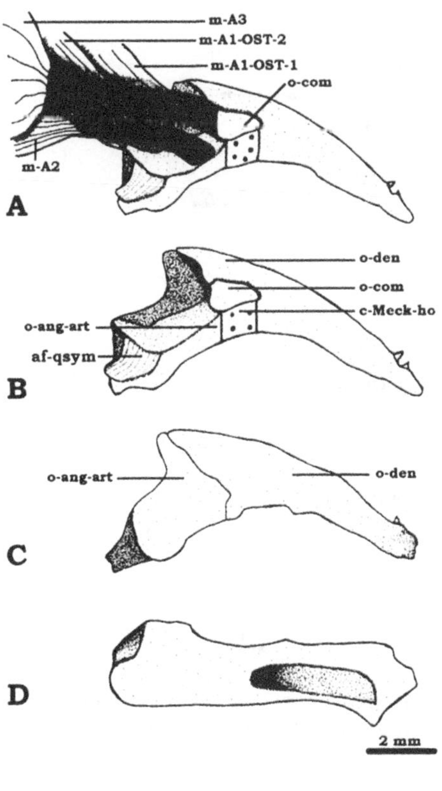

Fig. 3-19 Amphiliidae: *Phractura brevicauda*. **A)** Mandible and adductor mandibulae, medial view. **B)** Mandible, medial view. **C)** Mandible, lateral view. **D)** Mandible, ventral view.

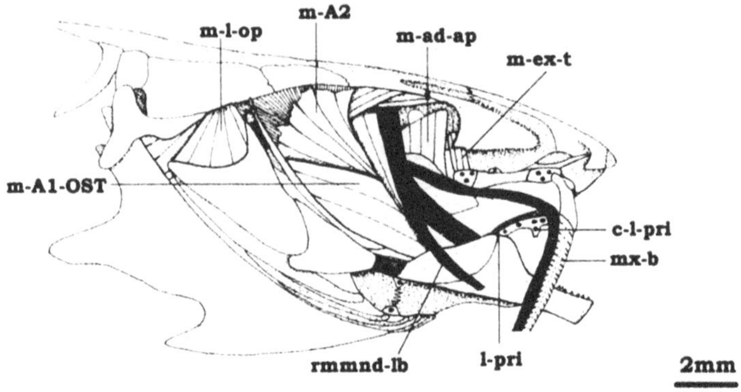

Fig. 3-20 Ariidae: Lateral view of cephalic musculature of *Arius heudeloti*, all muscles exposed.

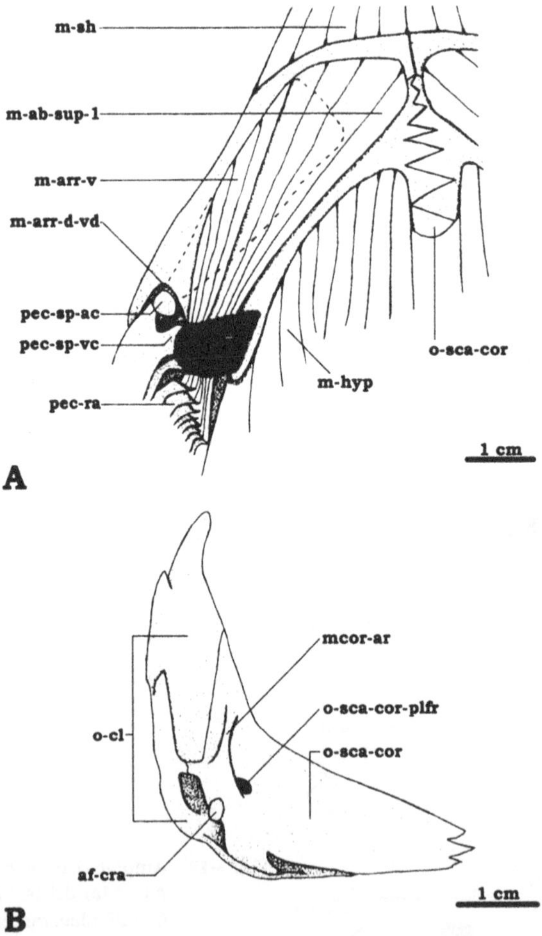

Fig. 3-21 Ariidae: pectoral girdle of *Arius heudeloti*. **A)** Ventral view, all muscles exposed. **B)** Posterior view, posttemporo-supracleithrum and all muscules removed.

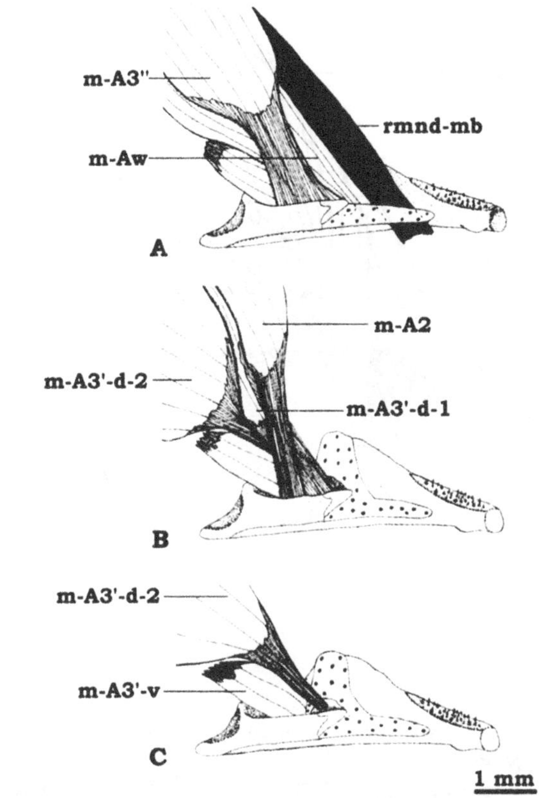

Fig. 3-22 Ariidae: mesial view of mandible and adductor mandibulae muscle of *Arius heudeloti*. **A)** Adductor mandibulae complex exposed. **B)** A3″, Aω and ramus mandibularis removed. **C)** A2 and A-3′-d-1 removed.

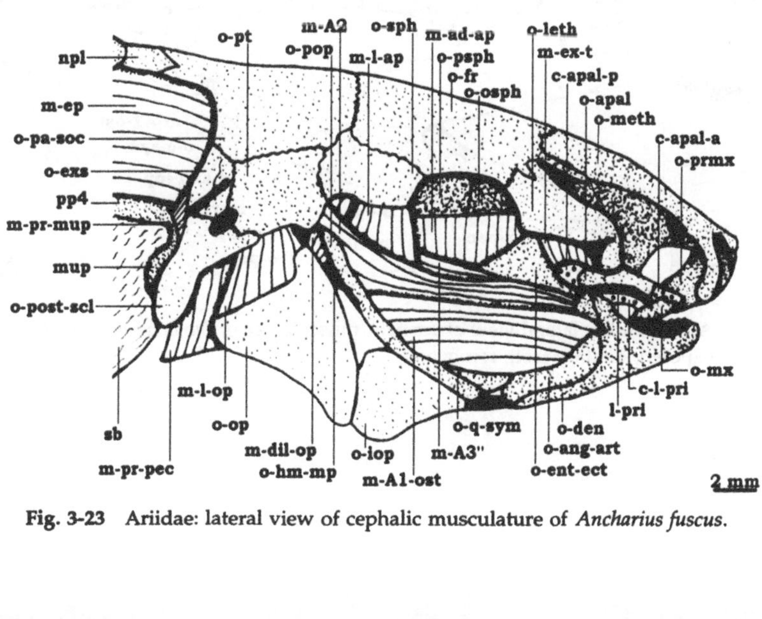

Fig. 3-23 Ariidae: lateral view of cephalic musculature of *Ancharius fuscus*.

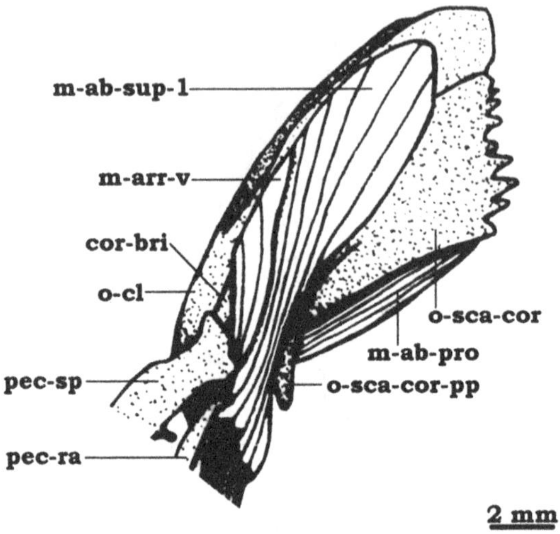

Fig. 3-24 Ariidae: ventral view of pectoral girdle of *Ancharius fuscus*, all muscles exposed.

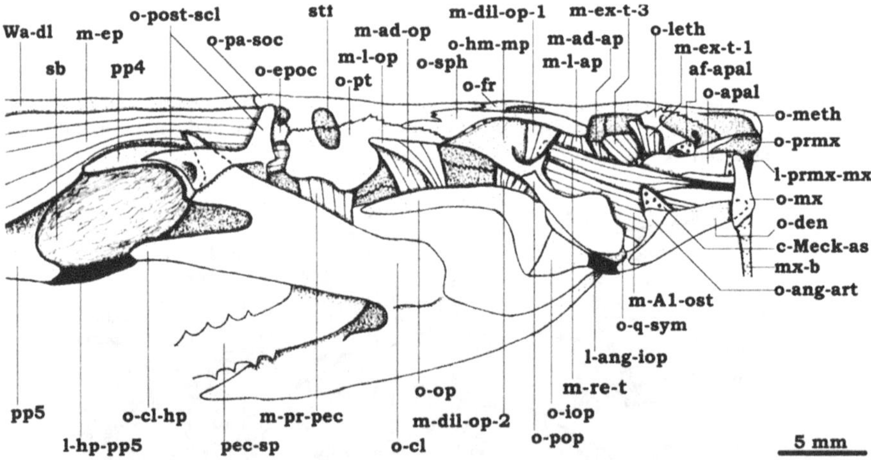

Fig. 3-25 Aspredinidae: lateral view of cephalic musculature of *Bunocephalus knerii*, all muscles exposed, dentary teeth removed.

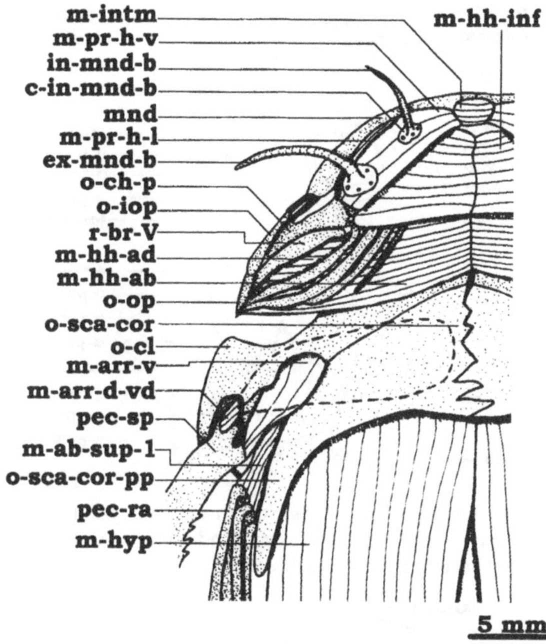

Fig. 3-26 Aspredinidae: ventral view of cephalic region and pectoral girdle of *Bunocephalus knerii*, all muscles exposed.

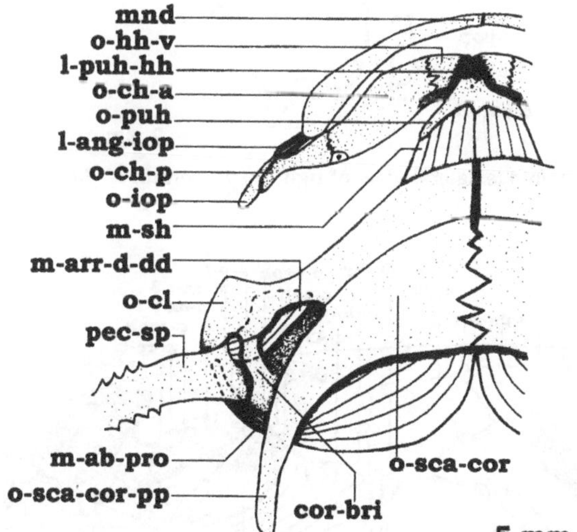

Fig. 3-27 Aspredinidae: ventral view of cephalic region and pectoral girdle of *Bunocephalus knerii*, mandibular barbels, opercle, branchiostegal rays, pectoral rays, protractor hyoidei, intermandibularis, hyohyoideus inferior, hyohyoideus adductor, hyohyoideus abductor, arrector ventralis, abductor superficialis, hypoaxialis and ventral division of arrector dorsalis were removed.

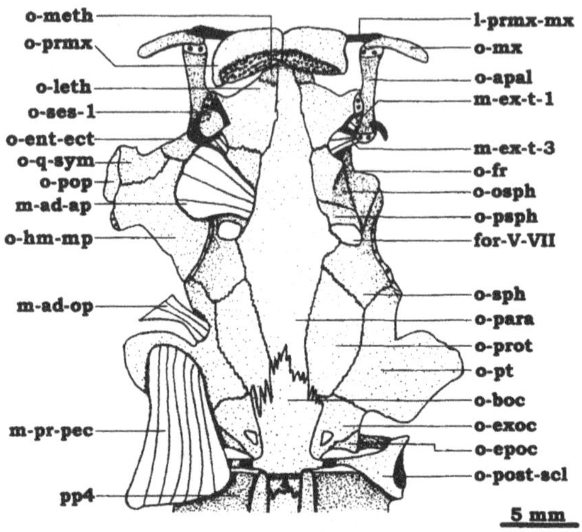

Fig. 3-28 Aspredinidae: ventral view of neurocranium and palatine-maxillary system of *Bunocephalus knerii*. Left side: suspensorium, adductor arcus palatini, adductor operculi and protractor pectoralis, also illustrated.

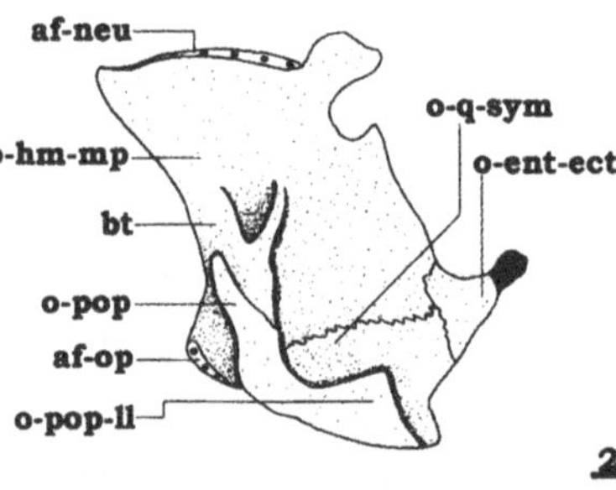

Fig. 3-29 Aspredinidae: lateral view of right suspensorium of *Bunocephalus knerii.*

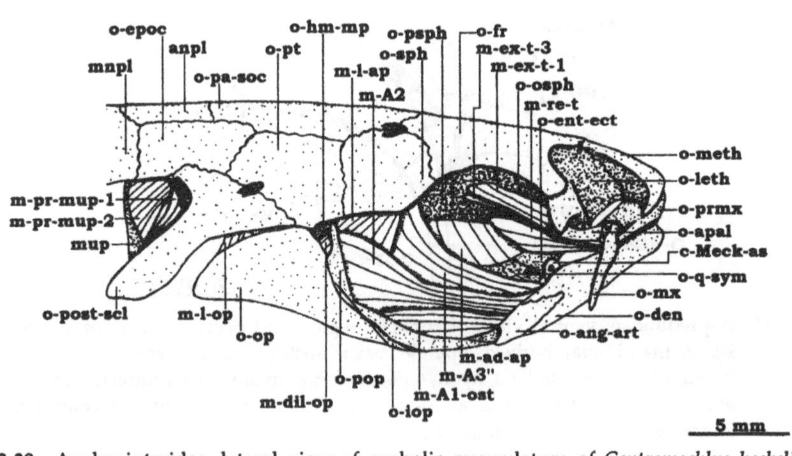

Fig. 3-30 Auchenipteridae: lateral view of cephalic musculature of *Centromochlus heckelii*, all muscles exposed.

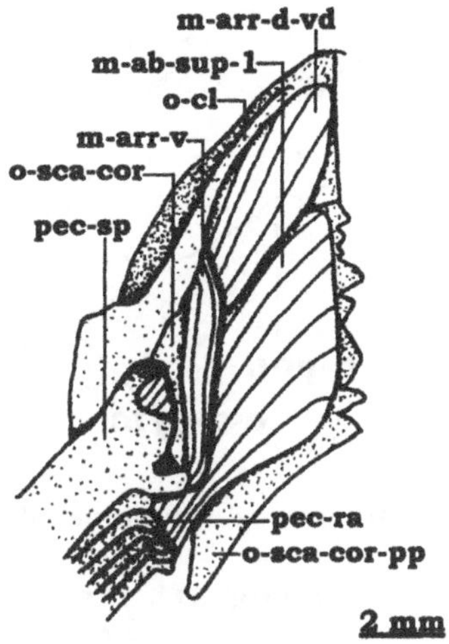

Fig. 3-31 Auchenipteridae: ventral view of pectoral girdle musculature of *Centromochlus heckelii*, all muscles exposed.

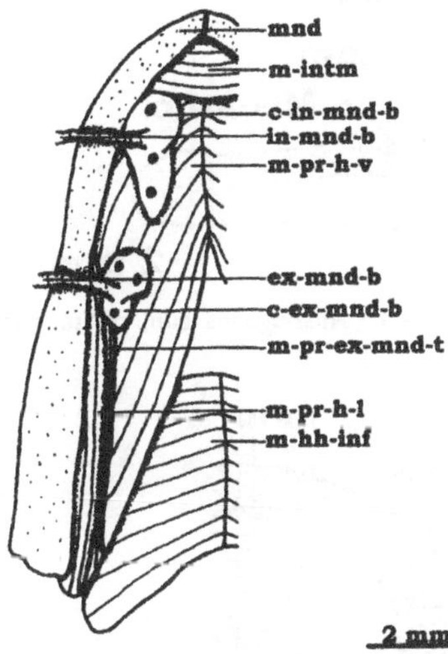

Fig. 3-32 Auchenipteridae: ventral view of ventral cephalic musculature of *Centromochlus heckelii*, all muscles exposed.

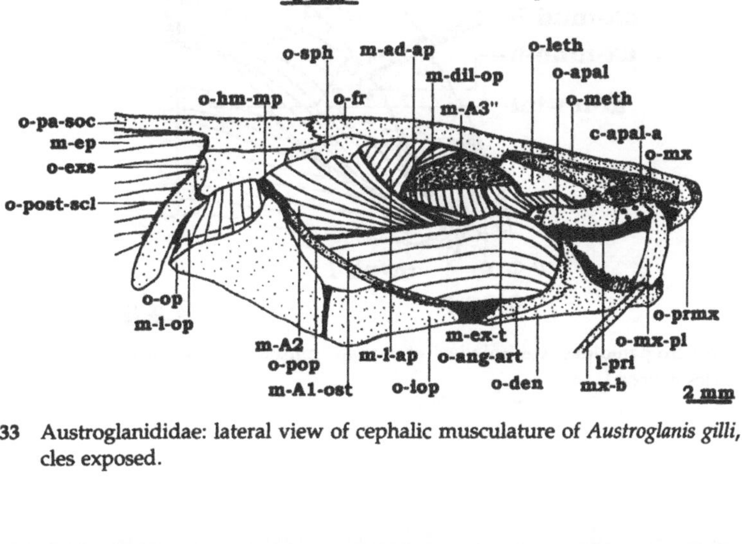

Fig. 3-33 Austroglanididae: lateral view of cephalic musculature of *Austroglanis gilli*, all muscles exposed.

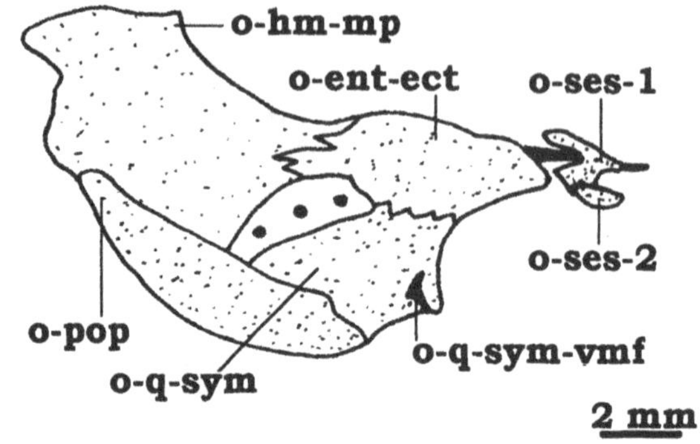

Fig. 3-34 Austroglanididae: mesial view of suspensorium of *Austroglanis gilli*.

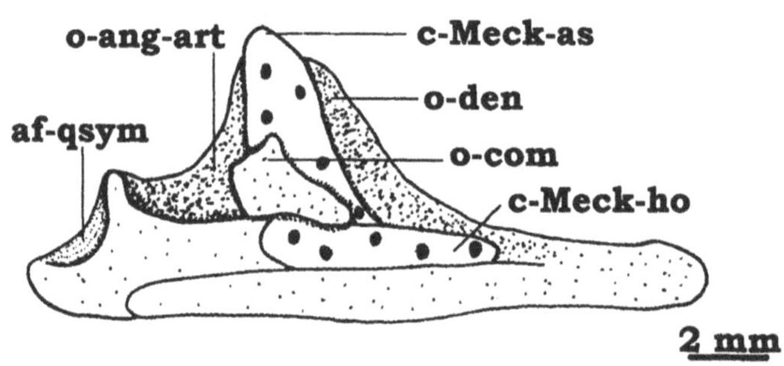

Fig. 3-35 Austroglanididae: mesial view of mandible of *Austroglanis gilli*, dentary teeth removed.

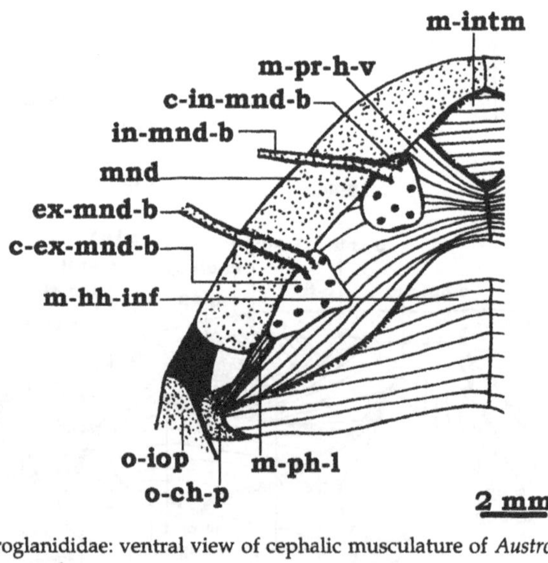

Fig. 3-36 Austroglanididae: ventral view of cephalic musculature of *Austroglanis gilli*, all muscles exposed.

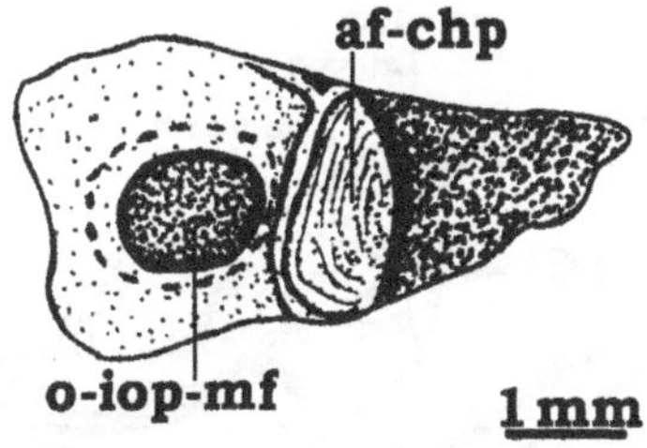

Fig. 3-37 Austroglanididae: mesial view of interopercular bone of *Austroglanis gilli*. Anterior and posterior surfaces correspond to right and left sides respectively.

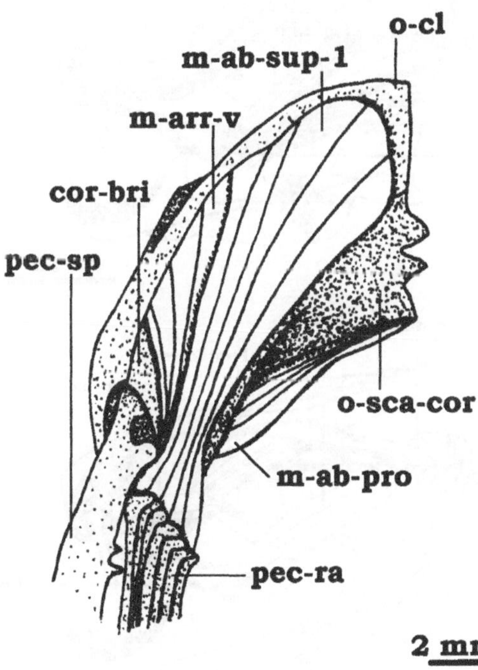

Fig. 3-38 Austroglanididae: ventral view of pectoral girdle musculature of *Austroglanis gilli*, all muscles exposed.

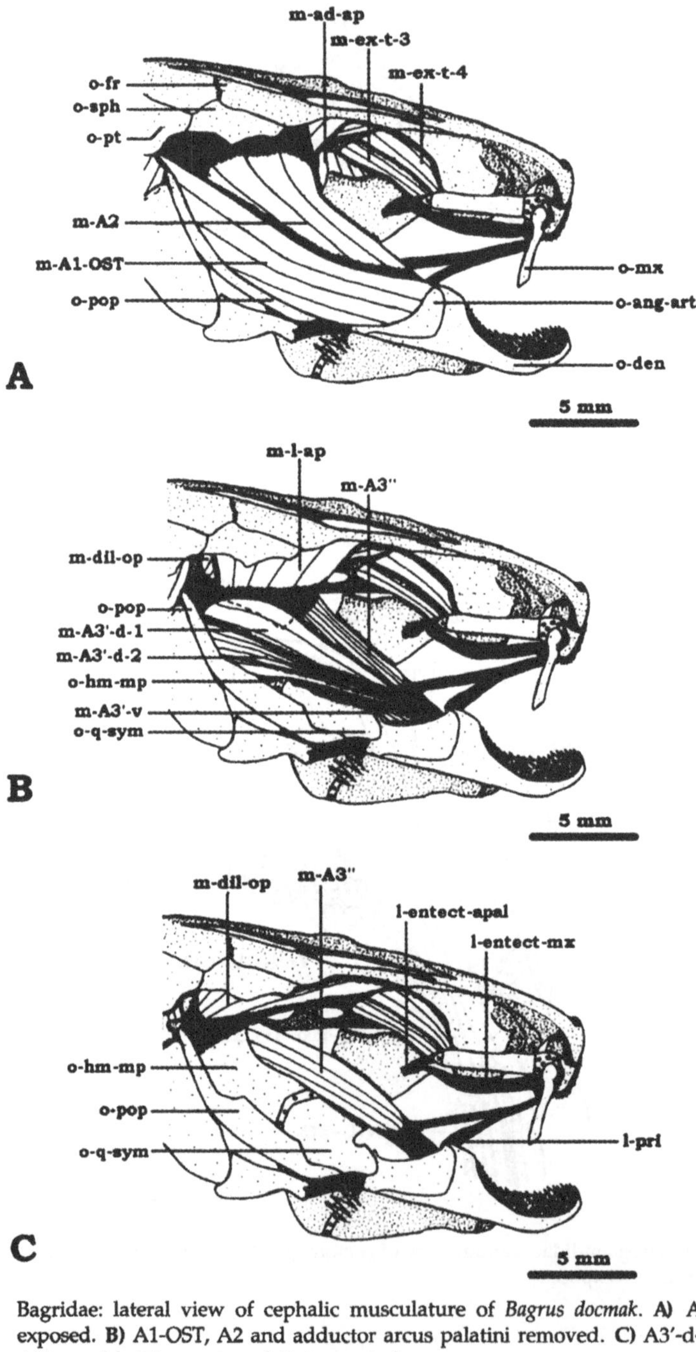

Fig. 3-39 Bagridae: lateral view of cephalic musculature of *Bagrus docmak*. **A)** All muscles exposed. **B)** A1-OST, A2 and adductor arcus palatini removed. **C)** A3′-d-1, A3′-d-2, A3′-v and levator arcus palatini removed.

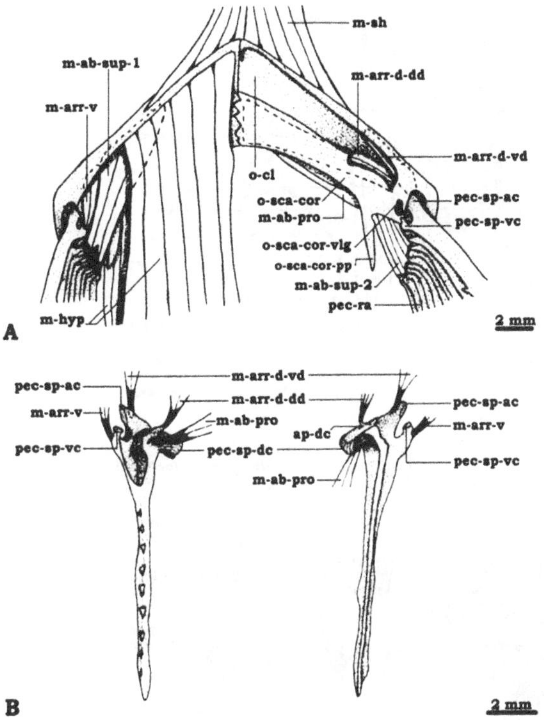

Fig. 3-40 Bagridae *Bagrus docmak*. **A)** Ventral view of pectoral girdle. Left side: all muscles exposed. Right side: hypoaxialis, arrector ventralis, section 1 of abductor superficialis and posttemporo-supracleithrum removed. **B)** Median (left side) and lateral (right side) view of pectoral spine, showing insertions of muscles responsible for its movement.

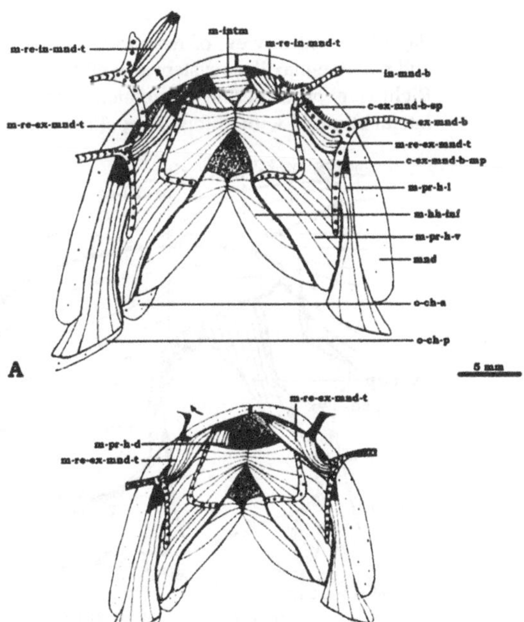

Fig. 3-41 Bagridae: *Bagrus docmak*. **A)** Ventral view of ventral cephalic musculature with, on left side, internal mandibular barbel, its retractor muscle and associated cartilage pulled laterally. **B)** Same view, but internal mandibular barbels, their retractors and supporting parts of associated cartilages, as well as intermandibularis muscle removed. Right side: pars dorsalis of protractor hyoidei also removed.

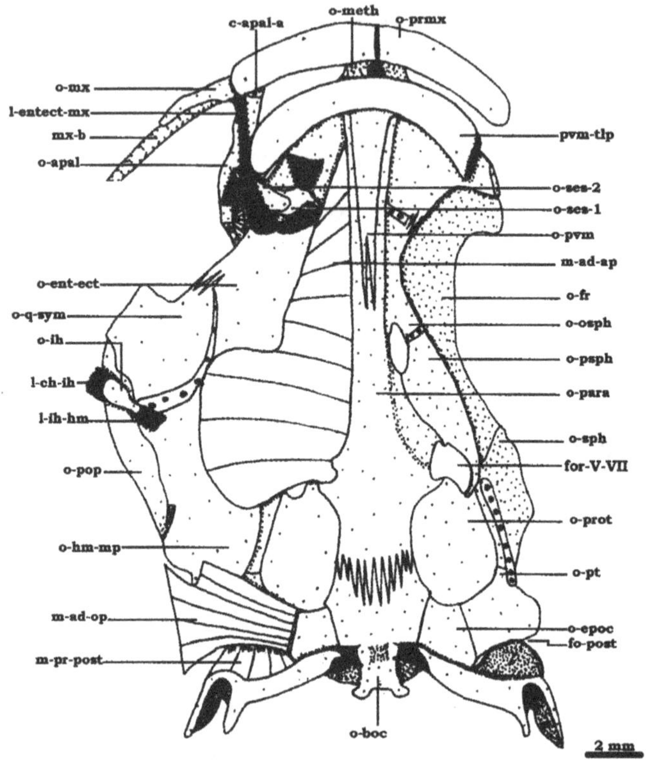

Fig. 3-42 Bagridae: ventral view of neurocranium of *Bagrus docmak*. Left side: suspensorium and palatine-maxillary system as well as associated muscles and ligaments illustrated. Right side: prevomerine anterolateral arm partially cut to show articulatory facet of lateral-ethmoid with autopalatine. Prevomerine and premaxillary teeth removed.

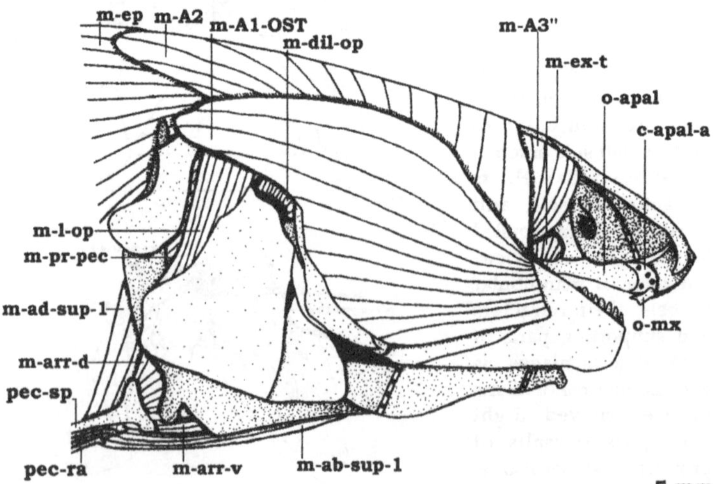

Fig. 3-43 Cetopsidae: lateral view of cephalic and pectoral girdle musculature of *Cetopsis coecutiens*, all muscles exposed.

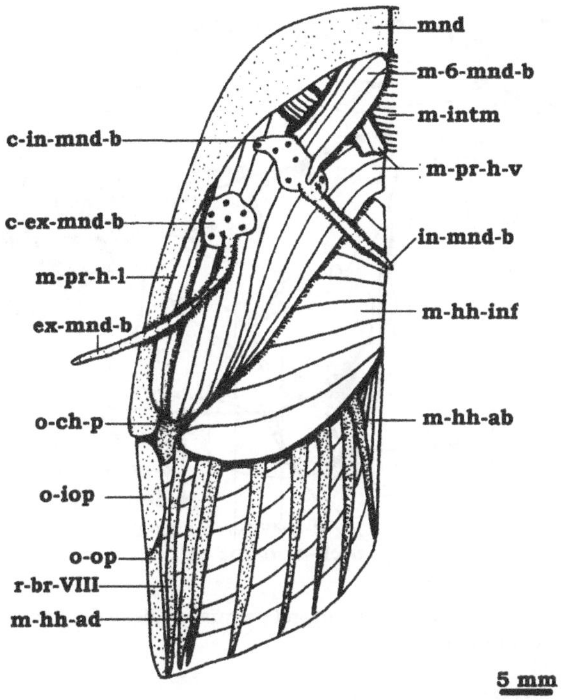

Fig. 3-44 Cetopsidae: ventral view of the cephalic and pectoral girdle musculature of *Cetopsis coecutiens*, all muscles exposed.

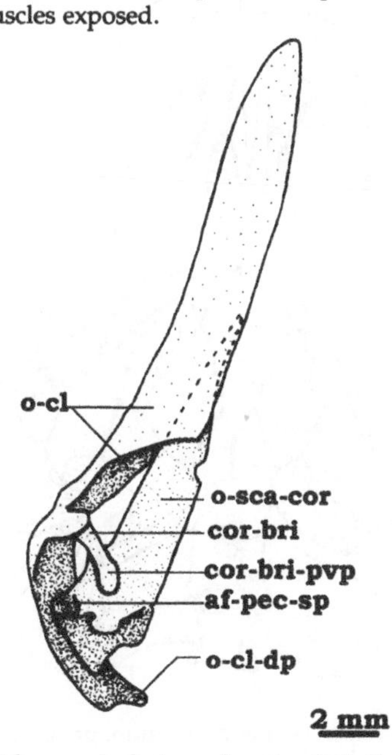

Fig. 3-45 Cetopsidae: ventral view of pectoral girdle of *Cetopsis coecutiens*.

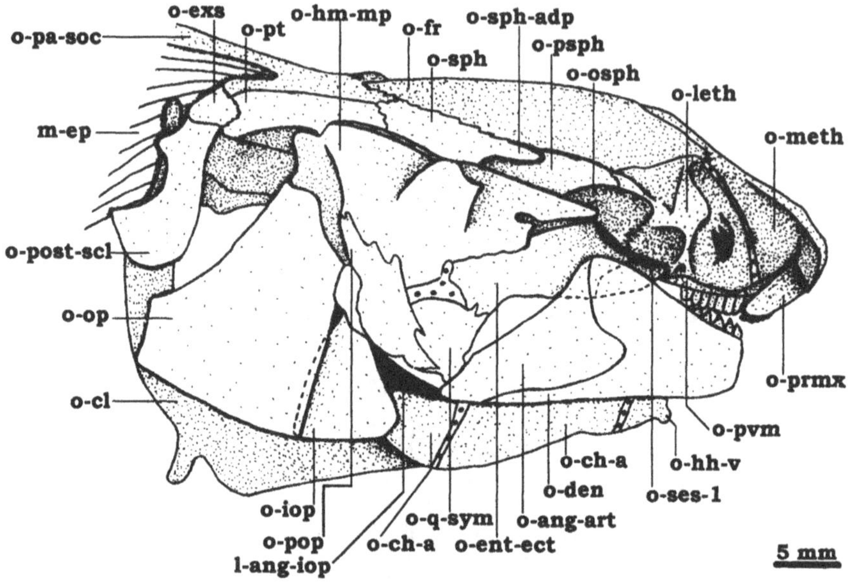

Fig. 3-46 Cetopsidae: lateral view of cranium and pectoral girdle of *Cetopsis coecutiens*. Muscle epaxialis also illustrated.

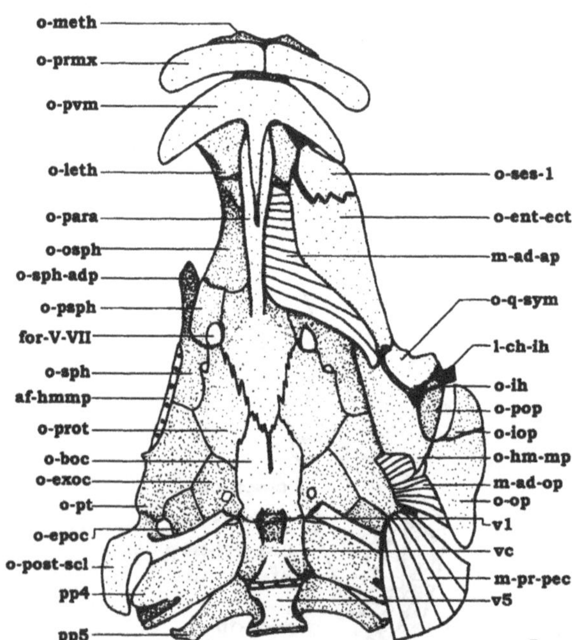

Fig. 3-47 Cetopsidae: ventral view of neurocranium of *Cetopsis coecutiens*. Right side: suspensorium, as well as muscles adductor arcus palatini, adductor operculi and protractor pectoralis, also illustrated. Both premaxillary and prevomerine teeth removed.

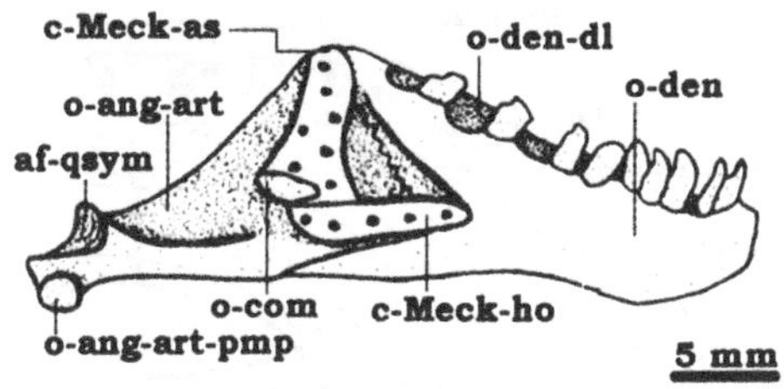

Fig. 3-48 Cetopsidae: mesial view of left lower jaw of *Cetopsis coecutiens*.

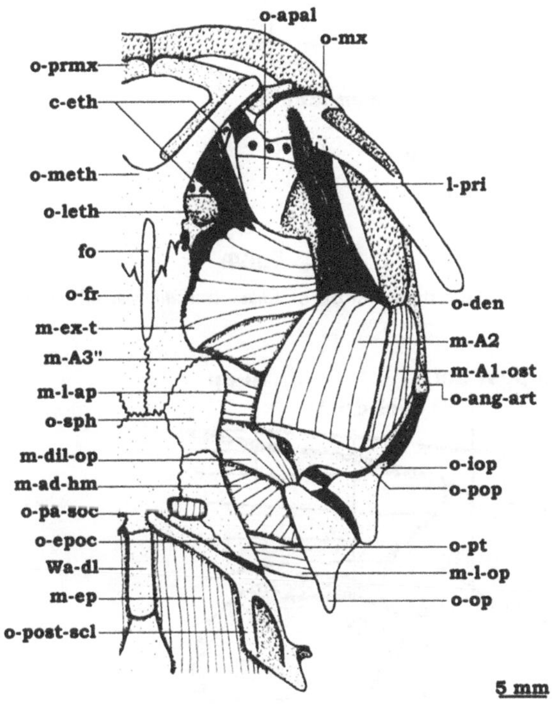

Fig. 3-49 Chacidae: dorsal view of cephalic musculature of *Chaca bankanensis*, all muscles exposed.

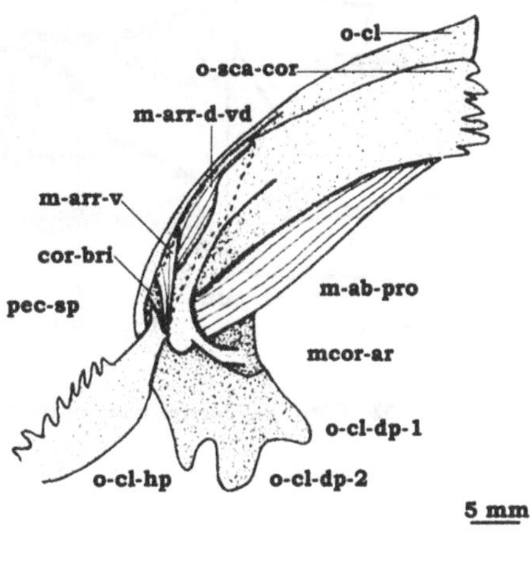

Fig. 3-50 Chacidae: ventral view of pectoral girdle musculature of *Chaca bankanensis*, all muscles exposed.

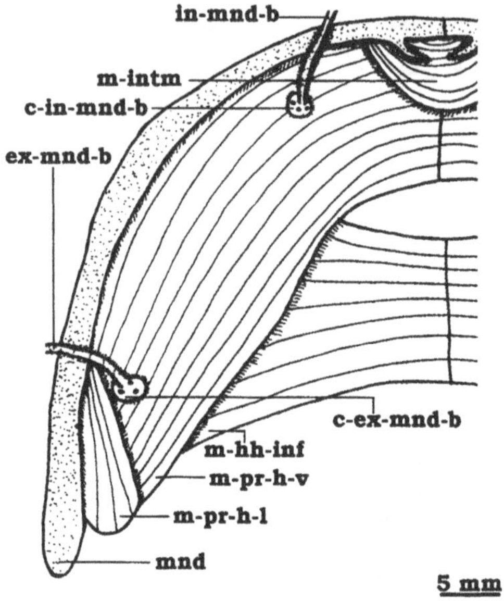

Fig. 3-51 Chacidae: ventral view of cephalic musculature of *Chaca bankanensis*, all muscles exposed.

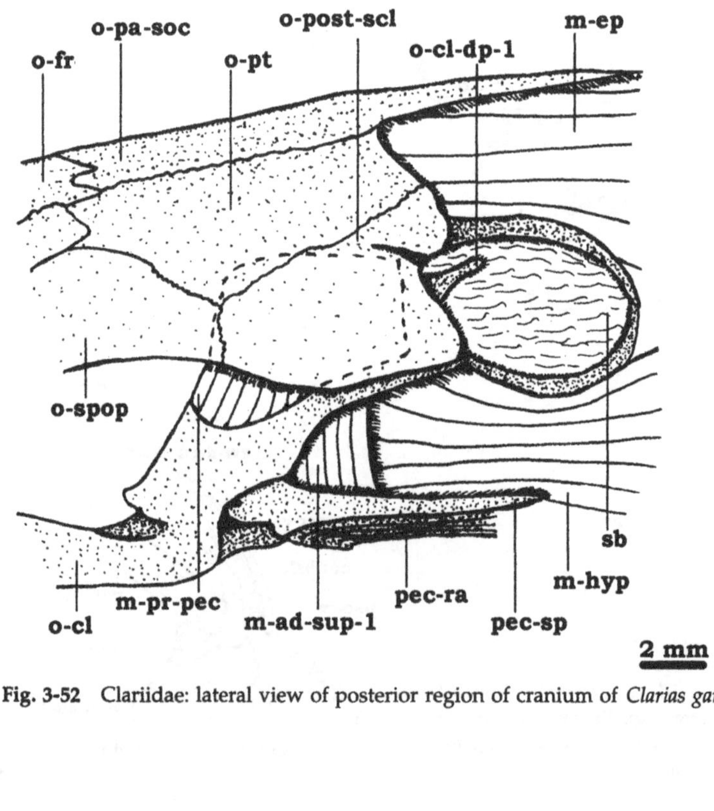

Fig. 3-52 Clariidae: lateral view of posterior region of cranium of *Clarias gariepinus*.

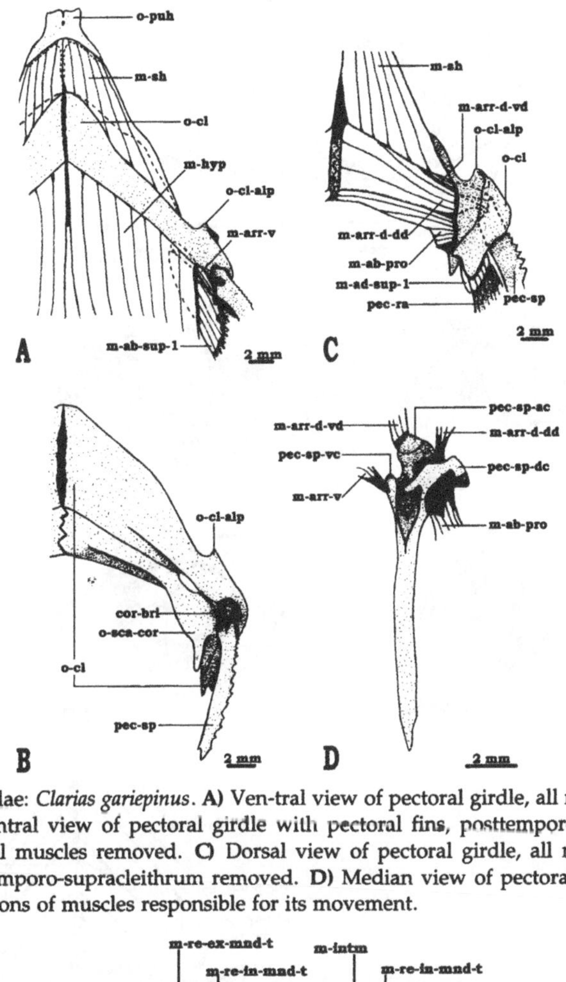

Fig. 3-53 Clariidae: *Clarias gariepinus*. **A)** Ven-tral view of pectoral girdle, all muscles exposed. **B)** Ventral view of pectoral girdle with pectoral fins, posttemporo-supracleithrum and all muscles removed. **C)** Dorsal view of pectoral girdle, all muscles exposed, posttemporo-supracleithrum removed. **D)** Median view of pectoral spine, showing insertions of muscles responsible for its movement.

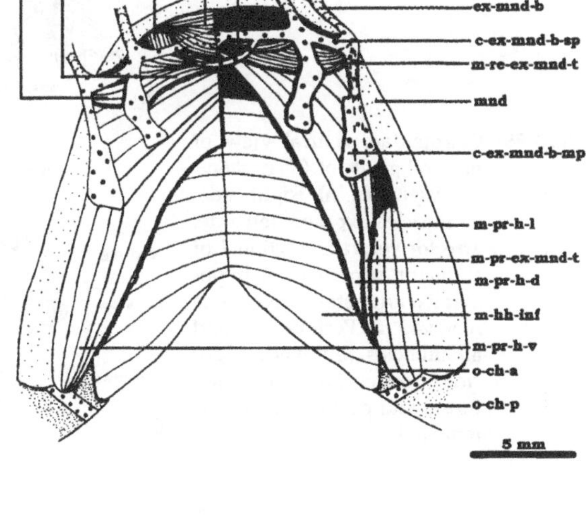

Fig. 3-54 Clariidae: ventral view of cephalic musculature of *Clarias gariepinus*. Right side: pars ventralis of protractor hyoidei, intermandibularis, muscle 2 of mandibular barbels and depressor interni mandibularis tentaculi removed.

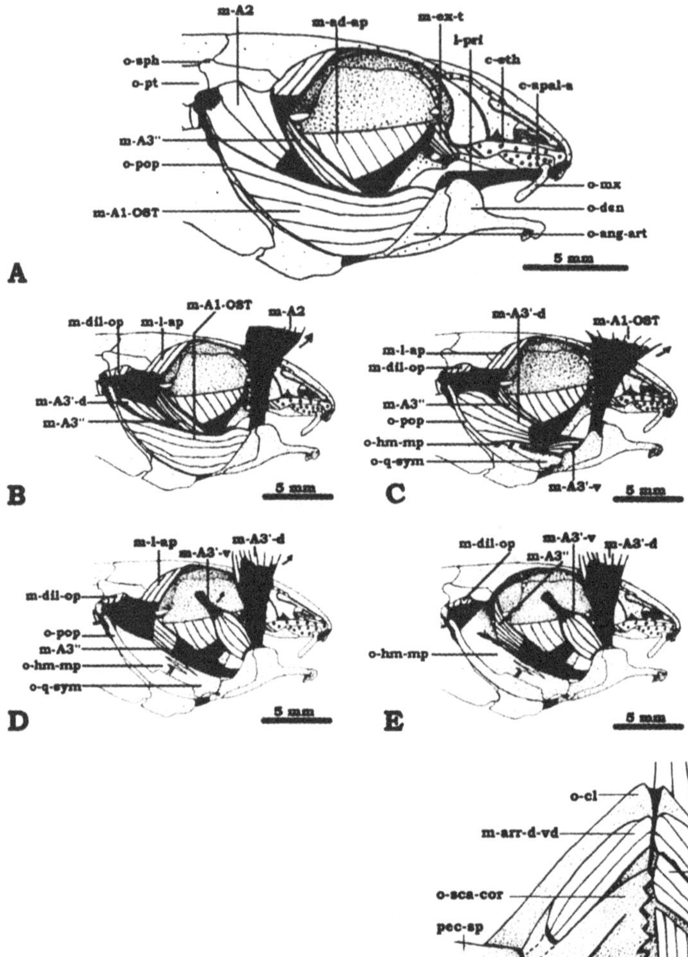

Fig. 3-55 Claroteidae: lateral view of cephalic musculature of *Chrysichthys nigrodigitatus*, mandibular teeth removed. **A)** All muscles exposed. **B)** Ligamentum primordium removed and A2 folded back. **C)** A2 removed and A1-OST folded back. **D)** A3'-d and A3'-v folded back and A1-OST removed. **E)** Levator arcus palatini removed.

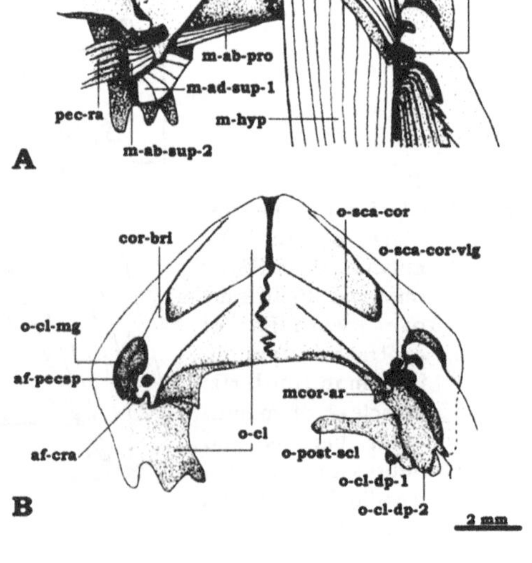

Fig. 3-56 Claroteidae: ventral view of pectoral girdle of *Chrysichthys nigrodigitatus*. **A)** Right side: all muscles exposed; left side: arrector ventralis, section 1 of abductor superficialis, hypoaxialis and sternohyoideus removed. **B)** Pectoral fins and all muscles removed. Right side: posttemporo-supracleithrum and pectoral spine also removed.

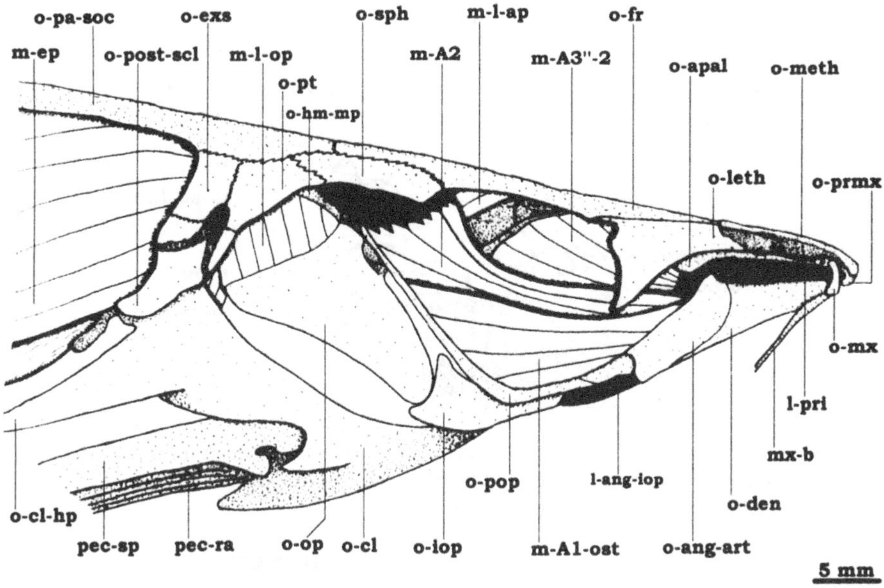

Fig. 3-57 Cranoglanididae: lateral view of cephalic musculature of *Cranoglanis bouderius*, all muscles exposed.

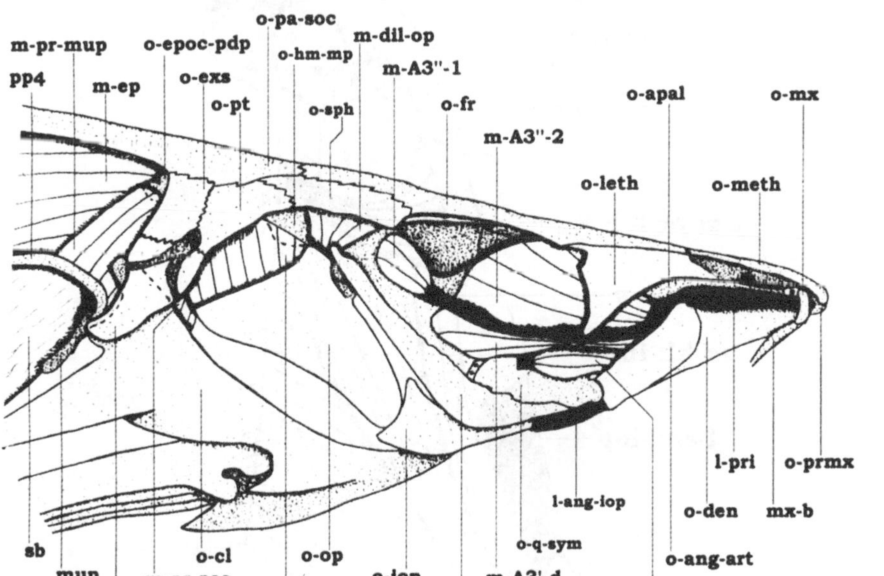

Fig. 3-58 Cranoglanididae: lateral view of cephalic musculature of *Cranoglanis bouderius*. Levator operculi, A1-ost, A2, levator arcus palatini, and anterolateral fibres of epaxialis removed.

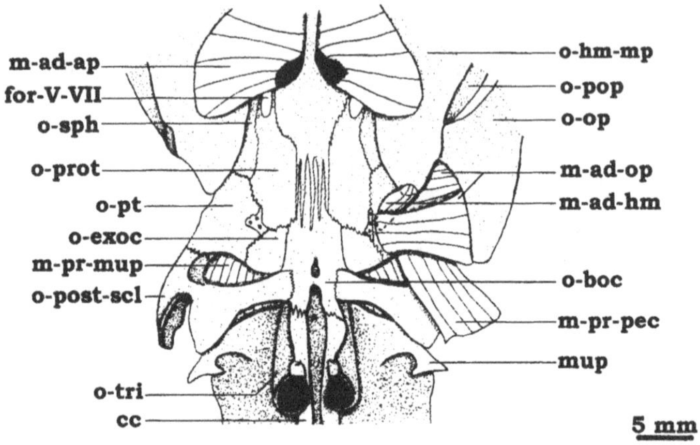

Fig. 3-59 Cranoglanididae: ventral view of posterior region of neurocranium and suspensorium of *Cranoglanis bouderius*. Left side: opercular bone, adductor hyomandibularis, adductor operculi and protractor pectoralis removed.

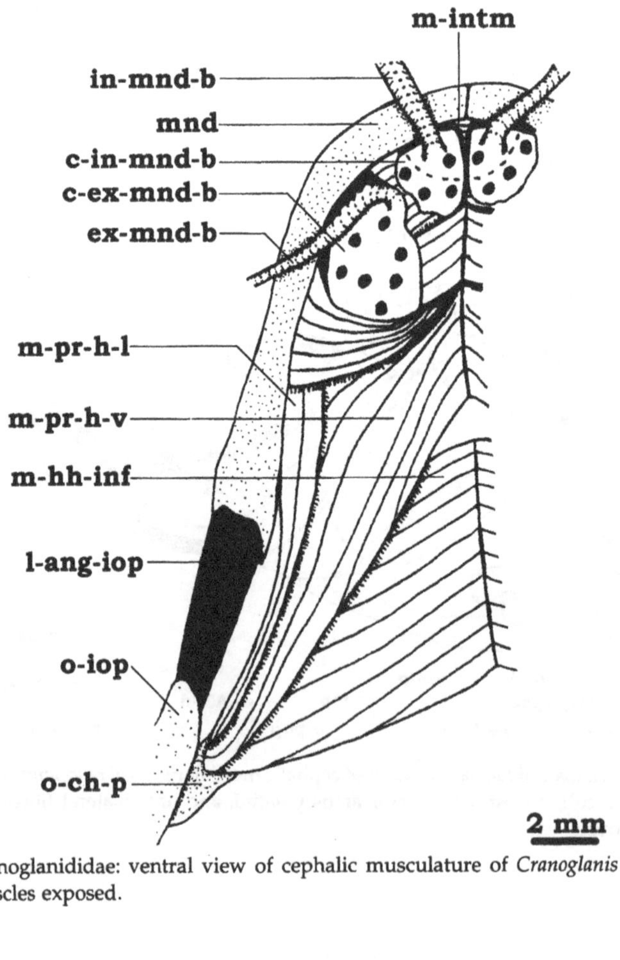

Fig. 3-60 Cranoglanididae: ventral view of cephalic musculature of *Cranoglanis bouderius*, all muscles exposed.

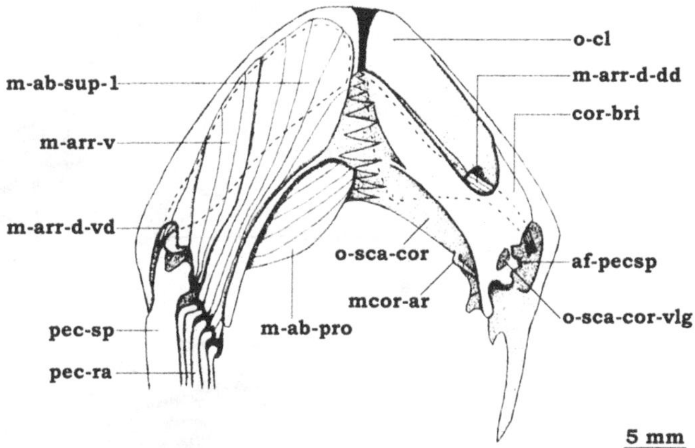

Fig. 3-61 Cranoglanididae: ventral view of the pectoral girdle of *Cranoglanis bouderius*. Left side: all muscles exposed. Right side: arrector ventralis, abductor superficialis, abductor profundus and dorsal division of arrector dorsalis removed.

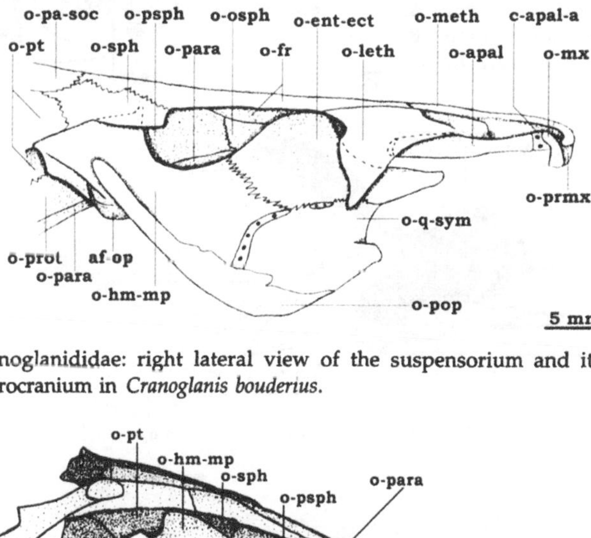

Fig. 3-62 Cranoglanididae: right lateral view of the suspensorium and its attachment to neurocranium in *Cranoglanis bouderius*.

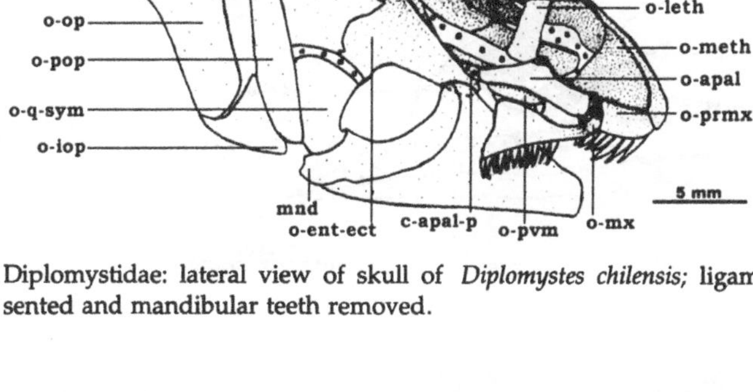

Fig. 3-63 Diplomystidae: lateral view of skull of *Diplomystes chilensis;* ligaments not represented and mandibular teeth removed.

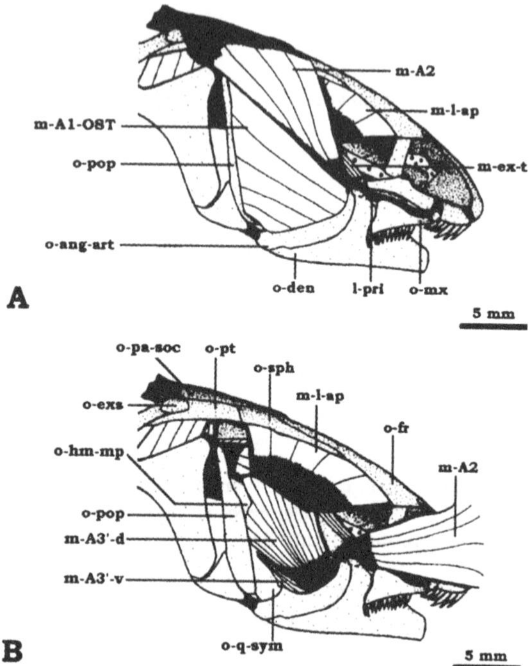

Fig. 3-64 Diplomystidae: lateral view of cheek musculature of *Diplomystes chilensis*. **A)** All muscles exposed. **B)** A1-OST and A2 folded back.

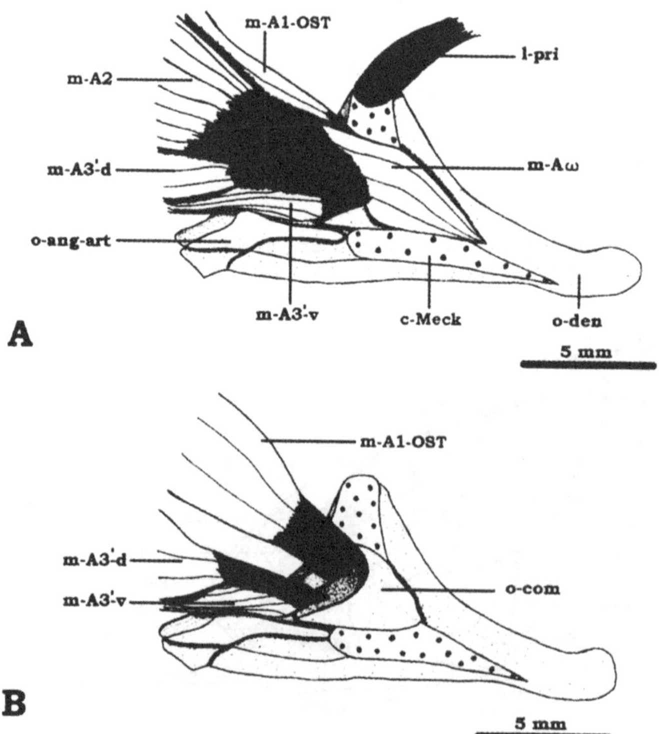

Fig. 3-65 Diplomystidae: medial view of left lower jaw of *Diplomystes chilensis*. **A)** Adductor mandibulae complex exposed. **B)** A1-OST insertion on lateral face of mandible, Aw, A2 and ligamentum primordium removed.

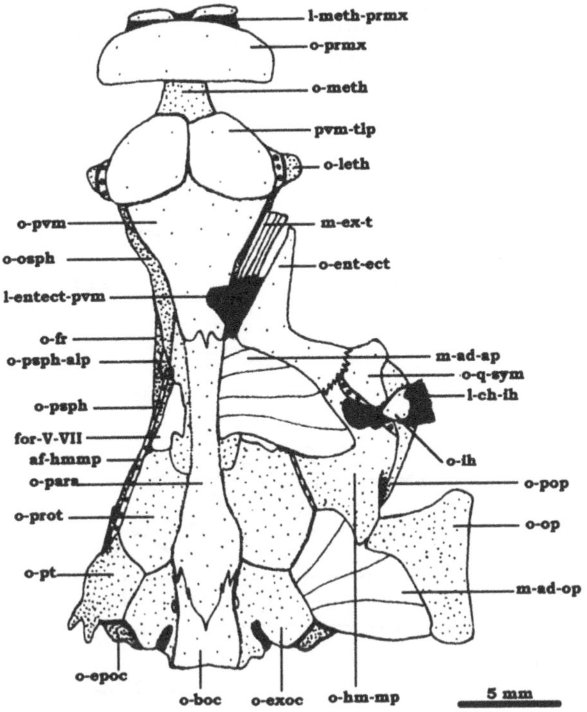

Fig. 3-66 Diplomystidae: ventral view of neurocranium of *Diplomystes chilensis*, premaxillary and prevomerine teeth removed. Right side: suspensorium and associated muscles and ligaments also represented.

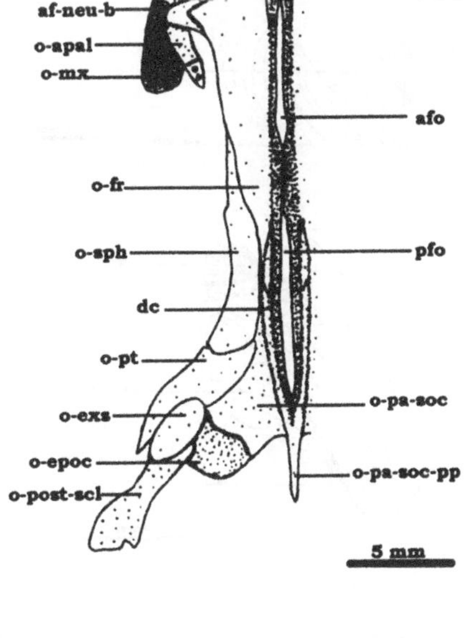

Fig. 3-67 Diplomystidae: dorsal view of neurocranium of *Diplomystes chilensis*. Right side: autopalatine and maxilla also represented.

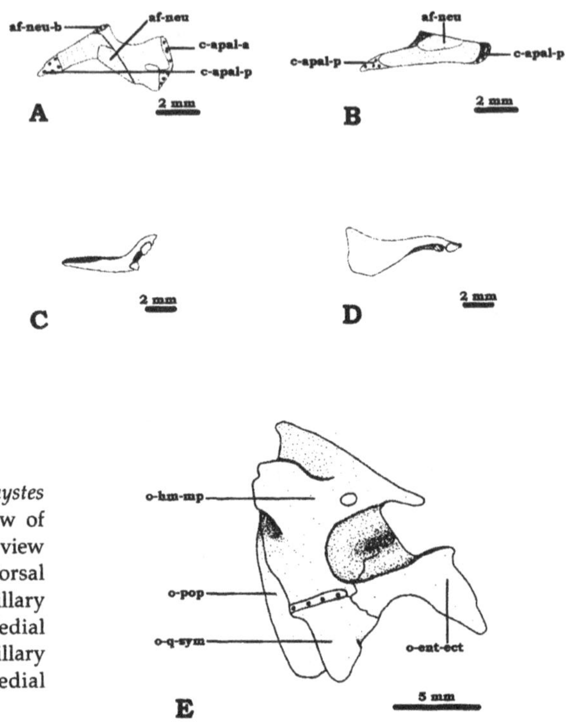

Fig. 3-68 Diplomystidae: *Diplomystes chilensis*. **A)** Dorsal view of autopalatine. **B)** Medial view of autopalatine. **C)** Dorsal view of maxilla, maxillary teeth removed. **D)** Medial view of maxilla, maxillary teeth removed. **E)** Medial view of suspensorium.

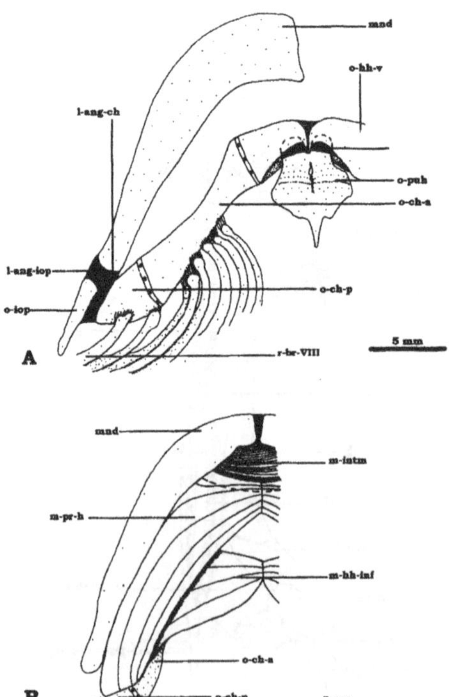

Fig. 3-69 Diplomystidae: *Diplomystes chilensis*. **A)** Ventral view of hyoid arch and mandible, musculature not illustrated. **B)** Ventral view of cephalic musculature, all muscles exposed.

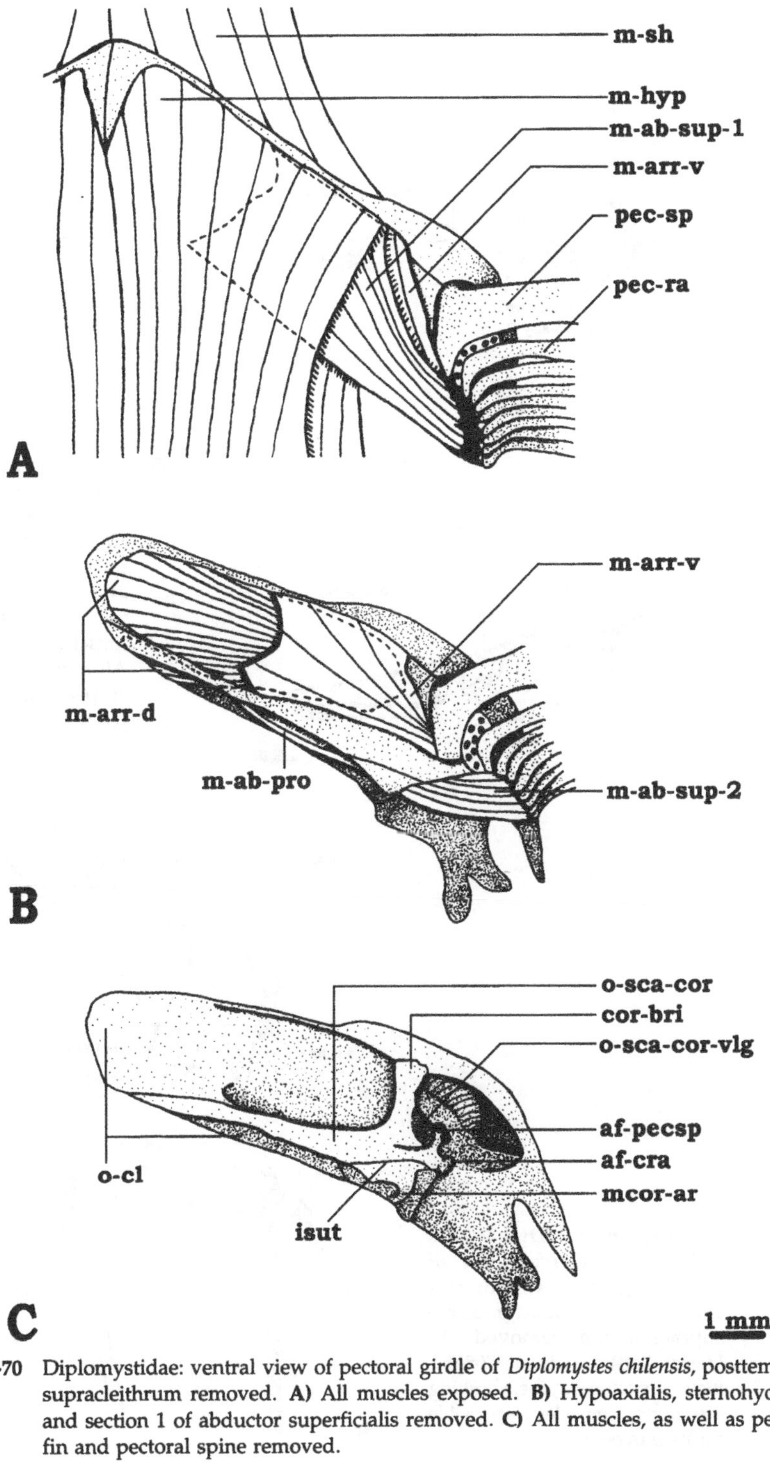

Fig. 3-70 Diplomystidae: ventral view of pectoral girdle of *Diplomystes chilensis*, posttemporo-supracleithrum removed. **A)** All muscles exposed. **B)** Hypoaxialis, sternohyoideus and section 1 of abductor superficialis removed. **C)** All muscles, as well as pectoral fin and pectoral spine removed.

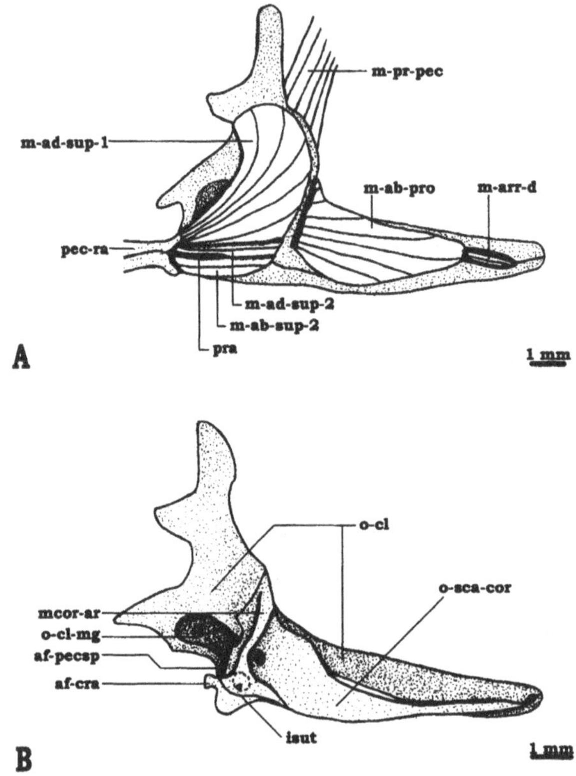

Fig. 3-71 Diplomystidae: posterior view of pectoral girdle of *Diplomystes chilensis*, posttemporo-supracleithrum removed. **A)** All muscles exposed. **B)** All muscles, as well as pectoral fin and pectoral spine removed.

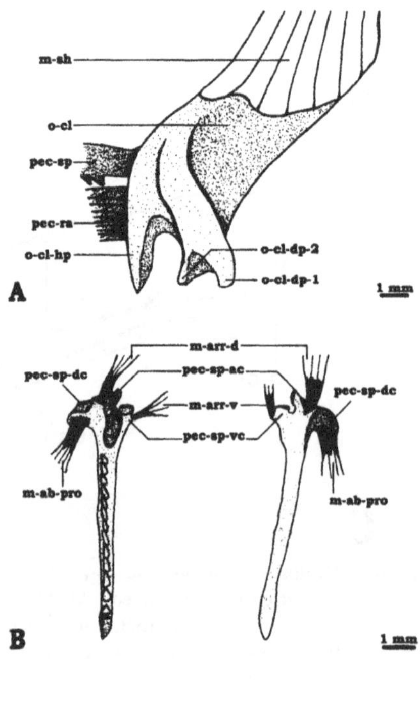

Fig. 3-72 Diplomystidae: *Diplomystes chilensis*. **A)** Dorsal view of pectoral girdle. All muscles exposed; posttemporo-supracleithrum removed. **B)** Mesial (left) and lateral (right) view of pectoral spine, showing insertions of muscles responsible for its movement.

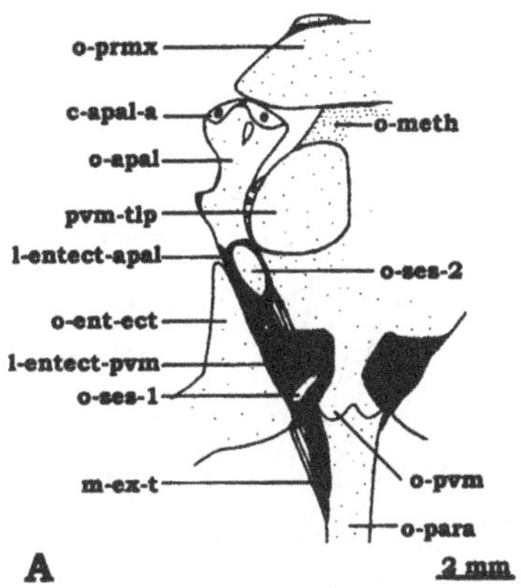

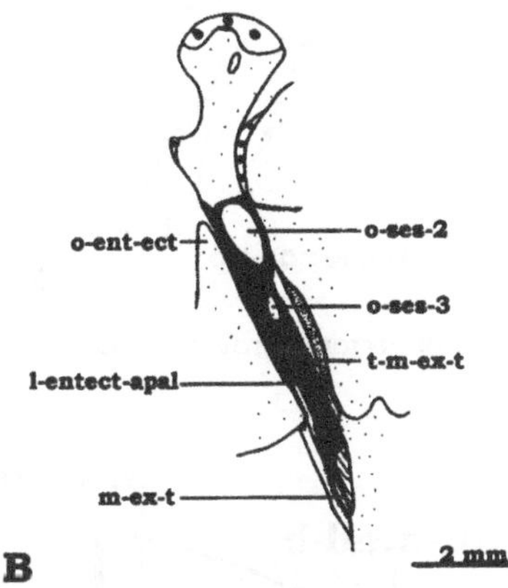

Fig. 3-73 Diplomystidae: ventral view of anterior region of suspensorium of *Diplomystes chilensis*. **A)** Prevomerine and premaxillary teeth removed. **B)** Sesamoid bone 1 and ligament between entoectopterygoid and prevomer removed.

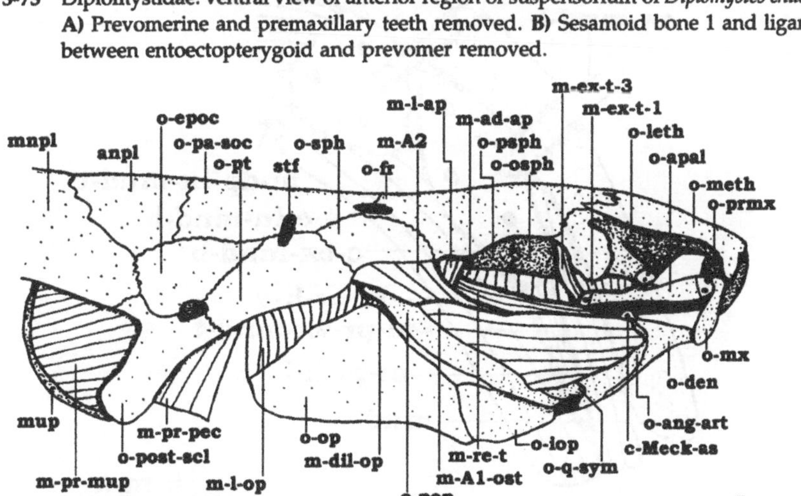

Fig. 3-74 Doradidae: lateral view of cephalic musculature of *Franciscodoras marmoratus*, all muscles exposed.

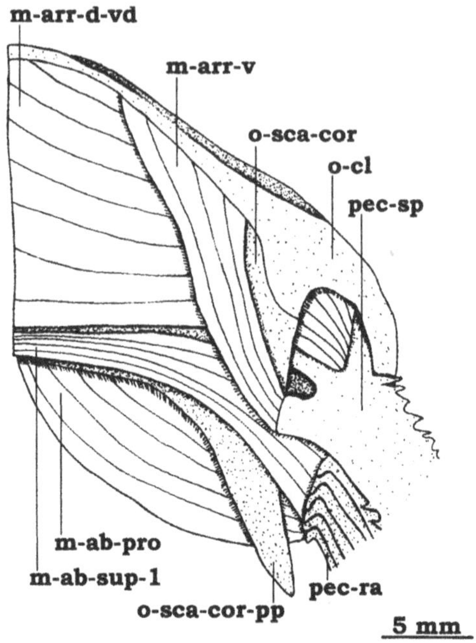

Fig. 3-75 Doradidae: ventral view of pectoral girdle musculature of *Franciscodoras marmoratus*, all muscles exposed.

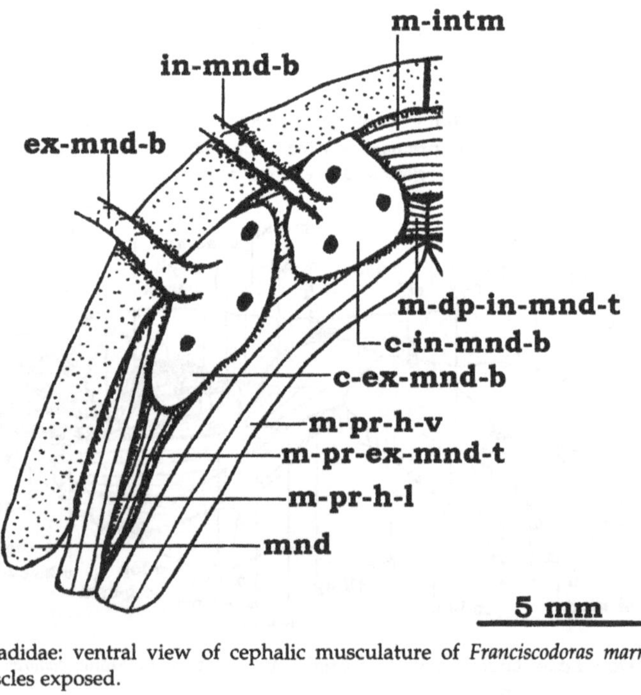

Fig. 3-76 Doradidae: ventral view of cephalic musculature of *Franciscodoras marmoratus*, all muscles exposed.

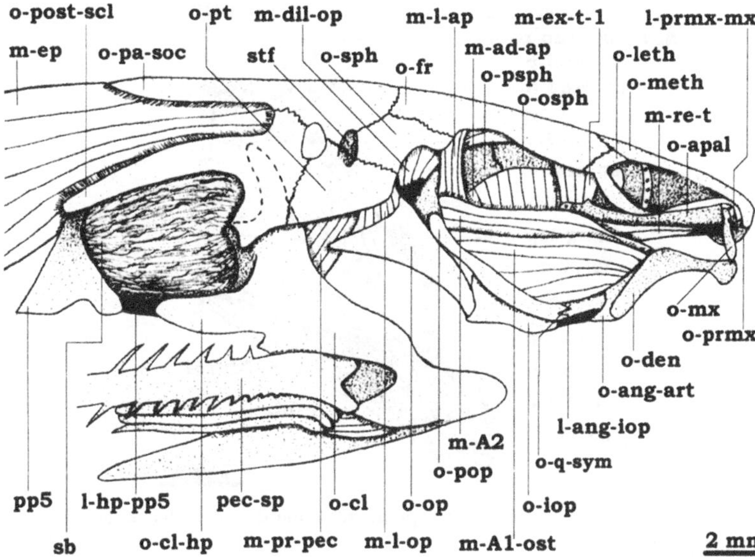

Fig. 3-77 Erethistidae: lateral view of cephalic musculature of *Erethistes pusillus*, all muscles exposed; dentary and premaxillary teeth removed.

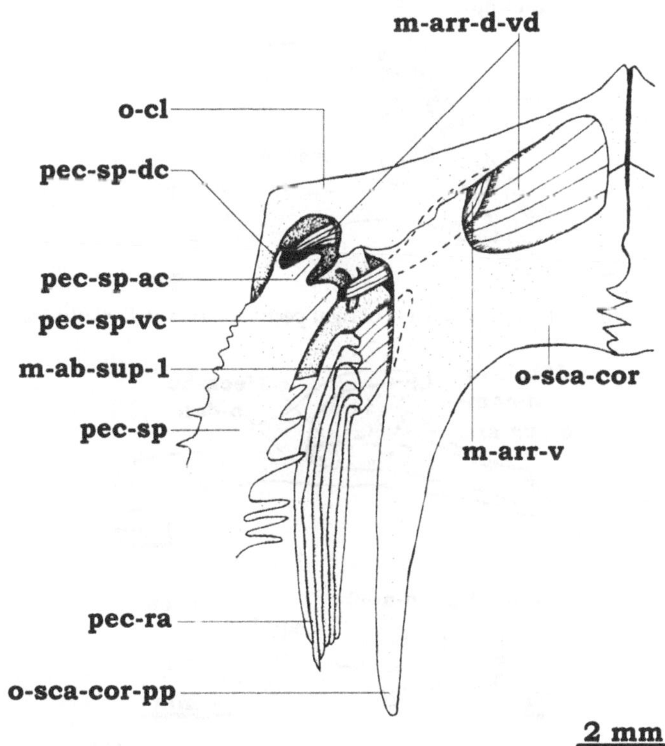

Fig. 3-78 Erethistidae: ventral view of pectoral girdle of *Erethistes pusillus*, all muscles exposed.

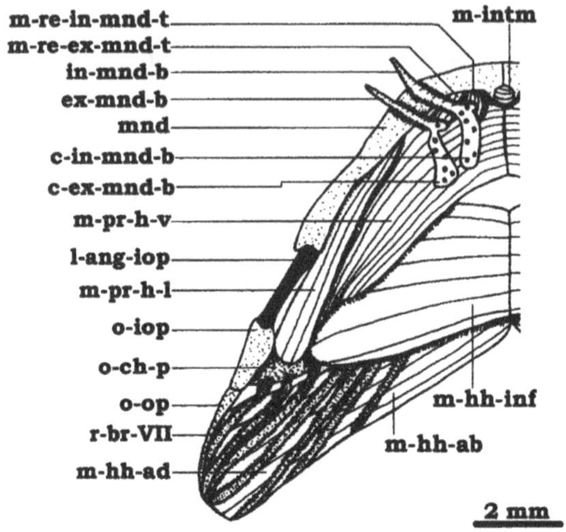

Fig. 3-79 Erethistidae: ventral view of cephalic musculature of *Erethistes pusillus*, all muscles exposed.

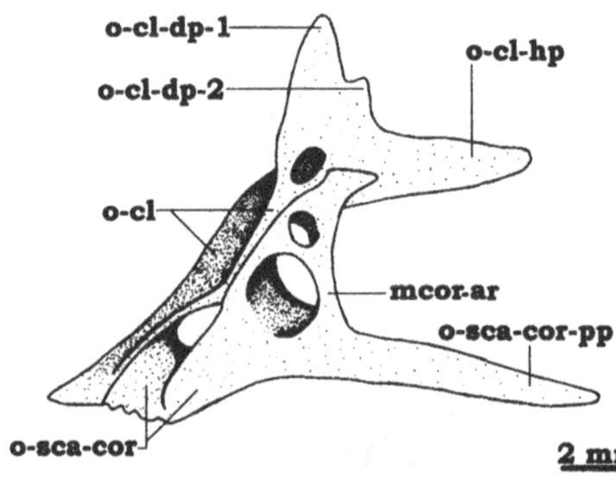

Fig. 3-80 Erethistidae: medial view of right pectoral girdle of *Erethistes pusillus*.

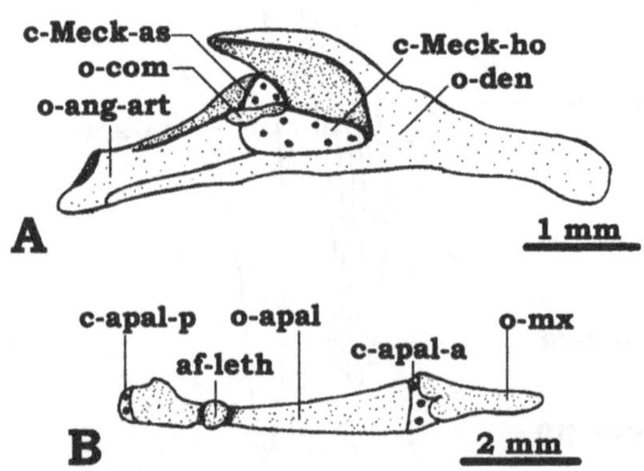

Fig. 3-81 Erethistidae: *Erethistes pusillus*. **A)** Medial view of left mandible, mandibular teeth removed. **B)** Medial view of left autopalatine and maxilla.

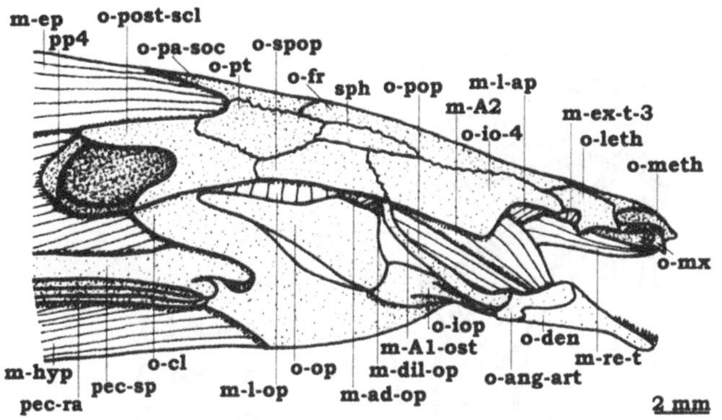

Fig. 3-82 Heteropneustidae: lateral view of cephalic musculature of *Heteropneustes fossilis*, all muscles exposed.

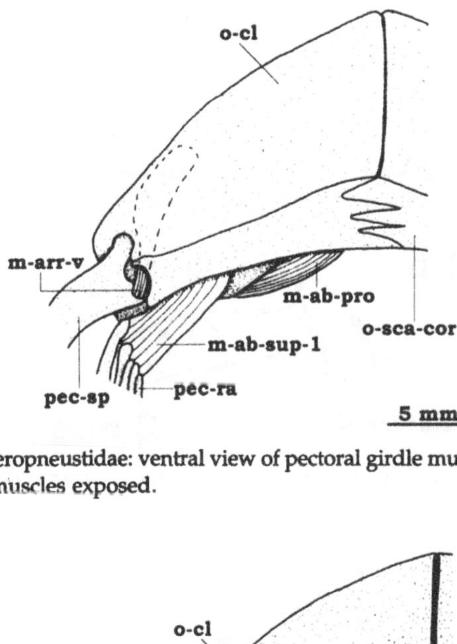

Fig. 3-83 Heteropneustidae: ventral view of pectoral girdle musculature of *Heteropneustes fossilis*, all muscles exposed.

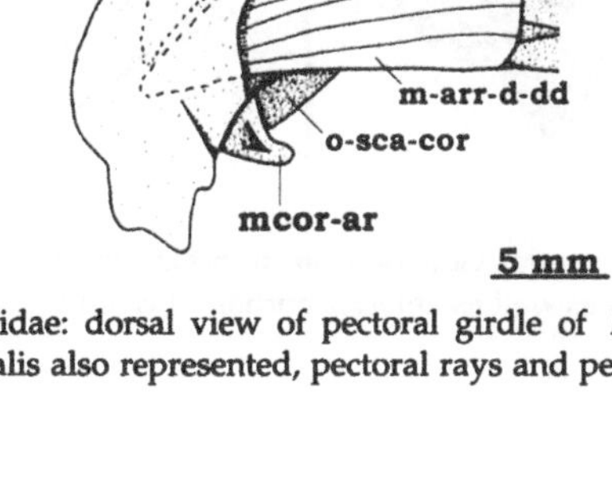

Fig. 3-84 Heteropneustidae: dorsal view of pectoral girdle of *Heteropneustes fossilis*. Muscle arrector dorsalis also represented, pectoral rays and pectoral spine removed.

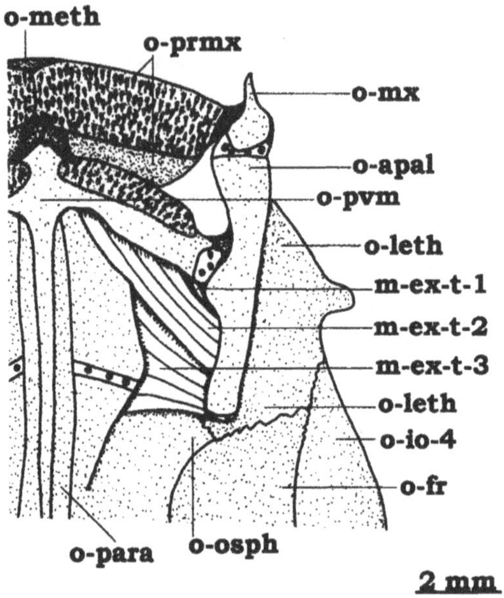

Fig. 3-85 Heteropneustidae: ventral view of anterior region of neurocranium of *Heteropneustes fossilis*. Autopalatine, maxilla and the muscle extensor tentaculi also illustrated.

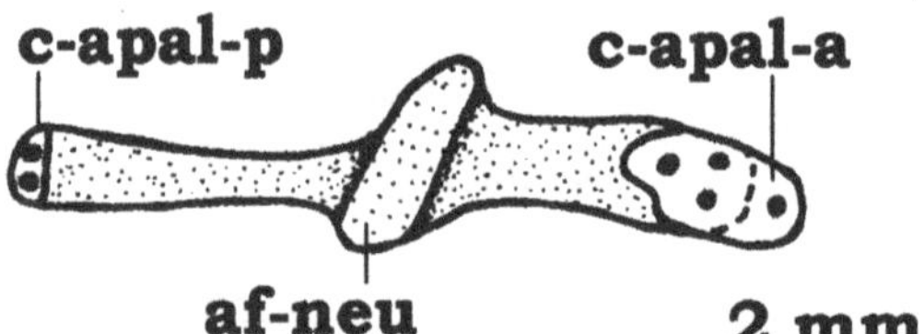

Fig. 3-86 Heteropneustidae: mesial view of left autopalatine of *Heteropneustes fossilis*.

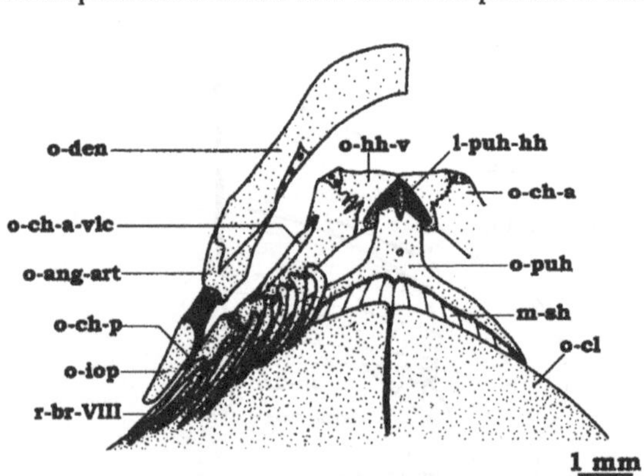

Fig. 3-87 Heteropneustidae: ventral view of splanchnocranium of *Heteropneustes fossilis*. Muscle sternohyoideus as well as anterior portion of cleithrum, also represented.

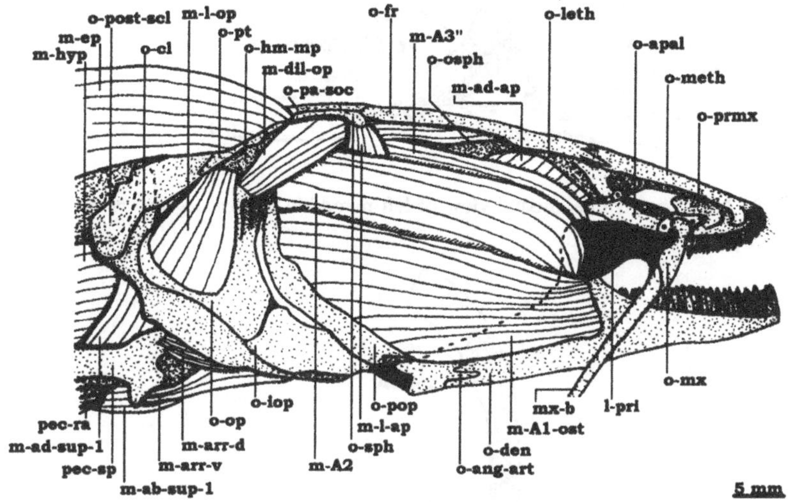

Fig. 3-88 Nematogenyidae: lateral view of cephalic musculature of *Nematogenys inermis*, all muscles exposed.

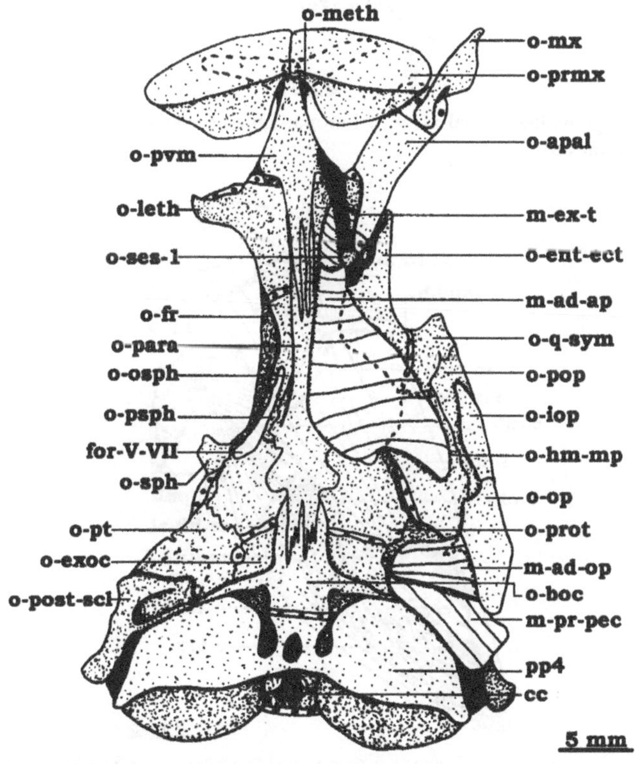

Fig. 3-89 Nematogenyidae: ventral view of neurocranium of *Nematogenys inermis*. Right side: suspensorium, as well as autopalatine, maxilla, adductor arcus palatini, extensor tentaculi, adductor operculi and protractor pectoralis, also illustrated. Premaxillary teeth removed.

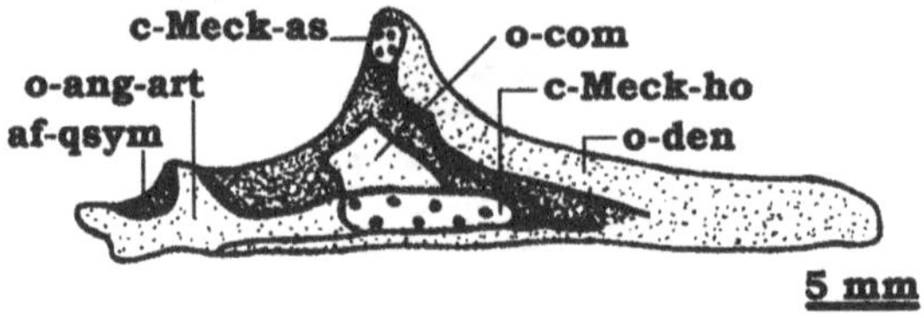

Fig. 3-90 Nematogenyidae: medial view of left mandible of *Nematogenys inermis*, mandibular teeth removed.

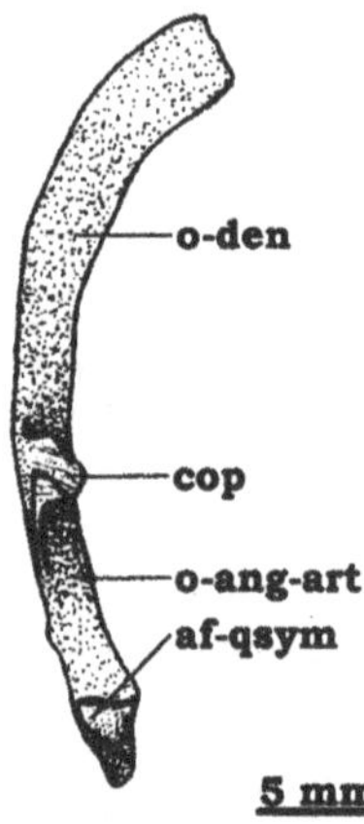

Fig. 3-91 Nematogenyidae: dorsal view of left mandible of *Nematogenys inermis*, mandibular teeth removed.

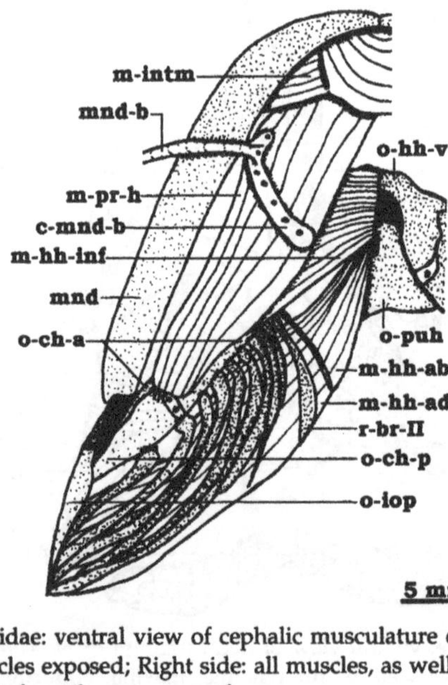

Fig. 3-92 Nematogenyidae: ventral view of cephalic musculature of *Nematogenys inermis*. Left side: all muscles exposed; Right side: all muscles, as well as mandibular barbels and their associated cartilages, removed.

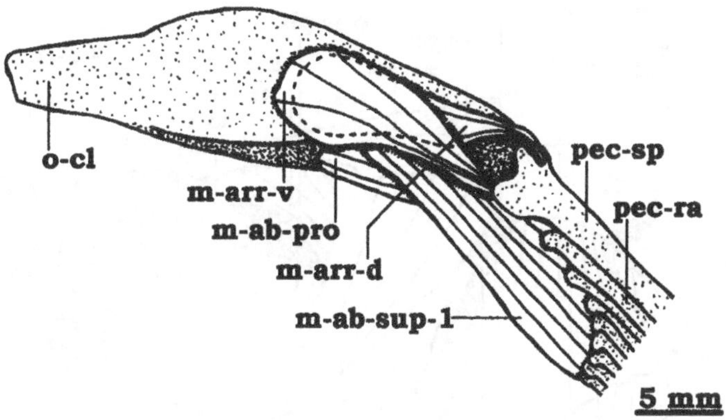

Fig. 3-93 Nematogenyidae: ventral view of pectoral girdle musculature of *Nematogenys inermis*, all muscles exposed.

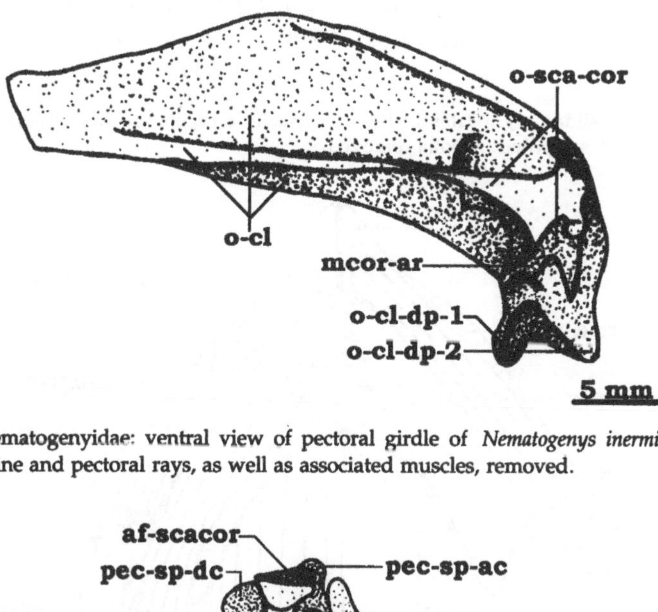

Fig. 3-94 Nematogenyidae: ventral view of pectoral girdle of *Nematogenys inermis*. Pectoral spine and pectoral rays, as well as associated muscles, removed.

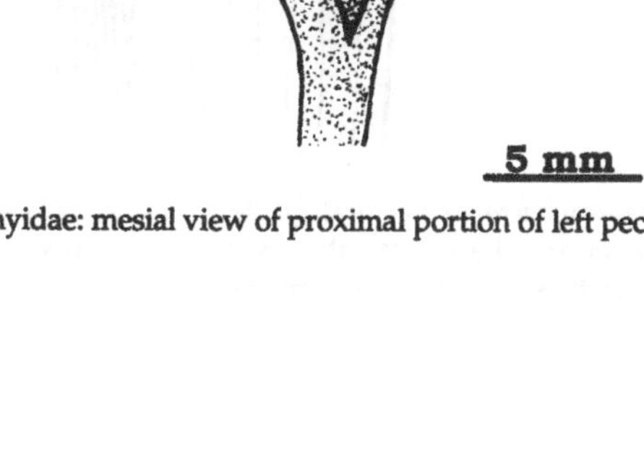

Fig. 3-95 Nematogenyidae: mesial view of proximal portion of left pectoral spine of *Nematogenys inermis*.

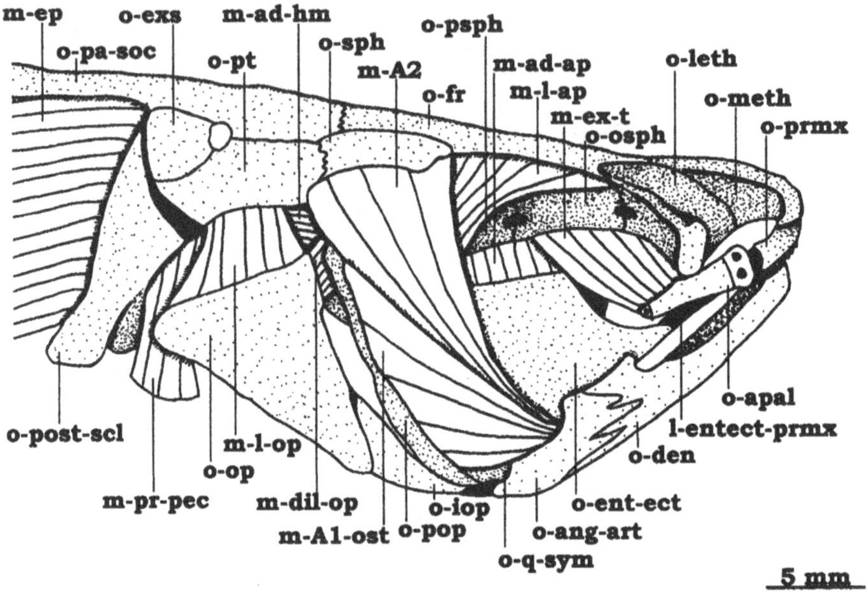

Fig. 3-96 Pangasiidae: right lateral view of cephalic musculature of *Pangasius siamensis*, all muscles exposed, maxilla removed.

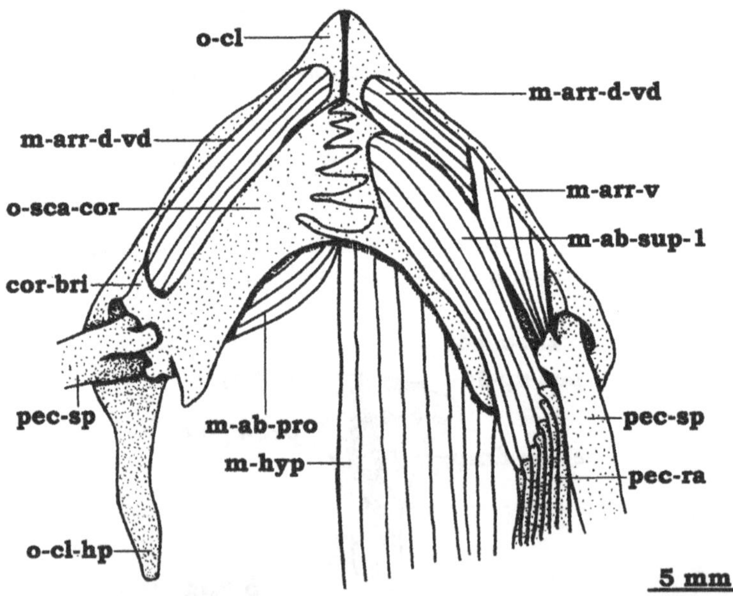

Fig. 3-97 Pangasiidae: ventral view of pectoral girdle of *Pangasius siamensis*. Right side: all musculature exposed; Left side: abductor superficialis, arrector ventralis and hypoaxialis, as well as pectoral rays, removed.

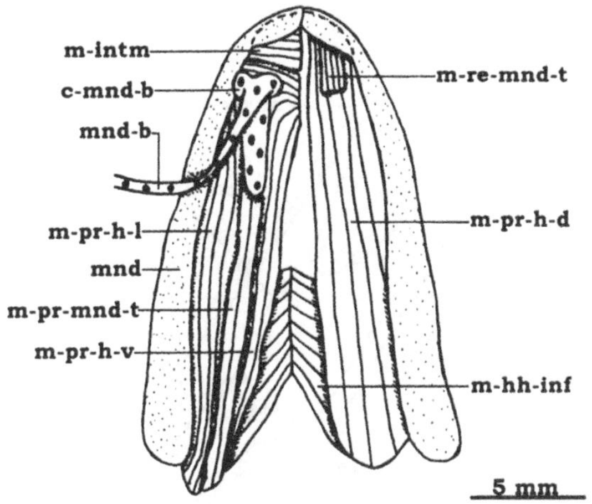

Fig. 3-98 Pangasiidae: ventral view of cephalic musculature of *Pangasius siamensis*. Left side: all musculature exposed; Right side: intermandibularis, ventral and lateral parts of the protractor hyoidei and protractor mandibularis tentaculi, as well as mandibular barbel and its respective cartilage, removed.

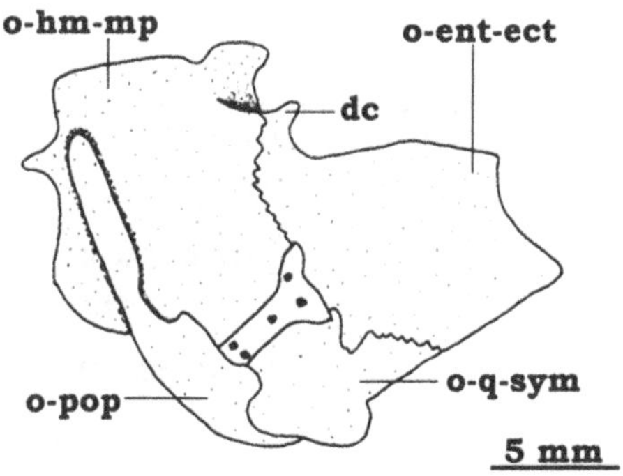

Fig. 3-99 Pangasiidae: lateral view of right suspensorium of *Pangasius siamensis*.

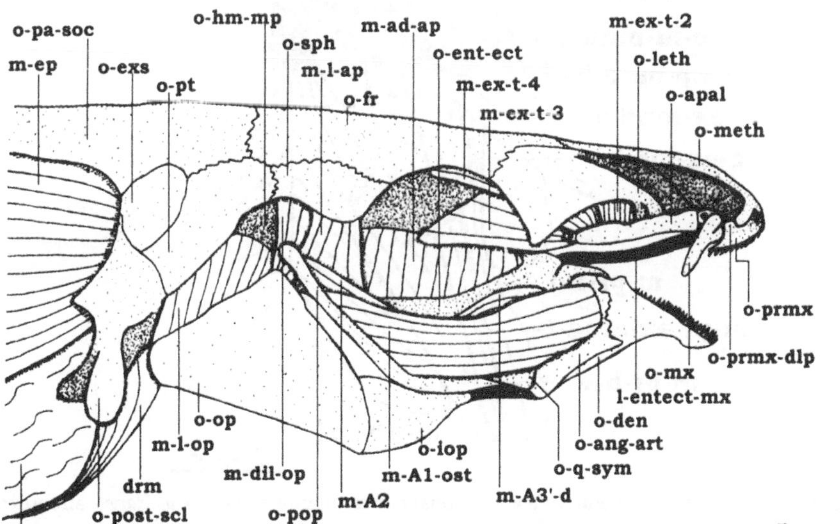

Fig. 3-100 Pimelodidae: lateral view of cephalic musculature of *Pimelodus clarias*, all muscles exposed.

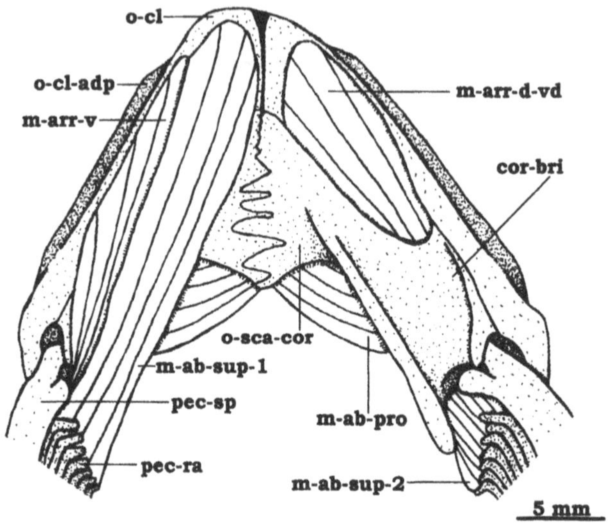

Fig. 3-101 Pimelodidae: ventral view of pectoral girdle musculature of *Pimelodus clarias*. Left side: all muscles exposed; Right side: arrector ventralis and section 1 of abductor superficialis removed.

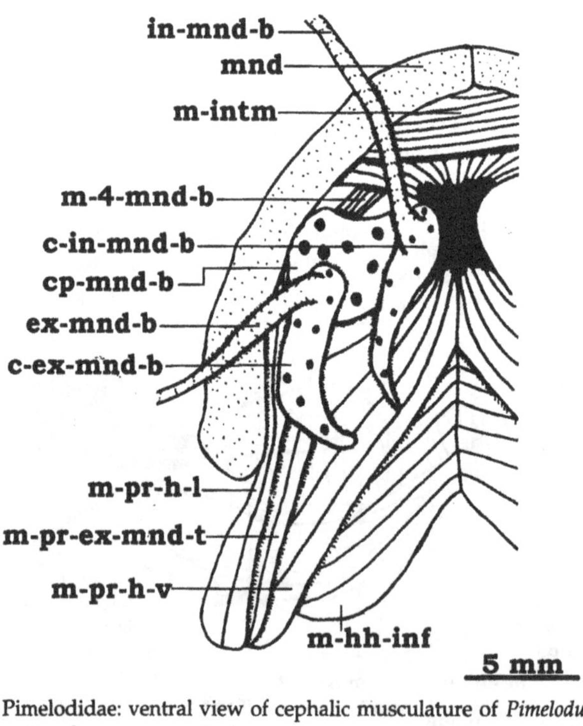

Fig. 3-102 Pimelodidae: ventral view of cephalic musculature of *Pimelodus clarias*, all muscles exposed.

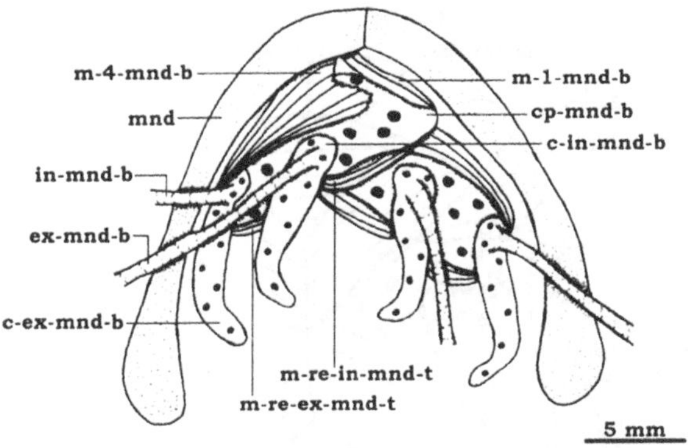

Fig. 3-103 Pimelodidae: ventral view of cephalic musculature of *Pimelodus clarias*. Intermandibularis, protractor hyoidei, hyoideus inferioris and protractor externi mandibularis tentaculi removed. Part of the plates carrying the mandibular barbels also removed to expose muscle 4 of the mandibular barbels.

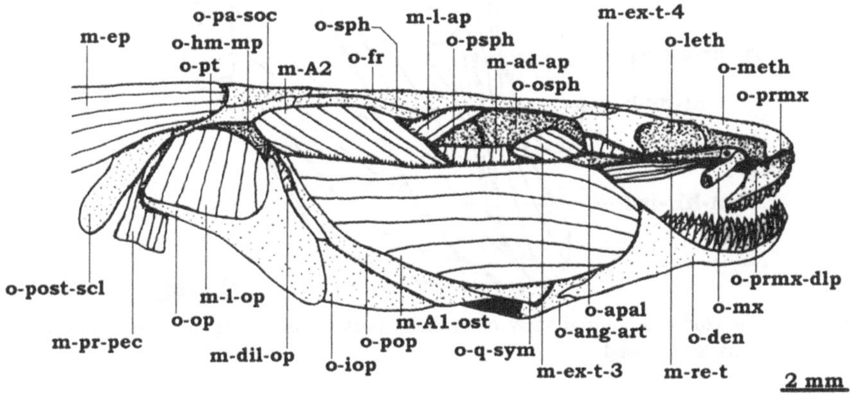

Fig. 3-104 Pimelodidae: lateral view of cephalic musculature of *Heptapterus mustelinus*, all muscles exposed.

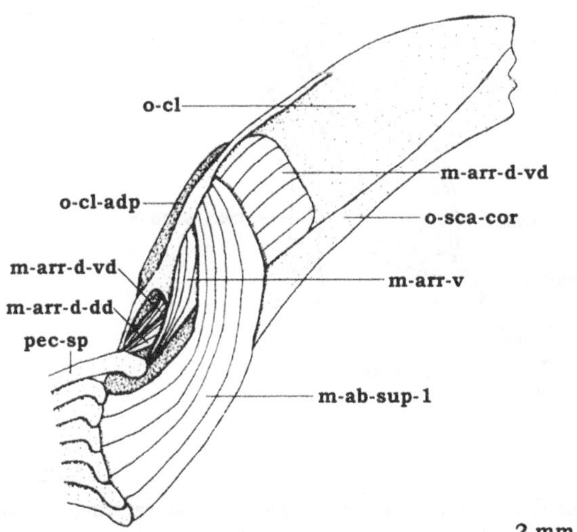

Fig. 3-105 Pimelodidae: ventral view of pectoral girdle musculature of *Heptapterus mustelinus*, all muscles exposed.

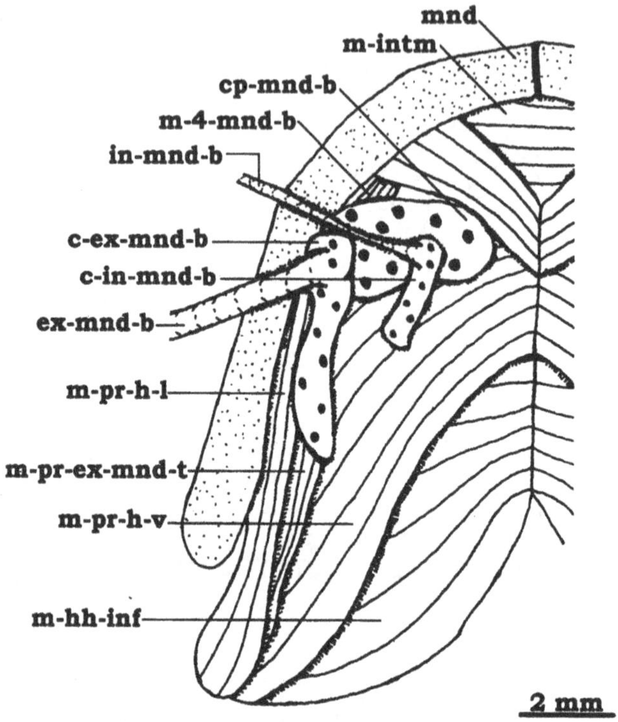

Fig. 3-106 Pimelodidae: ventral view of cephalic musculature of *Heptapterus mustelinus*, all muscles exposed.

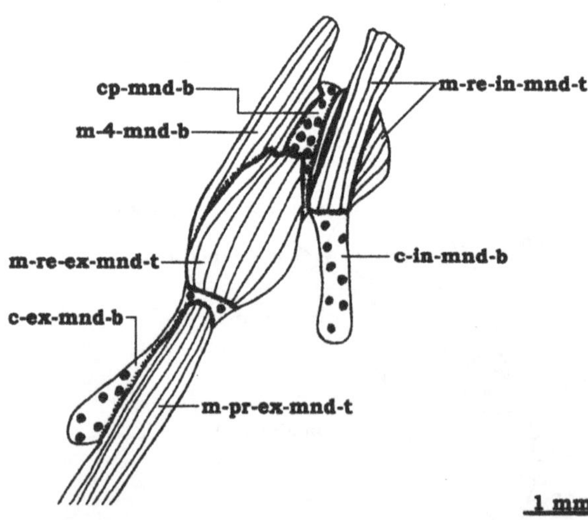

Fig. 3-107 Pimelodidae: dorsal view of left cartilaginous plate carrying mandibular barbels and associated structures in *Heptapterus mustelinus*.

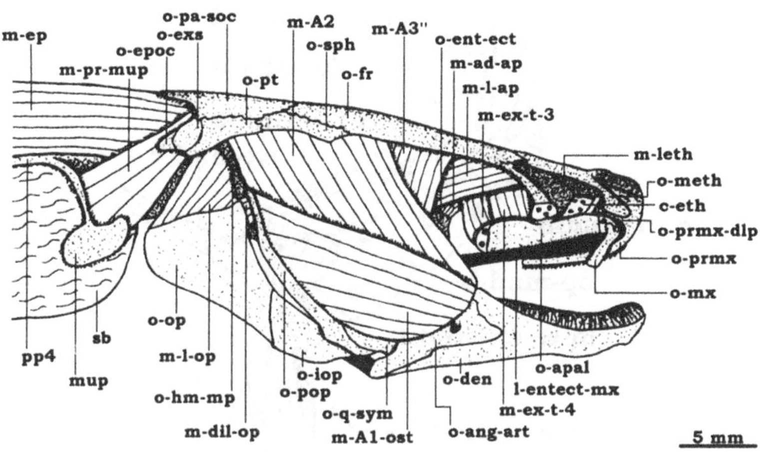

Fig. 3-108 Pimelodidae: lateral view of cephalic musculature of *Pseudopimelodus raninus*. All muscles exposed, but ligament between maxilla and premaxilla, as well as posttemporo-supracleithrum, removed.

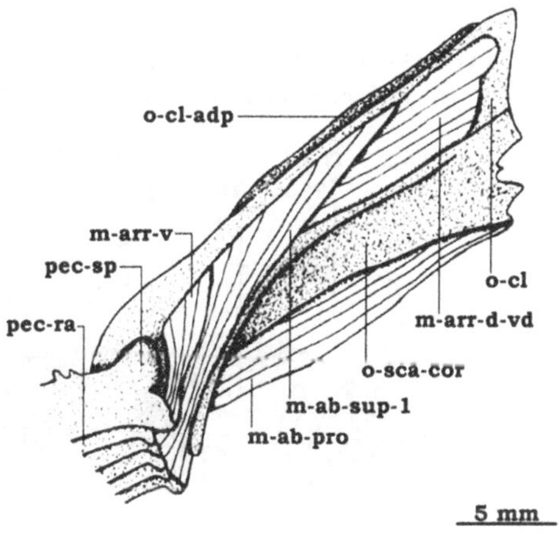

Fig. 3-109 Pimelodidae: ventral view of pectoral girdle musculature of *Pseudopimelodus raninus*, all muscles exposed.

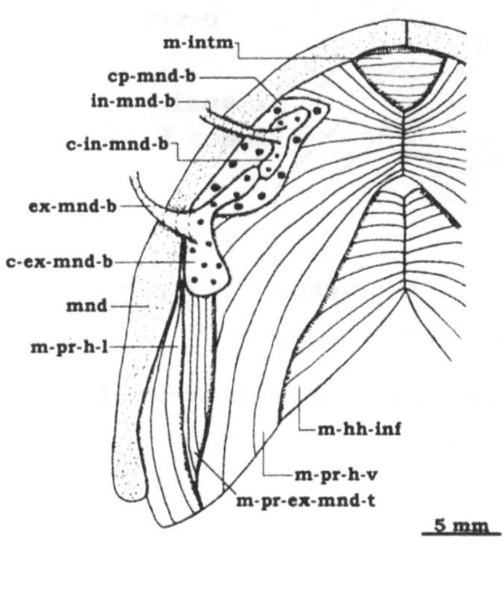

Fig. 3-110 Pimelodidae: ventral view of cephalic musculature of *Pseudopimelodus raninus*, all muscles exposed.

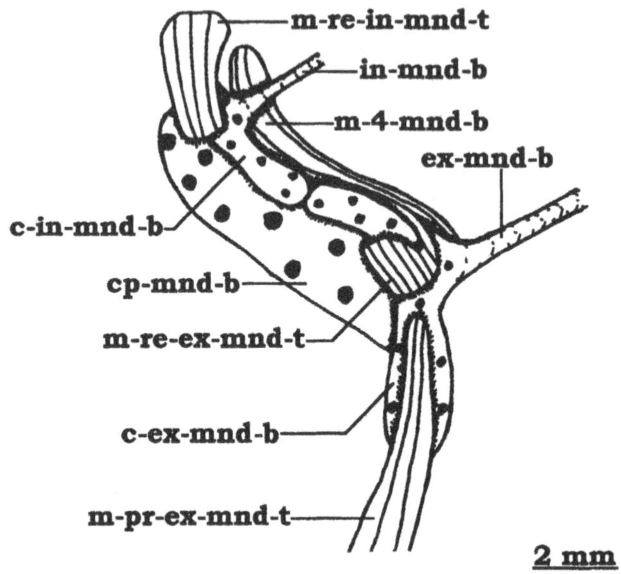

Fig. 3-111 Pimelodidae: dorsal view of right cartilaginous plate carrying mandibular barbels and associated structures in *Pseudopimelodus raninus*.

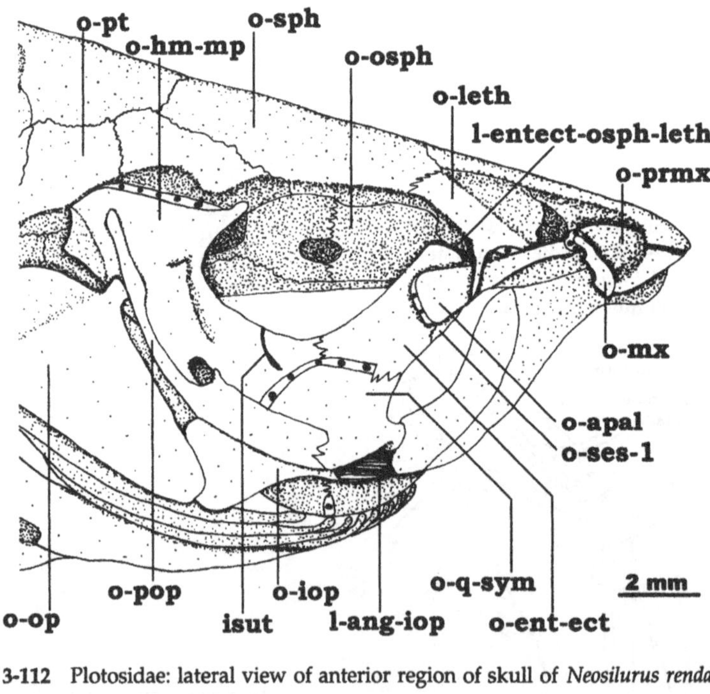

Fig. 3-112 Plotosidae: lateral view of anterior region of skull of *Neosilurus rendahli*.

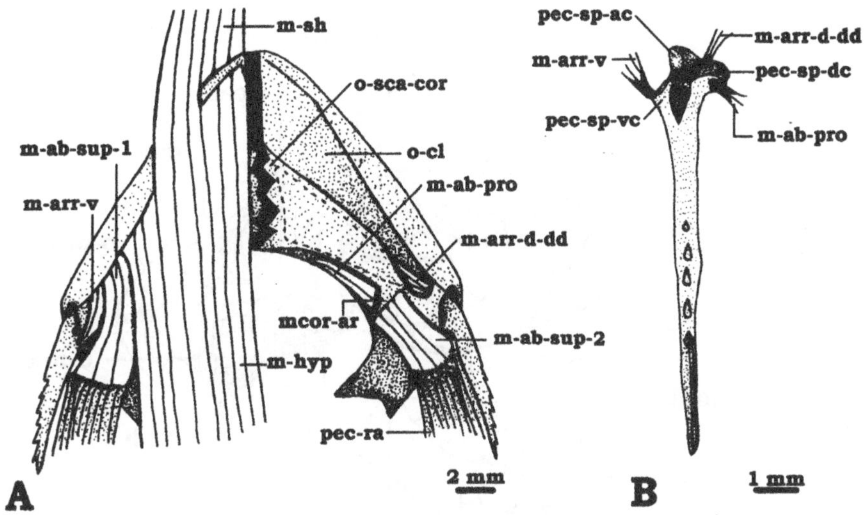

Fig. 3-113 Plotosidae: *Plotosus lineatus*. **A)** Ventral view of pectoral girdle. Left side: all muscles exposed; Right side: hypoaxialis, sternohyoideus, arrector ventralis, section 1 of abductor superficialis and posttemporo-supracleithrum removed. **B)** Median view of pectoral spine, showing insertions of muscles responsible for its movement.

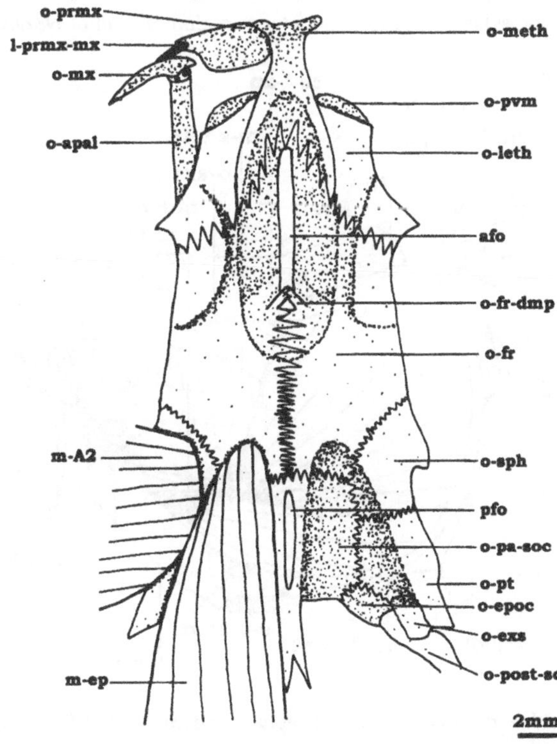

Fig. 3-114 Plotosidae: dorsal view of neurocranium of *Plotosus lineatus*. Left side: palatine-maxillary system, as the adductor mandibulae and epaxialis muscles, are also represented.

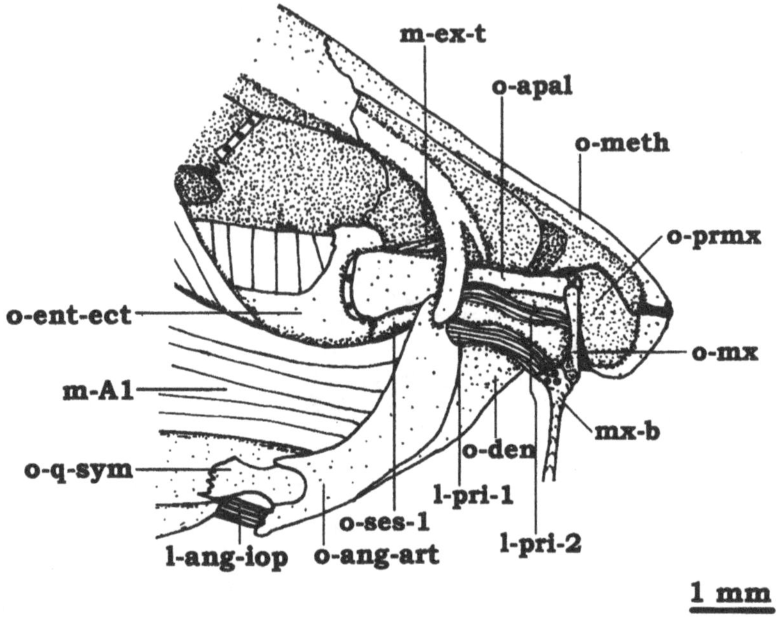

Fig. 3-115 Plotosidae: lateral view of anterior region of skull of *Neosilurus rendahli*.

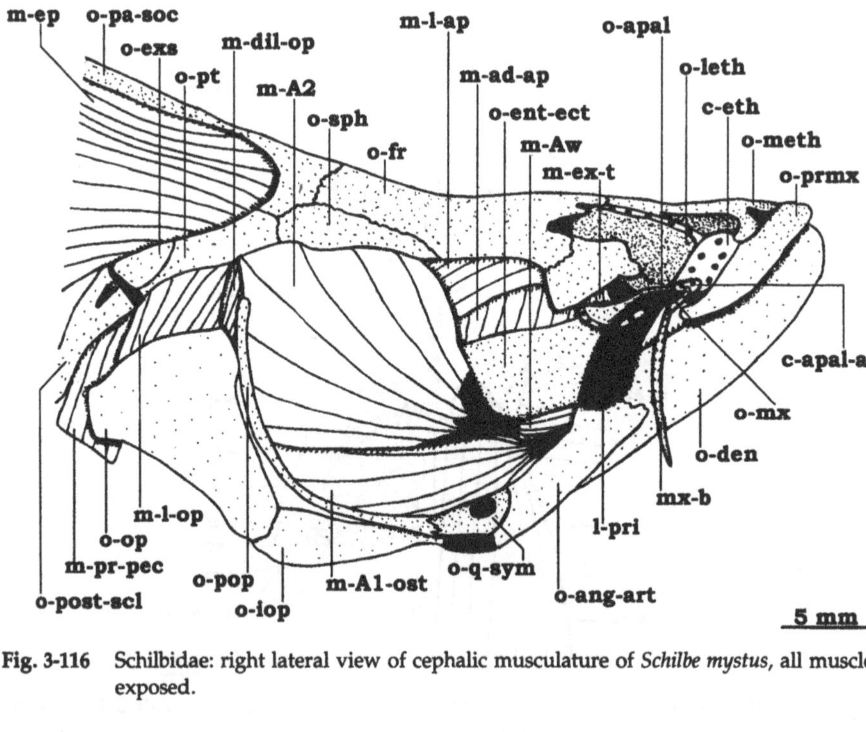

Fig. 3-116 Schilbidae: right lateral view of cephalic musculature of *Schilbe mystus*, all muscles exposed.

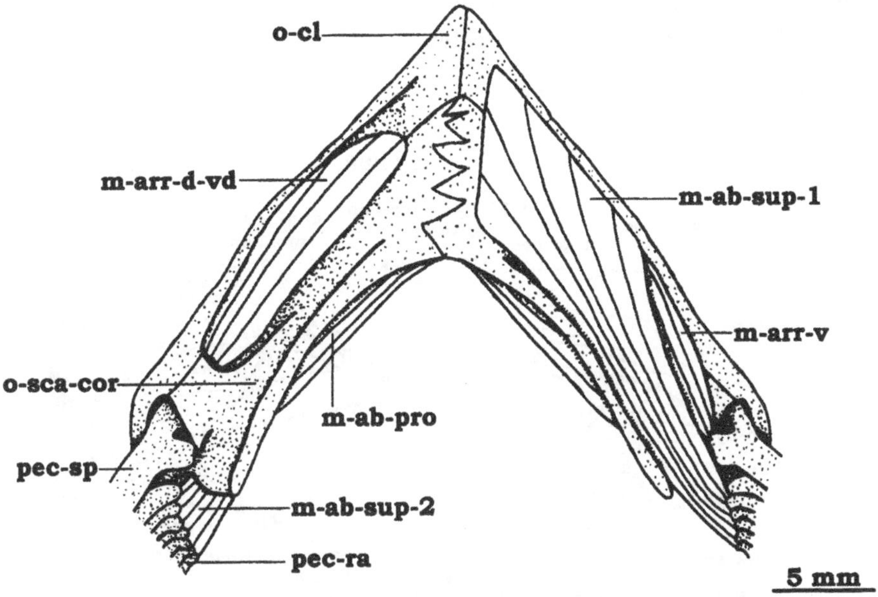

Fig. 3-117 Schilbidae: ventral view of pectoral girdle of *Schilbe mystus*. Left side: all muscles exposed; Right side: both abductor superficialis 1 and arrector ventralis removed.

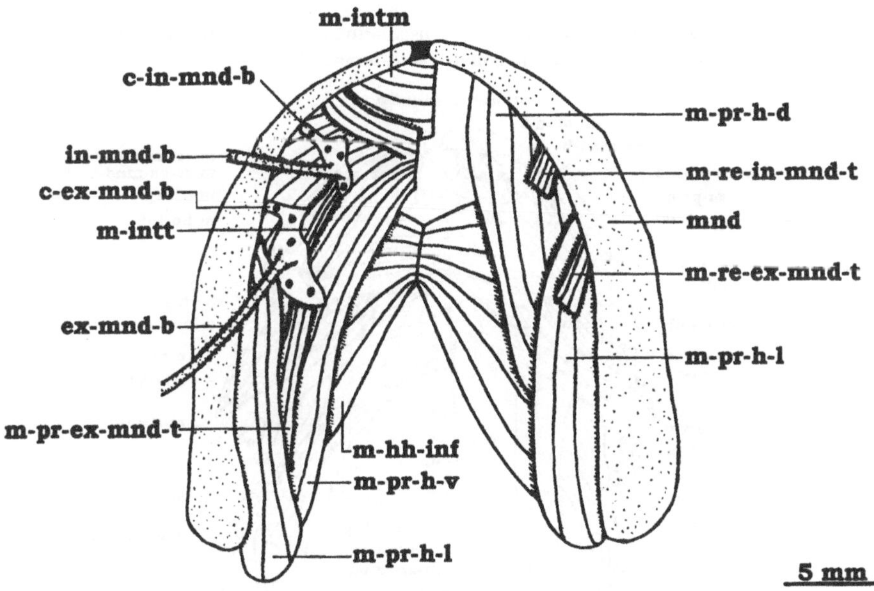

Fig. 3-118 Schilbidae: ventral view of cephalic musculature of *Schilbe mystus*. Left side: all the muscles exposed; Right side: intermandibularis, protractor hyoideus ventral, intertentacularis, protractor externi mandibularis tentaculi, as well as mandibular barbels and their respective cartilages removed.

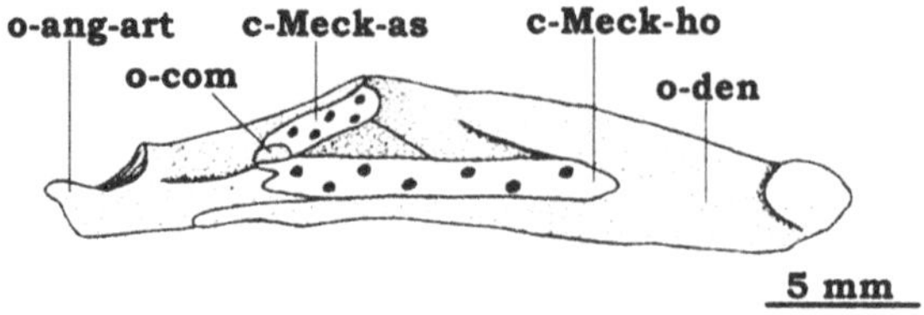

Fig. 3-119 Schilbidae: mesial view of left lower jaw of *Schilbe mystus*

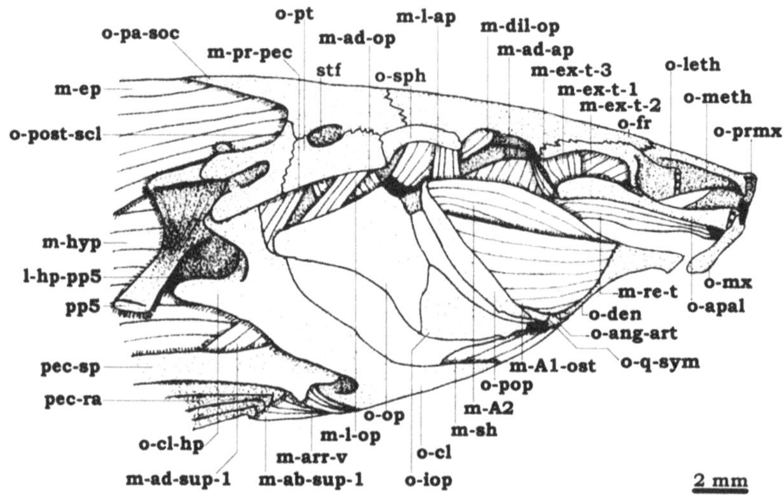

Fig. 3-120 Sisoridae: lateral view of cephalic musculature of *Glyptothorax fukiensis*, all muscles exposed, dentary and premaxillary teeth removed.

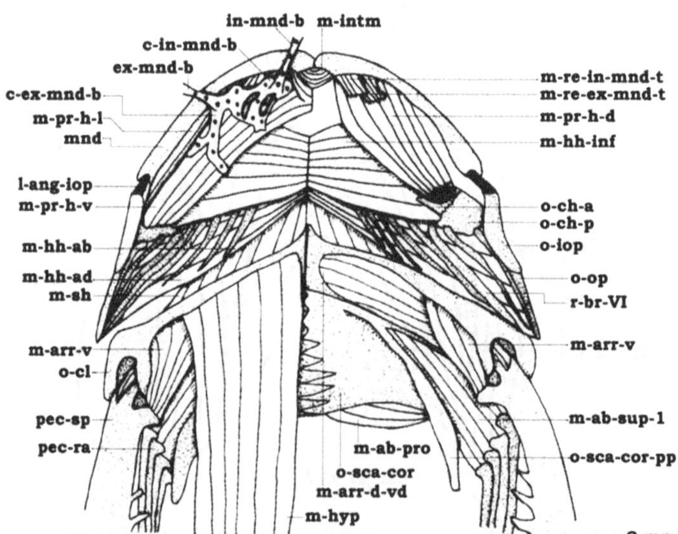

Fig. 3-121 Sisoridae: ventral view of the cephalic region and pectoral girdle of *Glyptothorax fukiensis*. Left side: all muscles exposed; right side: mandibular barbels, their associated cartilages hypoaxialis, and ventral and lateral parts of protractor hyoidei removed. On both sides, ligament between posterior ceratohyal and angulo-articular removed.

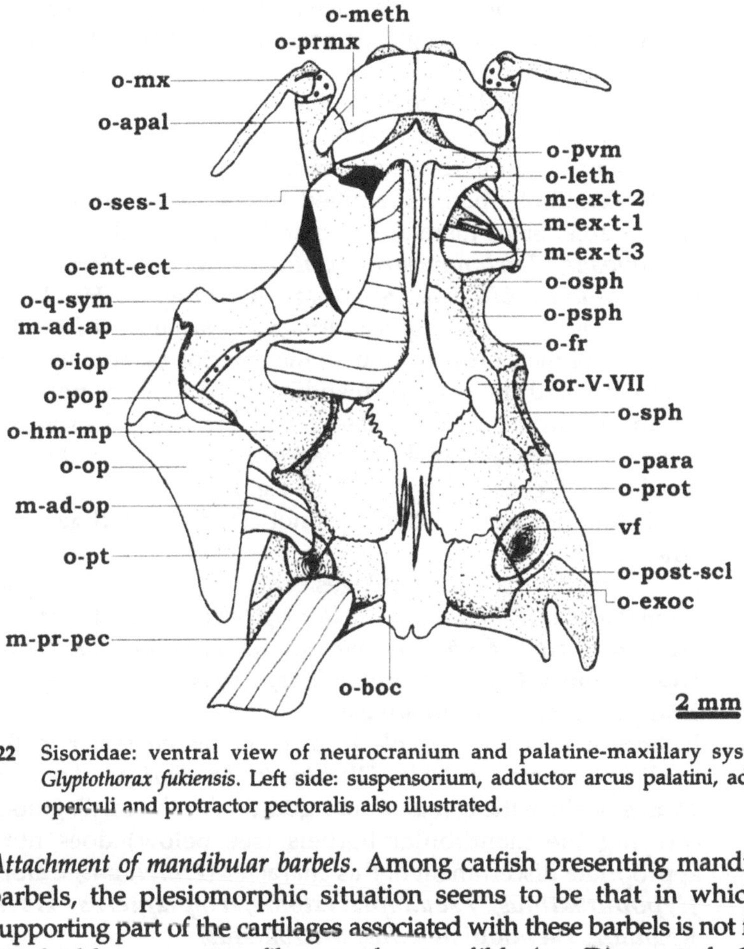

Fig. 3-122 Sisoridae: ventral view of neurocranium and palatine-maxillary system of *Glyptothorax fukiensis*. Left side: suspensorium, adductor arcus palatini, adductor operculi and protractor pectoralis also illustrated.

2. *Attachment of mandibular barbels*. Among catfish presenting mandibular barbels, the plesiomorphic situation seems to be that in which the supporting part of the cartilages associated with these barbels is not firmly attached by numerous fibres to the mandible (see Diogo et al., 2003a) [State 0: e.g. Fig. 3.26]. In the derived condition the supporting part of these cartilages is firmly attached to the mandible [State 1: e.g. Fig. 3.5].
   - CS–0: Supporting part of cartilages not firmly associated with mandible (***Nematogenys, Silurus, Wallago, Cetopsis, Hemicetopsis, Bunocephalus, Aspredo, Xyliphius, Chaca***)
   - CS–1: Supporting part of cartilages firmly associated with mandible **(all genera not in other CS)**
   - Inapplicable: Since there are no mandibular barbels (***Diplomystes, Ageneiosus, Trichomycterus, Callichthys, Corydoras, Loricaria, Hypoptopoma, Lithoxus, Scoloplax, Astroblepus, Hatcheria***) or since there is a large additional cartilaginous complex connecting the mandibular barbels and the mandible (see below) (***Pimelodus, Calophysus, Hypophthalmus, Pseudoplatystoma, Heptapterus, Goeldiella, Rhamdia, Pseudopimelodus, Microglanis***)

3. *Ramification of mandibular barbels (character inspired from Regan, 1911b).* Among the catfish genera studied presenting mandibular barbels the plesiomorphic configuration clearly seems to be that in which these barbels are not ramified [State 0: e.g. Fig. 3.26]. The derived condition of this character is thus the ramification of these barbels [State 1].
   — CS–0: Mandibular barbels not ramified **(all genera not in other CS)**
   — CS–1: Mandibular barbels ramified **(*Synodontis, Mochokus*)**
   — Inapplicable: Since there are no mandibular barbels **(*Diplomystes, Ageneiosus, Trichomycterus, Callichthys, Corydoras, Loricaria, Hypoptopoma, Lithoxus, Scoloplax, Astroblepus, Hatcheria*)**
4. *Contact between anterior part of the cartilages of mandibular barbels.* In the vast majority of the catfish examined, the anterior parts of the cartilages of the internal and the external mandibular barbels are not in contact (Diogo and Chardon, 2000b) [State 0: e.g. Fig. 3.26]. In some catfish, however, the anterior part of the cartilages of the internal and the external mandibular contact each other [State 1: e.g. Fig. 3.5].
   — CS–0: Anterior parts of cartilages not contacting **(all genera not in other CS)**
   — CS–1: Anterior parts of cartilages contacting **(*Bagrus, Bagrichthys, Hemibagrus, Clarias, Uegitglanis, Heterobranchus, Heteropneustes, Plotosus, Cnidoglanis, Paraplotosus, Neosilurus, Glyptothorax, Glyptosternon, Liobagrus, Amblyceps, Synodontis, Ictalurus, Amiurus, Anadoras*)**
   — ?: Since it was not possible to discern this character in the very small specimens available **(*Parakysis, Mochokus, Acanthodoras, Doras*)** or since the complex configuration of the cartilaginous plate carrying the mandibular barbels (see below) does not allow appropriate discernment of this character **(*Pimelodus, Calophysus, Hypophthalmus, Pseudoplatystoma, Heptapterus, Goeldiella, Rhamdia, Pseudopimelodus, Microglanis*)**
   — Inapplicable: Since there are no mandibular barbels **(*Diplomystes, Ageneiosus, Trichomycterus, Callichthys, Corydoras, Loricaria, Hypoptopoma, Lithoxus, Scoloplax, Astroblepus, Hatcheria*)** or since there is only one pair of these barbels **(*Nematogenys, Pangasius, Helicophagus, Silurus, Wallago, Rita*)**
5. *Contact between posterior part of the cartilages of mandibular barbels.* In all catfish genera analysed except *Gagata*, the posterior parts of the cartilages of the internal and the external mandibular barbels are not in contact [State 0: e.g. Fig. 3.26]. However, in this genus the moving parts (see terminology of Diogo and Chardon, 2000b) of these cartilages contact each other [State 1: Fig. 3.102].
   — CS–0: Posterior parts of cartilages not contacting **(all genera not in other CS)**
   — CS–1: Posterior parts of cartilages contacting **(*Gagata*)**

— Inapplicable: Since there are no mandibular barbels (***Diplomystes, Ageneiosus, Trichomycterus, Callichthys, Corydoras, Loricaria, Hypoptopoma, Lithoxus, Scoloplax, Astroblepus, Hatcheria***) or since there is only one pair of these barbels (***Nematogenys, Pangasius, Helicophagus, Silurus, Wallago, Rita***)

6. *Separation of cartilages of mandibular barbels*. As seen above in those catfish exhibiting mandibular barbels the cartilages of the external and internal barbels may be in contact or be separated [State 0: e.g. Figs. 3.5, 3.26]. However, among the catfish examined, *Chaca* is unique in having a markedly extreme separation between the basal cartilages of the external and internal barbels of the same side of the fish [Sate 1: e.g. Fig. 3.51].
   — CS–0: Cartilages of external and internal barbels contacting or loosely separated **(all genera not in other CS)**
   — CS–1: Marked separation of cartilages of external and internal barbels (***Chaca***)
   — Inapplicable: Since there are no mandibular barbels (***Diplomystes, Ageneiosus, Trichomycterus, Callichthys, Corydoras, Loricaria, Hypoptopoma, Lithoxus, Scoloplax, Astroblepus, Hatcheria***) or since there is only one pair of these barbels (***Nematogenys, Pangasius, Helicophagus, Silurus, Wallago, Rita***)
7. *Moving and supporting parts of cartilages of mandibular barbels* (*unordered multistate character*). Plesiomorphically in those catfish exhibiting mandibular barbels there is a differentiation of the cartilages of the mandibular barbels into a small anterior (supporting) and a somewhat longer posterior (moving) part (see Diogo et al., 2003a) [State 0: e.g. Fig. 3.69]. However, these two parts are not recognisable in the cetopsid genera examined, in which the base of the mandibular barbels is situated near the posterior, and not the anterior extremity of these cartilages [State 1: e.g. Fig. 3.44], nor in the aspredinid genera analysed, in which the cartilages of the mandibular barbels are extremely reduced in size [State 2: e.g. Fig. 3.26].
   — CS–0: Cartilages differentiated into supporting and moving parts **(all genera not in other CS)**
   — CS–1: Cartilages not differentiated into supporting and moving parts, with base of barbels near their posterior margin (***Cetopsis, Hemicetopsis, Helogenes***)
   — CS–2: Cartilages not differentiated into supporting and moving parts, very reduced in size (***Bunocephalus, Aspredo, Xyliphius***)
   — Inapplicable: Since there are no mandibular barbels (***Diplomystes, Ageneiosus, Trichomycterus, Callichthys, Corydoras, Loricaria, Hypoptopoma, Lithoxus, Scoloplax, Astroblepus, Hatcheria***)
8. *Anteroposterior length of cartilages of mandibular barbels*. Plesiomorphically in those catfish exhibiting mandibular barbels, the cartilages of these barbels are not markedly elongated anteroposterorly [State 0: e.g. Fig. 3.32]. In a few catfish genera examined, however, these cartilages are

extremely elongated anteroposteriorly, with their total length being approximately similar to that of the muscle protractor hyoideus [State 1].

- — CS–0: Cartilages not markedly elongated anteroposteriorly **(all genera not in other CS)**
- — CS–1: Cartilages markedly elongated anteroposteriorly **(*Siluranodon, Ailia, Pseudeutropius, Laides, Synodontis*)**
- — Inapplicable: Since there are no mandibular barbels **(*Diplomystes, Ageneiosus, Trichomycterus, Callichthys, Corydoras, Loricaria, Hypoptopoma, Lithoxus, Scoloplax, Astroblepus, Hatcheria*)**

9. *Shape of cartilages of internal mandibular barbels (ordered multistate character).* Contrary to the plesiomorphic situation in those catfish with mandibular barbels [State 0: e.g. Fig. 3.26], in *Gagata* the cartilage of the internal mandibular barbel exhibits a highly irregular shape, seeming at first sight to be pierced by a small foramen, due to the fact that the posteromesial projection of this cartilage nearly contacts its main body [State 1]. In genera of state 2 the posteromesial projection is still more developed, contacting the main body of the cartilage, delimiting a true, small foramen [State 2: e.g. Fig. 3.121].
   - — CS–0: Cartilage of internal barbel without apparent or true foramen **(all genera not in other CS)**
   - — CS–1: Cartilage of internal barbel with apparent foramen **(*Gagata*)**
   - — CS–2: Cartilage of internal barbel with true foramen **(*Glyptothorax, Glyptosternon, Amblyceps*)**
   - — ?: Since there is only one pair of these barbels and it is not possible to discern whether they correspond to the internal pair (in this case, the corresponding option would be 'CS–0') or the external (in this case, the corresponding option would be 'Inapplicable') **(*Nematogenys, Pangasius, Helicophagus, Silurus, Wallago, Rita*)**
   - — Inapplicable: Since there are no mandibular barbels **(*Diplomystes, Ageneiosus, Trichomycterus, Callichthys, Corydoras, Loricaria, Hypoptopoma, Lithoxus, Scoloplax, Astroblepus, Hatcheria*)**
10. *Anterior bifurcation of cartilages of mandibular barbels.* Contrary to the plesiomorphic situation in those catfish with mandibular barbels (Diogo and Chardon, 2000b) [State 0: e.g. Fig. 3.26], in the taxa of CS–1 the cartilages associated with the mandibular barbels are anteriorly bifurcate [State 1: e.g. Fig. 3.98].
   - — CS–0: Cartilages anteriorly not bifurcate **(all genera not in other CS)**
   - — CS–1: Cartilages anteriorly bifurcate **(*Pangasius, Helicophagus, Schilbe, Laides, Pseudeutropius*)**
   - — ?: Since it was not possible to appraise this character due to the highly peculiar configuration of the cartilages of the mandibular barbels **(*Siluranodon, Ailia*)**
   - — Inapplicable: Since there are no mandibular barbels **(*Diplomystes, Ageneiosus, Trichomycterus, Callichthys, Corydoras, Loricaria, Hypoptopoma, Lithoxus, Scoloplax, Astroblepus, Hatcheria*)**

11. *Enlargement of cartilage of mandibular barbels.* In those catfish with mandibular barbels the cartilages of these barbels could exhibit different shapes and sizes [State 0: see characters above]. However, markedly enlarged, voluminous, circular cartilages are present among all the catfish examined, only in the specimens of genera *Cranoglanis* and *Franciscodoras* [State 1: e.g. Fig. 3.60].
    — CS–0: Cartilages not markedly enlarged **(all genera not in other CS)**
    — CS–1: Presence of markedly enlarged, voluminous, circular cartilages **(*Cranoglanis, Franciscodoras, Austroglanis*)**
    — Inapplicable: Since there are no mandibular barbels **(*Diplomystes, Ageneiosus, Trichomycterus, Callichthys, Corydoras, Loricaria, Hypoptopoma, Lithoxus, Scoloplax, Astroblepus, Hatcheria*)**
12. *Presence of cartilaginous plates carrying the mandibular barbels (ordered multistate character) (character inspired from Ghiot, 1978).* Plesiomorphically those catfish exhibiting mandibular barbels lack "cartilaginous plates carrying mandibular barbels" (see terminology of Diogo and Chardon, 2000b) [State 0: e.g. Fig. 3.92]. In all the pimelodids examined, in addition to the basal cartilages supporting the mandibular barbels usually present in most catfish (see above), there are also 'cartilaginous plates carrying mandibular barbels' (see terminology of Diogo and Chardon, 2000b). These cartilages constitute a very unique, readily recognised feature, although they are somewhat less developed in *Pseudopimelodus* and *Microglanis* [State 1: e.g. Fig. 3.110] than in the other six pimelodid genera analysed [State 2: e.g. Fig. 3.103].
    — CS–0: Cartilaginous plates absent **(all genera not in other CS)**
    — CS–1: Cartilaginous plates thinner and less developed than in CS–1 **(*Pseudopimelodus, Microglanis*)**
    — CS–2: Presence of voluminous, robust cartilaginous plates **(*Pimelodus, Calophysus, Hypophthalmus, Pseudoplatystoma, Heptapterus, Goeldiella, Rhamdia*)**
    — Inapplicable: Since there are no mandibular barbels **(*Diplomystes, Ageneiosus, Trichomycterus, Callichthys, Corydoras, Loricaria, Hypoptopoma, Lithoxus, Scoloplax, Astroblepus, Hatcheria*)**
13. *Posterior bifurcation of cartilages of external mandibular barbels.* Contrary to the plesiomorphic situation in those catfish with mandibular barbels [State 0: e.g. Fig. 3.26], in those few taxa of CS–1 the cartilages associated with the external mandibular barbels are posteriorly bifurcate [State 1: e.g. Fig. 3.121].
    — CS–0: Cartilages of external barbels not posteriorly bifurcate **(all genera not in other CS)**
    — CS–1: Cartilages of external barbels posteriorly bifurcate **(*Glyptothorax, Akysis, Liobagrus, Amblyceps*)**
    — ?: Since there is only one pair of mandibular barbels and it is not possible to discern whether they correspond to the internal pair (in

this case, the corresponding option would be 'Inapplicable') or the external (in this case, the corresponding option would be 'CS–0') **(*Nematogenys, Pangasius, Helicophagus, Silurus, Wallago, Rita*)**
- Inapplicable: Since there are no mandibular barbels **(*Diplomystes, Ageneiosus, Trichomycterus, Callichthys, Corydoras, Loricaria, Hypoptopoma, Lithoxus, Scoloplax, Astroblepus, Hatcheria*)**

14. *Anteroposterior length of supporting part of cartilages of mandibular barbels.* Plesiomorphically in those catfish presenting mandibular barbels the anterior, supporting parts of these cartilages are typically short [State 0: e.g. Fig. 3.92]. In *Synodontis* the supporting parts of the cartilages of the mandibular barbels are, however, markedly elongated anteroposteriorly, with their length approximately similar to that of the moving parts of these barbels, which are already particularly elongated in this same genus (see above) [State 1].
    - CS–0: Supporting parts of cartilages not markedly elongated anteroposteriorly **(all genera not in other CS)**
    - CS–1: Supporting parts of cartilages markedly elongated anteroposteriorly **(*Synodontis*)**
    - Inapplicable: Since there are no mandibular barbels **(*Diplomystes, Ageneiosus, Trichomycterus, Callichthys, Corydoras, Loricaria, Hypoptopoma, Lithoxus, Scoloplax, Astroblepus, Hatcheria*)**
15. *Contact between internal mandibular barbels.* Contrary to the situation found in other catfish presenting mandibular barbels [State 0: e.g. Fig. 3.5], in *Cranoglanis* the cartilages of the internal mandibular barbels are deeply in contact with each other on the midline [State 1: e.g. Fig. 3.60].
    - CS–0: Cartilages not in deep contact with each other **(all genera not in other CS)**
    - CS–1: Cartilages in deep contact with each other on the midline **(*Cranoglanis*)**
    - ?: Since there is only one pair of mandibular barbels and it is not possible to discern whether they correspond to the internal pair (in this case, the corresponding option would be 'CS–0') or the external (in this case, the corresponding option would be 'Inapplicable') **(*Nematogenys, Pangasius, Helicophagus, Silurus, Wallago, Rita*)**
    - Inapplicable: Since there are no mandibular barbels **(*Diplomystes, Ageneiosus, Trichomycterus, Callichthys, Corydoras, Loricaria, Hypoptopoma, Lithoxus, Scoloplax, Astroblepus, Hatcheria*)**
16. *Anterior bifurcation of retractor of external mandibular barbel.* Contrary to all other catfish presenting muscles of mandibular barbels, in which the retractor of the external mandibular barbels is not bifurcate [State 0: e.g. Fig. 3.79], in *Bagrus* this muscle exhibits an anterior bifurcation, with its two anterior tendons being separated by the muscles protractor hyoideus and intermandibularis [State 1: e.g. Fig. 3.41].
    - CS–0: Not anteriorly bifurcate **(all genera not in other CS)**
    - CS–1: Anteriorly bifurcate **(*Bagrus*)**

— ?: Since there is only one pair of mandibular barbels and it is not possible to discern whether they correspond to the external pair (in this case, the corresponding option would be 'CS–0') or the internal (in this case, the corresponding option would be 'Inapplicable') (***Nematogenys, Pangasius, Helicophagus, Silurus, Wallago, Rita***)

— Inapplicable: Since there are no mandibular barbels (***Diplomystes, Ageneiosus, Trichomycterus, Callichthys, Corydoras, Loricaria, Hypoptopoma, Lithoxus, Scoloplax, Astroblepus, Hatcheria***) or since these barbels are present but the ventral cephalic musculature is not differentiated in small muscles of mandibular barbels (***Nematogenys, Bunocephalus, Aspredo, Xyliphius, Wallago, Silurus, Chaca***)

17. *Presence of mesial cartilaginous complex connecting the internal mandibular barbels on the midline.* Contrary to all the other catfish presenting mandibular barbels [State 0: e.g. Fig. 3.92], in the three clariid genera examined, as well as in *Heteropneustes*, there is a well-developed, mesial cartilaginous complex connecting the internal mandibular barbels on the midline (see terminology of Diogo and Chardon, 2000b) [State 1: e.g. Fig. 3.54].

— CS–0: Absence of cartilaginous complex connecting the internal mandibular barbels on the midline **(all genera not in other CS)**

— CS–1: Presence of cartilaginous complex connecting the internal mandibular barbels on the midline (***Clarias, Uegitglanis, Heterobranchus, Heteropneustes***)

— Inapplicable: Since there are no mandibular barbels (***Diplomystes, Ageneiosus, Trichomycterus, Callichthys, Corydoras, Loricaria, Hypoptopoma, Lithoxus, Scoloplax, Astroblepus, Hatcheria***)

18. *Presence of depressor of internal mandibular barbels (ordered multistate character).* Among those catfish presenting muscles of mandibular barbels, the plesiomorphic condition seems to be that in which the depressor of the internal mandibular barbels is absent (i.e., the retractors of the barbels are present, but the depressor of the internal barbels is missing: Diogo and Vandewalle, 2003) [State 0: e.g. Fig. 3.5]. In those catfish genera of CS–1, this muscle is present [State 1: e.g. Fig. 3.76] but not as developed as in *Amphilius* and *Paramphilius* [State 2: e.g. Fig. 3.8].

— CS–0: Depressor of internal mandibular barbels absent **(all genera not in other CS)**

— CS–1: Depressor of internal mandibular barbels present (***Chrysichthys, Clarotes, Auchenoglanis, Malapterurus, Phractura, Leptoglanis, Andersonia, Clarias, Heterobranchus, Heteropneustes, Arius, Genidens, Doumea, Belonoglanis, Trachyglanis, Franciscodoras***)

— CS–2: Depressor of internal mandibular barbels highly developed (***Amphilius, Paramphilius***)

— ?: Since there is only one pair of mandibular barbels, and it is not possible to discern whether they correspond to the internal pair (in this case, the corresponding option would be 'CS–0') or the external (in this case, the corresponding option would be 'Inapplicable') (***Nematogenys, Pangasius, Helicophagus, Silurus, Wallago, Rita***) or since it was not possible to discern this character, due to the very small size and/or the poor condition of the specimens examined (***Zaireichthys, Uegitglanis, Anadoras, Acanthodoras, Doras, Centromochlus, Auchenipterus, Helogenes, Cetopsis, Hemicetopsis, Ancharius***)

— Inapplicable: Since there are no mandibular barbels (***Diplomystes, Ageneiosus, Trichomycterus, Callichthys, Corydoras, Loricaria, Hypoptopoma, Lithoxus, Scoloplax, Astroblepus, Hatcheria***) or since these barbels are present but the ventral cephalic musculature is not differentiated in small muscles of mandibular barbels (***Nematogenys, Bunocephalus, Aspredo, Xyliphius, Wallago, Silurus, Chaca***)

19. *Presence of intertentacularis*. As with the depressor of the internal mandibular barbels, among those catfish presenting muscles of mandibular barbels the plesiomorphic condition seems to be that in which the muscle intertentacularis is absent (see above) [State 0: e.g. Fig. 3.5]. In those catfish of CS–1 the intertentacularis is present [State 1: e.g. Fig. 3.118].

— CS–0: Intertentacularis absent **(all genera not in other CS)**

— CS–1: Intertentacularis present (***Chrysichthys, Schilbe, Laides, Pseudeutropius, Siluranodon, Ailia, Clarotes, Phractura, Leptoglanis, Andersonia, Paramphilius, Doumea, Belonoglanis, Trachyglanis, Amphilius, Arius, Genidens***)

— ?: Since it was not possible to discern this character due to very small size and/or poor condition of the specimens examined (***Zaireichthys, Ancharius***)

— Inapplicable: Since there are no mandibular barbels (***Diplomystes, Ageneiosus, Trichomycterus, Callichthys, Corydoras, Loricaria, Hypoptopoma, Lithoxus, Scoloplax, Astroblepus, Hatcheria***), since these barbels are present but the ventral cephalic musculature is not differentiated in small muscles of mandibular barbels (***Nematogenys, Bunocephalus, Aspredo, Xyliphius, Wallago, Silurus, Chaca***), or since there is only one pair of mandibular barbels, and, thus, it cannot exist a muscle intertentacularis connecting the cartilages of the internal and the external mandibular barbels (***Nematogenys, Pangasius, Helicophagus, Silurus, Wallago, Rita***)

20. *Presence of muscle 1 of mandibular barbels (character inspired from Ghiot, 1978)*. As with the depressor of the internal mandibular barbels and the intertentacularis, among those catfish presenting muscles of mandibular barbels, the plesiomorphic condition clearly seems to be that in which

the muscle 1 of these barbels is absent (Diogo and Chardon, 2000b) [State 0: e.g. Fig. 3.79], with this muscle found only in catfish of CS–1 [State 1: e.g. Fig. 3.103].

— CS–0: Muscle 1 of mandibular barbels absent **(all genera not in other CS)**

— CS–1: Muscle 1 of mandibular barbels present ***(Pimelodus, Calophysus, Hypophthalmus, Pseudoplatystoma)***

— Inapplicable: Since there are no mandibular barbels ***(Diplomystes, Ageneiosus, Trichomycterus, Callichthys, Corydoras, Loricaria, Hypoptopoma, Lithoxus, Scoloplax, Astroblepus, Hatcheria)*** or since these barbels are present but the ventral cephalic musculature is not differentiated in small muscles of mandibular barbels ***(Nematogenys, Bunocephalus, Aspredo, Xyliphius, Wallago, Silurus, Chaca)***

21. *Presence of muscle 2 of mandibular barbels*. Plesiomorphically among those catfish presenting muscles of mandibular barbels, muscle 2 of these barbels is also absent (Diogo and Chardon, 2000b) [State 0: e.g. Fig. 3.79], with this muscle found only in the two genera of CS–1 [State 1: e.g. Fig. 3.54].

— CS–0: Muscle 2 of mandibular barbels absent **(all genera not in other CS)**

— CS–1: Muscle 2 of mandibular barbels present ***(Clarias, Heterobranchus)***

— ?: Since it was not possible to discern this character due to the poor condition of the specimens examined ***(Uegitglanis)***

— Inapplicable: Since there are no mandibular barbels ***(Diplomystes, Ageneiosus, Trichomycterus, Callichthys, Corydoras, Loricaria, Hypoptopoma, Lithoxus, Scoloplax, Astroblepus, Hatcheria)*** or since these barbels are present but the ventral cephalic musculature is not differentiated in small muscles of mandibular barbels ***(Nematogenys, Bunocephalus, Aspredo, Xyliphius, Wallago, Silurus, Chaca)***

22. *Presence of muscle 3 of mandibular barbels*. Plesiomorphically among those catfish presenting muscles of mandibular barbels, muscle 3 of these barbels is absent (Diogo and Chardon, 2000b) [State 0: e.g. Fig. 3.79]. In *Amphilius* this muscle is present [State 1: e.g. Fig. 3.8].

— CS–0: Muscle 3 of mandibular barbels absent **(all genera not in other CS)**

— CS–1: Muscle 3 of mandibular barbels present ***(Amphilius)***

— Inapplicable: Since there are no mandibular barbels ***(Diplomystes, Ageneiosus, Trichomycterus, Callichthys, Corydoras, Loricaria, Hypoptopoma, Lithoxus, Scoloplax, Astroblepus, Hatcheria)*** or since these barbels are present but the ventral cephalic musculature is not differentiated in small muscles of mandibular barbels ***(Nematogenys, Bunocephalus, Aspredo, Xyliphius, Wallago, Silurus, Chaca)***

23. *Presence of muscle 4 of mandibular barbels (character inspired from Ghiot, 1978)*. As noted above for muscles 1, 2 and 3 of the mandibular barbels, plesiomorphically those Siluriformes presenting muscles of mandibular barbels lack a muscle 4 of these barbels (Diogo and Chardon, 2000b) [State 0: e.g. Fig. 3.79]. In those catfish of CS–1 this muscle is present [State 1: e.g. Fig. 3.103].
    - CS–0: Muscle 4 of mandibular barbels absent **(all genera not in other CS)**
    - CS–1: Muscle 4 of mandibular barbels present **(*Pimelodus, Calophysus, Hypophthalmus, Pseudoplatystoma, Heptapterus, Goeldiella, Rhamdia, Pseudopimelodus, Microglanis*)**
    - Inapplicable: Since there are no mandibular barbels **(*Diplomystes, Ageneiosus, Trichomycterus, Callichthys, Corydoras, Loricaria, Hypoptopoma, Lithoxus, Scoloplax, Astroblepus, Hatcheria*)** or since these barbels are present but the ventral cephalic musculature is not differentiated in small muscles of mandibular barbels **(*Nematogenys, Bunocephalus, Aspredo, Xyliphius, Wallago, Silurus, Chaca*)**
24. *Presence of muscle 6 of mandibular barbels*. Plesiomorphically those catfishes presenting small muscles associated with the mandibular barbels lack a muscle 6 of these barbels [State 0: e.g. Fig. 3.79]. The only three catfish genera examined presenting a muscle 6 of the mandibular barbels running from the anteroventromesial surface of the cartilage of the internal barbel to the mandible are those mentioned in CS–1 [State 1: e.g. Fig. 3.44].
    - CS–0: Muscle 6 of mandibular barbels absent **(all genera not in other CS)**
    - CS–1: Muscle 6 of mandibular barbels present **(*Cetopsis, Hemicetopsis, Helogenes*)**
    - Inapplicable: Since there are no mandibular barbels **(*Diplomystes, Ageneiosus, Trichomycterus, Callichthys, Corydoras, Loricaria, Hypoptopoma, Lithoxus, Scoloplax, Astroblepus, Hatcheria*)** or since these barbels are present but the ventral cephalic musculature is not differentiated in small muscles of mandibular barbels **(*Nematogenys, Bunocephalus, Aspredo, Xyliphius, Wallago, Silurus, Chaca*)**
25. *Anterior extension of cartilages of internal mandibular barbels (ordered multistate character)*. Contrary to the other catfishes presenting mandibular barbels [State 0: e.g. Fig. 3.92], in *Malapterurus* there is an anterior extension of the cartilages of the internal mandibular barbels, with the anterior margin of these cartilages situating at about the same level of the anterior margin of the mandible [State 1]. Such an anterior extension of the cartilages of the internal mandibular barbels is still more pronounced in *Auchenoglanis*, where the anterior margin of these cartilages extends anteriorly beyond the anterior margin of the mandible [State 2].
    - CS–0: Absence of anterior extension of cartilage of mandibular barbel **(all genera not in other CS)**

— CS–1: Presence of anterior extension of cartilage of mandibular barbel (***Malapterurus***)
— CS–2: Anterior extension of cartilage of mandibular barbel more pronounced than in CS–1, with anterior margin of this cartilage extending anteriorly beyond the anterior margin of mandible (***Auchenoglanis***)
— ?: Since there is only one pair of mandibular barbels, and, thus, it is not possible to discern if it corresponds to the internal pair (in this case, the corresponding option would be 'CS–0') or the external one (in this case, the corresponding option would be 'Inapplicable') (***Nematogenys, Pangasius, Helicophagus, Silurus, Wallago, Rita***)
— Inapplicable: Since there are no mandibular barbels (***Diplomystes, Ageneiosus, Trichomycterus, Callichthys, Corydoras, Loricaria, Hypoptopoma, Lithoxus, Scoloplax, Astroblepus, Hatcheria***)

26. *Presence of muscle intermandibularis.* Plesiomorphically catfish present a muscle intermandibularis [State 0: e.g. Fig. 3.69], but in those catfishes of CS–1 this muscle is absent [State 1: e.g. Fig. 3.9].
— CS–0: Muscle intermandibularis present **(all genera not in other CS)**
— CS–1: Muscle intermandibularis absent (***Phractura, Doumea, Belonoglanis, Trachyglanis, Andersonia***)
— ?: Since it was not possible to discern this character due to the small size and/or poor condition of the specimens examined (***Malapterurus, Leptoglanis, Paramphilius, Zaireichthys***)

27. *Development of muscle intermandibularis.* The plesiomorphic condition for those catfishes presenting a muscle intermandibularis seems to be that in which these fishes present a well-developed, broad muscle intermandibularis (Diogo and Vandewalle, 2003) [State 0: e.g. Fig. 3.69]. In catfishes of CS–1, this muscle is significantly more reduced in size than in catfishes of CS–0 [State 1: e.g. Fig. 3.8].
— CS–0: Well-developed, broad muscle intermandibularis (***Diplomystes, Nematogenys, Trichomycterus, Hatcheria, Bagrus, Pseudopimelodus, Microglanis, Bagrichthys, Hemibagrus, Silurus, Wallago, Helogenes, Cetopsis, Hemicetopsis, Arius, Genidens***)
— CS–1: Muscle intermandibularis significantly more reduced in size than in CS–0 **(all genera not in other CS)**
— ?: Since, as noted above, it was not possible to appropriately discern whether the muscle intermandibularis is present (in this case, the corresponding option would be 'CS–0') or not (in this case, the corresponding option would be 'Inapplicable') (***Malapterurus, Leptoglanis, Paramphilius, Zaireichthys***)
— Inapplicable: Since the muscle intermandibularis is missing (***Phractura, Doumea, Belonoglanis, Trachyglanis, Andersonia***)

28. *Shape of muscle intermandibularis (character inspired from Schaefer and Lauder, 1986).* Contrary to the situation found in other catfishes presenting a

muscle intermandibularis [State 0: e.g. Fig. 3.69], in *Astroblepus* the fibres of this muscle are not continuous on the midline [State 1].

— CS–0: Muscle intermandibularis continuous on the midline **(all genera not in other CS)**

— CS–1: Muscle intermandibularis not continuous on the midline **(*Astroblepus*)**

— ?: Since, as mentioned above, it was not possible to discern whether the intermandibularis is present (in this case, the corresponding option would be 'CS–0') or not (in this case, the corresponding option would be 'Inapplicable') **(*Malapterurus, Leptoglanis, Paramphilius, Zaireichthys*)**

— Inapplicable: Since there is no muscle intermandibularis **(*Phractura, Doumea, Belonoglanis, Trachyglanis, Andersonia*)**

29. *Attachment of protractor hyoideus*. Contrary to the situation found in all other catfishes examined, in which the protractor hyoideus originates exclusively on the hyoid arch and inserts on the mandible [State 0: e.g. Fig. 3.69], in the specimens of *Synodontis* examined this muscle runs not only from the hyoid arch, but also from the mesial surface of the suspensorium, to the mandible [State 1].

— CS–0: Protractor hyoideus originates exclusively on the hyoid arch **(all genera not in other CS)**

— CS–1: Protractor hyoideus originates on the hyoid arch and suspensorium **(*Synodontis*)**

30. *Differentiation of protractor hyoideus*. Plesiomorphically in catfish the muscle protractor hyoideus is not differentiated into pars ventralis, pars lateralis and pars dorsalis (Diogo and Vandewalle, 2003) [State 0: e.g. Fig. 3.69]. In the derived condition, this muscle is differentiated into these three different parts [State 1: e.g. Fig. 3.98]

— CS–0: Protractor hyoideus not differentiated into pars dorsalis, ventralis and lateralis **(*Diplomystes, Nematogenys, Trichomycterus, Hatcheria, Callichthys, Corydoras, Loricaria, Hypoptopoma, Lithoxus, Scoloplax, Astroblepus*)**

— CS–1: Protractor hyoideus differentiated into pars dorsalis, ventralis and lateralis **(all genera not in other CS)**

31. *Position of pars lateralis of protractor hyoideus*. Contrary to the situation found in all other catfishes examined presenting a protractor hyoideus differentiated into pars lateralis, ventralis and dorsalis, in which the pars lateralis lies mainly lateral to the pars ventralis (Diogo and Vandewalle, 2003) [State 0: e.g. Fig. 3.32], in the four genera of CS–1 the pars lateralis lies completely dorsal to the pars ventralis [State 1: e.g. Fig. 3.54].

— CS–0: Pars lateralis lying mainly lateral to pars ventralis **(all genera not in other CS)**

— CS–1: Pars lateralis completely dorsal to pars ventralis **(*Clarias, Uegitglanis, Heterobranchus, Heteropneustes*)**

— Inapplicable: Since the protractor hyoideus is not differentiated into pars lateralis, dorsalis and ventralis ***(Diplomystes, Nematogenys, Trichomycterus, Hatcheria, Callichthys, Corydoras, Loricaria, Hypoptopoma, Lithoxus, Scoloplax, Astroblepus)***

32. *Mesial aponeurosis of left and right pars dorsalis of protractor hyoideus.* Contrary to the situation found in all other catfishes examined presenting a protractor hyoideus differentiated into pars lateralis, ventralis and dorsalis, in which the pars dorsalis of each side do not contact on the midline (Diogo and Vandewalle, 2003) [State 0: e.g. Fig. 3.98], in the three genera of CS–1 the pars dorsalis of each side meet mesially in a well-developed mesial aponeurosis [State 1: e.g. Fig. 3.54].
    — CS–0: Pars dorsalis of each side not meeting on the midline **(all genera not in other CS)**
    — CS–1: Pars dorsalis of each side meeting mesially on mesial aponeurosis **(*Clarias, Heterobranchus, Cranoglanis, Austroglanis*)**
    — Inapplicable: Since the protractor hyoideus is not differentiated into pars lateralis, dorsalis and ventralis **(*Diplomystes, Nematogenys, Trichomycterus, Hatcheria, Callichthys, Corydoras, Loricaria, Hypoptopoma, Lithoxus, Scoloplax, Astroblepus*)**
33. *Relation between pars lateralis and pars ventralis of protractor hyoideus.* Contrary to the situation found in all other catfishes examined presenting a protractor hyoideus differentiated into pars lateralis, ventralis and dorsalis, in which these three parts are well separated from each other [State 0: e.g. Fig. 3.8], in those few catfishes of CS–1 a large portion of the fibres of the pars lateralis is deeply mixed with those of the pars ventralis [State 1: e.g. Fig. 3.60].
    — CS–0: Large portion of fibres of pars lateralis not mixed with those of pars ventralis **(all genera not in other CS)**
    — CS–1: Large portion of fibres of pars lateralis mixed with those of pars ventralis **(*Plotosus, Neosilurus, Paraplotosus, Cnidoglanis, Chaca, Cranoglanis*)**
    — Inapplicable: Since the protractor hyoideus is not differentiated into pars lateralis, dorsalis and ventralis **(*Diplomystes, Nematogenys, Trichomycterus, Hatcheria, Callichthys, Corydoras, Loricaria, Hypoptopoma, Lithoxus, Scoloplax, Astroblepus*)**
34. *Mesial aponeurosis of protractor hyoideus (character inspired from Howes, 1983a; Schaefer and Lauder, 1986).* The plesiomorphic siluriform configuration is that in which the right and left portions of the protractor hyoideus (or its pars ventralis in those catfish where the muscle in differentiated into pars ventralis, dorsalis and lateralis: see above), do come into contact on a medial aponeurosis [State 0: e.g. Fig. 3.69]. This configuration changed only among all the catfish examined, in the four genera of CS–1 [State 1].
    — CS–0: Presence of mesial aponeurosis of protractor hyoideus **(all genera not in other CS)**

— CS–1: Left and right parts of whole protractor hyoideus completely separated on the midline (***Loricaria, Hypoptopoma, Lithoxus, Astroblepus***)

35. *Shape of hyohyoideus inferior (ordered multistate character)*. Plesiomorphically in catfish the hyohyoideus inferior is a well-developed, somewhat triangular muscle with a marked median aponeurosis [State 0: e.g. Fig. 3.69]. In those siluriforms of CS–1 this is a hypertrophied, somewhat transversal muscle with a barely marked median aponeurosis [State 1], with its hypertrophy even more pronounced and its median aponeurosis even less distinct in catfish of CS–2 [State 2: e.g. Fig. 3.9].
    — CS–0: Hyohyoideus inferior with fibres running essentially anteromesially to meet in a marked median aponeurosis **(all genera not in other CS)**
    — CS–1: Hyohyoideus inferior hypertrophied, with fibres running essentially mesially to meet in a barely defined median aponeurosis (***Mochokus, Auchenoglanis, Malapterurus, Franciscodoras, Anadoras, Doras***)
    — CS–2: Hyohyoideus inferior even more hypertrophied, with median aponeurosis almost indistinguishable (***Phractura, Doumea, Belonoglanis, Trachyglanis, Leptoglanis, Zaireichthys, Andersonia, Bagrichthys***)
36. *Bilateral bifurcation of hyohyoideus inferior (character inspired from Schaefer and Lauder, 1986)*. Contrary to all other catfishes [State 0: e.g. Fig. 3.69], in the loricariid, astroblepid and scoloplacid genera examined the hyohyoideus inferior exhibits a marked bilateral bifurcation [State 1].
    — CS–0: Hyohyoideus inferior not bifurcate **(all genera not in other CS)**
    — CS–1: Hyohyoideus inferior with bilateral bifurcation (***Loricaria, Hypoptopoma, Lithoxus, Astroblepus, Scoloplax***)
37. *Contact between hyohyoideus abductor and pectoral girdle (ordered multistate character)*. Plesiomorphically the hyohyoideus abductor does not attach to the pectoral girdle (Diogo and Vandewalle, 2003) [State 0: e.g. Fig. 3.92]. In catfish of CS–1, this muscle is hypertrophied, with its median aponeurosis firmly attached to the ventral surface of the pectoral girdle [State 1]. In catfish of CS–2, the hypertrophy of this muscle is even more pronounced, and some of its fibres inserted directly on the anteroventral surface of the pectoral girdle [State 2: e.g. Fig. 3.26].
    — CS–0: Hyohyoideus abductor does not come into contact with pectoral girdle **(all genera not in other CS)**
    — CS–1: Hyohyoideus abductor hypertrophied, with median aponeurosis firmly attached to pectoral girdle (***Synodontis, Mochokus, Franciscodoras, Anadoras, Acanthodoras, Doras, Centromochlus, Ageneiosus, Auchenipterus, Loricaria, Lithoxus***)
    — CS–2: Hyohyoideus abductor even more hypertrophied, with some of its fibres inclusively inserted directly on ventral surface of pectoral girdle (***Bunocephalus, Aspredo, Xyliphius, Callichthys, Corydoras***)

Musculature Associated with Pectoral Girdle and Fins

38. *Posteroventral attachment of sternohyoideus.* Plesiomorphically in catfish the posteroventral fibres of the sternohyoideus do not cover a significant part of the anteroventral surface of the pectoral girdle [State 0: e.g. Fig. 13.8], as is the case in the specimens examined of genera *Clarias* and *Heterobranchus* [State 1: e.g. Fig. 3.53].
    — CS–0: Sternohyoideus does not cover a significant part of anteroventral surface of pectoral girdle **(all genera not in other CS)**
    — CS–1: Sternohyoideus covers a significant part of anteroventral surface of pectoral girdle **(*Clarias, Heterobranchus*)**
    — ?: Since it is not possible to discern this character due to the fact that the posteroventral fibres of sternohyoideus are completely mixed with and hence indistinguishable from the anteroventral fibres of the hypaxialis **(*Cetopsis, Hemicetopsis, Helogenes, Nematogenys, Trichomycterus, Hatcheria, Plotosus*)**
39. *Posterodorsal attachment of sternohyoideus.* In catfish the posterodorsal fibres of the sternohyoideus usually attach essentially on the anterodorsomesial surface of the pectoral girdle [State 0], but in the two genera of CS–1 the posterodorsal fibres of this muscle also attach on a significant part of the dorsolateral surface of this girdle [State 1].
    — CS–0: Fibres of sternohyoideus attach on anterodorsomesial surface of pectoral girdle **(all genera not in other CS)**
    — CS–1: Fibres of sternohyoideus attach not only on anterodorsomesial, but also on dorsolateral surface of pectoral girdle **(*Neosilurus, Plotosus*)**
40. *Position of hypoaxialis.* Contrary to all other catfishes examined [State 0: e.g. Fig. 3.70], in *Siluranodon* the hypoaxialis is not visible in a ventral view of the anteroventral region of the body at the level of the pectoral girdle [State 1].
    — CS–0: Hypoaxialis visible in ventral view of anteroventral region of body at the level of pectoral girdle **(all genera not in other CS)**
    — CS–1: Hypoaxialis not visible in ventral view of anteroventral region of body at the level of pectoral girdle **(*Siluranodon*)**
41. *Absence of abductor superficialis.* Abductor superficialis is plesiomorphically present in Siluriformes [State 0: e.g. Fig. 3.70]; uniquely in specimens examined of genus *Chaca* this muscle is absent [State 1: e.g. Fig. 3.50].
    — CS–0: Abductor superficialis present **(all genera not in other CS)**
    — CS–1: Abductor superficialis present **(*Chaca*)**
42. *Anterior attachment of part 1 of abductor superficialis.* The plesiomorphic situation for catfish is seemingly that in which part 1 of the abductor superficialis reaches the anteroventral surface of the cleithrum (Diogo et al., 2001c) [State 0: e.g. Fig. 3.70]. In the derived situation, part 1 of this muscle originates on the posteroventral, and not on the anteroventral surface of the pectoral girdle [State 1: e.g. Fig. 3.7].
    — CS–0: Abductor superficialis 1 reaches anteroventral surface of the cleithrum **(all genera not in other CS)**

— CS–1: Abductor superficialis 1 does not reach anteroventral surface of cleithrum (***Chrysichthys, Pangasius, Clarotes, Auchenoglanis, Phractura, Doumea, Belonoglanis, Trachyglanis, Akysis, Parakysis, Amphilius, Leptoglanis, Zaireichthys, Andersonia, Paramphilius, Nematogenys, Trichomycterus, Hatcheria, Gagata, Glyptosternon, Glyptothorax, Bagarius, Erethistes, Hara, Bunocephalus, Aspredo, Xyliphius, Rita, Synodontis, Mochokus, Franciscodoras, Anadoras, Acanthodoras, Doras, Centromochlus, Ageneiosus, Loricaria, Hypoptopoma, Lithoxus, Scoloplax, Astroblepus***)
— Inapplicable: Since the abductor superficialis is absent (***Chaca***)

43. *Development of part 1 of abductor superficialis.* Contrary to all other catfish examined [State 0: e.g. Fig. 3.70], *Siluranodon* and *Laides* present a markedly hypertrophied abductor superficialis 1 completely covering all the ventral surface of the pectoral girdle [State 1].
— CS–0: Abductor superficialis 1 not hypertrophied **(all genera not in other CS)**
— CS–1: Hypertrophied abductor superficialis 1, covering all the ventral surface of pectoral girdle (***Siluranodon, Laides***)
— Inapplicable: Since the abductor superficialis is absent (***Chaca***)

44. *Position of part 1 of abductor superficialis.* Contrary to the catfish plesiomorphic situation [State 0: e.g. Fig. 3.70], in the three genera of CS–1 the abductor superficialis 1 situates on the posterior, and not on the ventral, surface of the pectoral girdle [State 1].
— CS–0: Abductor superficialis 1 situated on ventral surface of pectoral girdle **(all genera not in other CS)**
— CS–1: Abductor superficialis 1 situated on posterior surface of pectoral girdle (***Loricaria, Hypoptopoma, Lithoxus***)
— Inapplicable: Since abductor superficialis is absent (***Chaca***)

45. *Development of part 2 of abductor superficialis.* Contrary to the situation found in all other catfish examined [State 0: e.g. Fig. 3.70], in *Centromochlus* the abductor superficialis 2 is markedly hypertrophied and not confined to the posteroventrolateral surface of the pectoral girdle, reaching its ventromesial surface [State 1].
— CS–0: Abductor superficialis 2 not hypertrophied **(all genera not in other CS)**
— CS–1: Hypertrophied abductor superficialis 2 reaching ventromesial surface of pectoral girdle (***Centromochlus***)
— Inapplicable: Since abductor superficialis is absent (***Chaca***)

46. *Subdivision of abductor profundus.* Contrary to all other catfish examined [State 0: e.g. Fig. 3.71], in *Amphilius* the abductor profundus is subdivided into two well-developed, well-distinguished sections [State 1: e.g. Fig. 3.10].
— CS–0: Abductor profundus not differentiated into two sections **(all genera not in other CS)**
— CS–1: Abductor profundus differentiated into two well-developed sections (***Amphilius***)

47. *Development of abductor profundus.* The plesiomorphic siluriform situation is seemingly that found in *Diplomystes* and many other catfish, in which the abductor profundus originates somewhat far from the mesial symphysis of the pectoral girdle, a situation that seems to be related to the plesiomorphically poorly-developed scapulo-coracoid (Diogo et al., 2001c) [State 0: e.g. Fig. 3.70]. In the derived condition, this muscle almost reaches, or does reach, the mesial symphysis of this girdle [State 0: e.g. Fig. 3.27].
    - — CS–0: Abductor profundus originates far from midline (***Diplomystes, Malapterurus, Parakysis, Plotosus, Silurus, Wallago, Cnidoglanis, Paraplotosus, Neosilurus, Nematogenys, Callichthys, Corydoras, Loricaria, Hypoptopoma, Lithoxus, Scoloplax, Astroblepus, Cetopsis, Hemicetopsis, Helogenes, Heptapterus, Goeldiella, Rhamdia, Liobagrus***)
    - — CS–1: Abductor profundus almost reaching, or indeed reaching, mesial symphysis of pectoral girdle **(all genera not in other CS)**
48. *Development of arrector ventralis (unordered multistate character).* Plesiomorphically in catfish the arrector ventralis is a well-developed muscle with its fibres running essentially transversally or slightly obliquely (Diogo et al., 2001b) [State 0: Fig. 3.70]. In catfish of CS–1, this is a relatively thin muscle with fibres running in a much more marked anteroposterior axis [State 1: Fig. 3.31]. In a completely different, peculiar and unique configuration, the specimens examined of genus *Glyptosternon* present a markedly hypertrophied arrector ventralis covering almost all the ventral surface of the pectoral girdle and reaching mesially the mesial symphysis of pectoral girdle [State 2].
    - — CS–0: Arrector ventralis well developed, fibres extending mainly transversally or slightly obliquely (***Diplomystes, Nematogenys, Trichomycterus, Hatcheria, Callichthys, Corydoras, Loricaria, Hypoptopoma, Lithoxus, Scoloplax, Astroblepus, Phractura, Doumea, Andersonia, Belonoglanis, Trachyglanis, Zaireichthys, Leptoglanis, Amphilius, Paramphilius***)
    - — CS–1: Relatively thin arrector ventralis with fibres running in a more marked anteroposterior axis **(all genera not in other CS)**
    - — CS–2: Markedly hypertrophied arrector ventralis covering all ventral surface and reaching mesial symphysis of pectoral girdle (***Glyptosternon***)
    - — ?: Since it was not possible to discern this character due to the highly peculiar configuration of the pectoral girdle (see below) (***Cetopsis, Hemicetopsis, Helogenes***)
49. *Subdivision of arrector ventralis (ordered multistate character).* Plesiomorphically in catfish the arrector ventralis is constituted by a single mass of fibres [State 0: Fig. 3.70]. In the amphiliin and leptoglanidin Amphiliidae, however, this muscle is differentiated into two quite distinct sections, which are associated posterolaterally [State 1: e.g. Fig. 3.7].

This differentiation is still more pronounced in the doumein amphiliids, in which the arrector ventralis is fully differentiated in an additional muscle of the pectoral girdle, whose fibres are completely separated from those of the arrector ventralis [State 2: e.g. Fig. 3.11].

— CS–0: Arrector ventralis not mesially bifurcate **(all genera not in other CS)**

— CS–1: Arrector ventralis markedly bifurcate mesially **(*Amphilius, Leptoglanis, Zaireichthys, Paramphilius*)**

— CS–2: Differentiation still more pronounced, with arrector ventralis fully differentiated into an additional muscle **(*Phractura, Andersonia, Doumea, Belonoglanis, Trachyglanis*)**

50. *Development of protractor pectoralis* (*ordered multistate character*). Protractor pectoralis is plesiomorphically a well-developed muscle inserting mainly on the anterodorsal surface of the pectoral girdle [State 0: e.g. Fig. 3.71], but in *Chaca* [State 1], and especially in the four genera of CS–1 [State 2: e.g. Fig. 3.52], this is a markedly hypertrophied, anteroposteriorly elongated muscle inserting in an also anteroposteriorly elongated, deep concavity of the anterolateral surface of the pectoral girdle.

— CS–0: Protractor pectoralis inserting on anterodorsal surface of pectoral girdle **(all genera not in other CS)**

— CS–1: Protractor pectoralis hypertrophied, markedly elongated anteroposteriorly **(*Chaca*)**

— CS–2: Protractor pectoralis hypertrophy, as well as its anteroposterior elongation, even more pronounced than in CS–1 **(*Clarias, Uegitglanis, Heterobranchus, Heteropneustes*)**

51. *Differentiation of arrector dorsalis*. The plesiomorphic condition for catfish is that in which the scapulo-coracoid does not present a posterodorsal, large laminar projection subdividing the arrector dorsalis into two well-developed, well-distinguished divisions [State 0: e.g. Fig. 3.70), as present in the derived condition (Diogo et al., 2001c) [State 1: e.g. Fig. 3.40].

— CS–0: Scapulo-coracoid without horizontal lamina subdividing arrector dorsalis into well-developed, well-distinguished dorsal and ventral divisions **(*Diplomystes, Cetopsis, Hemicetopsis, Helogenes, Callichthys, Corydoras, Loricaria, Hypoptopoma, Lithoxus, Scoloplax, Astroblepus, Nematogenys, Trichomycterus, Hatcheria*)**

— CS–1: Horizontal lamina of scapulo-coracoid present, similar to that of cleithrum **(all genera not in other CS)**

52. *Development of ventral division of arrector dorsalis*. The plesiomorphic condition for those catfish having an arrector dorsalis subdivided into two divisions (see above) is to present a well-developed, voluminous ventral division (Diogo et al., 2001c) [State 0: e.g. Fig. 3.56]. Exceptionally, the ventral division of the arrector dorsalis is markedly reduced in size in *Bagrus* and *Hemibagrus* [State 1: e.g. Fig. 3.40].

— CS–0: Voluminous ventral division of arrector dorsalis **(all genera not in other CS)**

— CS–1: Ventral division of arrector dorsalis considerably reduced in size ***(Bagrus, Hemibagrus)***

— Inapplicable: Since the arrector dorsalis is not subdivided into dorsal and ventral divisions ***(Diplomystes, Cetopsis, Hemicetopsis, Helogenes, Callichthys, Corydoras, Loricaria, Hypoptopoma, Lithoxus, Scoloplax, Astroblepus, Nematogenys, Trichomycterus, Hatcheria)***

53. *Position of ventral division of arrector dorsalis* (*ordered multistate character*). The plesiomorphic condition for those catfish with the arrector dorsalis subdivided into two divisions is that the ventral division situates, as its name indicates, on the ventral surface of the pectoral girdle (Diogo et al., 2001c) [State 0: e.g. Fig. 3.56]. In catfish of CS–1, part of the ventral division of the arrector dorsalis is dorsal to the pectoral girdle [State 1: e.g. Fig. 3.50] while in those of CS–2, the vast majority or even the totality of this division lies on the dorsal surface of this girdle [State 2: e.g. Fig. 3.53].

— CS–0: Ventral division of arrector dorsalis lying on ventral surface of pectoral girdle **(all genera not in other CS)**

— CS–1: Part of ventral division of arrector dorsalis lying on dorsal surface of pectoral girdle ***(Chaca)***

— CS–2: Vast majority or even totality of ventral division of arrector dorsalis lying on dorsal surface of pectoral girdle ***(Heteropneustes, Plotosus, Cnidoglanis, Paraplotosus, Neosilurus, Laides, Siluranodon, Ailia, Clarias, Heterobranchus, Bunocephalus, Aspredo, Xyliphius, Hypophthalmus)***

— ?: Since distinct discernment of this character was not possible due to the poor condition of the specimens examined ***(Uegitglanis)***

— Inapplicable: Since the arrector dorsalis is not subdivided into dorsal and ventral divisions ***(Diplomystes, Cetopsis, Hemicetopsis, Helogenes, Callichthys, Corydoras, Loricaria, Hypoptopoma, Lithoxus, Scoloplax, Astroblepus, Nematogenys, Trichomycterus, Hatcheria)***

54. *Development of dorsal division of arrector dorsalis*. The plesiomorphic condition for those catfish with the arrector dorsalis subdivided into two divisions is to present a well-developed, voluminous dorsal division [State 0: e.g. Fig. 3.40], but in the three aspredinid genera examined, the dorsal division of the arrector dorsalis is markedly reduced in size, being lodged on the dorsolateral surface of the pectoral girdle and thus not visible in dorsal view [State 1: e.g. Fig. 3.27].

— CS–0: Voluminous dorsal division of arrector dorsalis **(all genera not in other CS)**

— CS–1: Dorsal division of arrector dorsalis considerably reduced in size ***(Bunocephalus, Aspredo, Xyliphius)***

— Inapplicable: Since the arrector dorsalis is not subdivided into dorsal and ventral divisions ***(Diplomystes, Cetopsis, Hemicetopsis,***

***Helogenes, Callichthys, Corydoras, Loricaria, Hypoptopoma, Lithoxus, Scoloplax, Astroblepus, Nematogenys, Trichomycterus, Hatcheria*)**

55. *Position of dorsal division of arrector dorsalis* (*unordered multistate character*). The plesiomorphic condition for those catfish with the arrector dorsalis subdivided into two divisions is that the dorsal division lies, as its name indicates, on the dorsal surface of the pectoral girdle [State 0: e.g. Fig. 3.40]. In catfish of CS–1, the dorsal division of the arrector dorsalis is mesially bifurcate into a well-developed dorsomesial and a well developed ventromesial part, situated respectively on the dorsal and ventral surface of the pectoral girdle [State 1]. In catfish of CS–2, the dorsal division of the arrector dorsalis is also mesially bifurcated with, however, the ventromesial portion situated ventral to the pectoral girdle and considerably more developed than the dorsomedial portion situated dorsal to this girdle [State 2]. A different configuration of the dorsal division of the arrector dorsalis is found in catfish of CS–3. Here the totality of this division lies on the ventral surface of the pectoral girdle and the lateral foramen between the scapulo coracoid and the cleithrum where this division usually passes is completely closed [State 3: e.g. Fig. 3.7].
    — CS–0: Dorsal division of arrector dorsalis lying on dorsal surface of pectoral girdle **(all genera not in other CS)**
    — CS–1: Dorsal division of arrector dorsalis mesially bifurcate, with part of it lying dorsal and part ventral to the pectoral girdle **(*Leptoglanis, Synodontis, Mochokus, Franciscodoras, Doras, Acanthodoras, Anadoras, Centromochlus, Ageneiosus, Auchenipterus*)**
    — CS–2: Dorsal division of arrector dorsalis mesially bifurcate, with most of its fibres lying ventral to the pectoral girdle **(*Doumea, Zaireichthys*)**
    — CS–3: Totality of dorsal division of arrector dorsalis lying ventral to the pectoral girdle **(*Amphilius, Phractura, Belonoglanis, Trachyglanis, Andersonia, Malapterurus*)**
    — Inapplicable: Since the arrector dorsalis is not subdivided into dorsal and ventral divisions **(*Diplomystes, Cetopsis, Hemicetopsis, Helogenes, Callichthys, Corydoras, Loricaria, Hypoptopoma, Lithoxus, Scoloplax, Astroblepus, Nematogenys, Trichomycterus, Hatcheria*)**

Neurocranium and Anterior Vertebrae

56. *Ventral projection of mesethmoid cornua*. Contrary to all other catfish examined [State 0: e.g. Fig. 3.63], in specimens of genera *Cetopsis* and *Hemicetopsis* examined the mesethmoid cornua present a well-developed, ventral, vertical projection of laminar bone [State 1: e.g. Fig. 3. 46].
    — CS–0: Mesethmoid cornua not presenting ventral, vertical projection of laminar bone **(all genera not in other CS)**

— CS–1: Mesethmoid cornua with well-developed, ventral, vertical projection of laminar bone **(*Cetopsis, Hemicetopsis*)**

57. *Cartilage of mesethmoid (inspired from Friel, 1994).* Contrary to all other catfish examined [State 0: e.g. Fig. 3.67], the three genera of CS–1 present a small, Y-shaped cartilage on the posteromesial surface of the mesethmoid [State 1].
    — CS–0: Mesethmoid not presenting Y-shaped cartilage on posteromesial surface **(all genera not in other CS)**
    — CS–1: Mesethmoid presenting small, Y-shaped cartilage on posteromesial surface **(*Bunocephalus, Aspredo, Xyliphius*)**
58. *Ventral processes of mesethmoid (character inspired from Lundberg and McDade, 1986).* As stated by Lundberg and McDade (1986), plesiomorphically catfish lack major ventral processes on the lateral surface of the mesethmoid [State 0: e.g. Fig. 3.66]. In genera of CS–1, this bone exhibits two well-developed, ventrolateral processes mainly directed laterally, but also anteriorly, which are significantly posterior to the mesethmoid cornua [State 1].
    — CS–0: Mesethmoid lacks ventrolateral processes **(all genera not in other CS)**
    — CS–1: Mesethmoid exhibits ventrolateral projections **(*Heptapterus, Goeldiella, Rhamdia*)**
59. *Dorsomesial concavity of mesethmoid.* Contrary to the plesiomorphic situation for catfish, in which the mesethmoid lacks major, deep dorsal concavities [State 0: e.g. Fig. 3.67], in specimens of *Pangasius* examined, this bone exhibits two well-developed, dorsolateral crests, which delimit a well-defined, deep dorsomesial concavity [State 1].
    — CS–0: Mesethmoid lacks major, deep dorsomesial concavitiy **(all genera not in other CS)**
    — CS–1: Mesethmoid exhibits well-defined, deep dorsomesial concavity **(*Pangasius*)**
60. *Dorsolateral projections of mesethmoid.* Plesiomorphically catfish lack major dorsolateral projections of the mesethmoid [State 0: e.g. Fig. 3.63] but in genera of CS–1, this bone exhibits prominent dorsolateral projections of laminar bone, with its dorsal surface thus markedly larger than its ventral surface [State 1: e.g. Fig. 3.25].
    — CS–0: Mesethmoid lacks major dorsolateral projections of laminar bone **(all genera not in other CS)**
    — CS–1: Mesethmoid with prominent dorsolateral projections of laminar bone **(*Bunocephalus, Aspredo, Xyliphius, Doras, Acanthodoras, Anadoras, Centromochlus, Ageneiosus, Auchenipterus*)**
    — ?: Since distinct discernment of this character was not possible in the specimens examined **(*Ancharius*)**
61. *Highly developed ventromesial crest of mesethmoid (character inspired from Schaefer, 1997).* Contrary to the plesiomorphic situation for catfish in

which the mesethmoid lacks major ventral crests [State 0: e.g. Fig. 3.89], in specimens examined of the four genera of CS–1 this bone exhibits a highly developed, somewhat roundish, prominent ventromesial crest [State 1].

— CS–0: Mesethmoid lacking highly developed ventromesial crest **(all genera not in other CS)**

— CS–1: Mesethmoid exhibiting highly developed ventromesial crest **(*Loricaria, Hypoptopoma, Lithoxus, Astroblepus*)**

62. *Well-developed cartilage between ventral surface of mesethmoid and premaxilla (character inspired from Schaefer, 1997)*. Contrary to the plesiomorphic situation for catfish in which there is no well-developed cartilage between the mesethmoid and the premaxilla [State 0: e.g. Fig. 3.64], siluriforms of CS–1 exhibit a large, well-developed cartilaginous structure between these two bones for articulation between them [State 1].

— CS–0: No well-developed cartilage between mesethmoid and premaxilla **(all genera not in other CS)**

— CS–1: Well-developed cartilage between mesethmoid and premaxilla **(*Loricaria, Hypoptopoma, Lithoxus, Astroblepus*)**

63. *Ligamentous connection between ventral surface of mesethmoid and premaxilla (character inspired from Schaefer, 1997)*. Contrary to the plesiomorphic single ligamentous connection between the mesethmoid and the premaxilla [State 0: e.g. Fig. 3.66], the loricariids examined present two well-defined ligaments between these two structures [State 1].

— CS–0: Single ligamentous connection between ventral surface of mesethmoid and premaxilla **(all genera not in other CS)**

— CS–1: Presence of two well-defined ligaments between ventral surface of mesethmoid and premaxilla **(*Loricaria, Hypoptopoma, Lithoxus*)**

64. *Prominent dorsomesial crest of mesethmoid*. In the plesiomorphic siluriform configuration, the mesethmoid lacks major dorsomesial crests [State 0: e.g. Fig. 3.67] but in *Leptoglanis* this bone exhibits a prominent, anteroposteriorly elongated, dorsomesial crest [State 1: e.g. Fig. 3.17].

— CS–0: Mesethmoid lacks major dorsomesial crests **(all genera not in other CS)**

— CS–1: Mesethmoid exhibits prominent dorsomesial crest **(*Leptoglanis*)**

65. *Layer of cartilage-like tissue (character inspired from de Pinna and Vari, 1995)*. Siluriforms plesiomorphically lack a 'layer of cartilage-like tissue' (see terminology of de Pinna and Vari, 1995) protecting the olfactory capsule. Such a 'layer of cartilage-like tissue' is present, however, in the cetopsins examined [State 1].

— CS–0: Absence of 'layer of cartilage-like tissue' protecting olfactory capsule **(all genera not in other CS)**

— CS–1: Presence of 'layer of cartilage-like tissue' protecting olfactory capsule **(*Cetopsis, Hemicetopsis*)**

66. *Enclosure of olfactory foramen* (*character inspired from Mo, 1991*). Contrary to other catfish examined, in which the olfactory foramen is enclosed by the lateral ethmoid or by both this bone and the mesethmoid [State 0: e.g. Fig. 3.43], in *Auchenoglanis* the foramen is exclusively enclosed by the mesethmoid [State 1].
    - CS–0: Olfactory foramen not completely enclosed by mesethmoid **(all genera not in other CS)**
    - CS–1: Olfactory foramen completely enclosed by mesethmoid **(*Auchenoglanis*)**
67. *Ossification of mesethmoid* (*character inspired from de Pinna and Vari, 1995*). Primitively the mesethmoid is well ossified in adult catfishes [State 0: e.g. Fig. 3.63] but in adult *Helogenes*, a large, mesial part of this bone remains unossified, cartilaginous, with only the anterior cornua well ossified.
    - CS–0: Mesethmoid well ossified **(all genera not in other CS)**
    - CS–1: Much of mesethmoid remains cartilaginous **(*Helogenes*)**
68. *Dorsal surface of mesethmoid*. Contrary to the plesiomorphic condition for siluriforms [State 0: e.g. Fig. 3.67], in *Belonoglanis* and *Trachyglanis* the dorsal surface of the mesethmoid is markedly compressed transversally at about midpoint the length of this bone, thus conferring a somewhat Y-shaped aspect to its posterodorsal surface in dorsal view [State 1: e.g. Fig. 3.12].
    - CS–0: Posterodorsal surface of mesethmoid not Y-shaped **(all genera not in other CS)**
    - CS–1: Posterodorsal surface of mesethmoid Y-shaped **(*Belonoglanis, Trachyglanis*)**
69. *Anterior bifurcation of mesethmoid* (*ordered multistate character*). Plesiomorphically in catfish the anterior margin of the mesethmoid is not significantly bifurcate (Diogo, 2003b] [State 0: e.g. Fig. 3.67]. In the catfish of CS–1, the mesethmoid is markedly bifurcate anteriorly [State 1: e.g. Fig. 3.108], while in *Amphilius* [State 2], and especially in *Chaca* [State 3: e.g. Fig. 3.49], this bone exhibits a remarkable bifurcate anterior margin.
    - CS–0: Mesethmoid not significantly bifurcate anteriorly **(all genera not in other CS)**
    - CS–1: Mesethmoid markedly bifurcate anteriorly **(*Chrysichthys, Schilbe, Clarotes, Pseudopimelodus, Microglanis, Auchenoglanis, Silurus, Wallago, Helogenes, Ictalurus, Amiurus, Ageneiosus, Ancharius*)**
    - CS–2: Anterior bifurcation of mesethmoid more pronounced than in CS–1 **(*Amphilius*)**
    - CS–3: Anterior bifurcation of mesethmoid even more prominent, with anterior cornua of mesethmoid inclusively longer than main body of this bone **(*Chaca*)**
70. *Anteromesial process of mesethmoid* (*character inspired from Schaefer, 1990*). Contrary to all other catfish [State 0: e.g. Fig. 3.67], in specimens of

genus *Scoloplax* examined the mesethmoid is prolonged anteriorly by a prominent, anteriorly pointed, anteromesial process [State 1].

— CS–0: Mesethmoid without prominent anteromesial process **(all genera not in other CS)**

— CS–1: Mesethmoid exhibiting prominent anteromesial process **(*Scoloplax*)**

71. *Premaxilla markedly elongated anteroposteriorly*. Although the anteroposterior length of the premaxilla is relatively variable among catfish [State 0: see characters below], this bone is particularly elongated anteroposteriorly in *Wallago*, with its anteroposterior length about half the total length of the neurocranium [State 1].

— CS–0: Premaxilla not markedly elongated anteroposteriorly **(all genera not in other CS)**

— CS–1: Premaxilla markedly elongated anteroposteriorly **(*Wallago*)**

72. *Dorsolateral process of premaxilla (ordered multistate character)*. Plesiomorphically catfish lack major dorsal processes of the premaxilla for insertion of premaxillo-maxillary ligament (see, e.g., de Pinna, 1993) [State 0: e.g. Fig. 3.63]. Such a process is present in siluriforms of CS–1 [State 1: e.g. Fig. 3.104] but particularly developed in siluriforms of CS–2 [State 2: e.g. Fig. 3.100].

— CS–0: Absence of well-developed dorsolateral process of premaxilla **(all genera not in other CS)**

— CS–1: Presence of well-developed dorsolateral process of premaxilla for insertion of premaxillo-maxillary ligament **(*Bagrus, Rita, Bagrichthys, Hemibagrus, Heptapterus, Goeldiella, Rhamdia, Pseudopimelodus, Microglanis*)**

— CS–2: Dorsolateral process of premaxilla more developed than in CS–1 **(*Pimelodus, Calophysus, Hypophthalmus, Pseudoplatystoma, Chaca*)**

73. *Width of premaxilla (ordered multistate character)*. Primitively in catfish the premaxillae are well-developed, broad structures [State 0: e.g. Fig. 3.67]. In catfish of CS–1, they are considerably compressed transversally [State 1: e.g. Fig. 3.17]. The transversal compression of the premaxilla is still more pronounced in genera of CS–2 [State 2: e.g. Fig. 3.12].

— CS–0: Premaxilla not compressed transversally **(all genera not in other CS)**

— CS–1: Premaxilla compressed transversally **(*Loricaria, Hypoptopoma, Lithoxus, Bagrichthys, Phractura, Doumea, Leptoglanis, Andersonia, Auchenoglanis*)**

— CS–2: Premaxilla more compressed transversally than in CS–1 **(*Belonoglanis, Trachyglanis, Callichthys, Corydoras*)**

74. *Posterolateral process of premaxilla (character inspired from Chen and Lundberg, 1994)*. Contrary to all other catfish examined [State 0], in *Liobagrus* and

Amblyceps the premaxilla exhibits a long, thin, posterolaterally oriented, prominent posterodorsolateral process [State 1: e.g. Fig. 3.3].

— CS–0: Absence of prominent posterolateral process of premaxilla **(all genera not in other CS)**

— CS–1: Presence of prominent posterolateral process of premaxilla **(*Liobagrus, Amblyceps*)**

75. *Lateral bifurcation of premaxilla.* Plesiomorphically in catfish the premaxilla is not significantly laterally bifurcate [State 0: Fig. 3.66] but in the specimens examined of the six genera of CS–1, this bone exhibits a marked lateral bifurcation [State 1: Fig. 3.20].

— CS–0: Premaxilla not significantly laterally bifurcate **(all genera not in other CS)**

— CS–1: Marked lateral bifurcation of premaxilla **(*Arius, Ancharius, Genidens, Pangasius, Helicophagus, Ageneiosus*)**

76. *Posteroventral extension of premaxilla.* Plesiomorphically catfish lack major posteroventral extensions of the premaxilla [State 0: e.g. Fig. 3.66]. In catfish of CS–1, the premaxilla exhibits a large, somewhat roundish, posteroventral horizontal extension of laminar bone [State 1: e.g. Fig. 3.89].

— CS–0: Absence of posteroventral extension of premaxilla **(all genera not in other CS)**

— CS–1: Presence of posteroventral extension of premaxilla **(*Paraplotosus, Plotosus, Cnidoglanis, Neosilurus, Nematogenys, Heteropneustes, Uegitglanis*)**

77. *Posterior process of premaxilla (character inspired from Brown and Ferraris, 1988).* Contrary to other catfish examined [State 0: e.g. Fig. 3.67], in *Chaca* the premaxilla exhibits a highly developed, long and thin posterior process oriented posteromesially [State 1: e.g. Fig. 3.49].

— CS–0: Absence of prominent posterolateral process of premaxilla **(all genera not in other CS)**

— CS–1: Presence of prominent posterolateral process of premaxilla **(*Chaca*)**

78. *Relation between nasal barbels and premaxilla (character inspired from Mo, 1991).* When catfish present nasal barbels, these barbels are usually situated on the posterior nostrils (see Mo, 1991) [State 0]. However, in the nematogenyid and trichomycterid genera examined, the nasal barbels are situated on the anterior nostrils, being supported mainly by the anterodorsal surface of the premaxillae [State 1].

— CS–0: Nasal barbels, if present, situated on posterior nostrils **(all genera not in other CS)**

— CS–1: Nasal barbels on anterior nostrils, supported by anterodorsal surface of premaxillae **(*Nematogenys, Trichomycterus, Hatcheria*)**

79. *Anterior projection of premaxilla (character inspired from Friel, 1994).* Contrary to other catfish examined [State 0: e.g. Fig. 3.1.13.4], in *Aspredo* the

premaxilla exhibits a large, broad anterior projection of laminar bone [State 1].

— CS–0: Absence of anterior projection of premaxilla **(all genera not in other CS)**

— CS–1: Presence of large anterior projection of premaxilla **(*Aspredo*)**

80. *Subdivision of premaxilla* (*character inspired from de Pinna, 1996*). Each premaxilla in siluriforms is primitively composed of a single bony piece [State 0: e.g. Fig. 3.89]. In specimens examined of the two genera of CS–1, each premaxilla is subdivided into two or more segments [State 1: e.g. Fig. 3.122].

— CS–0: Each premaxilla composed of a single bony piece **(all genera not in other CS)**

— CS–1: Each premaxilla subdivided into two or more segments **(*Glyptothorax, Bagarius*)**

81. *Presence of premaxillary teeth*. Contrary to the plesiomorphic situation found in all other catfish examined [State 0: e.g. Fig. 3.63], in specimens of genera of CS–1, the premaxilla is completely edentulous [State 1].

— CS–0: Presence of premaxillary teeth **(all genera not in other CS)**

— CS–1: Absence of premaxillary teeth **(*Gagata, Xyliphius, Siluranodon, Hypophthalmus, Callichthys, Corydoras*)**

82. *Connection between premaxilla and mesethmoid* (*character inspired from Howes, 1983a*). Plesiomorphically in catfish the ligaments between the premaxilla and the mesethmoid are very short [State 0: e.g. Fig. 3.66] but in siluriforms of CS–1, these ligaments are long, conferring a great mobility to the premaxilla [State 1].

— CS–0: Short ligaments between premaxilla and mesethmoid **(all genera not in other CS)**

— CS–1: Long ligaments between premaxilla and mesethmoid **(*Callichthys, Corydoras, Loricaria, Hypoptopoma, Lithoxus, Astroblepus*)**

83. *Dorsal facet of lateral ethmoid* (*character inspired from de Pinna, 1996*). Contrary to all other catfish examined [State 0: e.g. Fig. 3.63], in *Gagata* the posterior region of the dorsal surface of the lateral ethmoid has a well-defined arced facet abutting the ventral surface of the frontal [State 1].

— CS–0: Absence of arced dorsal facet of lateral ethmoid **(all genera not in other CS)**

— CS–1: Presence of arced dorsal facet of lateral ethmoid **(*Gagata*)**

84. *Shape of lateral ethmoid*. The plesiomorphic condition for siluriforms is that in which the anterior surface of the lateral ethmoid is somewhat rounded or irregularly shaped in dorsal view (see, e.g., Mo, 1991; Diogo, 2003b) [State 0: e.g. Fig 3.67]. In specimens examined of genera of CS–1, the lateral ethmoid is markedly truncated anteriorly [State 1: e.g. Fig 3.89].

— CS–0: Lateral ethmoid not markedly truncated anteriorly **(all genera not in other CS)**

— CS–1: Lateral ethmoid clearly truncated anteriorly ***(Amphilius, Phractura, Doumea, Belonoglanis, Trachyglanis, Leptoglanis, Zaireichthys, Andersonia, Paramphilius, Nematogenys, Loricaria, Lithoxus, Astroblepus, Glyptothorax, Glyptosternon, Pseudopimelodus, Microglanis)***

85. *Articulatory facet of neurocranium for autopalatine (character inspired from Lundberg et al., 1991b).* Plesiomorphically in catfish the articulatory facet of the neurocranium for the autopalatine is not considerably elongated anteroposteriorly [State 0: e.g. Fig. 3.66] but in pimelodin catfishes examined, this facet is developed into a large, alate lateral projection remarkably elongated anteroposteriorly [State 1].
   — CS–0: Articulatory facet of neurocranium for autopalatine not markedly elongated posteriorly **(all genera not in other CS)**
   — CS–1: Articulatory facet of neurocranium for autopalatine remarkably elongated anteroposteriorly ***(Pimelodus, Calophysus, Hypophthalmus, Pseudoplatystoma)***

86. *Dorsal concavity between lateral ethmoid and frontal (ordered multistate character).* The plesiomorphic condition for catfish, found in *Diplomystes* and the vast majority of other siluriforms, is likely that in which there is no well-defined concavity between the posterodorsal surface of the lateral ethmoid and the anterodorsal surface of the frontal [State 0: e.g. Fig. 3.67]. The specimens examined in genera of CS–1 exhibit a deep dorsal concavity between the frontal and lateral ethmoid [State 1: e.g. Fig. 3.49], while specimens examined in genera of CS–2 exhibit a complete foramen between these two bones [State 2: e.g. Fig. 3.116].
   — CS–0: Absence of well-defined dorsal concavity between lateral ethmoid and frontal **(all genera not in other CS)**
   — CS–1: Presence of well-defined and deep dorsal concavity between lateral ethmoid and frontal ***(Silurus, Wallago, Cranoglanis, Goeldiella, Synodontis, Ictalurus, Amiurus, Franciscodoras, Anadoras, Acanthodoras, Doras, Ageneiosus, Chaca, Austroglanis)***
   — CS–2: Complete foramen between dorsal surfaces of lateral ethmoid and frontal ***(Pangasius, Schilbe, Laides, Pseudeutropius, Siluranodon, Ailia, Helicophagus, Arius, Ancharius, Genidens)***
   — ?: Since it was not possible to discern this character due to the poor condition of the specimens examined ***(Auchenipterus)***

87. *Presence of ramified projections of lateral ethmoid near olfactory foramen.* Uniquely in *Cetopsis*, and contrary to all other catfish examined [State 0: e.g. Fig. 3.63], the lateral ethmoid exhibits ramified projections near the olfactory foramen [State 1: e.g. Fig. 3.43].
   — CS–0: Absence of ramified projections of lateral ethmoid near olfactory foramen **(all genera not in other CS)**
   — CS–1: Presence of ramified projections of lateral ethmoid near olfactory foramen ***(Cetopsis)***

88. *Lateral laminar projection of lateral ethmoid (ordered multistate character).* The plesiomorphic condition for catfish clearly seems to be that present in *Diplomystes* and most siluriforms in which the lateral ethmoid presents no major well-developed lateral projections of laminar bone [State 0: e.g. Fig. 3.63]. In catfish of CS–1, each lateral ethmoid peculiarly exhibits a well-developed, broad dorsolateral projection of laminar bone, which surrounds a significant part of the anterodorsolateral surface of the eye [State 1: e.g. Fig. 3.55], with this projection even more developed in specimens of genus *Cranoglanis* [State 1: e.g. Fig. 3.57].
    — CS–0: Lateral ethmoid lacking major lateral projection of laminar bone **(all genera not in other CS)**
    — CS–1: Lateral ethmoid exhibiting well-developed, broad dorsolateral projection of laminar bone, surrounding anterodorsolateral surface of the eye **(*Chrysichthys, Clarotes, Arius, Ancharius, Genidens, Ictalurus, Amiurus, Austroglanis*)**
    — CS–2: Remarkably developed dorsolateral projection of lateral ethmoid **(*Cranoglanis*)**
89. *Contact between lateral ethmoid and sphenotic (character inspired from Bornbusch, 1991b).* Contrary to other catfish [State 0: e.g. Fig. 3.67], in the silurids examined the dorsal surface of the lateral ethmoid contacts the dorsal surface of the sphenotic by means of a thin, elongated lateral laminar projection [State 1].
    — CS–0: Absence of thin, elongated lateral laminar projection of lateral ethmoid contacting sphenotic **(all genera not in other CS)**
    — CS–1: Presence of thin, elongated lateral laminar projection of lateral ethmoid contacting sphenotic **(*Silurus, Wallago*)**
90. *Foramen on dorsomesial surface of lateral ethmoid.* Plesiomorphically in catfish the lateral ethmoid does not present major foramens near its dorsomesial surface [State 0: e.g. Fig. 3.67] but in siluriforms of CS–1, a well-developed foramen is present on the lateral ethmoid near the dorsomesial surface of this bone and the posterolateral surface of the mesethmoid [State 1: e.g. Fig. 3.13].
    — CS–0: Absence of foramen on dorsomesial surface of lateral ethmoid **(all genera not in other CS)**
    — CS–1: Presence of foramen on dorsomesial surface of lateral ethmoid **(*Leptoglanis, Zaireichthys, Glyptothorax, Glyptosternon, Liobagrus, Amblyceps*)**
    — ?: Since distinct discernment of this character was not possible in the specimens examined **(*Gagata*)**
91. *Posterodorsal projection of lateral ethmoid (character inspired from de Pinna, 1996).* Contrary to all other catfish examined [State 0: e.g. Fig. 3.63], in the specimens analysed of the three genera of CS–1, the lateral ethmoid exhibits a long, thin posterolateral projection surrounding a significant part of the dorsolateral surface of the frontal [State 1: e.g. Fig. 3.120].
    — CS–0: Lateral ethmoid without long, thin posterolateral projection **(all genera not in other CS)**

— CS–1: Lateral ethmoid exhibiting long, thin posterolateral projection **(*Glyptothorax, Glyptosternon, Bagarius*)**

92. *Thin anterior projection of ethmoid cartilage.* Peculiarly in *Chaca*, and contrary to all other catfish examined [State 0: e.g. Fig. 3.63], the ethmoid cartilage exhibits a long, thin anterior projection, which is situated ventral to the also long and thin anterior cornua of the mesethmoid [State 1: e.g. Fig. 3.49].
    — CS–0: Absence of thin anterior projection of ethmoid cartilage **(all genera not in other CS)**
    — CS–1: Presence of thin anterior projection of ethmoid **cartilage (*Chaca*)**
93. *Presence of prevomer.* Plesiomorphically in catfish the prevomer is present [State 0: e.g. Fig. 3.66] but in the specimens examined of the six genera of CS–1, this bone is absent or undifferentiated [State 1: e.g. Fig. 3.28].
    — CS–0: Prevomer present **(all genera not in other CS)**
    — CS–1: Prevomer absent or undifferentiated **(*Bunocephalus, Aspredo, Xyliphius, Chaca, Microglanis, Scoloplax*)**
94. *Length of prevomer (character inspired from de Pinna and Vari, 1995).* Contrary to all other catfish examined [State 0: e.g. Fig. 3.66], in *Helogenes* the prevomer is markedly compressed anteroposteriorly and almost reduced to its anterior arms carrying the prevomeine tooth-plates [State 1].
    — CS–0: Prevomer not markedly compressed anteroposteriorly **(all genera not in other CS)**
    — CS–1: Prevomer markedly compressed anteroposteriorly **(*Helogenes*)**
    — Inapplicable: Since the prevomer is absent or undifferentiated **(*Bunocephalus, Aspredo, Xyliphius, Chaca, Microglanis, Scoloplax*)**
95. *Position of anterior margin of prevomer (ordered multistate character).* Plesiomorphically in catfish the anterior margin of the prevomer lies relatively far from the anterior margin of the mesethmoid [State 0: e.g. Fig. 3.66]. In specimens of genera *Trichomycterus* and *Hatcheria* examined, the prevomer extends anteriorly, with its anterior margin lying at about the same level as the anterior margin of the premaxilla [State 1]. In specimens of *Nematogenys* examined, the prevomer is extended further anteriorly, its anterior margin lying at about the same level of the anterior margin of the mesethmoid [State 2: e.g. Fig. 3.89].
    — CS–0: Anterior margin of prevomer markedly posterior to anterior margin of mesethmoid **(all genera not in other CS)**
    — CS–1: Prevomer with anterior extension, its anterior margin lying at level of anterior margin of premaxilla **(*Trichomycterus, Hatcheria*)**
    — CS–2: Anterior margin of prevomer in markedly anterior position, lying at level of anterior margin of mesethmoid **(*Nematogenys*)**
    — Inapplicable: Since the prevomer is absent or undifferentiated **(*Bunocephalus, Aspredo, Xyliphius, Chaca, Microglanis, Scoloplax*)**
96. *Anterodorsomesial projection of frontal.* Contrary to all other catfish examined [State 0], in specimens of the six genera of CS–1 analysed, the

frontal exhibits a well-developed, thin, extensive anterodorsomesial projection, which almost separates, or completely separates, the dorsomesial surface of the lateral ethmoid and the dorsolateral surface of the mesethmoid in dorsal view [State 1].

— CS–0: Absence of well-developed, thin, extensive anterodorsomesial projection of frontal **(all genera not in other CS)**

— CS–1: Presence of well-developed, thin, extensive anterodorsomesial projection of frontal **(*Clarias, Heterobranchus, Paraplotosus, Neosilurus, Cnidoglanis, Heteropneustes*)**

97. *Contact between frontal and pterotic*. Plesiomorphically in catfish the frontal does not come into contact with the pterotic (see, e.g., Chardon, 1968; Mo, 1991) [State 0: e.g. Fig. 3.67]. In specimens examined of genera listed in CS–1, the posterolateral margin of the frontal comes into contact with the anteromesial margin of the pterotic in dorsal view [State 1: e.g. Fig. 3.17].

— CS–0: Frontal and pterotic not in contact **(all genera not in other CS)**

— CS–1: Frontal and pterotic in contact in dorsal view **(*Pimelodus, Hypophtalmus, Calophysus, Phractura, Doumea, Belonoglanis, Trachyglanis, Andersonia, Clarias, Uegitglanis, Heterobranchus, Heteropneustes, Silurus, Leptoglanis, Auchenipterus, Malapterurus*)**

— ?: Since it was not possible to discern the limits between frontal, pterotic and sphenotic in the specimens examined **(*Zaireichthys*)**

98. *Dorsomesial process of frontal (ordered multistate character)*. Contrary to all other catfish examined [State 0: e.g. Fig. 3.24], in *Paraplotosus* there is a well-developed dorsomesial process of the frontal meeting its counterpart on the midline, thus forming a well-developed V-shaped structure on the dorsal surface of the neurocranium [State 1], which is still more prominent in *Plotosus* [State 2: e.g. Fig. 3.114].

— CS–0: Frontal not presenting dorsomesial process **(all genera not in other CS)**

— CS–1: Frontal presenting dorsomesial process **(*Paraplotosus*)**

— CS–2: Dorsomesial process of frontal more developed than in CS–1 **(*Plotosus*)**

99. *Dorsal salience of frontal and lateral ethmoid*. Contrary to the other siluriforms examined [State 0: e.g. Fig. 3.67], in the specimens of genera *Plotosus* and *Cnidoglanis* analysed, there is a well-developed dorsal salience of the neurocranium, at the level of the posterodorsal region of the lateral ethmoid and the anterodorsal region of the frontal [State 1: e.g. Fig. 3.114].

— CS–0: Absence of dorsal salience of frontal and lateral ethmoid **(all genera not in other CS)**

— CS–1: Presence of dorsal salience of frontal and lateral ethmoid **(*Plotosus, Cnidoglanis*)**

100. *Width of frontal.* Plesiomorphically in catfish the frontal is very large and hence clearly visible in both dorsal and ventral views of the neurocranium [State 0: e.g. Fig. 3.66]. However, in specimens of genera *Cetopsis* and *Hemicetopsis* examined, this bone is so markedly compressed transversally that it is almost completely confined to the neurocranial dorsomesial crests for attachment of the adductor mandibulae and is not visible in a ventral view of the neurocranium [State 1: e.g. Fig. 3. 46].
    - CS–0: Frontal not markedly compressed transversally **(all genera not in other CS)**
    - CS–1: Frontal markedly compressed transversally ***(Cetopsis, Hemicetopsis)***
101. *Foramen on dorsolateral surface of frontal (character inspired from de Pinna and Vari, 1995).* Plesiomorphically catfish lack any large openings other than the anterior and posterior median cranial fontanels [State 0: e.g. Fig. 3.67] but in *Helogenes* there is a large foramen on the dorsolateral surface of each frontal [State 1].
    - CS–0: Frontal without large foramen on dorsolateral surface **(all genera not in other CS)**
    - CS–1: Frontal exhibiting large foramen on dorsolateral surface ***(Helogenes)***
102. *Presence of anterior fontanel.* Primitively in adult catfish the anterior fontanel is present and usually well developed (see, e.g., Regan, 1911b) [State 0: e.g. Fig. 3.67]. In the adult specimens examined of genera of CS–1, the anterior fontanel is absent [State 1: e.g. Fig. 3.13].
    - CS–0: Anterior fontanel present **(all genera not in other CS)**
    - CS–1: Anterior fontanel absent ***(Zaireichthys, Cetopsis, Loricaria, Hypoptopoma, Lithoxus, Scoloplax, Astroblepus)***
103. *Additional median cranial fontanel.* Contrary to all other catfish (see above) [State 0: e.g. Fig. 3.67], specimens examined of genera listed in CS–1 present an additional median cranial fontanel, for a total of three median cranial fontanels [State 1].
    - CS–0: Absence of additional median cranial fontanel **(all genera not in other CS)**
    - CS–1: Presence of additional median cranial fontanel ***(Clarias, Ailia, Uegitglanis, Heterobranchus, Heteropneustes)***
104. *Foramen between sphenotic, frontal and parieto-supraoccipital (character inspired from de Pinna, 1996).* As mentioned above, plesiomorphically catfish lack any large openings other than the anterior and posterior median cranial fontanels [State 0: e.g. Fig. 3.67], but peculiarly in *Gagata* there is a well-developed foramen between the dorsal surfaces of the frontal, sphenotic and parieto-supraoccipital on each side of the head [State 1].
    - CS–0: Absence of well-developed foramen between sphenotic, frontal and parieto-supraoccipital **(all genera not in other CS)**
    - CS–1: Presence of well-developed foramen between sphenotic, frontal and parieto-supraoccipital ***(Gagata)***

105. *Contact between lateral ethmoid and pterosphenoid.* Contrary to the plesiomorphic situation found in other catfish, in which the anterodorsal margin of the pterosphenoid and posterodorsal margin of the lateral ethmoid lie relatively far from each other [State 0: e.g. Fig. 3.63], in *Cetopsis* and *Hemicetopsis* the anterodorsal surface of the pterosphenoid is markedly expanded anteriorly and the posterodorsal surface of the lateral ethmoid is markedly expanded posteriorly, with these bones being almost in contact, or inclusively in contact [State 1: e.g. Fig. 3.46].
    — CS–0: Anterodorsal margin of pterosphenoid and posterodorsal margin of lateral ethmoid relatively far from each other **(all genera not in other CS)**
    — CS–1: Anterodorsal margin of pterosphenoid and posterodorsal margin of lateral ethmoid near to each other, or inclusively in contact **(*Cetopsis, Hemicetopsis*)**
    — ?: Since discerning the limits of these bones was not possible in the specimens examined **(*Helogenes*)**
106. *Anteroventrolateral process of pterosphenoid.* The plesiomorphic configuration for the siluriforms is seemingly that present in other ostariophysans and in the vast majority of the siluriforms, in which the pterosphenoid lacks a well-developed anteroventrolateral process for articulation with the hyomandibulo-metapterygoid (see, e.g., Diogo et al., 2001a) [State 0: e.g. Fig. 3.89]. However, such a process is present in the specimens examined of genus *Diplomystes* and genus *Trichomycterus* [State 1: e.g. Fig. 3.63].
    — CS–0: Absence of anteroventrolateral process of pterosphenoid **(all genera not in other CS)**
    — CS–1: Presence of anteroventrolateral process of pterosphenoid **(*Diplomystes, Trichomycterus*)**
    — ?: Since discerning this character was not possible in the specimens examined **(*Hatcheria*)**
107. *Lateral process of pterosphenoid.* Peculiarly in *Plotosus* and *Paraplotosus*, and contrary to all other catfish examined [State 0: e.g. Fig. 3.46], the pterosphenoid exhibits a prominent, anterolaterally pointed, lateral process [State 1].
    — CS–0: Absence of lateral process of pterosphenoid **(all genera not in other CS)**
    — CS–1: Presence of lateral process of pterosphenoid **(*Plotosus, Paraplotosus*)**
108. *Posterior portion of parasphenoid.* Contrary to all other catfish in which the posterior portion of the parasphenoid is markedly large [State 0: e.g. Fig. 3.66], in the plotosids examined the posterior portion of this bone is notably compressed transversally [State 1].
    — CS–0: Posterior portion of parasphenoid not compressed transversally **(all genera not in other CS)**

— CS–1: Posterior portion of parasphenoid markedly compressed transversally (***Plotosus, Paraplotosus, Cnidoglanis, Neosilurus***)

109. *Anterodorsolateral salience of sphenotic* (*ordered multistate character*). The plesiomorphic condition for catfish is seemingly that found in *Diplomystes* and the vast majority of siluriforms, as well as of ostariophysans, in which the sphenotic lacks major well-developed processes or saliences [State 0: e.g. Fig. 3.67]. However, in catfish of CS–1 there is a well-developed, large anterodorsolateral salience of the sphenotic [State 1: e.g. Fig. 3.89], which is particularly pronounced in catfish of CS–2 [State 2].
 — CS–0: Absence of well-developed anterodorsolateral salience of sphenotic **(all genera not in other CS)**
 — CS–1: Presence of well-developed anterodorsolateral salience of sphenotic (***Chrysichthys, Pangasius, Clarotes, Nematogenys, Trichomycterus, Hatcheria, Bagarius, Amiurus, Plotosus, Neosilurus, Paraplotosus, Pimelodus, Calophysus, Rhamdia, Austroglanis***)
 — CS–2: Anterodorsolateral salience of sphenotic even more pronounced than in CS–1 (***Malapterurus, Auchenoglanis***)

110. *Lateral bifurcation of anterodorsolateral salience of sphenotic*. Contrary to other catfish [State 0: e.g. Fig. 3.89], in *Austroglanis* there is a marked bifurcation of the anterodorsolateral salience of the sphenotic (see above) [State 1: e.g. Fig. 3.33].
 — CS–0: Absence of marked bifurcation of anterodorsolateral salience of sphenotic (***Chrysichthys, Pangasius, Clarotes, Nematogenys, Trichomycterus, Hatcheria, Bagarius, Amiurus, Plotosus, Neosilurus, Paraplotosus, Pimelodus, Calophysus, Rhamdia, Malapterurus, Auchenoglanis***)
 — CS–1: Presence of marked bifurcation of anterodorsolateral salience of sphenotic (***Austroglanis***)
 — Inapplicable: Since there is no well-developed anterodorsolateral salience of sphenotic **(all genera not in other CS)**

111. *Ventromesial projection of sphenotic*. Contrary to all other catfish examined [State 0: e.g. Fig. 3.66], in specimens of *Callichthys* analysed the sphenotic exhibits a well-developed, thin ventromesial projection that extends mesially to contact a similar well-developed, thin ventrolateral projection of the prootic [State 1].
 — CS–0: Absence of well-developed ventromesial projection of sphenotic **(all genera not in other CS)**
 — CS–1: Presence of well-developed ventromesial projection of sphenotic (***Callichthys***)

112. *Anterodorsal process of sphenotic*. As noted above, plesiomorphically in siluriforms the sphenotic lacks major well-developed processes or saliences [State 0: e.g. Fig. 3.67], but in the four genera of CS–1 this bone exhibits a prominent, anteriorly directed anterodorsal process [State 1: Fig. 3.46].
 — CS–0: Absence of prominent anterodorsal process of sphenotic **(all genera not in other CS)**

— CS–1: Presence of prominent anterodorsal process of sphenotic (***Cetopsis, Hemicetopsis, Helogenes, Ageneiosus***)

113. *Fusion between sphenotic, prootic and pterosphenoid (character inspired from de Pinna, 1992)*. Contrary to other catfish examined [State 0: e.g. Fig. 3.67], in the specimens of genera *Trichomycterus* and *Hatcheria* examined the sphenotic, prootic and pterosphenoid are seemingly fused into a single element [State 1].
    — CS–0: Sphenotic, prootic and pterosphenoid not fused **(all genera not in other CS)**
    — CS–1: Sphenotic, prootic and pterosphenoid seemingly fused into a single element (***Trichomycterus, Hatcheria***)

114. *Size of utricular otolith (ordered multistate character) (character inspired from, e.g., Chardon, 1968; Mo, 1991; de Pinna, 1993; Oliveira et al., 2001)*. Contrary to the plesiomorphic siluriform condition present in other catfish examined, in which the utriculus is not a particularly conspicuous element and is confined within the central area of the prootic [State 0: e.g. Fig. 3.66], in genera of CS–1 [State 1], and especially in genera of CS–2 [State 2], the utricular otolith is a greatly enlarged element profoundly inflating the ventral surface of the neurocranium.
    — CS–0: Utricular otolith not markedly developed **(all genera not in other CS)**
    — CS–1: Utricular otolith markedly developed, profoundly inflating ventral surfaces of prootic and pterotic (***Plotosus, Neosilurus, Cnidoglanis, Paraplotosus***)
    — CS–2: Utricular otolith more developed than in CS–1, profoundly inflating ventral surfaces of prootic and pterotic, and also the exoccipital (***Arius, Genidens***)

115. *Fossa between ventromesial surface of pterotic and ventrolateral surface of exoccipital (ordered multistate character)*. Contrary to all other catfish examined [State 0: e.g. Fig. 3.66], in *Bagarius* and *Gagata* a large, somewhat deep fossa occurs on the neurocranial floor between the ventromesial surface of the pterotic and the ventrolateral surface of the exoccipital [State 1], which is still more enlarged and deeper in Glyptosternon and Glyptothorax [State 2: e.g. Fig. 3.122].
    — CS–0: Absence of large fossa between ventral surfaces of pterotic and exoccipital **(all genera not in other CS)**
    — CS–1: Presence of large, somewhat deep fossa between ventral surfaces of pterotic and exoccipital (***Bagarius, Gagata***)
    — CS–2: Fossa between ventral surfaces of pterotic and exoccipital more developed, enlarged and deeper than in CS–1 (***Glyptosternon, Glyptothorax***)

116. *Posterior process of exoccipital*. Contrary to all other catfish examined [State 0: e.g. Fig. 3.66], in specimens of genera of CS–1 analysed, the exoccipital

exhibits a prominent, thin posterior process, which comes into contact posteriorly with the anterior surface of the fourth parapophysis [State 1].

— CS–0: Absence of prominent posterior process of exoccipital **(all genera not in other CS)**

— CS–1: Presence of prominent posterior process of exoccipital **(*Arius, Genidens*)**

117. *Fusion between exoccipital and basioccipital* (*character inspired from Chardon, 1968*). Contrary to all other catfish examined [State 0: e.g. Fig. 3.66], in specimens of the five genera of CS–1 analysed, the exoccipital and basioccipital are seemingly fused into a single element [State 1].

— CS–0: Exoccipital and basioccipital not fused **(all genera not in other CS)**

— CS–1: Exoccipital and basioccipital fused in a single element **(*Trichomycterus, Hatcheria, Callichthys, Corydoras, Scoloplax*)**

118. *Emergence of epioccipital on cranial roof* (*character inspired from Mo, 1991; de Pinna, 1993; Lundberg, 1993*). Contrary to other catfish examined, in which the epioccipital situates essentially on the posterior wall of the cranium and does not constitute a significant part of the cranial roof [State 0: e.g. Fig. 3.67], in siluriforms of CS–2 the epioccipital constitutes a significant part of the dorsal surface of the cranial roof [State 1: e.g. Fig. 3.74].

— CS–0: Epioccipital not constituting significant part of cranial roof **(all genera not in other CS)**

— CS–1: Epioccipital constituting significant part of cranial roof **(*Franciscodoras, Anadoras, Acanthodoras, Doras, Centromochlus, Ageneiosus, Auchenipterus*)**

— ?: Since distinct discernment of this character was not possible in the specimens examined **(*Ancharius*)**

119. *Presence of prominent, roundish posterodorsal process of epioccipital*. Contrary to all other catfish examined [State 0: e.g. Fig. 3.67], in *Cranoglanis* there is a prominent, stout, roundish posterodorsal process of the epioccipital, from which originate a goodly part of the fibres of the protractor of the müllerian process [State 1: e.g. Fig. 3.58].

— CS–0: Absence of prominent, stout, roundish posterodorsal process of epioccipital **(all genera not in other CS)**

— CS–1: Presence of prominent, stout, roundish posterodorsal process of epioccipital **(*Cranoglanis*)**

120. *Presence of well-developed, laterally opened, 'fossa post-temporalis'* (*character inspired from Mo, 1991*). Contrary to all other catfish examined [State 0: e.g. Fig. 3.63], in the four bagrid genera analysed and in *Malapterurus* there is a well-developed posterodorsal projection of the pterotic which, together with the anterodorsolateral margin of the post-temporo-supracleithrum, delimits a well-developed, laterally opened, 'fossa post-temporalis' (see terminology of Mo, 1991) [State 1: e.g. Fig. 3.42].

— CS–0: Absence of well-developed posterodorsolateral projection of pterotic delimiting well-developed, laterally opened, 'fossa post-temporalis' **(all genera not in other CS)**

— CS–1: Presence of well-developed posterodorsolateral projection of pterotic delimiting well-developed, laterally opened, 'fossa post-temporalis' **(*Bagrus, Bagrichthys, Hemibagrus, Rita, Malapterurus*)**

121. *Foramen between pterotic and parieto-supraoccipital.* As noted above, plesiomorphically catfish lack any large openings other than the anterior and posterior median cranial fontanels [State 0: e.g. Fig. 3.67]. Uniquely in *Paramphilius* a small foramen lies between the dorsal surfaces of the pterotic and parieto-supraoccipital on each side of the head [State 1].
   — CS–0: Absence of small foramen between pterotic and parieto-supraoccipital **(all genera not in other CS)**
   — CS–1: Presence of small foramen between pterotic and parieto-supraoccipital **(*Paramphilius*)**

122. *Presence of prominent ventrolateral process of pterotic* (*character inspired from Lundberg, 1982*). Contrary to all other catfish examined [State 0: e.g. Fig. 3.66], in *Ictalurus* and *Amiurus* a prominent, ventrolateral process of the pterotic occurs, which receives the 'subpterotic process' (see terminology of Lundberg, 1982) of the post-temporo-supracleithrum [State 1].
   — CS–0: Absence of prominent ventrolateral process of pterotic **(all genera not in other CS)**
   — CS–1: Presence of prominent ventrolateral process of pterotic **(*Ictalurus, Amiurus*)**

123. *Presence of 'supratemporal fossa'* (*ordered multistate character*) (*character inspired from de Pinna, 1996*). As noted above, plesiomorphically catfish lack any large openings or fossas other than the anterior and posterior median cranial fontanels [State 0: e.g. Fig. 3.67]. In siluriforms of CS–1, a well-defined, deep 'supratemporal fossa' (see terminology of de Pinna, 1996) lies between the dorsomesial surface of the pterotic and the dorsolateral surface of the parieto-supraoccipital on each side of the head [State 1: e.g. Fig. 3.74]. In the two genera of CS–2 a complete foramen occurs between the dorsomesial surface of the pterotic and the dorsolateral surface of the parieto-supraoccipital on each side of the head [State 2].
   — CS–0: Absence of supratemporal fossa **(all genera not in other CS)**
   — CS–1: Presence of supratemporal fossa between dorsal surfaces of pterotic and parieto-supraoccipital **(*Glyptothorax, Akysis, Parakysis, Glyptosternon, Erethistes, Hara, Bunocephalus, Aspredo, Xyliphius, Franciscodoras, Anadoras, Acanthodoras, Doras*)**
   — CS–2: Presence of complete supratemporal foramen between dorsal surfaces of pterotic and parieto-supraoccipital **(*Bagarius, Gagata*)**

124. *Parietal as separate ossification on cranial roof* (*character inspired from Lundberg, 1975b*). As emphasised by Lundberg (1975b), contrary to the plesiomorphic condition found in all other catfish examined, in which the parietal and the supraoccipital are likely fused into a single element [State 0: e.g. Fig. 3.67], in both young and adult *Helogenes* the parietal

seems to be present as a separate element on the dorsal surface of the cranial roof [State 1].

— CS–0: Parietal not a separate element on the cranial roof **(all genera not in other CS)**

— CS–1: Parietal seemingly a separate element on the cranial roof **(*Helogenes*)**

125. *Presence of fossa between dorsomesial limb of posttemporo-supracleithrum, extrascapular and pterotic (ordered multistate character).* As noted above, plesiomorphically catfish lack any well-developed openings or fossas other than the anterior and posterior median cranial fontanels [State 0: e.g. Fig. 3.67]. In catfish of CS–1 a well-developed, deep fossa occurs between the dorsomesial limb of the posttemporo-supracleithrum, extrascapular and pterotic [State 1: e.g. Fig. 3.33], which is remarkably developed in specimens of Arius and Genidens examined [State 2: e.g. Fig. 3.20].

— CS–0: Absence of well-developed, deep fossa between dorsomesial limb of posttemporo-supracleithrum, extrascapular and pterotic **(all genera not in other CS)**

— CS–1: Presence of well-developed, deep fossa between dorsomesial limb of posttemporo-supracleithrum, extrascapular and pterotic **(*Chrysichthys, Schilbe, Laides, Pseudeutropius, Siluranodon, Pangasius, Helicophagus, Clarotes, Auchenoglanis, Cranoglanis, Ictalurus, Amiurus, Austroglanis, Ancharius*)**

— CS–2: Fossa between dorsomesial limb of posttemporo-supracleithrum, extrascapular and pterotic remarkably developed **(*Arius, Genidens*)**

— Inapplicable: Since there is no distinct extrascapular **(*Clarias, Silurus, Wallago, Uegitglanis, Heterobranchus, Phractura, Belonoglanis, Trachyglanis, Doumea, Amphilius, Andersonia, Leptoglanis, Zaireichthys, Paramphilius, Nematogenys, Trichomycterus, Hatcheria, Helogenes, Glyptothorax, Glyptosternon, Bagarius, Liobagrus, Akysis, Parakysis, Amblyceps, Gagata, Erethistes, Hara, Bunocephalus, Aspredo, Xyliphius, Silurus, Wallago, Ailia, Hypophthalmus, Synodontis, Mochokus, Franciscodoras, Anadoras, Acanthodoras, Doras, Chaca, Centromochlus, Ageneiosus, Auchenipterus, Callichthys, Corydoras, Loricaria, Hypoptopoma, Lithoxus, Scoloplax, Astroblepus*)** or since the configuration of both the pterotic and the posttemporo-supracleithrum is too modified to appropriately apply this character **(*Malapterurus*)**

126. *Presence of fossa between dorsomesial limb of posttemporo-supracleithrum and parieto-supraoccipital.* Plesiomorphically catfish lack any well-developed openings or fossas other than the anterior and posterior median cranial fontanels (see above) [State 0: e.g. Fig. 3.67]. In siluriforms of CS–1, however, there is a well-developed, deep fossa between the anteromesial

surface of the dorsomesial limb of posttemporo-supracleithrum, parieto-supraoccipital and, eventually, epioccipital [State 1: e.g. Fig. 3.25].
- CS–0: Absence of well-developed, deep fossa between posttemporo-supracleithrum, parieto-supraoccipital and, eventually, epioccipital **(all genera not in other CS)**
- CS–1: Presence of well-developed, deep fossa between posttemporo-supracleithrum, parieto-supraoccipital and, eventually, epioccipital **(*Erethistes, Hara, Bunocephalus, Aspredo, Xyliphius, Chaca*)**

127. *Size of basioccipital*. Contrary to the catfish examined [State 0: e.g. Fig. 3.66], in *Andersonia* the basioccipital is a peculiarly developed structure even larger than the broad prootic [State 1] (see Diogo, 2003b).
- CS–0: Basioccipital not peculiarly developed **(all genera not in other CS)**
- CS–1: Peculiarly developed basioccipital larger than broad prootic **(*Andersonia*)**

128. *Ventral process of basioccipital*. Contrary to all other catfish examined [State 0: e.g. Fig. 3.89], in specimens of *Arius* and *Genidens* analysed a well-developed ventral process of the basioccipital occurs, which comes into contact with a similar ventral process of the first vertebra, forming a prominent, V-shaped ventral salience on the posterior surface of the neurocranial floor [State 1].
- CS–0: Absence of well-developed ventral process of basioccipital **(all genera not in other CS)**
- CS–1: Presence of well-developed ventral process of basioccipital **(*Arius, Genidens*)**

129. *Anteroposterior compression of parieto-supraoccipital (character inspired from de Pinna and Vari, 1995)*. Contrary to all other catfish examined [State 0: e.g. Fig. 3.67], *Helogenes* exhibits a markedly anteroposteriorly compressed parieto-supraoccipital [State 1].
- CS–0: Parieto-supraoccipital not markedly compressed anteroposteriorly **(all genera not in other CS)**
- CS–1: Parieto-supraoccipital markedly compressed anteroposteriorly **(*Helogenes*)**

130. *Presence of lateral projection of parieto-supraoccipital*. Contrary to all other catfish examined [State 0: e.g. Fig. 3.67], in the two genera of CS–1 a well-developed lateral projection of the parieto-supraoccipital occurs that is markedly extended laterally, thus surrounding a significant part of the posterior margin of the pterotic and which, together with the main body and the posterior process of the parieto-supraoccipital, forms a somewhat T-shaped parieto-supraoccipital in dorsal view [State 1: e.g. Fig. 3.46].
- CS–0: Absence of well-developed lateral projection of parieto-supraoccipital **(all genera not in other CS)**
- CS–1: Presence of well-developed lateral projection of parieto-supraoccipital **(*Cetopsis, Hemicetopsis*)**

131. *Posterior margin of parieto-supraoccipital (unordered multistate character) (character inspired from Chardon, 1968)*. Plesiomorphically catfish present a well-developed posterior process of the parieto-supraoccipital [State 0: e.g. Fig. 3.67] but in specimens of genera of CS–1 examined, this process is particularly conspicuous [State 1: e.g. Fig. 3.77]. A different configuration is found in those catfish of CS–2, in which the parieto-supraoccipital is markedly truncated posteriorly [State 2: e.g. Fig. 3.25].
   — CS–0: Well-developed posterior process of parieto-supraoccipital **(all genera not in other CS)**
   — CS–1: Particularly conspicuous posterior process of parieto-supraoccipital **(*Bagrichthys, Pangasius, Helicophagus, Chrysichthys, Schilbe, Laides, Pseudeutropius, Siluranodon, Phractura, Doumea, Andersonia, Belonoglanis, Trachyglanis, Arius, Genidens, Cranoglanis, Austroglanis, Erethistes, Hara, Pimelodus, Ictalurus*)**
   — CS–2: Parieto-supraoccipital truncated posteriorly **(*Bunocephalus, Chaca, Synodontis, Mochokus, Franciscodoras, Anadoras, Acanthodoras, Doras, Centromochlus, Ageneiosus, Auchenipterus, Trichomycterus, Hatcheria, Malapterurus*)**

132. *Lateral laminae of parieto-supraoccipital (ordered multistate character) (character inspired from He et al., 1999)*. Plesiomorphically in catfish the parieto-supraoccipital lacks major lateral laminae [State 0: e.g. Fig. 3.67]. Uniquely in *Andersonia, Belonoglanis* and *Trachyglanis* the parieto-supraoccipital exhibits a well-developed, broad, dorsolateral laminar extension which, however, is considerably less developed in *Andersonia* and *Trachyglanis* [State 1: e.g. Fig. 3.12] than in *Belonoglanis* [State 2].
   — CS 0: Absence of lateral laminae of parieto-supraoccipital **(all genera not in other CS)**
   — CS–1: Presence of well-developed lateral laminae of parieto-supraoccipital **(*Andersonia, Trachyglanis*)**
   — CS–2: Lateral laminae of parieto-supraoccipital even more developed than in CS–1 **(*Belonoglanis*)**

133. *Presence of well-developed foramen between anterodorsal surface of extrascapular, dorsomesial surface of pterotic and dorsolateral surface of parieto-supraoccipital*. As noted above, plesiomorphically catfish lack any well-developed openings other than the anterior and posterior median cranial fontanels [State 0: e.g. Fig. 3.67], but in *Pangasius* and *Helicophagus* a well-developed foramen lies between the anterodorsal surface of the extrascapular dorsomesial surface of the pterotic and dorsolateral surface of the parieto-supraoccipital [State 1: e.g. Fig. 3.96].
   — CS–0: Absence of foramen between extrascapular, pterotic and parieto-supraoccipital **(all genera not in other CS)**
   — CS–1: Presence of well-developed foramen between extrascapular, pterotic and parieto-supraoccipital **(*Pangasius, Helicophagus*)**

— Inapplicable: Since there is no distinct extrascapular (***Clarias, Silurus, Wallago, Uegitglanis, Heterobranchus, Phractura, Belonoglanis, Trachyglanis, Doumea, Amphilius, Andersonia, Leptoglanis, Zaireichthys, Paramphilius, Nematogenys, Trichomycterus, Hatcheria, Helogenes, Glyptothorax, Glyptosternon, Bagarius, Liobagrus, Akysis, Parakysis, Amblyceps, Gagata, Erethistes, Hara, Bunocephalus, Aspredo, Xyliphius, Silurus, Wallago, Ailia, Hypophthalmus, Synodontis, Mochokus, Franciscodoras, Anadoras, Acanthodoras, Doras, Chaca, Centromochlus, Ageneiosus, Auchenipterus, Callichthys, Corydoras, Loricaria, Hypoptopoma, Lithoxus, Scoloplax, Astroblepus***)

134. *Development of infraorbital 4 and suprapreopercle (ordered multistate character) (character inspired from, e.g., Regan, 1911b).* Contrary to all other catfish studied [State 0], in specimens of *Uegitglanis* examined the infraorbital 4 and the suprapreopercle are well-developed bones, with their connection to the neurocranium relatively feeble however, [State 1]. In specimens examined of genera *Heteropneustes, Clarias* and *Heterobranchus* both the infraorbital 4 and the suprapreopercle are highly developed, markedly enlarged bones firmly ankylosed to the neurocranium [State 2: e.g. Fig. 3.82].
— CS–0: Infraorbital 4 and suprapreopercle not enlarged **(all genera not in other CS)**
— CS–1: Infraorbital 4 and suprapreopercle enlarged, feebly attached to neurocranium (***Uegitglanis***)
— CS–2: Infraorbital 4 and suprapreopercle markedly enlarged and firmly ankylosed to neurocranium (***Clarias, Heteropneustes, Heterobranchus***)

135. *Ventral superficial ossification of complex centrum (character inspired from Chardon, 1968; Mo, 1991).* Contrary to other catfish examined [State 0: e.g. Fig. 3.59], in *Arius* and *Genidens* there is a superficial, ventral ossification of the complex centrum, which completely covers the aortic groove in ventral view [State 1].
— CS–0: Absence of ventral superficial ossification of complex centrum completely covering aortic groove **(all genera not in other CS)**
— CS–1: Presence of ventral superficial ossification of complex centrum completely covering aortic groove (***Arius, Genidens***)

136. *Ventral ossification covering separation between complex centrum and basioccipital (character inspired from de Pinna, 1992).* Contrary to other catfish examined [State 0: e.g. Fig. 3.89], in specimens analysed of genera of CS–1, a ventral ossification seemingly covers separation between the complex centrum and basioccipital; hence this separation is not visible in a ventral view [State 1].
— CS–0: Absence of ventral ossification covering separation between complex centrum and basioccipital **(all genera not in other CS)**

— CS–1: Presence of well-developed ventral ossification apparently covers separation between complex centrum and basioccipital (***Trichomycterus, Hatcheria, Callichthys, Corydoras, Loricaria, Hypoptopoma, Lithoxus, Scoloplax, Astroblepus***)

137. *Bony tube linking bony capsules of swim bladder (character inspired from Chardon, 1968; He et al., 1999).* Plesiomorphically catfish lack a bony tube linking the parapophyses of each side of the Weberian apparatus [State 0: e.g. Fig. 3.47], but in *Andersonia, Leptoglanis* and *Zaireichthys* there is a well-developed, thin bony tube linking the parapophyses of each side of the Weberian apparatus, which are modified in bony capsules enclosing the swim bladder [State 1].
    - — CS–0: Absence of well-developed, thin bony tube linking parapophyses of Weberian apparatus **(all genera not in other CS)**
    - — CS–1: Presence of well-developed, thin bony tube linking parapophyses of Weberian apparatus (***Leptoglanis, Zaireichthys, Andersonia***)
138. *Development of parapophysis 5 (character inspired from de Pinna, 1996).* Contrary to all other catfish examined, in which the parapophysis of the fifth vertebra is a narrow laminar structure [State 0: e.g. Fig. 3.47], in specimens of genera of CS–1 analysed, parapophysis 5 is a strong structure directed immediately perpendicular to the vertebral axis [State 1: e.g. Fig. 3.120].
    - — CS–0: Parapophysis 5 not markedly strong and perpendicular to vertebral axis **(all genera not in other CS)**
    - — CS–1: Parapophysis 5 markedly strong and perpendicular to vertebral axis (***Glyptothorax, Glyptosternon, Bagarius, Liobagrus, Akysis, Parakysis, Amblyceps, Gagata, Erethistes, Hara, Bunocephalus, Aspredo, Xyliphius***)
139. *Lateral expansion of parapophysis 5 (character inspired from de Pinna, 1996).* Contrary to all other catfish examined [State 0: e.g. Fig. 3.47], in siluriforms of CS–1 this parapophysis is markedly extended laterally, almost or quite reaching the lateral surface of the body [State 1: e.g. Fig. 3.120].
    - — CS–0: Parapophysis 5 not markedly expanded laterally **(all genera not in other CS)**
    - — CS–1: Parapophysis 5 markedly expanded laterally (***Glyptothorax, Bagarius, Gagata, Erethistes, Hara, Bunocephalus, Aspredo, Xyliphius***)
140. *Presence of lateral ramification of parapophysis 5 (character inspired from de Pinna, 1996).* As described by de Pinna (1996), in the two genera of CS–1, contrary to the other siluriforms examined [State 0: e.g. Fig. 3.47], the parapophysis of the fifth vertebra is markedly ramified laterally [State 1: e.g. Fig. 3.77].
    - — CS–0: Absence of lateral ramification of parapophysis 5 **(all genera not in other CS)**

— CS–1: Presence of lateral ramification of parapophysis 5 (***Erethistes, Hara***)

141. *Dorsal lamina of Weberian apparatus* (*character inspired from Chardon, 1968; Friel, 1994*). Contrary to other catfish examined [State 0], in siluriforms of CS–1 a well-developed, anteroposteriorly elongated dorsal lamina is observed in the Weberian apparatus contacting the dorsal surface of the body [State 1: e.g. Fig. 3.74].
    — CS–0: Absence of dorsal lamina of Weberian apparatus **(all genera not in other CS)**
    — CS–1: Presence of dorsal lamina of Weberian apparatus contacting dorsal surface of body (***Bunocephalus, Aspredo, Xyliphius, Chaca***)

142. *Configuration of nuchal plates* (*character inspired from Lundberg, 1993*). As described by Lundberg (1993), catfish may/may not present well-developed nuchal plates on the posterodorsal region of the cranium [State 0]. But the nuchal plates forming a markedly enlarged nuchal shield, with a large median X-shaped nuchal plate enclosing almost completely, or even completely, an also large anterior nuchal plate is a peculiar condition found only in those specimens examined of genera of CS–1 [State 1: e.g. Fig. 3.74].
    — CS–0: Nuchal plates, if present, not forming markedly enlarged nuchal shield in which the median X-shaped nuchal plate encloses almost completely, or even completely, the anterior nuchal plate **(all genera not in other CS)**
    — CS–1: Nuchal plates forming markedly enlarged nuchal shield in which the median X-shaped nuchal plate encloses almost completely, or even completely, the anterior nuchal plate (***Franciscodoras, Anadoras, Acanthodoras, Doras, Mochokus, Synodontis, Centromochlus, Auchenipterus***)

Pectoral Girdle and Fins

143. *Contact between dorsomedial surface of posttemporo-supracleithrum and neurocranium*. As pointed out in Diogo (2003b), in *Paramphilius*, contrary to all other catfish examined [State 0: e.g. Fig. 3.67], there is almost no contact between the dorsomesial limb of the posttemporo-supracleithrum and the neurocranium [State 1].
    — CS–0: Dorsomesial surface of posttemporo-supracleithrum loosely or firmly attached to neurocranium **(all genera not in other CS)**
    — CS–1: Almost no contact between dorsomesial limb of posttemporo-supracleithrum and neurocranium (***Paramphilius***)

144. *Dorsomesial limb of posttemporo-supracleithrum* (*character inspired from Chardon, 1968*). Contrary to all other catfish examined [State 0: e.g. Fig. 3.67], in the four genera of CS–1 the dorsomesial limb of the posttemporo-supracleithrum is very thin and markedly extended mesially, its mesial

margin almost reaching the midline dorsal to the posterodorsal surface of the neurocranium [State 1: e.g. Fig. 3.49].

— CS–0: Dorsomesial limb of posttemporo-supracleithrum not markedly thin and mesially extended **(all genera not in other CS)**

— CS–1: Dorsomesial limb of posttemporo-supracleithrum markedly thin and mesially extended **(*Bunocephalus, Aspredo, Xyliphius, Parakysis, Chaca*)**

145. *Anteroventrolateral process of posttemporo-supracleithrum* (*ordered multistate character*). In the plesiomorphic condition, the posttemporo-supracleithrum lacks major anterolateral processes [State 0: e.g. Fig. 3.67] but in *Leptoglanis* and *Zaireichthys*, a well-developed, anteroventrolateral process of the posttemporo-supracleithrum pointed ventrally occurs which, however, is less developed in *Leptoglanis* [State 1: e.g. Fig. 3.17] than in *Zaireichthys* [State 2: e.g. Fig. 3.13].

— CS–0: Absence of anteroventrolateral process of posttemporo-supracleithrum **(all genera not in other CS)**

— CS–1: Presence of anteroventrolateral process of posttemporo-supracleithrum **(*Leptoglanis*)**

— CS–2: Anteroventrolateral process of posttemporo-supracleithrum more developed than in CS–1 **(*Zaireichthys*)**

146. *Shape of posttemporo-supracleithrum*. Usually in siluriforms the posttemporo-supracleithrum, when seen in lateral view, has a somewhat rectangular, roughly oblique shape [State 0: e.g. Fig. 3.63], but specimens of genera of CS–1 present a somewhat L-shaped posttemporo-supracleithrum in lateral view, with the posteroventrolateral limb (see terminology of Diogo et al., 2001c) of this bone being markedly oriented posteriorly [State 1: e.g. Fig. 3.17].

— CS–0: Posttemporo-supracleithrum not L-shaped in lateral view **(all genera not in other CS)**

— CS–1: Posttemporo-supracleithrum somewhat L-shaped in lateral view **(*Leptoglanis, Zaireichthys, Akysis, Parakysis, Glyptothorax, Glyptosternon, Bagarius, Gagata*)**

— Inapplicable: Due to the highly modified shape of the posttemporo-supracleithrum in these catfish (see above) **(*Bunocephalus, Aspredo, Xyliphius, Chaca*)**

147. *Anterodorsal foramen of posttemporo-supracleithrum*. In the plesiomorphic condition the posttemporo-supracleithrum lacks major foramen on its dorsal surface [State 0: e.g. Fig. 3.67], but in specimens of the four genera of CS–1, this bone exhibits a well-developed foramen on its anterodorsal surface, near the posterior margin of the pterotic [State 1].

— CS–0: Absence of anterodorsal foramen of posttemporo-supracleithrum **(all genera not in other CS)**

— CS–1: Presence of anterodorsal foramen of posttemporo-supracleithrum **(*Paraplotosus, Neosilurus, Cnidoglanis, Plotosus*)**

148. *Transcapular process of posttemporo-supracleithrum (ordered multistate character) (character inspired from Chardon, 1968; de Pinna, 1996).* Plesiomorphically catfish lack major posterodorsomesial processes attaching to the dorsolateral surface of the fourth parapophysis [State 0: e.g. Fig. 3.89]. Specimens examined of genera of CS–1, however, present a well-developed posterior projection of the posterodorsomesial surface of the posttemporo-supracleithrum ('transcapular' process), which is loosely attached to the dorsolateral surface of the fourth parapophysis [State 1]. In siluriforms of CS–2 [State 2: e.g. Fig. 3.82], and particularly in *Erethistes* and *Hara* [State 3: e.g. Fig. 3.77], the transcapular process is markedly developed and firmly attached to the dorsolateral margin of the fourth parapophysis.
   — CS–0: Absence of well-developed transcapular process of posttemporo-supracleithrum **(all genera not in other CS)**
   — CS–1: Presence of well-developed transcapular process of posttemporo-supracleithrum loosely attached to parapophysis 4 **(*Uegitglanis, Akysis, Parakysis*)**
   — CS–2: Transcapular process of posttemporo-supracleithrum more developed than in CS–1 and firmly attached to parapophysis 4 **(*Clarias, Heterobranchus, Heteropneustes, Chaca, Glyptothorax, Bagarius, Gagata, Bunocephalus, Aspredo, Xyliphius, Scoloplax*)**
   — CS–3: Remarkably developed transcapular process of posttemporo-supracleithrum firmly attached to parapophysis 4 **(*Erethistes, Hara*)**
149. *Anterolateral extension of posttemporo-supracleithrum.* In the plesiomorphic condition the posttemporo-supracleithrum lacks major anterolateral extensions (Diogo et al., 2001c) [State 0: e.g. Fig. 3.63], but in specimens examined of genera *Genidens* and *Arius* this bone exhibits a well-developed anterolateral extension, which is markedly extended anteriorly and surrounds a significant part of the lateral margin of the pterotic [State 1: e.g. Fig. 3.20].
   — CS–0: Absence of well-developed anterolateral extension of posttemporo-supracleithrum **(all genera not in other CS)**
   — CS–1: Presence of well-developed anterolateral extension of posttemporo-supracleithrum **(*Genidens, Arius*)**
   — ?: Since proper discernment of this character was not possible in the specimens examined **(*Ancharius*)**
150. *Development of mesial limb of posttemporo-supracleithrum (unordered multistate character) (character inspired from Chardon, 1968).* Plesiomorphically in catfish the posttemporo-supracleithrum exhibits a well-developed, stout mesial limb attaching to the posteroventral surface of the neurocranium [State 0: e.g. Fig. 3.59]. In catfish of CS–1, however, the mesial limb of this bone is a markedly thin, anteroposteriorly compressed structure [State 1: e.g. Fig. 3.89]. A different situation is found in catfish of CS–2, in which the mesial limb of the posttemporo-supracleithrum is completely missing [State 2].
   — CS–0: Mesial limb of posttemporo-supracleithrum a well-developed, stout structure **(all genera not in other CS)**

— CS–1: Mesial limb of posttemporo-supracleithrum markedly compressed anteroposteriorly (***Helogenes, Nematogenys, Amphilius, Phractura, Doumea, Belonoglanis, Trachyglanis, Plotosus, Neosilurus, Paraplotosus, Cnidoglanis, Andersonia, Chaca, Paramphilius, Bagarius, Silurus, Wallago, Liobagrus, Akysis, Parakysis, Amblyceps, Gagata, Erethistes, Hara, Bunocephalus, Aspredo, Xyliphius, Pseudopimelodus, Microglanis, Malapterurus***)

— CS–2: Mesial limb of posttemporo-supracleithrum totally absent (***Clarias, Uegitglanis, Heterobranchus, Heteropneustes, Trichomycterus, Hatcheria, Zaireichthys, Glyptosternon, Ailia, Callichthys, Corydoras, Loricaria, Hypoptopoma, Lithoxus, Scoloplax, Astroblepus***)

151. *Ventral process of mesial limb of posttemporo-supracleithrum.* Contrary to other catfish [State 0: e.g. Fig. 3.89], in specimens of the two genera of CS–2 examined, the mesial limb of this bone exhibits a well-developed, ventromesially pointed ventral process [State 1].

— CS–0: Absence of well-developed ventral process of mesial limb of posttemporo-supracleithrum **(all genera not in other CS)**

— CS–1: Presence of well-developed ventral process of mesial limb of posttemporo-supracleithrum (***Franciscodoras, Anadoras***)

— Inapplicable: Since there is no mesial limb of posttemporo-supracleithrum (***Clarias, Uegitglanis, Heterobranchus, Heteropneustes, Trichomycterus, Hatcheria, Zaireichthys, Glyptosternon, Ailia, Callichthys, Corydoras, Loricaria, Hypoptopoma, Lithoxus, Scoloplax, Astroblepus***)

152. *Connection between dorsal surface of cleithrum and posttemporo-supracleithrum (character inspired from Schaefer, 1990).* Contrary to other catfish examined, in which the dorsal surface of the cleithrum is lodged in a space delimited by both the posttemporo-supracleithrum and the anterior vertebrae, and thus is usually visible in a dorsal view of the cephalic region [State 0: e.g. Fig. 3.25], in the loricariids and scoloplacids studied the dorsal portion of the cleithrum is completely lodged in a ventral concavity of the posttemporo-supracleithrum and hence is not visible in dorsal view [State 1].

— CS–0: Dorsal portion of cleithrum not lodged in ventral concavity of posttemporo-supracleithrum **(all genera not in other CS)**

— CS–1: Dorsal portion of cleithrum lodged in ventral concavity of posttemporo-supracleithrum (***Loricaria, Hypoptopoma, Lithoxus, Scoloplax***)

153. *Posteroventrolateral bifurcation of posttemporo-supracleithrum.* Plesiomorphically in catfish the posttemporo-supracleithrum exhibits a slight or well-developed bifurcation on its posteroventrolateral surface to receive the dorsal margin of the cleithrum (Diogo et al., 2001) [State 0: e.g. Fig. 3.56]. However, this bifurcation is peculiarly developed in

catfish of CS–1, in which the excavation to receive the cleithrum almost extends to the anterior margin of the posttemporo-supracleithrum [State 1: e.g. Fig. 3.28].

— CS–0: Bifurcation of posttemporo-supracleithrum for cleithrum not peculiarly developed **(all genera not in other CS)**

— CS–1: Bifurcation of posttemporo-supracleithrum for cleithrum peculiarly developed **(*Amphilius, Paramphilius, Phractura, Andersonia, Trachyglanis, Belonoglanis, Doumea, Leptoglanis, Zaireichthys, Glyptothorax, Glyptosternon, Bagarius, Liobagrus, Akysis, Parakysis, Amblyceps, Gagata, Erethistes, Hara, Bunocephalus, Aspredo, Xyliphius, Chaca*)**

— Inapplicable: Since dorsal portion of cleithrum is peculiarly lodged in ventral concavity of posttemporo-supracleithrum (see below) **(*Loricaria, Hypoptopoma, Lithoxus, Scoloplax*)**

154. *Posteroventromesial projection of mesial limb of posttemporo-supracleithrum.* As noted above, in the plesiomorphic condition the mesial limb of the posttemporo-supracleithrum lacks major ventral processes or projections [State 0: e.g. Fig. 3.89] but in *Synodontis* the mesial limb of this bone exhibits a well-developed posteroventromesial projection, which is firmly connected with the posteroventral surface of the neurocranium [State 1].

— CS–0: Absence of well-developed posteroventromesial projection of mesial limb of posttemporo-supracleithrum **(all genera not in other CS)**

— CS–1: Presence of well-developed posteroventromesial projection of mesial limb of posttemporo-supracleithrum **(*Synodontis*)**

— Inapplicable: Since there is no mesial limb of posttemporo-supracleithrum **(*Clarias, Uegitglanis, Heterobranchus, Heteropneustes, Trichomycterus, Hatcheria, Zaireichthys, Glyptosternon, Ailia, Callichthys, Corydoras, Loricaria, Hypoptopoma, Lithoxus, Scoloplax, Astroblepus*)**

155. *Suture between mesial limb of posttemporo-supracleithrum and neurocranium (character inspired from Chardon, 1968*). Plesiomorphically in catfish the mesial limb of the posttemporo-supracleithrum is only connected to the neurocranium by a ligament [State 0: e.g. Fig. 3.89]. In specimens examined of genera of CS–1, however, these structures are firmly sutured to each other [State 1: e.g. Fig. 3.59].

— CS–0: Absence of suture between mesial limb of posttemporo-supracleithrum and neurocranium **(all genera not in other CS)**

— CS–1: Presence of suture between mesial limb of posttemporo-supracleithrum and neurocranium **(*Arius, Ancharius, Genidens, Cranoglanis , Akysis, Parakysis, Leptoglanis, Synodontis, Mochokus, Franciscodoras, Anadoras, Acanthodoras, Doras, Centromochlus, Ageneiosus, Auchenipterus, Pangasius, Helicophagus*)**

— ?: Since proper discernment of this character was not possible in specimens examined (***Glyptothorax, Bagarius, Gagata, Hypophthalmus***)
— Inapplicable: Since there is no mesial limb of the posttemporo-supracleithrum (***Clarias, Uegitglanis, Heterobranchus, Heteropneustes, Trichomycterus, Hatcheria, Zaireichthys, Glyptosternon, Ailia, Callichthys, Corydoras, Loricaria, Hypoptopoma, Lithoxus, Scoloplax, Astroblepus***)

156. *Connection between mesial limb of posttemporo-supracleithrum and neurocranium (character inspired from Chardon, 1968).* Contrary to the plesiomorphic condition found in other catfish examined, in which the mesial limb of the posttemporo-supracleithrum is mainly associated with the basioccipital [State 0: e.g. Fig. 3.89], in *Cetopsis* and *Hemicetopsis* the mesial limb of the former bone is mainly connected to the exoccipital [State 1: e.g. Fig. 3.47].
— CS–0: Mesial limb of posttemporo-supracleithrum not mainly associated with exoccipital **(all genera not in other CS)**
— CS–1: Mesial limb of posttemporo-supracleithrum mainly associated with exoccipital (***Cetopsis, Hemicetopsis***)
— Inapplicable: Since there is no mesial limb of the posttemporo-supracleithrum (***Clarias, Uegitglanis, Heterobranchus, Heteropneustes, Trichomycterus, Hatcheria, Zaireichthys, Glyptosternon, Ailia, Callichthys, Corydoras, Loricaria, Hypoptopoma, Lithoxus, Scoloplax, Astroblepus***)

157. *Posterior laminar projection of mesial limb of posttemporo-supracleithrum.* As mentioned above, in the plesiomorphic siluriform condition the mesial limb of the posttemporo-supracleithrum lacks major ventral processes or projections [State 0: e.g. Fig. 3.89]. In catfish of CS–1, the mesial limb of this bone, however, exhibits a well-developed, large, posterior projection of laminar bone posteroventrally pointed, which is roughly triangular in ventral view [State 1: e.g. Fig. 3.59].
— CS–0: Absence of well-developed posterior laminar projection of mesial limb of posttemporo-supracleithrum **(all genera not in other CS)**
— CS–1: Presence of well-developed posterior laminar projection of mesial limb of posttemporo-supracleithrum (***Cranoglanis, Chrysichthys, Schilbe, Laides, Pseudeutropius, Siluranodon, Ailia, Pangasius, Helicophagus, Clarotes, Ictalurus, Amiurus, Austroglanis, Arius, Genidens, Ancharius***)
— Inapplicable: Since there is no mesial limb of the posttemporo-supracleithrum (***Clarias, Uegitglanis, Heterobranchus, Heteropneustes, Trichomycterus, Hatcheria, Zaireichthys, Glyptosternon, Ailia, Callichthys, Corydoras, Loricaria, Hypoptopoma, Lithoxus, Scoloplax, Astroblepus***)

158. *Fusion between posttemporo-supracleithrum and pterotic (character inspired from Howes, 1983b)*. Contrary to all other catfish examined [State 0: e.g. Fig. 3.63], in siluriforms of CS–1 the posttemporo-supracleithrum seemingly completely fused with the pterotic [State 1].
   — CS–0: Posttemporo-supracleithrum not fused with pterotic **(all genera not in other CS)**
   — CS–1: Posttemporo-supracleithrum and pterotic seemingly fused into a single element **(*Callichthys, Corydoras, Loricaria, Hypoptopoma, Lithoxus, Scoloplax, Astroblepus*)**

159. *Dorsal processes of cleithrum*. Plesiomorphically catfish exhibit two well-developed dorsal processes of the cleithrum for articulation with the posttemporo-supracleithrum (Diogo et al., 2001c) [State 0: e.g. Fig. 3.72], but in siluriforms of CS–1 the cleithrum exhibits only a well-developed dorsal process for articulation with this latter bone [State 1: e.g. Fig. 3.7].
   — CS–0: Presence of two well-developed dorsal processes of cleithrum for articulation with posttemporo-supracleithrum **(all genera not in other CS)**
   — CS–1: Presence of one single well-developed dorsal process of cleithrum for articulation with posttemporo-supracleithrum **(*Amphilius, Paramphilius, Cetopsis, Hemicetopsis, Helogenes, Aspredo, Doras, Acanthodoras, Anadoras, Malapterurus, Loricaria, Hypoptopoma, Lithoxus, Scoloplax, Astroblepus*)**

160. *Ventrolateral foramen of cleithrum*. Plesiomorphically catfish lack major foramens on the ventrolateral surface of the cleithrum [State 0: e.g. Fig. 3.72] but in specimens of *Hypophthalmus* examined, this bone exhibits a remarkably large, roundish ventrolateral foramen, which allows the dorsal condyle of the pectoral spine to be seen in a dorsal view of the pectoral girdle [State 1].
   — CS–0: Absence of large ventrolateral foramen of cleithrum **(all genera not in other CS)**
   — CS–1: Presence of large ventrolateral foramen of cleithrum **(*Hypophthalmus*)**

161. *Orientation of dorsal processes of cleithrum*. The plesiomorphic condition for siluriforms is that in which the dorsal processes of the cleithrum are essentially oriented posterodorsally [State 0: e.g. Fig. 3.72]. In *Amphilius* and *Paramphilius*, however, the single dorsal process of each cleithrum (see above) is essentially oriented anterodorsally, and not posterodorsally [State 1: e.g. Fig. 3.7]
   — CS–0: Dorsal processes of cleithra essentially oriented posterodorsally **(all genera not in other CS)**
   — CS–1: Dorsal processes of cleithra essentially oriented anterodorsally **(*Amphilius, Paramphilius*)**

162. *Cartilage between dorsal processes of cleithrum (character inspired from Mo, 1991)*. Contrary to other catfish examined [State 0: e.g. Fig. 3.72], in

siluriforms of CS–1 a well-developed cartilage lies between the two dorsal processes of the cleithrum [State 1].

- CS–0: Absence of well-developed cartilage between dorsal processes of cleithrum **(all genera not in other CS)**
- CS–1: Presence of well-developed cartilage between dorsal processes of cleithrum **(*Clarotes, Chrysichthys, Auchenoglanis, Callichthys, Corydoras, Bunocephalus, Chaca*)**
- Inapplicable: Since there is only one well-developed process of cleithrum (see above) **(*Amphilius, Paramphilius, Cetopsis, Hemicetopsis, Helogenes, Aspredo, Doras, Acanthodoras, Anadoras, Malapterurus, Loricaria, Hypoptopoma, Lithoxus, Scoloplax, Astroblepus*)**

163. *Posterior projection of second dorsal process of cleithrum.* Contrary to other catfish examined [State 0: e.g. Fig. 3.72], in *Chaca* the second dorsal process of the cleithrum (see terminology of Diogo et al., 2001c) is remarkably developed, enlarged and posteriorly projected [State 1: e.g. Fig. 3.50].
    - CS–0: Second dorsal process of cleithrum not markedly developed **(all genera not in other CS)**
    - CS–1: Second dorsal process of cleithrum markedly developed **(*Chaca*)**
    - Inapplicable: Since there is only one well-developed process of cleithrum (see above) **(*Amphilius, Paramphilius, Cetopsis, Hemicetopsis, Helogenes, Aspredo, Doras, Acanthodoras, Anadoras, Malapterurus, Loricaria, Hypoptopoma, Lithoxus, Scoloplax, Astroblepus*)**
164. *Ligamentous connection between humeral process of cleithrum and anterior vertebrae (unordered multistate character) (character inspired from de Pinna, 1996).* As emphasised by de Pinna (1996), plesiomorphically catfish lack well-defined ligamentous connections between humeral process of cleithrum and anterior vertebrae [State 0: e.g. Fig. 3.58]. In specimens examined of genera of CS–1, there is a ligament attaching posteriorly on either the parapophysis or rib of the sixth vertebra and anteriorly ending in soft tissue without actually reaching the humeral process of the cleithrum or any other bone [State 1]. In *Akysis* a well-defined ligament runs from the humeral process to the sixth vertebra and/or its rib [State 2]. In catfish of CS–3 a well-defined, strong ligament runs from humeral process to anterior vertebrae, but attaches posteriorly on the parapophysis of the fifth vertebra, rather than on the sixth vertebra or its rib [State 3: e.g. Fig. 3.25].
    - CS–0: Absence of well-defined humero-vertebral ligament **(all genera not in other CS)**
    - CS–1: Humero-vertebral ligament attaching posteriorly on sixth vertebra and anteriorly ending in soft tissue **(*Liobagrus, Parakysis, Amblyceps*)**

— CS–2: Humero-vertebral ligament attaching posteriorly on sixth vertebra and anteriorly on humeral process (***Akysis***)
— CS–3: Humero-vertebral ligament attaching posteriorly on fifth parapophysis and anteriorly on humeral process (***Glyptothorax, Bagarius, Gagata, Erethistes, Hara, Bunocephalus, Aspredo, Xyliphius***)
— Inapplicable: Since the humeral process is undifferentiated (***Amphilius, Paramphilius, Nematogenys, Trichomycterus, Hatcheria, Cetopsis, Hemicetopsis, Helogenes, Parakysis, Malapterurus, Glyptosternon, Ailia, Wallago, Calophysus, Hypophthalmus, Heptapterus, Ageneiosus***)

165. *Position of humeral process of cleithrum.* Plesiomorphically in catfish dorsal surface of humeral process of cleithrum considerably ventral to dorsal processes of this bone [State 0: e.g. Fig. 3.72], but in siluriforms of genera of CS–1 examined, this process lies on posterodorsal surface of cleithrum, at about the same level of the dorsal processes [State 1: e.g. Fig. 3.58] (see Diogo et al., 2001c; Diogo, 2003b).
— CS–0: Humeral process considerably ventral to dorsal processes of cleithrum **(all genera not in other CS)**
— CS–1: Humeral process at about the same level as dorsal processes of cleithrum (***Doumea, Zaireichthys, Synodontis, Mochokus, Callichthys, Corydoras, Trachyglanis, Belonoglanis, Phractura, Andersonia, Leptoglanis, Pimelodus, Calophysus, Hypophthalmus, Cranoglanis***)
— Inapplicable: Since there is no distinct humeral process of the cleithrum (***Amphilius, Paramphilius, Nematogenys, Trichomycterus, Hatcheria, Cetopsis, Hemicetopsis, Helogenes, Parakysis, Malapterurus, Glyptosternon, Ailia, Wallago, Calophysus, Hypophthalmus, Heptapterus, Ageneiosus***)

166. *Anterolateral process of cleithrum.* Contrary to other catfish examined [State 0: e.g. Fig. 3.72], siluriforms of CS–1 present a well-developed, anteriorly pointed anterolateral process of the cleithrum [State 1: e.g. Fig. 3.53].
— CS–0: Absence of well-developed anterolateral process of cleithrum **(all genera not in other CS)**
— CS–1: Presence of well-developed anterolateral process of cleithrum (***Clarias, Heterobranchus, Xyliphius***)

167. *Anteromesial process of cleithrum (ordered multistate character).* Contrary to other catfish examined [State 0: e.g. Fig. 3.72], in siluriforms of CS–1 the cleithrum exhibits a well-developed anteromesial process [State 1: e.g. Fig. 3.4], which is especially developed in *Xyliphius* [State 2].
— CS–0: Absence of well-developed anteromesial process of cleithrum **(all genera not in other CS)**
— CS–1: Presence of well-developed anteromesial process of cleithrum (***Glyptothorax, Glyptosternon, Bagarius, Liobagrus, Akysis, Parakysis, Amblyceps, Gagata, Erethistes, Hara***)

— CS–2: Anteromesial process of cleithrum even more developed than in CS–1 **(*Xyliphius*)**

168. *Anterodorsal laminar projection of cleithrum.* Contrary to the other catfish examined [State 0: e.g. Fig. 3.70], specimens analysed of genera of CS–1 present a markedly developed, anterodorsal laminar projection [State 1: e.g. Fig. 3.101].
 — CS–0: Absence of markedly developed anterodorsal laminar projection of cleithrum **(all genera not in other CS)**
 — CS–1: Presence of markedly developed anterolateral laminar projection of cleithrum **(*Pimelodus, Calophysus, Pseudoplatystoma, Heptapterus, Goeldiella, Rhamdia, Pseudopimelodus, Microglanis, Synodontis, Mochokus, Ictalurus, Amiurus, Franciscodoras, Centromochlus, Ageneiosus, Auchenipterus, Ancharius*)**
169. *Posteroventral projection of cleithrum.* Contrary to all other siluriforms examined [State 0: e.g. Fig. 3.70], in the cetopsids studied the cleithrum exhibits a prominent posteroventral projection to receive the anterior surface of the coracoid bridge [State 1: e.g. Fig. 3.45].
 — CS–0: Absence of posteroventral projection of cleithrum **(all genera not in other CS)**
 — CS–1: Presence of posteroventral projection of cleithrum **(*Cetopsis, Hemicetopsis, Helogenes*)**
170. *Length of dorsolateral portion of cleithrum.* Contrary to the plesiomorphic situation found in other catfish [State 0: e.g. Fig. 3.71], in genera of CS–1 the dorsolateral portion of the cleithrum is markedly elongated anteroposteriorly [State 1: e.g. Fig. 3.52].
 — CS–0: Dorsolateral portion of cleithrum not markedly elongated anteroposteriorly **(all genera not in other CS)**
 — CS–1: Dorsolateral portion of cleithrum markedly elongated anteroposteriorly **(*Clarias, Uegitglanis, Heterobranchus, Heteropneustes*)**
171. *Anteroventromesial laminar projection of cleithrum.* Contrary to other siluriforms examined [State 0: e.g. Fig. 3.70], in the cetopsin cetopsids analysed the cleithrum exhibits a prominent anteroventromesial projection of laminar bone, which covers, in ventral view, the anterior parts of scapulo-coracoid, abductor superficialis 1 and arrector ventralis [State 1: e.g. Fig. 3.45].
 — CS–0: Absence of prominent anteroventromesial projection of laminar bone of cleithrum **(all genera not in other CS)**
 — CS–1: Presence of prominent anteroventromesial projection of laminar bone of cleithrum **(*Cetopsis, Hemicetopsis*)**
172. *Posteroventral lamina of cleithrum.* Contrary to other siluriforms [State 0: e.g. Fig. 3.70], *Heteropneustes* exhibits a large, markedly developed, posteroventral horizontal lamina covering the region of the coracoid

bridge, as well as almost all the muscle arrector ventralis, in ventral view [State 1: e.g. Fig. 3.83].

— CS–0: Absence of large posteroventral lamina of cleithrum **(all genera not in other CS)**

— CS–1: Presence of large posteroventral lamina of cleithrum **(*Heteropneustes*)**

173. *Interdigitation between cleithra.* In the plesiomorphic condition, each cleithrum meets its counterpart mesially by means of ligamentous tissue [State 0: e.g. Fig. 3.70], but in the amphiliid and scoloplacid catfish examined, each cleithrum meets its counterpart medially in an interdigitation of several strong serrations [State 1: e.g. Fig. 3.7].

— CS–0: Absence of interdigitation between cleithra **(all genera not in other CS)**

— CS–1: Presence of interdigitation between cleithra **(*Amphilius, Phractura, Doumea, Belonoglanis, Trachyglanis, Zaireichthys, Andersonia, Leptoglanis, Paramphilius, Scoloplax*)**

174. *Ankylosis between cleithrum and scapulo-coracoid.* As referred above, one of the main morphological differences between the plesiomorphic siluriform configuration of the scapulo-coracoid [State 0: e.g. Fig. 3.45] and the derived configuration present in a great number of catfish, is a more pronounced ankylosis in the latter between anterior margin of scapulo-coracoid and posterior margin of cleithrum. Among the siluriforms examined, this derived configuration is present in catfish of CS–1 [State 1: e.g. Fig. 3.56].

— CS–0: Absence of pronounced ankylosis between cleithrum and scapulo-coracoid **(*Diplomystes, Cetopsis, Hemicetopsis, Silurus, Wallago, Helogenes*)**

— CS–1: Presence of pronounced ankylosis between cleithrum and scapulo-coracoid **(all genera not in other CS)**

— ?: Since, due to the very reduced size of the scapulo-coracoid, and in particular the notably reduced size of its mesial arm, it is not possible to judge this character in a proper manner **(*Nematogenys, Trichomycterus, Hatcheria*)**

175. *Firm association between posterolateral processes of scapulo-coracoid and cleithrum (character inspired from Reis, 1998a).* Contrary to all other catfish examined [State 0: e.g. Fig. 3.70], in the callichthyids analysed the cleithrum exhibits a prominent posterolateral process, which is ankylosed posteriorly with an also prominent posterolateral process of the scapulo-coracoid [State 1].

— CS–0: Absence of firm connection between posterolateral processes of scapulo-coracoid and cleithrum **(all genera not in other CS)**

— CS–1: Presence of firm connection between posterolateral processes of scapulo-coracoid and cleithrum **(*Callichthys, Corydoras*)**

176. *Dorsolateral foramen of cleithrum* (*Character inspired from Schaefer, 1990*). Peculiarly in specimens of *Scoloplax* examined, and contrary to all other siluriforms studied [State 0: e.g. Fig. 3.72], a large, somewhat circular foramen occurs on the dorsolateral surface of the cleithrum, ventral to the dorsal process of this bone [State 1].
    — CS–0: Absence of large dorsolateral foramen of cleithrum **(all genera not in other CS)**
    — CS–1: Presence of large dorsolateral foramen of cleithrum **(*Scoloplax*)**
177. *Anteroventrolateral expansion of cleithrum*. Plesiomorphically catfish lack major anterior processes of the cleithrum [State 0: e.g. Fig. 3.70], but in *Trachyglanis* [State 1] there is a well-developed anteroventrolateral process of the cleithrum.
    — CS–0: Absence of well-developed anteroventrolateral expansion of cleithrum **(all genera not in other CS)**
    — CS–1: Presence of well-developed anteroventrolateral expansion of cleithrum **(*Trachyglanis*)**
178. *Mesial laminar expansion of dorsomesial surface of cleithrum* (*character inspired from de Pinna, 1996*). Contrary to all other catfish examined [State 0: e.g. Fig. 3.70], in *Glyptosternon* [State 1] the dorsomesial surface of the cleithrum exhibits a well-developed, large mesial laminar expansion.
    — CS–0: Absence of well-developed mesial expansion of dorsomesial surface of cleithrum **(all genera not in other CS)**
    — CS–1: Presence of well-developed mesial expansion of dorsomesial surface of cleithrum **(*Glyptosternon*)**
179. *Foramen on dorsal margin of cleithrum*. Peculiarly in specimens of *Hypoptopoma* examined, and contrary to other catfish studied [State 0: e.g. Fig. 3.72], a very large, circular foramen occurs on the dorsal surface of the cleithrum [State 1].
    — CS–0: Absence of large foramen on dorsal surface of cleithrum **(all genera not in other CS)**
    — CS–1: Presence of large foramen on dorsal surface of cleithrum **(*Hypoptopoma*)**
180. *Development of scapulo-coracoid*. Plesiomorphically in catfish the scapulo-coracoid is a slender structure with a thin median process, which does not suture with its counterpart medially (see, e.g., Diogo et al., 2001c) [State 0: e.g. Fig. 3.70]. In siluriforms of CS–1, the scapulo-coracoid is a well-developed structure visible in dorsal view and meeting its counterpart in a strong median interdigitation [State 1: e.g. Fig. 3.56].
    — CS–0: Absence of interdigitation between scapulo-coracoids ***(Diplomystes, Nematogenys, Trichomycterus, Hatcheria, Cetopsis, Hemicetopsis, Silurus, Wallago, Helogenes, Astroblepus)***
    — CS–1: Presence of interdigitation between well-developed scapulo-coracoids **(all genera not in other CS)**
181. *Posterior process of scapulo-coracoid* (*ordered multistate character*) (*character inspired from Mo, 1991; de Pinna, 1996*). Plesiomorphically catfish lack

major posterior processes of the scapulo-coracoid [State 0: e.g. Fig. 3.70]. In the catfish of CS–1, the scapulo-coracoid exhibits a well-developed posterior process [State 1], whose development is more pronounced in catfish of CS–2 [State 2: e.g. Fig. 3.121], and especially in genera of CS–3 [State 3: e.g. Fig. 3.1.15.2].

- CS–0: Absence of well-developed posterior process of scapulo-coracoid **(all genera not in other CS)**
- CS–1: Presence of well-developed posterior process of scapulo-coracoid **(*Bagrichthys, Bagarius, Rhamdia, Ancharius*)**
- CS–2: Posterior process of scapulo-coracoid more developed than in CS–1 **(*Bagrus, Hemibagrus, Glyptothorax, Bunocephalus, Xyliphius, Franciscodoras, Acanthodoras, Centromochlus, Callichthys, Loricaria, Hypoptopoma, Lithoxus, Scoloplax*)**
- CS–3: Posterior process of scapulo-coracoid even more developed than in CS–2 **(*Erethistes, Hara, Aspredo, Anadoras, Doras, Auchenipterus, Corydoras*)**

182. *Posteroventromesial process of scapulo-coracoid* (*ordered multistate character*). As noted above, plesiomorphically catfish lack major posterior processes of the scapulo-coracoid [State 0]. But in *Belonoglanis*, *Trachyglanis* and *Andersonia* there is a prominent posteroventromesial process of the scapulo-coracoid, which differs from the posterior process of this bone described above, and is less developed in *Trachyglanis* [State 1] than in *Belonoglanis* and *Andersonia* [State 2: e.g. Fig. 3.14].

- CS–0: Absence of well-developed posteroventromesial process of scapulo-coracoid **(all genera not in other CS)**
- CS–1: Presence of well-developed posteroventromesial process of scapulo-coracoid **(*Trachyglanis*)**
- CS–2: Posteroventromesial process of scapulo-coracoid more developed than in CS–1 **(*Belonoglanis, Andersonia*)**

183. *Posteroventral projection of coracoid bridge* (*ordered multistate character*). Plesiomorphically in catfish the coracoid bridge is not significantly expanded posteroventrally [State 0: e.g. Fig. 3.70]. In catfish of CS–1 the coracoid bridge exhibits a well-developed posteroventral projection of laminar bone that covers a great part of the muscle arrector ventralis in ventral view [State 1: e.g. Fig. 3.45], which is particularly well developed in genus *Belonoglanis* [State 2: e.g. Fig. 3.1.2.9].

- CS–0: Absence of posteroventral projection of coracoid bridge **(all genera not in other CS)**
- CS–1: Presence of well-developed posteroventral projection of coracoid bridge **(*Trachyglanis, Cetopsis, Hemicetopsis, Helogenes*)**
- CS–2: Posteroventral projection of coracoid bridge even better developed than in CS–1 **(*Belonoglanis*)**
- Inapplicable: Since there is no coracoid bridge **(*Paramphilius, Malapterurus, Nematogenys, Glyptosternon, Siluranodon, Ailia, Astroblepus*)**

184. *Foramen on posteroventrolateral surface of scapulo-coracoid.* Contrary to other catfish examined, in which the scapulo-coracoid does not present a posteroventrolateral foramen [State 0: e.g. Fig. 3.70], in *Doumea* the posteroventrolateral surface of this bone is pierced by a large, circular foramen [State 1].
   — CS–0: Absence of posteroventrolateral foramen of scapulo-coracoid **(all genera not in other CS)**
   — CS–1: Presence of posteroventrolateral foramen of scapulo-coracoid **(*Doumea*)**

185. *Presence of mesocoracoid arch (ordered multistate character) (character inspired from Mo, 1991).* Plesiomorphically catfish present a mesocoracoid arch separated from, and well-distinguished from, the posterior surface of the main body of the scapulo-coracoid [State 0: e.g. Fig. 3.71]. However, in catfish of CS–1 the mesocoracoid arch is reduced to a thin, laminar structure completely fused with the posterior surface of the main body of the scapulo-coracoid [State 1: e.g. Fig. 3.21], with these two structures being completely indistinguishable from each other in siluriforms of CS–2 [State 2].
   — CS–0: Mesocoracoid arch separated from main body of scapulo-coracoid **(all genera not in other CS)**
   — CS–1: Mesocoracoid arch reduced to thin structure fused with main body of scapulo-coracoid **(*Arius, Genidens*)**
   — CS–2: Mesocoracoid arch and main body of scapulo-coracoid completely indistinguishable from each other **(*Ancharius, Bunocephalus, Parakysis, Aspredo, Xyliphius, Rita, Franciscodoras, Anadoras, Acanthodoras, Doras, Centromochlus, Ageneiosus, Auchenipterus*)**

186. *Dorsal projection of mesocoracoid arch (character inspired from Mo, 1991).* Contrary to other catfish examined [State 0: e.g. Fig. 3.71], in *Auchenoglanis* the mesocoracoid arch exhibits a prominent dorsal projection, with its dorsal extremity extending dorsally to the main body of the scapulo-coracoid [State 1].
   — CS–0: Absence of dorsal projection of mesocoracoid arch **(all genera not in other CS)**
   — CS–1: Presence of dorsal projection of mesocoracoid arch **(*Auchenoglanis*)**
   — Inapplicable: Due to the absence of a distinct mesocoracoid arch **(*Bunocephalus, Parakysis, Aspredo, Xyliphius, Rita, Franciscodoras, Anadoras, Acanthodoras, Doras, Centromochlus, Ageneiosus, Auchenipterus, Ancharius*)**

187. *Dorsal bifurcation of mesocoracoid arch.* Contrary to other catfish examined [State 0: e.g. Fig. 3.71], in the two genera of CS–1 the mesocoracoid arch is markedly bifurcate dorsally [State 1: e.g. Fig. 3.80].
   — CS–0: Absence of dorsal bifurcation of mesocoracoid arch **(all genera not in other CS)**

— CS–1: Mesocoracoid arch markedly bifurcate dorsally (***Erethistes, Hara***)
— Inapplicable: Due to the absence of a distinct mesocoracoid arch (***Bunocephalus, Parakysis, Aspredo, Xyliphius, Rita, Franciscodoras, Anadoras, Acanthodoras, Doras, Centromochlus, Ageneiosus, Auchenipterus, Ancharius***)

188. *Shape of articulatory facet of scapulo-coracoid for pectoral spine* (*unordered multistate character*). Plesiomorphically in siluriforms, the articulatory facet of the scapulo-coracoid for the pectoral spine (or first pectoral ray) is essentially elongated ventrodorsally (Diogo et al., 20001c) [State 0: e.g. Fig. 3.71]. In *Zaireichthys* and *Leptoglanis* this articulatory facet is essentially elongated anteroposteriorly [State 1], while in catfish of CS–2 it is a highly modified structure with a markedly globular aspect [State 2: e.g. Fig. 3.7]. In catfish of CS–3 the facet has, in turn, a markedly concave aspect [State 3: e.g. Fig. 3.11].
— CS–0: Articulatory facet essentially elongated ventrodorsally **(all genera not in other CS)**
— CS–1: Articulatory facet essentially elongated anteroposteriorly (***Leptoglanis, Zaireichthys***)
— CS–2: Highly modified articulatory facet with markedly globular aspect (***Amphilius, Paramphilius, Nematogenys, Chaca, Astroblepus***)
— CS–3: Articulatory facet with markedly concave aspect (***Phractura, Doumea, Belonoglanis, Trachyglanis, Andersonia, Malapterurus***)

189. *Marked elongation of articulatory facet of scapulo-coracoid for pectoral spine.* As described above, the articulatory facet of the scapulo-coracoid for the pectoral spine may present different configurations [State 0: see description above]; however, a markedly elongated, thin articulatory facet is uniquely present among the catfish studied, in the doradid and auchenipterid specimens [State 1].
— CS–0: Articulatory facet not markedly elongated **(all genera not in other CS)**
— CS–1: Articulatory facet being a markedly elongated, thin structure (***Franciscodoras, Anadoras, Acanthodoras, Doras, Centromochlus, Ageneiosus, Auchenipterus***)

190. *Markedly enlarged mesocoracoid arch.* Contrary to the plesiomorphic situation found in other catfish, in which the mesocoracoid arch is a thin structure [State 0: e.g. Fig. 3.71], in siluriforms of CS–1 the mesocoracoid arch is markedly expanded transversally [State 1: e.g. Fig. 3.11].
— CS–0: Mesocoracoid arch not markedly expanded transversally **(all genera not in other CS)**
— CS–1: Mesocoracoid arch markedly expanded transversally (***Phractura, Belonoglanis, Trachyglanis, Doumea, Amphilius, Andersonia, Leptoglanis, Zaireichthys, Paramphilius, Glyptosternon, Glyptothorax***)

— Inapplicable: Due to the absence of a distinct mesocoracoid arch (***Bunocephalus, Parakysis, Aspredo, Xyliphius, Rita, Franciscodoras, Anadoras, Acanthodoras, Doras, Centromochlus, Ageneiosus, Auchenipterus, Ancharius***)

191. *Shape of mesocoracoid arch* (*character inspired from Oliveira et al., 2001*). Plesiomorphically in catfish the mesocoracoid arch has a roughly tubular shape in posterior view [State 0: e.g. Fig. 3.71], but in catfish of CS–1 the mesocoracoid arch is a uniqually shaped, somewhat triangular, mesially pointed structure [State 1: e.g. Fig. 3.84].
   - — CS–0: Mesocoracoid arch roughly tubular in posterior view **(all genera not in other CS)**
   - — CS–1: Mesocoracoid arch with somewhat triangular shape in posterior view (***Plotosus, Cnidoglanis, Neosilurus, Paraplotosus, Heteropneustes, Malapterurus***)
   - — ?: Since proper discernment of this character was not possible in the specimens examined (***Uegitglanis, Clarias, Heterobranchus***)
   - — Inapplicable: Due to the absence of a distinct mesocoracoid arch (***Bunocephalus, Parakysis, Aspredo, Xyliphius, Rita, Franciscodoras, Anadoras, Acanthodoras, Doras, Centromochlus, Ageneiosus, Auchenipterus, Ancharius***)

192. *Absence of coracoid bridge*. Plesiomorphically siluriforms present a coracoid bridge [State 0: e.g. Fig. 3.70] but in specimens examined of genera of CS–1, however, there is no distinct coracoid bridge [State 1: e.g. Fig. 3.94].
   - — CS–0: Presence of coracoid bridge **(all genera not in other CS)**
   - — CS–1: Absence of distinct coracoid bridge (***Paramphilius, Nematogenys, Glyptosternon, Siluranodon, Ailia, Laides, Malapterurus, Astroblepus***)
   - — ?: Since it was not possible to appraise this character due to the very peculiar configuration of the pectoral girdle (see above) (***Heteropneustes***)

193. *Presence of two bridges connecting ventral surfaces of scapulo-coracoid and cleithrum*. As seen above, catfish may/may not present a coracoid bridge (see above) [State 0: see descriptions above]. However, the specimens studied of *Erethistes*, *Hara* and *Loricaria* are unique in exhibiting two distinct, well-developed bridges connecting the anteroventral surface of the scapulo-coracoid to the ventral surface of the cleithrum [State 1: e.g. Fig. 3.78].
   - — CS–0: Absence of two well-developed bridges connecting ventral surfaces of scapulo-coracoid and cleithrum **(all genera not in other CS)**
   - — CS–1: Presence of two well-developed bridges connecting ventral surfaces of scapulo-coracoid and cleithrum (***Erethistes, Hara, Loricaria***)
   - — ?: Since it was not possible to appraise this character due to the very peculiar configuration of the pectoral girdle (see above) (***Heteropneustes***)

194. *Coracoid bridge markedly enlarged transversally.* Contrary to the plesiomorphic siluriform configuration, in which the coracoid bridge is not significantly enlarged transversally (Diogo et al., 2001c) [State 0: e.g. Fig. 3.70], catfish of CS–1 present a markedly enlarged coracoid bridge [State 1: e.g. Fig. 3.101].
    - CS–0: Coracoid bridge not markedly enlarged transversally **(all genera not in other CS)**
    - CS–1: Coracoid bridge markedly enlarged transversally **(*Chrysichthys, Clarotes, Arius, Genidens, Cranoglanis, Rita, Bagrichthys, Pimelodus, Pseudopimelodus, Ictalurus, Amiurus, Franciscodoras, Anadoras, Acanthodoras, Doras, Centromochlus, Auchenipterus, Austroglanis*)**
    - ?: Since it was not possible to appraise this character due to the very peculiar configuration of the pectoral girdle (see above) **(*Heteropneustes*)**
    - Inapplicable: Since there is no distinct coracoid bridge **(*Paramphilius, Nematogenys, Glyptosternon, Siluranodon, Ailia, Laides, Malapterurus, Astroblepus*)**
195. *Foramen situated posterior to coracoid bridge.* Contrary to all other catfish examined [State 0: e.g. Fig. 3.70], in specimens of *Silurus* there is a well-developed foramen on the scapulo-coracoid posterior to the coracoid bridge [State 1].
    - CS–0: Absence of foramen posterior to coracoid bridge **(all genera not in other CS)**
    - CS–1: Presence of foramen posterior to coracoid bridge **(*Silurus*)**
    - ?: Since it was not possible to appraise this character due to the very peculiar configuration of the pectoral girdle (see above) **(*Heteropneustes*)**
    - Inapplicable: Since there is no distinct coracoid bridge **(*Paramphilius, Nematogenys, Glyptosternon, Siluranodon, Ailia, Laides, Malapterurus, Astroblepus*)**
196. *Obtusion of tunnel delimited by coracoid bridge.* Contrary to all other catfish examined, in which the coracoid bridge, as indicated by its name, encloses a wide tunnel for the passage of the ventral division of the arrector dorsalis [State 0: e.g. Fig. 3.70], in specimens of *Clarias* and *Heterobranchus* the tunnel delimited by this bridge is completely obtuse [State 1: e.g. Fig. 3.53].
    - CS–0: Tunnel delimited by coracoid bridge not obtused **(all genera not in other CS)**
    - CS–1: Tunnel delimited by coracoid bridge completely obtuse **(*Clarias, Heterobranchus*)**
    - ?: Since it was not possible to appraise this character due to the very peculiar configuration of the pectoral girdle (see above) **(*Heteropneustes*)**

— Inapplicable: Since there is no coracoid bridge (***Paramphilius, Nematogenys, Glyptosternon, Siluranodon, Ailia, Laides, Malapterurus, Astroblepus***)

197. *Development of anterior condyle of pectoral spine*. Contrary to the plesiomorphic condition found in other catfish, in which the anterior condyle of the pectoral spine is not significantly developed [State 0: e.g. Fig. 3.72], in specimens of *Aspredo* and *Xyliphius* this condyle is a particularly large, stout structure [State 1].
 — CS–0: Anterior condyle of pectoral spine not significantly developed **(all genera not in other CS)**
 — CS–1: Anterior condyle of pectoral spine large and stout (***Aspredo, Xyliphius***)
 — Inapplicable: Since the first pectoral ray is not developed into a true pectoral spine presenting a dorsal, ventral and anterior condyle (***Amphilius, Paramphilius, Malapterurus, Phractura, Doumea, Belonoglanis, Trachyglanis, Andersonia, Cetopsis, Hemicetopsis, Helogenes, Nematogenys, Trichomycterus, Hatcheria, Glyptosternon, Calophysus, Hypophthalmus, Heptapterus***)
198. *Anterior process of dorsal condyle of pectoral spine* (*ordered multistate character*) (*character inspired from Mo, 1991*). Contrary to the plesiomorphic condition found in other catfish, in which the dorsal condyle of the pectoral spine does not present an anterior process [State 0: e.g. Fig. 3.72], siluriforms of CS–1 present a well-developed anterior process of the dorsal condyle of the pectoral spine [State 1], which is particularly developed in siluriforms of CS–2 [State 2: e.g. Fig. 3.40].
 — CS–0: Absence of anterior process of dorsal condyle of pectoral spine **(all genera not in other CS)**
 — CS–1: Presence of well-developed anterior process of dorsal condyle of pectoral spine (***Goeldiella, Rhamdia***)
 — CS–2: Anterior process of dorsal condyle of pectoral spine more developed than in CS–1 (***Bagrus, Bagrichthys, Hemibagrus, Rita, Pseudeutropius, Bunocephalus, Aspredo, Xyliphius, Pseudopimelodus, Microglanis, Synodontis, Mochokus, Franciscodoras, Anadoras, Acanthodoras, Doras, Centromochlus, Ageneiosus, Auchenipterus***)
 — Inapplicable: Since the first pectoral ray is not developed into a true pectoral spine presenting a dorsal, ventral and anterior condyle (***Amphilius, Paramphilius, Malapterurus, Phractura, Doumea, Belonoglanis, Trachyglanis, Andersonia, Cetopsis, Hemicetopsis, Helogenes, Nematogenys, Trichomycterus, Hatcheria, Glyptosternon, Calophysus, Hypophthalmus, Heptapterus***)
199. *Enlargement of first pectoral ray/pectoral spine* (*character inspired from Tilak, 1971*). Although, as described above, the first pectoral ray is frequently developed into a well-developed, stout structure in siluriforms [State 0:

e.g. Fig. 3.72], its remarkable enlargement in *Chaca* is a unique feature among all the catfish examined [State 1: e.g. Fig. 3.50].

— CS–0: No remarkable enlargement of first pectoral ray/pectoral spine **(all genera not in other CS)**

— CS–1: Pectoral spine remarkably enlarged **(*Chaca*)**

200. *Dorsomesial process of first pectoral ray/pectoral spine*. Contrary to the plesiomorphic condition found in other catfish examined, in which the first pectoral ray/pectoral spine lacks major dorsomesial processes (for argumentation, see Diogo et al., 2001c) [State 0: e.g. Fig. 3.72], in siluriforms of CS–1 this structure exhibits a prominent, somewhat rectangular dorsomesial process [State 1: e.g. Fig. 3.95].

— CS–0: Absence of prominent dorsomesial process of first pectoral ray/pectoral spine **(all genera not in other CS)**

— CS–1: Presence of prominent dorsomesial process of first pectoral ray/pectoral spine **(*Nematogenys, Trichomycterus, Hatcheria, Cetopsis, Hemicetopsis, Helogenes*)**

201. *Triangular process of first pectoral ray/pectoral spine* (*ordered multistate character*). Plesiomorphically in catfish the first pectoral ray/pectoral spine lacks triangular anteromedial processes [State 0: e.g. Fig. 3.72]. In *Doumea, Cranoglanis, Belonoglanis,* and *Trachyglanis* there is a prominent, somewhat triangular anteromedial process of the first pectoral ray/ pectoral spine, which, however, is less developed in *Doumea* and *Cranoglanis* [State 1] than in *Trachyglanis* and *Belonoglanis* [State 2: e.g. Fig. 3.15].

— CS–0: Absence of triangular process of first pectoral ray/pectoral spine **(all genera not in other CS)**

— CS–1: Presence of triangular process of first pectoral ray/pectoral spine **(*Doumea, Cranoglanis*)**

— CS–2: Triangular process of first pectoral ray/pectoral spine more developed than in CS–1 **(*Trachyglanis, Belonoglanis*)**

201. *Arced process of first pectoral ray/pectoral spine* (*ordered multistate character*). Contrary to all other catfishes examined [State 0: e.g. Fig. 3.72], in *Belonoglanis, Doumea* and *Trachyglanis* a well-developed, arced process occurs on the anterior surface of the first pectoral ray, which is less developed in *Doumea* [State 1] than in *Trachyglanis* and *Belonoglanis* [state 2: e.g. Fig. 3.15].

— CS–0: Absence of arced process of first pectoral ray/pectoral spine **(all genera not in other CS)**

— CS–1: Presence of arced process of first pectoral ray/pectoral spine **(*Doumea*)**

— CS–2: Markedly developed arced process of first pectoral ray/ pectoral spine **(*Trachyglanis, Belonoglanis*)**

203. *Size of internal support for pectoral fin rays* (*character inspired from de Pinna, 1996*). Contrary to other catfish examined [State 0], in specimens of the

five genera of CS–1 the internal support for the pectoral fin rays, composed of distal and proximal radials, is markedly small in overall size [State 1].

— CS–0: Internal support for pectoral fin rays not markedly small in size **(all genera not in other CS)**

— CS–1: Internal support for pectoral fin rays markedly small in size **(*Erethistes, Hara, Bunocephalus, Aspredo, Xyliphius*)**

204. *Complete fusion between proximal radials* (*character inspired from Mo, 1991*). As noted by Mo (1991), the plesiomorphic condition for Siluriformes is that in which the proximal radials of the pectoral fin are not fused [State 0] (Mo, 1991). In *Leptoglanis* and *Zaireichthys* the proximal radials are completely fused along their length [State 1].

— CS–0: Proximal radials not fused along their length **(all genera not in other CS)**

— CS–1: Proximal radials completely fused along their length **(*Leptoglanis, Zaireichthys*)**

205. *Shape of proximal radial* (*character inspired from de Pinna, 1996*). Contrary to other catfish examined [State 0], in which the first proximal radial is a compact and irregularly rounded cartilaginous structure, in the three genera of CS–1 the first radial is markedly elongate and roughly rectangular [State 1].

— CS–0: First proximal radial not markedly elongate **(all genera not in other CS)**

— CS–1: First proximal radial markedly elongate **(*Liobagrus, Amblyceps, Parakysis*)**

Facial Musculature

206. *Insertion of adductor mandibulae complex*. Plesiomorphically in catfish the different sections of the adductor mandibulae insert in different regions of the mandible [State 0: e.g. Fig. 3.65], but in *Malapterurus* all the sections of this muscle are inserted in a smooth, well-developed, almost rectangular surface on the posterodorsal surface of the mandible [State 1].

— CS–0: Adductor mandibulae sections inserting on different locations of mandible **(all genera not in other CS)**

— CS–1: All sections of adductor mandibulae inserted in smooth, well-developed, approximately rectangular surface on posterodorsal surface of mandible **(*Malapterurus*)**

207. *Differentiation of adductor mandibulae A1-OST*. Contrary to other catfish in which the adductor mandibulae A1-OST is constituted by a single mass of fibres [State 0: e.g. Fig. 3.64], in *Amphilius* and *Paramphilius* the hypertrophied A1-OST is differentiated into numerous well-developed sections [State 1: e.g. Fig. 3.6].

— CS–0: A1-OST not differentiated into numerous well-developed sections **(all genera not in other CS)**

— CS–1: A1-OST differentiated into numerous well-developed sections **(*Amphilius, Paramphilius*)**

208. *Origin of adductor mandibulae A1-OST*. Plesiomorphically in catfish the adductor mandibulae A1-OST does not originate on the cranial roof (Diogo and Chardon, 2000a) [State 0: e.g. Fig. 3.64], but in genera of CS–1 the A1-OST originates on the dorsal surface of the neurocranium [State 1: e.g. Fig. 3.6].
  — CS–0: A1-OST not originating on dorsal surface of neurocranium **(all genera not in other CS)**
  — CS–1: A1-OST originates on dorsal surface of neurocranium **(*Amphilius, Paramphilius, Cetopsis, Hemicetopsis, Liobagrus, Amblyceps*)**
  — ?: Since it was not possible to properly appraise this character in the specimens examined **(*Helogenes*)**
209. *Position of adductor mandibulae A2*. Contrary to other catfish examined [State 0: e.g. Fig. 3.64], in genera of CS–1 the A2 lies mostly lateral and not mesial to the A1-OST [State 1: e.g. Fig. 3.116].
  — CS–0: A2 essentially mesial to A1-OST **(all genera not in other CS)**
  — CS–1: A2 essentially lateral to A1-OST **(*Schilbe, Laides, Pseudeutropius, Siluranodon, Ailia*)**
210. *Insertion of adductor mandibulae A2*. Plesiomorphically in catfish the adductor mandibulae A2 does not insert directly on the mesial surface of the dentary bone (Diogo and Chardon, 2000a) [State 0: e.g. Fig. 3.65], but such is the case in genera of CS–1 [State 1].
  — CS–0: A2 not directly inserted on mesial surface of dentary bone **(all genera not in other CS)**
  — CS–1: A2 directly inserted on mesial surface of dentary bone **(*Amphilius, Paramphilius, Nematogenys, Trichomycterus, Hatcheria, Callichthys, Corydoras, Loricaria, Hypoptopoma, Lithoxus, Scoloplax, Astroblepus, Cetopsis, Hemicetopsis, Liobagrus, Amblyceps, Helogenes, Glyptothorax, Glyptosternon, Bagarius, Akysis, Parakysis, Gagata, Synodontis, Mochokus, Franciscodoras, Anadoras, Acanthodoras, Doras, Centromochlus, Ageneiosus, Auchenipterus, Erethistes, Hara, Bunocephalus, Aspredo, Xyliphius*)**
211. *Relation between adductor mandibulae A2 and dilatator operculi*. Contrary to all other catfish examined (Diogo and Vandewalle, 2003) [State 0: e.g. Fig. 3.64], in siluriforms of CS–1 the dilatator operculi are markedly lateral to the adductor mandibulae A2 [State 1: e.g. Fig. 3.88].
  — CS–0: Dilatator operculi not markedly lateral to A2 **(all genera not in other CS)**
  — CS–1: Dilatator operculi markedly lateral to A2 **(*Nematogenys, Trichomycterus, Hatcheria, Callichthys*)**
212. *Relation between adductor mandibulae A3′ and levator arcus palatini (ordered multistate character) (character inspired from Oliveira et al., 2001)*. Contrary to all other catfish examined [State 0: e.g. Fig. 3.64], in siluriforms of CS–1 the adductor mandibulae A3′ is partly mesial, and not lateral to

the levator arcus palatini [State 1], while in siluriforms of CS–2 the A3′ is completely mesial to the latter muscle [State 2].

— CS–0: A3′ lateral to levator arcus palatini **(all genera not in other CS)**
— CS–1: A3′ partly mesial to levator arcus palatini ***(Neosilurus, Chaca, Paraplotosus, Cranoglanis)***
— CS–2: A3′ completely mesial to levator arcus palatini ***(Clarias, Uegitglanis, Heterobranchus, Heteropneustes)***
— ?: Since it is very difficult to recognise the different sections of the adductor mandibulae because they all insert at the same location on the mandible (see above) ***(Malapterurus)***
— Inapplicable: Since there is no levator arcus palatini ***(Loricaria, Hypoptopoma, Lithoxus, Scoloplax, Astroblepus)***

213. *Differentiation of adductor mandibulae A3′-d (unordered multistate character).* Plesiomorphically in catfish the adductor mandibulae A3′-d is constituted by a single mass of fibres [State 0: e.g. Fig. 3.64]. In specimens of genera of CS–1 examined, the A3′-d is differentiated into two subdivisions, with the smallest attached to the posterodorsal surface of the coronomeckelian bone and the largest attached to the posterior end of this bone [State 1: e.g. Fig. 3.39]. A different situation is found in those specimens of genera of CS–2, in which the A3′-d is differentiated into a lateral division that inserts by means of a thick tendon in the posterodorsolateral surface of the coronomeckelian and a mesial division that inserts by a thin tendon in the posterodorsomesial margin of this bone [State 2: e.g. Fig. 3.22].

— CS–0: A3′-d not differentiated into two well-developed divisions **(all genera not in other CS)**
— CS–1: A3′-d differentiated into small and large divisions attached respectively to posterodorsal and posterior surfaces of coronomeckelian ***(Bagrus, Hemibagrus)***
— CS–2: A3′ differentiated into lateral and mesial divisions attached respectively to posterodorsolateral and posterodorsomesial surfaces of coronomeckelian ***(Arius, Ancharius, Genidens, Auchenoglanis, Austroglanis)***
— ?: Since it was not possible to discern this character due to the poor condition of the specimens examined ***(Rita)*** or since it is very difficult to recognise the different sections of the adductor mandibulae since all these sections insert at the same location on the mandible ***(Malapterurus)***

214. *Differentiation of adductor mandibulae A3′-v.* Plesiomorphically in catfish the adductor mandibulae A3′-v is constituted by a single mass of fibres [State 0: e.g. Fig. 3.64], but in *Amphilius* and *Paramphilius* the A3′-v is markedly subdivided into two well-developed divisions [State 1: e.g. Fig. 3.6].

— CS–0: A3′-v not differentiated into two well-developed divisions **(all genera not in other CS)**

— CS–1: A3′-v differentiated into small and large divisions attached respectively to posterodorsal and posterior surfaces of coronomeckelian **(*Amphilius, Paramphilius*)**

215. *Insertion of adductor mandibulae A3″ (ordered multistate character).* Plesiomorphically in those catfish presenting an adductor mandibulae A3″, the insertion of the A3′ on the mandible lies lateral to the A3″ and the A2 [State 0: e.g. Fig. 3.22]. In the derived condition the A3′ lies mesial to the A2 (but lateral to the A3″) [State 1], or even mesial to both the A2 and the A3″ [State 2].
   - — CS–0: Insertion of A3′ lateral to those of A2 and A3″ **(all genera not in other CS)**
   - — CS–1: Insertion of A3′ mesial to A2 and lateral to A3″ **(*Clarias, Uegitglanis, Heterobranchus, Heteropneustes*)**
   - — CS–2: Insertion of A3′ mesial to both A2 and A3″ **(*Chaca*)**
   - — ?: Since it was not possible to discern whether the A3″ is present (in this case, the situation would correspond to 'CS–0') or not (in this case, the situation would correspond to 'Inapplicable') in the specimens examined **(*Parakysis, Doras, Pimelodus, Calophysus, Loricaria, Hypoptopoma, Lithoxus, Scoloplax, Astroblepus*)**
   - — Inapplicable: Since there is no A3″ **(*Diplomystes, Amphilius, Doumea, Trachyglanis, Phractura, Pangasius, Helicophagus, Belonoglanis, Andersonia, Leptoglanis, Zaireichthys, Paramphilius, Auchenoglanis, Malapterurus, Glyptothorax, Liobagrus, Akysis, Parakysis, Amblyceps, Glyptosternon, Gagata, Erethistes, Hara, Bunocephalus, Aspredo, Xyliphius, Goeldiella, Synodontis, Mochokus, Amiurus*)**

216. *Origin of adductor mandibulae A3″ (character inspired from Oliveira et al., 2002).* Contrary to other catfish examined [State 0: e.g. Fig. 3.39], in specimens of genera of CS–2 the A3″ covers a great part of the dorsolateral margin of the hyomandibulo-metapterygoid [State 1: e.g. Fig. 3.20].
   - — CS–0: A3″ not covering great part of dorsolateral surface of hyomandibulo-metapterygoid **(all genera not in other CS)**
   - — CS–1: A3″ covering much of dorsolateral surface of hyomandibulo-metapterygoid **(*Arius, Genidens*)**
   - — ?: Since it was not possible to discern whether the A3″ is present (in this case, the situation would correspond to 'CS–0') or not (in this case, the situation would correspond to 'Inapplicable') in the specimens examined **(*Parakysis, Doras, Pimelodus, Calophysus, Loricaria, Hypoptopoma, Lithoxus, Scoloplax, Astroblepus*)**
   - — Inapplicable: Since there is no A3″ **(*Diplomystes, Amphilius, Doumea, Trachyglanis, Phractura, Pangasius, Helicophagus, Belonoglanis, Andersonia, Leptoglanis, Zaireichthys, Paramphilius, Auchenoglanis, Malapterurus, Glyptothorax,***

***Liobagrus, Akysis, Parakysis, Amblyceps, Glyptosternon, Gagata, Erethistes, Hara, Bunocephalus, Aspredo, Xyliphius, Goeldiella, Synodontis, Mochokus, Amiurus*)**

217. *Relation between adductor mandibulae A3'' and primordial ligament (unordered multistate character) (character inspired from Takahasi, 1925).* Plesiomorphically catfish lack a connection between the adductor mandibulae and the primordial ligament [State 0: e.g. Fig. 3.65]. In catfish of CS–1 the adductor mandibulae A3'' exhibits a marked anterior bifurcation, with part of it inserting on the mandible, the other part inserting on the anterior portion of the primordial ligament [State 1: e.g. Fig. 3.39]. A different configuration is found in catfish of CS–2, in which a large anterior tendon of the A3'' inserts on the mandible and on the posterior, and not the anterior, portion of the primordial ligament [State 2: e.g. Fig. 3.58].
   - CS–0: A3'' not connected with primordial ligament **(all genera not in other CS)**
   - CS–1: A3'' exhibiting marked anterior bifurcation, with part inserting on mandible and part inserting on anterior portion of primordial ligament **(*Bagrus, Hemibagrus*)**
   - CS–2: A3'' exhibiting large anterior tendon inserting on mandible and on posterior portion of primordial ligament **(*Cranoglanis, Ictalurus, Callichthys, Corydoras*)**
   - ?: Since it was not possible to discern whether the A3'' is present (in this case, the situation would correspond to 'CS–0') or not (in this case, the situation would correspond to 'Inapplicable') in the specimens examined **(*Parakysis, Doras, Pimelodus, Calophysus, Loricaria, Hypoptopoma, Lithoxus, Scoloplax, Astroblepus*)**
   - Inapplicable: Since there is no A3'' **(*Diplomystes, Amphilius, Doumea, Trachyglanis, Phractura, Pangasius, Helicophagus, Belonoglanis, Andersonia, Leptoglanis, Zaireichthys, Paramphilius, Auchenoglanis, Malapterurus, Glyptothorax, Liobagrus, Akysis, Parakysis, Amblyceps, Glyptosternon, Gagata, Erethistes, Hara, Bunocephalus, Aspredo, Xyliphius, Goeldiella, Synodontis, Mochokus, Amiurus*)**

218. *Differentiation of adductor mandibulae A3'' (character inspired from Takahasi, 1925).* Contrary to all other catfish examined [State 0: e.g. Fig. 3.55], in *Cranoglanis* the adductor mandibulae A3'' is subdivided into two well-developed, well-distinguished divisions [State 1: e.g. Fig. 3.58].
   - CS–0: A3'' not subdivided **(all genera not in other CS)**
   - CS–1: A3'' subdivided into two well-developed divisions **(*Cranoglanis*)**
   - ?: Since it was not possible to discern whether the A3'' is present (in this case, the situation would correspond to 'CS–0') or not (in this case, the situation would correspond to 'Inapplicable') in the

specimens examined (***Parakysis, Doras, Pimelodus, Calophysus, Loricaria, Hypoptopoma, Lithoxus, Scoloplax, Astroblepus***)

— Inapplicable: Since there is no A3″ (***Diplomystes, Amphilius, Doumea, Trachyglanis, Phractura, Pangasius, Helicophagus, Belonoglanis, Andersonia, Leptoglanis, Zaireichthys, Paramphilius, Auchenoglanis, Malapterurus, Glyptothorax, Liobagrus, Akysis, Parakysis, Amblyceps, Glyptosternon, Gagata, Erethistes, Hara, Bunocephalus, Aspredo, Xyliphius, Goeldiella, Synodontis, Mochokus, Amiurus***)

219. *Relation between adductor mandibulae A3″ and neurocranium* (*unordered multistate character*). Plesiomorphically catfish lack a connection between the adductor mandibulae A3″ and the neurocranium [State 0: e.g. Fig. 3.55], but in siluriforms of CS–1 the adductor mandibulae A3″ originates not only on the suspensorium, but also on the neurocranium, near its articulation with the hyomandibulo-metapterygoid [State 1: e.g. Fig. 3.30]. A different configuration is found in *Cetopsis* and *Hemicetopsis*, in which the A3″ does not originate on the suspensorium, but further anteriorly and dorsally than usual, namely on the dorsal surface of both the frontal and the lateral ethmoid [State 2: e.g. Fig. 3.43].

— CS–0: A3″ originating on suspensorium **(all genera not in other CS)**

— CS–1: A3″ originating on suspensorium and neurocranium near region of its articulation with hyomandibulo-metapterygoid (***Chaca, Plotosus, Cnidoglanis, Paraplotosus, Neosilurus, Clarias, Uegitglanis, Heterobranchus, Hypophthalmus, Pseudopimelodus, Microglanis, Centromochlus, Ageneiosus, Auchenipterus, Nematogenys***)

— CS–2: A3″ originating exclusively on neurocranium, on dorsal surfaces of frontal and lateral ethmoid (***Cetopsis, Hemicetopsis***)

— ?: Since it was not possible to discern whether the A3″ is present (in this case, the situation would correspond to 'CS–0') or not (in this case, the situation would correspond to 'Inapplicable') in the specimens examined (***Parakysis, Doras, Pimelodus, Calophysus, Loricaria, Hypoptopoma, Lithoxus, Scoloplax, Astroblepus***)

— Inapplicable: Since there is no A3″ (***Diplomystes, Amphilius, Doumea, Trachyglanis, Phractura, Pangasius, Helicophagus, Belonoglanis, Andersonia, Leptoglanis, Zaireichthys, Paramphilius, Auchenoglanis, Malapterurus, Glyptothorax, Liobagrus, Akysis, Parakysis, Amblyceps, Glyptosternon, Gagata, Erethistes, Hara, Bunocephalus, Aspredo, Xyliphius, Goeldiella, Synodontis, Mochokus, Amiurus***)

220. *Orientation of adductor mandibulae Aω* (*character inspired from Oliveira et al., 2002*). Plesiomorphically in siluriforms the adductor mandibulae Aω is small, anteroposteriorly oriented and lodged in the medial surface of the mandible [State 0: e.g. Fig. 3.65]. However, in specimens of genera of CS–1 examined the adductor mandibulae Aω is highly developed

and obliquely oriented, with its posterodorsal fibres significantly dorsal to the upper edge of the coronoid process [State 1: e.g. Fig. 3.22].

— CS–0: Aω not highly developed **(all genera not in other CS)**

— CS–1: Aω highly developed and obliquely oriented, with posterodorsal fibres significantly dorsal to coronoid process ***(Arius, Ancharius, Genidens, Cranoglanis, Ictalurus, Amiurus, Austroglanis, Auchenoglanis)***

— ?: Since it was not possible to discern whether the Aω is present (in this case, the situation would correspond to 'CS–0') or not (in this case, the situation would correspond to 'Inapplicable') in the specimens examined ***(Auchenipterus, Centromochlus)***

— Inapplicable: Since there is no Aω ***(Amphilius, Nematogenys, Trichomycterus, Hatcheria, Phractura, Doumea, Belonoglanis, Trachyglanis, Clarias, Uegitglanis, Heterobranchus, Heteropneustes, Andersonia, Leptoglanis, Zaireichthys, Paramphilius, Bagarius, Liobagrus, Akysis, Parakysis, Amblyceps, Gagata, Erethistes, Hara, Bunocephalus, Aspredo, Xyliphius, Synodontis, Mochokus, Franciscodoras, Anadoras, Acanthodoras, Doras, Malapterurus, Chaca, Callichthys, Corydoras, Loricaria, Hypoptopoma, Lithoxus, Scoloplax, Astroblepus)***

221. *Separation between extensor tentaculi and adductor arcus palatini.* The extensor tentaculi of catfish is the result of the differentiation of the adductor arcus palatini, with the fibres of both these muscles somewhat mixed in *Diplomystes and Nematogenys* [State 0], but the two muscles completely separated in the derived condition [State 1: e.g. Fig. 3.20] (see Diogo and Vandewalle, 2003).

— CS–0: Some fibres of extensor tentaculi mixed with those of the adductor arcus palatini ***(Diplomystes, Nematogenys)***

— CS–1: Extensor tentaculi completely separated from the adductor arcus palatini **(all genera not in other CS)**

222. *Development of extensor tentaculi.* Contrary to all other catfish examined [State 0: e.g. Fig. 3.66], the two genera of CS–1 present a markedly hypertrophied extensor tentaculi, originating not only on the ethmoid region, as usual, but also on the major part of both the frontal and the sphenotic [State 1].

— CS–0: Extensor tentaculi not markedly hypertrophied **(all genera not in other CS)**

— CS–1: Extensor tentaculi markedly hypertrophied ***(Ageneiosus, Auchenipterus)***

223. *Position of extensor tentaculi.(character inspired from Oliveira et al., 2002).* Contrary to all other catfish examined [State 0: e.g. Fig. 3.66], in the six genera of CS–1 a great part of the mesial fibres of the extensor tentaculi situate ventral to the adductor arcus palatini [State 1: e.g. Fig. 3.55].

— CS–0: Fibres of extensor tentaculi not ventral to adductor arcus palatini **(all genera not in other CS)**

— CS–1: Fibres of extensor tentaculi ventral to adductor arcus palatini (***Chrysichthys, Schilbe, Clarotes, Cranoglanis, Arius, Genidens***)
— ?: Since it was not possible to properly observe this character in the specimens examined (***Ancharius***)

224. *Length of extensor tentaculi (character inspired from Howes and Fumihito, 1991).* According to Howes and Fumihito (1991), the markedly anteroposteriorly elongated extensor tentaculi present in some silurids seemingly represents a derived character within Siluriformes. This is a delicate issue, since an elongated extensor tentaculi is also found in the diplomystids (see, e.g. Fig. 3.64), as well as in all the non-nematogenyid loricarioids examined, and in particular because this character cannot be compared with other ostariophysans, since in those fishes there is no extensor tentaculi. The codification was thus made under the assumption of Howes and Fumihito (1991), with a not markedly elongated extensor tentaculi coded as CS–0 [State 0: e.g. Fig. 3.89] and a markedly elongated extensor tentaculi as CS–1 [State 1: e.g. Fig. 3.66]. This character will be carefully discussed, however, in the following chapters.
— CS–0: Extensor tentaculi not markedly elongated anteroposteriorly **(all genera not in other CS)**
— CS–1: Extensor tentaculi markedly elongated anteroposteriorly (***Diplomystes, Trichomycterus, Hatcheria, Callichthys, Corydoras, Loricaria, Hypoptopoma, Lithoxus, Scoloplax, Astroblepus, Silurus, Wallago***)

225. *Origin of extensor tentaculi (character inspired from Oliveira et al., 2001).* Contrary to all other catfish examined [State 0: e.g. Fig. 3.64], in the plotosids and chacids examined the origin of the extensor tentaculi on the neurocranium extends farther anteriorly than its insertion on the autopalatine [State 1: e.g. Fig. 3.49].
— CS–0: Origin of extensor tentaculi not extending farther anteriorly than its insertion on autopalatine **(all genera not in other CS)**
— CS–1: Origin of extensor tentaculi extending farther anteriorly than its insertion on autopalatine (***Plotosus, Neosilurus, Cnidoglanis, Paraplotosus, Chaca***)

226. *Differentiation of extensor tentaculi (unordered multistate character) (character inspired from Alexander, 1965; Gosline, 1975; Howes, 1983a).* Plesiomorphically in catfish the extensor tentaculi is constituted by a single mass of fibres [State 0: e.g. Fig. 3.66]. Several different derived configurations can, however, be found in the Siluriformes. In catfish of CS–1 the extensor tentaculi is differentiated in at least four bundles promoting respectively adduction, abduction, dorsal rotation and ventral rotation of the maxillary barbel [State 1: e.g. Fig. 3.39]. Catfish of CS–2 present a different type of differentiation, with the most posterior of these bundles inserted, respectively, on the posterodorsal and posteroventral extremities of the autopalatine [State 2: e.g. Fig. 3.6]. Catfish of CS–3 present two bundles, with a thin bundle markedly

extended posteriorly and inserting on the posterior margin of the autopalatine, and the other bundle lying anterior and almost perpendicular to the first one and attaching on the posteromesial surface of the autopalatine [State 3]. *Silurus* and *Wallago* present an unusual configuration, in which the extensor tentaculi is markedly subdivided into two well-developed subdivisions attaching respectively on the posterior and posterolateral margin of the autopalatine [State 4]. Another rare configuration is seen in specimens of genera of CS–5, in which the extensor tentaculi is differentiated into two markedly elongated ventral bundles attaching on the posteroventrolateral surface of the autopalatine and a more dorsal bundle essentially oriented dorsoventrally and attaching on the posterodorsal surface of this bone, anterior to its articulation with the neurocranium [State 5]. But the most peculiar configuration found among the catfish examined is very likely that of genera of CS–6, in which the extensor tentaculi is differentiated into two well-developed, completely separated bundles attached respectively on the lateral and mesial surfaces of the autopalatine [State 6].

— CS–0: Extensor tentaculi presenting a single mass of fibres **(all genera not in other CS)**

— CS–1: Extensor tentaculi differentiated in at least four bundles promoting respectively adduction, abduction, dorsal rotation and ventral rotation of the maxillary barbel ***(Bagrus, Bagrichthys, Hemibagrus, Rita, Pimelodus, Hypophthalmus, Calophysus, Pseudoplatystoma, Heptapterus, Goeldiella, Rhamdia, Pseudopimelodus, Microglanis, Clarias, Uegitglanis, Heterobranchus, Heteropneustes)***

— CS–2: Extensor tentaculi with most posterior bundles inserted respectively on posterodorsal and posteroventral extremities of the autopalatine ***(Amphilius, Paramphilius, Phractura, Doumea, Andersonia, Belonoglanis, Trachyglanis, Leptoglanis, Zaireichthys, Glyptothorax, Glyptosternon, Bagarius, Liobagrus, Amblyceps, Gagata, Erethistes, Hara, Bunocephalus, Aspredo, Xyliphius, Synodontis, Mochokus, Franciscodoras, Anadoras, Acanthodoras, Doras, Centromochlus, Ageneiosus, Auchenipterus)***

— CS–3: Extensor tentaculi with two bundles, with thin bundle markedly extended posteriorly, inserting on posterior margin of autopalatine, the other bundle lying anterior and almost perpendicular to the first and attaching on posteromesial surface of the autopalatine ***(Auchenoglanis, Malapterurus)***

— CS–4: Extensor tentaculi subdivisions attaching respectively on posterior and posterolateral margin of autopalatine ***(Silurus, Wallago)***

— CS–5: Extensor tentaculi with two markedly elongated ventral bundles attaching on posteroventrolateral surface of autopalatine and a more dorsal bundle essentially oriented dorsoventrally and

attaching on posterodorsal surface of this bone ***(Callichthys, Corydoras, Scoloplax, Astroblepus)***
- — CS–6: Extensor tentaculi with two well-developed, completely separated bundles attaching, respectively on lateral and mesial surfaces of autopalatine ***(Loricaria, Hypoptopoma, Lithoxus)***
- — ?: Since it was not possible to properly observe this character in the specimens examined ***(Akysis, Parakysis, Pseudeutropius, Laides, Rita)***

227. *Insertion of extensor tentaculi.* Contrary to other catfish examined [State 0: e.g. Fig. 3.89], in siluriforms of CS–1 a significant part of fibres of the extensor tentaculi is inserted on the mesial and/or dorsal surfaces of the sesamoid bone 1 of the suspensorium [State 1: e.g. Fig. 4.21] (Diogo and Vandewalle, 2003).
    - — CS–0: Extensor tentaculi not inserting on mesial and/or dorsal surfaces of sesamoid bone 1 **(all genera not in other CS)**
    - — CS–1: Significant part of fibres of extensor tentaculi inserting on mesial and/or dorsal surfaces of sesamoid bone 1 ***(Cranoglanis, Chrysichthys, Schilbe, Pseudeutropius, Clarotes, Auchenoglanis, Genidens, Arius, Ancharius, Parakysis, Pimelodus, Calophysus, Hypophthalmus, Acanthodoras)***
    - — ?: Since it was not possible to properly observe this character in the specimens examined ***(Laides)***
    - — Inapplicable: Since there is no sesamoid bone 1 of the suspensorium ***(Synodontis, Mochokus, Rita, Trichomycterus, Hatcheria, Callichthys, Corydoras, Loricaria, Hypoptopoma, Lithoxus, Scoloplax, Astroblepus)***
228. *Posterior extension of retractor tentaculi.* Contrary to all other catfish examined presenting a muscle retractor tentaculi [State 0: e.g. Fig. 3.105], in siluriforms of CS–1 the retractor tentaculi is markedly extended posteriorly, with the posterior fibres of this muscle lying at about the same level as those of the levator arcus palatini [State 1: e.g. Fig. 3.25] (see Diogo and Vandewalle, 2003).
    - — CS–0: Retractor tentaculi not markedly extended posteriorly **(all genera not in other CS)**
    - — CS–1: Retractor tentaculi markedly extended posteriorly ***(Bunocephalus, Aspredo, Xyliphius, Franciscodoras, Malapterurus)***
    - — Inapplicable: Since there is no retractor tentaculi ***(Arius, Ancharius, Genidens, Bagrus, Bagrichthys, Hemibagrus, Cetopsis, Hemicetopsis, Helogenes, Chrysychthys, Shilbe, Pseudeutropius, Pangasius, Helicophagus, Clarotes, Auchenoglanis, Cranoglanis, Diplomystes, Nematogenys, Trichomycterus, Hatcheria, Plotosus, Paraplotosus, Neosilurus, Cnidoglanis, Pimelodus, Calophysus, Hypophthalmus, Pseudoplatystoma, Pseudopimelodus, Microglanis, Ictalurus, Amiurus, Ageneiosus, Callichthys, Corydoras, Auchenipterus, Anadoras, Acanthodoras, Doras, Austroglanis)***

229. *Relation between retractor tentaculi and levator arcus palatini.* Contrary to all other catfish examined presenting a muscle retractor tentaculi, in which this muscle lies anterior and/or mesial to the levator arcus palatini [State 0: e.g. Fig. 3.74], in the aspredinids studied the retractor tentaculi is lateral to the latter muscle [State 1: e.g. Fig. 3.25].
   - CS–0: Retractor tentaculi anterior and/or mesial to levator arcus palatini **(all genera not in other CS)**
   - CS–1: Retractor tentaculi lateral to levator arcus palatini **(*Bunocephalus, Aspredo, Xyliphius*)**
   - Inapplicable: Since there is no retractor tentaculi **(*Arius, Ancharius, Genidens, Bagrus, Bagrichthys, Hemibagrus, Cetopsis, Hemicetopsis, Helogenes, Chrysychthys, Shilbe, Pseudeutropius, Pangasius, Helicophagus, Clarotes, Auchenoglanis, Cranoglanis, Diplomystes, Nematogenys, Trichomycterus, Hatcheria, Plotosus, Paraplotosus, Neosilurus, Cnidoglanis, Pimelodus, Calophysus, Hypophthalmus, Pseudoplatystoma, Pseudopimelodus, Microglanis, Ictalurus, Amiurus, Ageneiosus, Callichthys, Corydoras, Auchenipterus, Anadoras, Acanthodoras, Doras, Austroglanis*)**

230. *Differentiation of retractor tentaculi.* Contrary to all other catfish examined presenting a muscle retractor tentaculi, in which this muscle is constituted by a single mass of fibres [State 0: e.g. Fig. 3.25], in the silurids analysed this muscle is markedly differentiated into two well-developed subdivisions [State 1].
   - CS–0: Retractor tentaculi not differentiated into two well-developed divisions **(all genera not in other CS)**
   - CS–1: Retractor tentaculi differentiated into two well-developed divisions **(*Silurus, Wallago*)**
   - Inapplicable: Since there is no retractor tentaculi **(*Arius, Ancharius, Genidens, Bagrus, Bagrichthys, Hemibagrus, Cetopsis, Hemicetopsis, Helogenes, Chrysychthys, Shilbe, Pseudeutropius, Pangasius, Helicophagus, Clarotes, Auchenoglanis, Cranoglanis, Diplomystes, Nematogenys, Trichomycterus, Hatcheria, Plotosus, Paraplotosus, Neosilurus, Cnidoglanis, Pimelodus, Calophysus, Hypophthalmus, Pseudoplatystoma, Pseudopimelodus, Microglanis, Ictalurus, Amiurus, Ageneiosus, Callichthys, Corydoras, Auchenipterus, Anadoras, Acanthodoras, Doras, Austroglanis*)**

231. *Origin of retractor tentaculi (character inspired from Howes, 1983a).* Contrary to all other catfish examined presenting a muscle retractor tentaculi, in which this muscle originates on the suspensorium [State 0: e.g. Fig. 3.6], in siluriforms of CS–1 the origin of this muscle is confined to the neurocranium [State 1: e.g. Fig. 3.30].
   - CS–0: Retractor tentaculi originated on suspensorium **(all genera not in other CS)**

— CS–1: Retractor tentaculi originated on neurocranium **(*Centromochlus, Loricaria, Hypoptopoma, Lithoxus, Scoloplax, Astroblepus*)**
— Inapplicable: Since there is no retractor tentaculi **(*Arius, Ancharius, Genidens, Bagrus, Bagrichthys, Hemibagrus, Cetopsis, Hemicetopsis, Helogenes, Chrysychthys, Shilbe, Pseudeutropius, Pangasius, Helicophagus, Clarotes, Auchenoglanis, Cranoglanis, Diplomystes, Nematogenys, Trichomycterus, Hatcheria, Plotosus, Paraplotosus, Neosilurus, Cnidoglanis, Pimelodus, Calophysus, Hypophthalmus, Pseudoplatystoma, Pseudopimelodus, Microglanis, Ictalurus, Amiurus, Ageneiosus, Callichthys, Corydoras, Auchenipterus, Anadoras, Acanthodoras, Doras, Austroglanis*)**

232. *Presence of retractor premaxillae (character inspired from Howes, 1983a).* Contrary to the plesiomorphic situation found in the other catfish examined [State 0: e.g. Fig. 3.63], specimens of the five genera of CS–1 present a retractor premaxillae running from the suspensorium to a ligament deeply associated with the premaxilla [State 1].
— CS–0: Absence of retractor premaxillae **(all genera not in other CS)**
— CS–1: Presence of retractor premaxillae **(*Loricaria, Hypoptopoma, Lithoxus, Scoloplax, Astroblepus*)**

233. *Presence of retractor palatini (character inspired from Howes, 1983a).* Contrary to the plesiomorphic situation found in other catfishes examined [State 0: e.g. Fig. 3.63], the loricariids studied present a retractor palatini attaching on connective tissue associated with the dorsomesial region of the primordial ligament [State 1].
— CS–0: Absence of retractor palatini **(all genera not in other CS)**
— CS–1: Presence of retractor palatini **(*Loricaria, Hypoptopoma, Lithoxus*)**

234. *Development of adductor arcus palatini.* Contrary to all other catfishes examined [State 0: e.g. Fig. 3.66], *Chaca* exhibits a highly developed, dorsally expanded, adductor arcus palatini [State 1].
— CS–0: Adductor arcus palatini not markedly developed **(all genera not in other CS)**
— CS–1: Highly developed, dorsally expanded adductor arcus palatini **(*Chaca*)**

235. *Insertion of adductor arcus palatini.* Contrary to all other catfish examined, in which the adductor arcus palatini inserts not only on the hyomandibulo-metapterygoid, but also on other bones of the suspensorium [State 0: e.g. Fig. 3.66], in the four genera of CS–1 this muscle inserts exclusively on the hyomandibulo-metapterygoid [State 1: e.g. Fig. 3.16].
— CS–0: Adductor arcus palatini not exclusively inserted on hyomandibulo-metapterygoid **(all genera not in other CS)**

— CS–1: Adductor arcus palatini exclusively inserted on hyomandibulo-metapterygoid **(*Phractura, Andersonia, Doumea, Belonoglanis*)**

236. *Relation between adductor arcus palatini and sesamoid bone 1 of the suspensorium.* Plesiomorphically in catfish the adductor arcus palatini is not associated with the sesamoid bone 1 of the suspensorium (Diogo and Vandewalle, 2003) [State 0: e.g. Fig. 3.42], but in siluriforms of CS–1 a significant part of the fibres of the adductor arcus palatini is attached on this bone [State 1: e.g. Fig. 3. 3].
    — CS–0: Adductor arcus palatini not associated with sesamoid bone 1 of suspensorium **(all genera not in other CS)**
    — CS–1: Significant part of fibres of adductor arcus palatini attached on sesamoid bone 1 of suspensorium **(*Clarias, Uegitglanis, Heterobranchus, Heteropneustes, Glyptothorax, Glyptosternon, Bagarius, Liobagrus, Akysis, Amblyceps, Gagata, Erethistes, Hara, Paraplotosus, Plotosus, Cnidoglanis, Neosilurus, Chaca*)**
    — Inapplicable: Since there is no sesamoid bone 1 of the suspensorium **(*Rita*)**
237. *Lateral insertion of adductor arcus palatini on suspensorium (character inspired from Oliveira et al., 2002).* Plesiomorphically in catfish the adductor arcus palatini is essentially inserted on the mesial surface of the suspensorium [State 0: e.g. Fig. 3.66], but in siluriforms of CS–1 a great part of the fibres of this muscle insert on the lateral surface of the entoectopterygoid [State 1: e.g. Fig. 3.20].
    — CS–0: Adductor arcus palatini not covering a significant part of lateral surface of entoectopterygoid **(all genera not in other CS)**
    — CS–1: Adductor arcus palatini covering a significant part of lateral surface of entoectopterygoid **(*Nematogenys, Genidens, Arius*)**
    — Inapplicable: Since there is no entoectopterygoid **(*Aspredo*)**
238. *Absence of levator arcus palatini (character inspired from Howes, 1983a).* Contrary to the plesiomorphic situation found in catfish examined [State 0: e.g. Fig. 3.64], in the specimens of *Loricaria* studied the levator arcus palatini is lacking [State 1].
    — CS–0: Presence of levator arcus palatini **(all genera not in other CS)**
    — CS–1: Absence of levator arcus palatini **(*Loricaria*)**
239. *Differentiation of levator arcus palatini.* In the plesiomorphic condition, the levator arcus palatini is constituted by a single mass of fibres [State 0: e.g. Fig. 3.64], but in catfish of CS–1 this muscle is differentiated into two well-developed sections [State 1: e.g. Fig. 3.6].
    — CS–0: Levator arcus palatini not differentiated into two well-developed sections **(all genera not in other CS)**
    — CS–1: Levator arcus palatini differentiated into two well-developed sections **(*Amphilius, Paramphilius, Cetopsis, Hemicetopsis, Helogenes*)**
    — Inapplicable: Since there is no levator arcus palatini **(*Loricaria*)**

240. *Insertion of levator arcus palatini (unordered multistate character).* Plesiomorphically in catfish the levator adductor arcus palatini inserts exclusively on the hyomandibulo-metapterygoid [State 0: e.g. Fig. 3.55]. In catfish of CS–1 this muscle inserts not only on the hyomandibulo-metapterygoid, but also on the quadrato-symplectic [State 1]. A different derived configuration is found in catfish of CS–2, in which the levator arcus, besides its insertion on the hyomandibulo-metapterygoid, inserts on the entoectopterygoid [State 2].
   - CS–0: Levator arcus palatini exclusively inserted on hyomandibulo-metapterygoid **(all genera not in other CS)**
   - CS–1: Levator arcus palatini inserting on hyomandibulo-metapterygoid and quadrato-symplectic **(*Clarias, Heterobranchus, Uegitglanis, Heteropneustes*)**
   - CS–2: Levator arcus palatini inserting on hyomandibulo-metapterygoid and entoectopterygoid **(*Pangasius, Helicophagus, Cetopsis, Hemicetopsis, Helogenes*)**
   - Inapplicable: Since there is no levator arcus palatini **(*Loricaria*)** or since there is no entoectopterygoid **(*Aspredo*)**

241. *Origin of levator arcus palatini (unordered multistate character) (character inspired from de Pinna and Vari, 1995; Oliveira et al., 2001).* Plesiomorphically in catfish the levator adductor arcus palatini essentially originates on the dorsolateral surface of the neurocranium [State 0: e.g. Fig. 3.55]. However, in siluriforms of CS–1 this muscle covers a significant part of the dorsal surface of the cranial roof, namely of the sphenotic and pterotic [State 1]. In catfish of CS–2 the levator arcus palatini also originates on the dorsal surface of the cranial roof, but significantly more anteriorly than in siluriforms of CS–1, originating on the dorsal surface of the frontal and the lateral ethmoid [State 2: e.g. Fig. 3.43].
   - CS–0: Levator arcus palatini not originating on dorsal surface of neurocranium **(all genera not in other CS)**
   - CS–1: Levator arcus palatini originating on dorsal surface of neurocranium, at level of its articulation with the suspensorium **(*Plotosus, Paraplotosus*)**
   - CS–2: Levator arcus palatini originating on dorsal surface of neurocranium, at level of ethmoid region **(*Cetopsis, Hemicetopsis*)**
   - Inapplicable: Since there is no levator arcus palatini **(*Loricaria*)**

242. *Relation between dilatator operculi and dorsal surface of neurocranium (character inspired from Oliveira et al., 2001).* Contrary to other catfish examined [State 0: e.g. Fig. 3.55], in siluriforms of CS–1 the dilatator operculi invades a significant part of the dorsal surface of the cranial roof, namely of the sphenotic and pterotic [State 1].
   - CS–0: Dilatator operculi not originating on dorsal surface of neurocranium **(all genera not in other CS)**
   - CS–1: Dilatator operculi originating on dorsal surface of neurocranium **(*Trichomycterus, Hatcheria, Plotosus*)**

243. *Origin of dilatator operculi (character inspired from Howes, 1983a)*. Contrary to all other catfish examined, in which the dilatator operculi originates on the neurocranium and often on the suspensorium [State 0: e.g. Fig. 3.55], in specimens of *Loricaria* studied the dilatator operculi is not in contact with the neurocranium, originating exclusively on the hyomandibulo-metapterygoid [State 1].
    — CS–0: Dilatator operculi in contact with neurocranium **(all genera not in other CS)**
    — CS–1: Dilatator operculi exclusively originating on hyomandibulo-metapterygoid **(*Loricaria*)**
244. *Relation between dilatator operculi and hyomandibulo-metapterygoid*. As seen above, in catfishes the dilatator operculi may originate not only on the neurocranium, but also on the hyomandibulo-metapterygoid, namely on the posterodorsal surface of this bone [State 0: e.g. Fig. 3.25]. However, uniquely in *Cetopsis* a great part of the fibres of the dilatator operculi also attaches anteriorly on the anterodorsal surface of the hyomandibulo-metapterygoid [State 1].
    — CS–0: Dilatator operculi not attaching on anterodorsal surface of hyomandibulo-metapterygoid **(all genera not in other CS)**
    — CS–1: Dilatator operculi, besides its usual attachments, also attaching on anterodorsal surface of hyomandibulo-metapterygoid **(*Cetopsis*)**
245. *Differentiation of dilatator operculi* (*unordered multistate character*) (*character inspired from Howes, 1983a*). Plesiomorphically in catfish the dilatator operculi is constituted by a single mass of fibres [State 0: e.g. Fig. 3.64]. However, in siluriforms of CS–1 this muscle is differentiated into a posterior bundle originating on the posterodorsal surface of the hyomandibulo-metapterygoid and an anterior bundle originating on the neurocranium [State 1: e.g. Fig. 3.25]. A different configuration is found in catfish of CS–2. Here the dilatator operculi is subdivided into a dorsal division originating on the dorsal surface of the neurocranium and a ventral division originating on both the lateral surface of the neurocranium and the posterior surface of the hyomandibulo-metapterygoid [State 2].
    — CS–0: Dilatator operculi not differentiated into two well-developed divisions **(all genera not in other CS)**
    — CS–1: Dilatator operculi with posterior bundle originating on posterodorsal surface of hyomandibulo-metapterygoid and anterior bundle originating on neurocranium **(*Xyliphius, Aspredo, Bunocephalus*)**
    — CS–2: Dilatator operculi with dorsal division originating on dorsal surface of neurocranium and ventral division originating on neurocranium and posterior surface of hyomandibulo-metapterygoid **(*Trichomycterus, Hatcheria*)**
246. *Origin of adductor operculi* (*unordered multistate character*) (*character inspired from Adriaens and Verraes, 1997a*). Contrary to all other catfish examined,

in which the adductor arcus palatini originates exclusively on the neurocranium [State 0: e.g. Fig. 3.66], in siluriforms of CS–1 this muscle originates on the neurocranium and on the posterodorsomesial surface of the hyomandibulo-metapterygoid [State 1], while in catfish of CS–2 it originates on the neurocranium and on the posterodorsolateral surface of the hyomandibulo-metapterygoid [State 2: e.g. Fig. 3.58].

— CS–0: Adductor operculi originating exclusively on neurocranium **(all genera not in other CS)**

— CS–1: Adductor operculi originating on neurocranium and posterodorsomesial surface of the hyomandibulo-metapterygoid **(*Auchenoglanis, Malapterurus*)**

— CS–2: Adductor operculi originating on neurocranium and posterodorsolateral surface of the hyomandibulo-metapterygoid **(*Clarias, Uegitglanis, Heterobranchus, Heteropneustes, Cranoglanis*)**

247. *Presence of adductor hyomandibularis*. Plesiomorphically catfish lack a muscle adductor hyomandibularis (sensu Diogo and Vandewalle, 2003) [State 0: e.g. Fig. 3.66]. An adductor hyomandibularis, completely differentiated from the adductor operculi and running from the ventrolateral surface of the neurocranium to the posterodorsal surface of the hyomandibulo-metapterygoid, is present in catfish of CS–1 [State 2: e.g. Fig. 3.59].

— CS–0: Absence of adductor hyomandibularis **(all genera not in other CS)**

— CS–1: Presence of well-developed adductor hyomandibularis **(*Arius, Ancharius, Genidens, Chrysichthys, Schilbe, Laides, Pseudeutropius, Siluranodon, Ailia, Pangasius, Helicophagus, Clarotes, Auchenoglanis, Liobagrus, Akysis, Parakysis, Amblyceps, Silurus, Wallago, Cranoglanis, Pimelodus, Calophysus, Hypophthalmus, Pseudoplatystoma, Heptapterus, Goeldiella, Rhamdia, Microglanis, Ictalurus, Amiurus, Callichthys, Corydoras, Astroblepus, Scoloplax, Pseudopimelodus, Chaca, Austroglanis*)**

— ?: Since it was not possible to properly observe this character in the specimens examined **(*Akysis, Parakysis*)**

248. *Insertion of adductor operculi on hyomandibulo-metapterygoid*. The plesiomorphic siluriform condition is most probably that found in *Diplomystes* and the great majority of the other catfish in which, besides its insertion on the opercle, the adductor operculi, or a part differentiated from it (adductor hyomandibularis: see above), also inserts on the hyomandibulo-metapterygoid [State 0: e.g. Fig. 3.66]. In catfish of CS–1 neither the adductor operculi nor a part differentiated from it insert on this latter bone [State 1: e.g. Fig. 3.25].

— CS–0: Adductor operculi or part differentiated from it inserted on hyomandibulo-metapterygoid **(all genera not in other CS)**

— CS–1: Neither adductor operculi nor part differentiated from it inserted on hyomandibulo-metapterygoid (***Phractura, Doumea, Belonoglanis, Trachyglanis, Andersonia, Glyptothorax, Glyptosternon, Bagarius, Gagata, Erethistes, Hara, Bunocephalus, Aspredo, Xyliphius, Synodontis, Mochokus, Franciscodoras, Anadoras, Acanthodoras, Doras, Centromochlus, Ageneiosus, Auchenipterus, Loricaria, Hypoptopoma, Lithoxus***)

— ?: Since it was not possible to properly observe this character in the specimens examined (***Nematogenys, Trichomycterus, Hatcheria, Akysis, Parakysis***)

249. *Absence of levator operculi (character inspired from Howes, 1983a)*. Contrary to the plesiomorphic situation found in other catfish examined [State 0: e.g. Fig. 3.64], in the specimens of *Loricaria* studied the levator operculi is lacking [State 1].

— CS–0: Presence of levator operculi **(all genera not in other CS)**

— CS–1: Absence of levator operculi (***Loricaria***)

250. *Insertion of levator operculi on opercle (character inspired from Howes, 1983a; de Pinna, 1998)*. Contrary to all other catfish examined, in which the levator operculi does not insert on the lateral surface of the opercle [State 0: e.g. Fig. 3.64], in *Nematogenys* and *Heptapterus* a significant part of this muscle inserts on the lateral surface of this bone [State 1: e.g. Fig. 3.88].

— CS–0: Levator operculi not attaching on significant part of lateral surface of opercle **(all genera not in other CS)**

— CS–1: Levator operculi attaching on significant part of lateral surface of opercle (***Nematogenys, Heptapterus, Austroglanis***)

— Inapplicable: Since there is no levator operculi (***Loricaria***)

251. *Origin of levator operculi*. Plesiomorphically in catfish the levator operculi originates exclusively on the neurocranium (Diogo and Vandewalle, 2003) [State 0: e.g. Fig. 3.64], but in siluriforms of CS–1 a great part of the fibres of this muscle also originate on the dorsolateral surface of the hyomandibulo-metapterygoid [State 1: e.g. Fig. 3.88].

— CS–0: Levator operculi originating on neurocranium **(all genera not in other CS)**

— CS–1: Levator operculi originating not only on neurocranium, but also on dorsolateral surface of hyomandibulo-metapterygoid (***Cranoglanis, Plotosus, Paraplotosus, Cnidoglanis, Neosilurus, Schilbe, Laides, Pseudeutropius, Siluranodon, Silurus, Wallago, Ailia, Pimelodus, Calophysus, Hypophthalmus, Pseudoplatystoma, Heptapterus, Goeldiella, Rhamdia, Pseudopimelodus, Microglanis, Ictalurus, Amiurus, Nematogenys, Trichomycterus, Hatcheria, Callichthys, Corydoras***)

— Inapplicable: Since there is no levator operculi (***Loricaria***)

252. *Presence of protractor operculi (character inspired from Howes, 1983a)*. Contrary to the plesiomorphic situation found in other catfish examined

[State 0: e.g. Fig. 3.64], the trichomycterids studied have a muscle protractor operculi running from the posterolateral surface of the hyomandibulo-metapterygoid to the anterolateral surface of the opercle [State 1].

— CS–0: Absence of protractor operculi **(all genera not in other CS)**

— CS–1: Presence of protractor operculi **(*Trichomycterus, Hatcheria*)**

253. *Presence of protractor posttemporalis* (*character inspired from Mo, 1991*). Contrary to the plesiomorphic configuration found in other catfish examined [State 0: e.g. Fig. 3.64], bagrids analysed have a muscle protractor posttemporalis (sensu Diogo et al., 1999) running from the neurocranium to the posttemporo-supracleithrum [State 1: e.g. Fig. 3.42].

— CS–0: Absence of protractor posttemporalis **(all genera not in other CS)**

— CS–1: Presence of protractor posttemporalis **(*Bagrus, Bagrichthys, Hemibagrus, Rita*)**

254. *Presence of protractor of Müllerian process* (*character inspired from Bridge and Haddon, 1894*). Contrary to plesiomorphic configuration found in other catfish examined [State 0: e.g. Fig. 3.88], in siluriforms of CS–1 there is a well-developed protractor of the Müllerian process attached to, and provoking movement of, the anterolateral surface of the parapophysis of the fourth vertebra (= Müllerian process) [State 1: e.g. Fig. 3.58].

— CS–0: Absence of protractor of Müllerian process **(all genera not in other CS)**

— CS–1: Presence of protractor of Müllerian process **(*Arius, Ancharius, Genidens, Pangasius, Cranoglanis, Malapterurus, Pseudopimelodus, Microglanis, Synodontis, Mochokus, Franciscodoras, Anadoras, Acanthodoras, Doras, Centromochlus, Ageneiosus, Auchenipterus*)**

255. *Differentiation of protractor of Müllerian process*. Contrary to other catfish presenting a protractor of the Müllerian process [State 0: e.g. Fig. 3.58], in *Centromochlus* this muscle is markedly subdivided into two well-developed, almost completely separated divisions [State 1: e.g. Fig. 3.30].

— CS–0: Protractor of Müllerian process not markedly subdivided into two well-developed divisions **(*Arius, Ancharius, Genidens, Pangasius, Cranoglanis, Malapterurus, Pseudopimelodus, Microglanis, Synodontis, Mochokus, Franciscodoras, Anadoras, Acanthodoras, Doras, Ageneiosus, Auchenipterus*)**

— CS–1: Protractor of Müllerian process markedly subdivided into two well-developed divisions **(*Centromochlus*)**

— Inapplicable: Since there is no protractor of Müllerian process **(all genera not in other CS)**

256. *Origin of protractor of Müllerian process* (*character inspired from Taverne and Aloulou-Triki, 1974*). Contrary to other catfish presenting a protractor of the Müllerian process [State 0: e.g. Fig. 3.58], in *Mochokus* and *Synodontis*

this muscle does not originates on the posterior surface of the cephalic region, but instead on the region of the dorsal fin and its support [State 1].

— CS–0: Protractor of Müllerian originates on posterior surface of cephalic region ***(Arius, Ancharius, Genidens, Pangasius, Cranoglanis, Malapterurus, Pseudopimelodus, Microglanis, Franciscodoras, Anadoras, Acanthodoras, Doras, Ageneiosus, Auchenipterus, Centromochlus)***

— CS–1: Protractor of Müllerian process originates on region of dorsal fin and its support ***(Synodontis, Mochokus)***

— Inapplicable: Since there is no protractor of Müllerian process **(all genera not in other CS)**

257. *Presence of muscle tensor tripodis* (*character inspired from Ladich, 2001*). Contrary to the plesiomorphic configuration found in other catfish examined [State 0: e.g. Fig. 3.88], in siluriforms of CS–1 a well-developed muscle 'tensor tripodis' (see terminology of Ladich, 2001) runs from the posterior surface of the neurocranium to the dorsal surface of the swim bladder near the tripus [State 1].

— CS–0: Absence of muscle tensor tripodis **(all genera not in other CS)**

— CS–1: Presence of muscle tensor tripodis ***(Pimelodus, Pseudoplatystoma, Calophysus)***

— Inapplicable: Since the marked encapsulation of the swim bladder does not allow a direct muscular connection between the neurocranium or the parapophysis of the fourth vertebra to the swim bladder ***(Hypophthalmus, Akysis, Parakysis, Amblyceps, Liobagrus, Erethistes, Hara, Gagata, Bagarius, Glyptosternon, Glyptothorax, Cetopsis, Nematogenys, Trichomycterus, Hatcheria, Amphilius, Paramphilius, Andersonia, Belonoglanis, Doumea, Phractura, Trachyglanis, Leptoglanis, Zaireichthys, Clarias, Uegitglanis, Heterobranchus, Heteropneustes)***

258. *Presence of drumming muscle of swim bladder* (*character inspired from Ladich, 2001*). Contrary to the plesiomorphic configuration found in other catfish examined [State 0: e.g. Fig. 3.88], in specimens of genera of CS–1 a well-developed 'drumming muscle' of the swim bladder runs from the parapophysis of the fourth vertebra and, eventually, from the posterior surface of the neurocranium to the anterior and anteroventral surfaces of the swim bladder [State 1: e.g. Fig. 3.100].

— CS–0: Absence of drumming muscle of swim bladder **(all genera not in other CS)**

— CS–1: Presence of drumming muscle of swim bladder ***(Pimelodus, Pseudoplatystoma, Calophysus)***

— Inapplicable: Since the marked encapsulation of the swim bladder does not allow a direct muscular connection between the neurocranium or the parapophysis of the fourth vertebra to the swim bladder ***(Hypophthalmus, Akysis, Parakysis, Amblyceps, Liobagrus, Erethistes, Hara, Gagata, Bagarius, Glyptosternon,***

***Glyptothorax, Cetopsis, Nematogenys, Trichomycterus, Hatcheria, Amphilius, Paramphilius, Andersonia, Belonoglanis, Doumea, Phractura, Trachyglanis, Leptoglanis, Zaireichthys, Clarias, Uegitglanis, Heterobranchus, Heteropneustes*)**

Splanchnocranium

259. *Presence of maxillary teeth (character inspired from, Regan, 1911b; Alexander, 1965; Gosline, 1975; Arratia, 1987; Grande and de Pinna 1998; and others).* Although the absence of maxillary teeth, a character found in all catfish except the diplomystids and fossil hypsidorids, has clearly been the most cited synapomorphy of recent non-diplomystid catfish, some authors have lately argued that it could, instead, constitute the plesiomorphic condition for the order Siluriformes. In this phylogenetic comparison I follow the opinion given in Grande and de Pinna's important recent paper (1998) in which the absence of maxillary teeth continues to be seen as a derived state among catfish [State 1: e.g. Fig. 3.39] and, thus, the configuration found in Diplomystes continues to be seen as the plesiomorphic state [State 0: e.g. Fig. 3.63]. This character is discussed in detail in Chapters 4 and 5.
    — CS–0: Presence of maxillary teeth **(*Diplomystes*)**
    — CS–1: Absence of maxillary teeth **(all genera not in other CS)**
260. *Articulatory facets of maxilla for autopalatine (character inspired from de Pinna and Vari, 1995).* Contrary to the plesiomorphic situation found in other catfish, in which the maxilla exhibits two proximal articulatory facets for articulation with the autopalatine [Sate 0: e.g. Fig. 3.68], in the cetopsids studied the former bone exhibits a single articulatory facet for the latter [State 1: e.g. Fig. 3.43].
    — CS–0: Maxilla with two articulatory facets for autopalatine **(all genera not in other CS)**
    — CS–1: Maxilla with one single articulatory facet for autopalatine **(*Cetopsis, Hemicetopsis, Helogenes*)**
261. *Foramen in proximal portion of maxilla.* Plesiomorphically in catfish the maxilla does not present a small foramen in its proximal region [State 0: e.g. Fig. 3.68], but in *Leptoglanis* the head of the maxilla is pierced by a small, circular foramen [State 1: e.g. Fig. 3.17].
    — CS–0: Proximal surface of maxilla not pierced by small foramen **(all genera not in other CS)**
    — CS–1: Proximal surface of maxilla pierced by small foramen **(*Leptoglanis*)**
262. *Foramen near distal tip of maxilla.* Plesiomorphically in catfish the maxilla does not present a small foramen near its distal tip [State 0: e.g. Fig. 3.68], but in *Paramphilius* the maxilla is pierced by a well-developed, circular foramen distally [State 1].
    — CS–0: Distal surface of maxilla not pierced by small foramen **(all genera not in other CS)**

— CS–1: Distal surface of maxilla pierced by small foramen **(*Paramphilius*)**

263. *Shape of maxilla (character inspired from Grande, 1987).* The plesiomorphic configuration for siluriforms is seemingly that in which the posterodistal margin of the maxilla is markedly concave, with the distal surface of this bone being larger than its proximal extremity. This is the case in catfish such as diplomystids, nematogenyids and cetopsins (Diogo et al., 2000a) [State 0: e.g. Fig. 3.63]. In the derived condition, the posterodistal margin of the maxilla is not markedly concave, with the maxillary bone essentially a simple, roughly tubular structure [State 1: e.g. Fig. 3.39].
   — CS–0: Posterodistal margin of maxilla markedly concave **(*Diplomystes, Nematogenys, Callichthys, Cetopsis, Hemicetopsis*)**
   — CS–1: Posterodistal margin of maxilla not markedly concave **(all genera not in other CS)**
   — ?: Since it was not possible to properly observe this character in the specimens examined **(*Helogenes, Trichomycterus, Hatcheria, Corydoras*)**
264. *Anteroventromesial expansion of maxilla.* Contrary to all other catfish examined [State 0: Fig. 3.68], in *Chaca* and *Paraplotosus* the maxilla exhibits a well-developed anteroventromesial expansion of laminar bone [State 1: e.g. Fig. 3.49].
   — CS–0: Absence of well-developed anteroventromesial expansion of maxilla **(all genera not in other CS)**
   — CS–1: Presence of well-developed anteroventromesial expansion of maxilla **(*Paraplotosus, Chaca*)**
265. *Markedly elongated maxilla* (*ordered multistate character*). Although primitively in catfish the maxilla is relatively elongated proximodistally [State 0: e.g. Fig. 3.63], it is not as long as the remarkably elongated maxilla found in the few siluriforms of CS–1 [State 1: e.g. Fig. 3.49] and especially in genera *Bagarius* and *Gagata* [State 2].
   — CS–0: Maxilla not remarkably elongated **(all genera not in other CS)**
   — CS–1: Remarkably elongated maxilla **(*Pangasius, Leptoglanis, Zaireichthys, Glyptosternon, Liobagrus, Aspredo, Wallago, Centromochlus, Ageneiosus, Auchenipterus, Chaca, Astroblepus, Scoloplax*)**
   — CS–2: Maxilla even more elongated than in CS–1 **(*Bagarius, Gagata*)**
266. *Markedly short maxilla* (*character inspired from Bornbusch, 1991b, 1995*). Contrary to all other catfish examined (see above) [State 0: see character above], in siluriforms of CS–1 the maxilla is peculiarly reduced to a very short, small structure [State 1: e.g. Fig. 3.82].
   — CS–0: Maxilla not reduced to very short, small structure **(all genera not in other CS)**
   — CS–1: Maxilla reduced to very short, small structure **(*Clarias, Schilbe, Laides, Pseudeutropius, Siluranodon, Ailia, Uegitglanis, Heterobranchus, Heteropneustes, Cetopsis, Hemicetopsis, Silurus, Wallago, Helogenes*)**

267. *Firm attachment between mesial surface of maxilla and lateral surface of mandible.* The plesiomorphic condition for catfishes, found in diplomystids and most other ostariophysans, seems to be that in which the mesial surface of the distal portion of the maxilla is firmly attached to the lateral surface of the mandible by short and strong ligamentous tissue (Alexander, 1965; Diogo et al., 2000a) [State 0: e.g. Figs. 3.63, 3.64]. In the derived condition the distal surface of the maxilla is not firmly attached to the mandible [State 1: e.g. Fig. 3.3].
   - CS–0: Mesial surface of maxilla firmly attached to mandible **(*Diplomystes*)**
   - CS–1: Mesial surface of maxilla not firmly attached to mandible **(all genera not in other CS)**
268. *Presence of smooth, roundish surface on proximoventral margin of maxilla.* Contrary to all other catfish examined [State 0: Fig. 3.68], in the loricariid specimens studied the maxilla exhibits a small, smooth, roundish surface on its proximoventral margin [State 1].
   - CS–0: Absence of smooth, roundish surface of maxilla **(all genera not in other CS)**
   - CS–1: Presence of smooth, roundish surface of maxilla **(*Loricaria, Hypoptopoma, Lithoxus*)**
269. *Presence of thin posterior laminar flange along posterior margin of maxilla (character inspired from Mo, 1991).* Contrary to all other catfish analysed [State 0: e.g. Fig. 3.68], in *Austroglanis* the maxilla exhibits a thin posterior laminar flange along its posterior margin [State 1: e.g. Fig. 3.33].
   - CS–0: Absence of thin posterior laminar flange along posterior margin of maxilla **(all genera not in other CS)**
   - CS–1: Presence of thin posterior laminar flange along posterior margin of maxilla **(*Austroglanis*)**
270. *Association between primordial ligament and cordlike tissue (character inspired from Mo, 1991).* Plesiomorphically catfish present a strong primordial ligament connecting the coronoid process of the mandible and the maxilla [State 0: e.g. Fig. 3.64]. However, in catfish of CS–1 this ligament is associated to, or differentiated into, a massive, somewhat cartilaginous or cordlike tissue [State 1: e.g. Fig. 3.64].
   - CS–0: Primordial ligament not associated to, or differentiated into, a massive, somewhat cartilaginous or cordlike tissue **(all genera not in other CS)**
   - CS–1: Primordial ligament associated to, or differentiated into, a massive, somewhat cartilaginous or cordlike tissue **(*Clarotes, Chrysichthys, Schilbe, Laides, Pseudeutropius, Siluranodon, Ailia, Arius, Ancharius, Genidens, Auchenoglanis*)**
   - ?: Since it was not possible to properly observe this character in the specimens examined **(*Cranoglanis, Austroglanis, Ictalurus, Amiurus, Pangasius, Silurus, Helicophagus, Malapterurus*)**

271. *Presence of two well-developed, well-distinguished ligaments connecting the coronoid process of the mandible and maxilla (character inspired from Oliveira et al., 2001).* Contrary to all other catfish examined (see above) [State 0: e.g. Fig. 3.64], in the plotosids analysed two well-developed, well distinguished primordial ligaments connect the coronoid process of the mandible to the maxilla and/or the maxillary barbels [State 1: e.g. Fig. 3.115].
    — CS–0: Absence of two well-developed, well-distinguished ligaments between maxilla and coronoid process **(all genera not in other CS)**
    — CS–1: Presence of two well-developed, well-distinguished ligaments between maxilla and coronoid process **(*Plotosus, Neosilurus, Cnidoglanis, Paraplotosus*)**
272. *Presence of enlarged base of maxillary barbels (character inspired from Oliveira et al., 2001).* Contrary to all other catfish examined [State 0: e.g. Fig. 3.25], in plotosids the maxillary barbels stem from a large, circular structure, which looks like a posterior extension of proximal part of these barbels [State 1: e.g. Fig. 3.115].
    — CS–0: Absence of enlarged base of maxillary barbels **(all genera not in other CS)**
    — CS–1: Presence of enlarged base of maxillary barbels **(*Plotosus, Neosilurus, Cnidoglanis, Paraplotosus*)**
273. *Presence of additional maxillary barbel (character inspired from Friel, 1994).* Contrary to other catfish [State 0: e.g. Fig. 3.20], *Aspredo* exhibits not just one, but two maxillary barbels associated with the maxilla [State 1].
    — CS–0: Presence of single maxillary barbel associated with maxilla **(all genera not in other CS)**
    — CS–1: Presence of two maxillary barbels associated with maxilla **(*Aspredo*)**
274. *Presence of strong ligament between premaxilla and proximal surface of maxilla.* Plesiomorphically catfish lack a well-developed, strong ligament linking the proximal surface of the maxilla and the premaxillary bone (Diogo et al., 2000a) [State 0: e.g. Fig. 3.67]. Such a ligament is present in siluriforms of CS–1 [State 1: e.g. Fig. 3.25].
    — CS–0: Absence of well-developed, strong ligament between proximal surface of maxilla and premaxilla **(*Diplomystes, Trichomycterus, Hatcheria, Nematogenys*)**
    — CS–1: Presence of well-developed, strong ligament between proximal surface of maxilla and premaxilla **(all genera not in other CS)**
    — ?: Since it was not possible to properly appraise this character in the specimens examined **(*Cetopsis, Hemicetopsis, Helogenes*)**
275. *Presence of rictal barbel (character inspired from de Pinna, 1998).* Contrary to other catfish examined [State 0], in the trichomycterids analysed a rictal barbel occurs, the proximal extremity of which is situated between the proximoventral margin of the maxilla and the dorsomesial margin of the premaxilla [State 1].
    — CS–0: Absence of rictal barbel **(all genera not in other CS)**
    — CS–1: Presence of rictal barbel **(*Trichomycterus, Hatcheria*)**

276. *Presence of triangular process of dorsal margin of autopalatine.* Plesiomorphically catfish lack major processes on the dorsal surface of the autopalatine [State 0: e.g. Fig. 3.68]. In *Helicophagus* and *Pseudeutropius* there is a well-developed, dorsally pointed triangular process on the dorsal surface of the autopalatine [State 1].
   — CS–0: Absence of triangular process on dorsal margin of autopalatine **(all genera not in other CS)**
   — CS–1: Presence of triangular process on dorsal margin of autopalatine **(*Helicophagus, Pseudeutropius*)**
277. *Development of anterior cartilage of autopalatine (character inspired from Mo, 1991).* Plesiomorphically in catfish the anterior cartilage of the autopalatine is a small, short structure [State 0: e.g. Fig. 3.68]. In siluriforms of CS–1 the anterior cartilage of the autopalatine is markedly elongated anteroposteriorly [State 1: e.g. Fig. 3.55].
   — CS–0: Anterior cartilage not markedly elongated anteroposteriorly **(all genera not in other CS)**
   — CS–1: Anterior cartilage markedly elongated anteroposteriorly **(*Arius, Ancharius, Genidens, Chrysichthys, Schilbe, Laides, Pseudeutropius, Siluranodon, Clarotes, Auchenoglanis, Cetopsis, Hemicetopsis, Helogenes, Austroglanis*)**
278. *Posteromesial extension of anterior cartilage of autopalatine (character inspired from de Pinna and Vari, 1995).* Contrary to other catfish examined [State 0: e.g. Fig. 3.68], in siluriforms of CS–1 the anterior cartilage of the autopalatine is markedly extended posteromesially, inclusively covering a significant part of the anterolateral surface of the main body of this bone [State 1: e.g. Fig. 3.55].
   — CS–0: Absence of posteromesial extension of anterior cartilage of autopalatine **(all genera not in other CS)**
   — CS–1: Presence of posteromesial extension of anterior cartilage of autopalatine **(*Chrysichthys, Cetopsis, Hemicetopsis, Clarotes, Helogenes*)**
279. *Markedly enlarged anterior cartilage of autopalatine (character inspired from de Pinna and Vari, 1995).* Contrary to other catfish, in which the anterior cartilage of the autopalatine exhibits a roughly roundish, rectangular or quadrangular shape and is not markedly expanded transversally [State 0: e.g. Fig. 3.63], in *Cetopsis* and *Hemicetopsis* this is a highly irregularly shaped, transversally enlarged structure [State 1: e.g. Fig. 3.43].
   — CS–0: Anterior cartilage of autopalatine not markedly enlarged transversally **(all genera not in other CS)**
   — CS–1: Anterior cartilage of autopalatine markedly enlarged transversally **(*Cetopsis, Hemicetopsis*)**
280. *Anterior portion of autopalatine with donut-like aspect (character inspired from Arratia, 1987).* Contrary to other catfish and other ostariophysans [State 0: e.g. Fig. 3.3], in *Diplomystes* a distinct anterior bifurcation of the autopalatine occurs, with the anterior surface of this bone, together with

the posterior margin of its anterior cartilage, conferring a somewhat donut-like shape to the anterior portion of this bone [State 1: e.g. Fig. 3.68].

— CS–0: Anterior portion of autopalatine not presenting a donut-like aspect **(all genera not in other CS)**

— CS–1: Anterior portion of autopalatine presenting a donut-like aspect **(*Diplomystes*)**

281. *Presence of prominent dorsolateral crest of autopalatine*. Plesiomorphically siluriforms lack major dorsolateral crests of the autopalatine [State 0: e.g. Fig. 3.68]. In *Auchenipterus* and *Ageneiosus* the autopalatine exhibits a prominent dorsolateral crest to receive the extensor tentaculi [State 1].

— CS–0: Absence of prominent dorsolateral crest of autopalatine **(all genera not in other CS)**

— CS–1: Presence of prominent dorsolateral crest of autopalatine **(*Auchenipterus, Ageneiosus*)**

282. *Presence of deep oval concavity of autopalatine*. Contrary to other catfish examined [State 0: e.g. Fig. 3.68], in *Ageneiosus* the autopalatine exhibits a deep, oval dorsal concavity [State 1].

— CS–0: Absence of deep, oval dorsal concavity of autopalatine **(all genera not in other CS)**

— CS–1: Presence of deep, oval dorsal concavity of autopalatine **(*Ageneiosus*)**

283. *Absence of posterior cartilage of autopalatine (character inspired from Mo, 1991; Arratia, 1992)*. Contrary to the plesiomorphic situation found in most catfish [State 0: e.g. Fig. 3.68], in siluriforms of CS–1 the posterior cartilage of the autopalatine is lacking [State 1: e.g. Fig. 3.17].

— CS–0: Presence of posterior cartilage of autopalatine **(all genera not in other CS)**

— CS–1: Absence of posterior cartilage of autopalatine **(*Phractura, Andersonia, Leptoglanis, Malapterurus, Zaireichthys, Paramphilius, Belonoglanis, Trachyglanis, Doumea, Amphilius, Trichomycterus, Hatcheria, Callichthys, Corydoras, Loricaria, Hypoptopoma, Lithoxus, Scoloplax, Astroblepus*)**

— ?: Since it was not possible to properly appraise this character in the specimens examined **(*Silurus, Wallago*)**

284. *Development of autopalatine (character inspired from Bornbusch, 1991b, 1995)*. Contrary to other catfish examined [State 0: e.g. Fig. 3.68], in the silurids analysed the autopalatine is reduced to a very short, nodular, irregularly shaped structure [State 1].

— CS–0: Autopalatine not reduced to a very short, nodular, irregularly shaped structure **(all genera not in other CS)**

— CS–1: Autopalatine reduced to a very short, nodular, irregularly shaped structure **(*Silurus, Wallago*)**

285. *Presence of prominent anteroventrolateral projection of autopalatine (character inspired from Mo, 1991)*. Uniquely in catfish of CS–1, and contrary to other siluriforms [State 0: e.g. Fig. 3.68], the autopalatine exhibits a

prominent anteroventrolateral projection of laminar bone [State 1: e.g. Fig. 3.55].

— CS–0: Absence of prominent anteroventrolateral projection of autopalatine **(all genera not in other CS)**

— CS–1: Presence of prominent anteroventrolateral projection of autopalatine ***(Chrysichthys, Schilbe, Pseudeutropius, Clarotes, Auchenoglanis)***

— ?: Since it was not possible to properly appraise this character in the specimens examined, due to the very peculiar shape of the autopalatine (see above) ***(Silurus, Wallago)***

286. *Shape of autopalatine* (*unordered multistate character*) (*character inspired from Gosline, 1975*). In the vast majority of catfish presenting a mesial articulation with the neurocranium (see below), this bone exhibits a somewhat spatulate shape, with the portion posterior to its mesial articulatory surface thinner than the portion situated anterior to this surface [State 0: e.g. Fig. 3.6]. In catfish of CS–1 the autopalatine is a roughly tubular structure with, however, its mesial articulation significantly salient mesially [State 1: e.g. Fig. 3.3]. A different configuration is found in siluriforms of CS–2, in which the autopalatine is a markedly simple tubular structure with no salient mesial articulatory surface and hence no major mesial projections or saliences [State 2: e.g. Fig. 3.39].

— CS–0: Spatulate autopalatine, with portion posterior to its mesial articulatory surface thinner than portion situated anterior to this surface **(all genera not in other CS)**

— CS–1: Autopalatine a roughly tubular structure with a markedly salient mesial articulatory surface for the neurocranium ***(Glyptothorax, Glyptosternon, Bagarius, Liobagrus, Akysis, Parakysis, Amblyceps, Gagata, Erethistes, Hara, Bunocephalus, Aspredo, Xyliphius, Clarias, Uegitglanis, Heterobranchus, Heteropneustes, Plotosus, Paraplotosus, Neosilurus, Cnidoglanis, Franciscodoras, Anadoras, Acanthodoras, Doras, Centromochlus, Ageneiosus, Auchenipterus)***

— CS–2: Autopalatine a markedly tubular, simple structure lacking major mesial projections or saliences ***(Bagrus, Bagrichthys, Hemibagrus, Rita, Arius, Genidens, Pimelodus, Calophysus, Hypophthalmus, Pseudoplatystoma, Heptapterus, Goeldiella, Rhamdia)***

— ?: Since it is very difficult to define this character due to the peculiar, irregular shape of the autopalatine in the specimens of these four genera ***(Silurus, Wallago, Pseudopimelodus, Chaca)***

— Inapplicable: Since in specimens of these genera the articulatory surface of the autopalatine for the neurocranium is mainly situated on the dorsal, and not the mesial surface of the autopalatine (see below) ***(Diplomystes, Nematogenys, Trichomycterus,***

***Hatcheria, Callichthys, Corydoras, Loricaria, Hypoptopoma, Lithoxus, Scoloplax, Astroblepus*)**

287. *Posterior bifurcation of autopalatine (character inspired from Mo, 1991; de Pinna, 1993).* Contrary to other catfish examined [State 0: e.g. Fig. 3.63], in the amphiliid specimens analysed the posterior tip of the bony portion of the autopalatine is markedly bifurcate [State 1: e.g. Fig. 3.17].
   — CS–0: Autopalatine not markedly bifurcate posteriorly **(all genera not in other CS)**
   — CS–1: Autopalatine markedly bifurcate posteriorly **(*Doumea, Belonoglanis, Trachyglanis, Andersonia, Zaireichthys, Phractura, Leptoglanis, Amphilius, Paramphilius*)**

288. *Dorsoventral compression of autopalatine (unordered multistate character) (character inspired from Brown and Ferraris, 1988; Arratia, 1990; Mo, 1991; de Pinna and Vari, 1995).* Plesiomorphically in catfish the autopalatine is not markedly compressed dorsoventrally [State 0: e.g. Fig. 3.63]. In siluriforms of CS–1 the autopalatine is markedly compressed dorsoventrally, with only a somewhat dorsal salience to form the articulatory surface for the neurocranium [State 1: e.g. Fig. 3.88]. A different configuration is found in *Chaca* in which, as described by Brown and Ferraris (1988), the autopalatine is markedly compressed dorsoventrally but exhibits two salient, oblique ventral arms [State 2]. A different derived configuration is also found in genera *Cetopsis* and *Hemicetopsis*, in which the autopalatine is a notably dorsoventrally compressed structure without major dorsal or ventral salience [State 3].
   — CS–0: Autopalatine not markedly compressed dorsoventrally **(all genera not in other CS)**
   — CS–1: Autopalatine markedly compressed dorsoventrally, with dorsal salience to form the articulatory surface for the neurocranium **(*Nematogenys, Trichomycterus, Hatcheria, Callichthys, Corydoras, Astroblepus, Scoloplax*)**
   — CS–2: Autopalatine markedly compressed dorsoventrally, with two salient, oblique ventral arms **(*Chaca*)**
   — CS–3: Autopalatine a notably dorsoventrally compressed structure without major dorsal or ventral salience **(*Cetopsis, Hemicetopsis*)**

289. *Presence of dorsal and ventral laminar expansions of posterior portion of autopalatine (character inspired from de Pinna, 1996).* Contrary to other siluriforms examined [State 0: e.g. Fig. 3.63], in amphiliids the posterior tip of the bony portion of the autopalatine is considerably broader dorsoventrally than the rest of the bone. This is seemingly associated with the presence of dorsal and ventral laminar expansions, and not an actual enlargement of the centre of ossification of the autopalatine (see de Pinna, 1996) [State 1: e.g. Fig. 3.17].
   — CS–0: Posterior portion of autopalatine not significantly expanded dorsoventrally due to presence of ventral and dorsal laminar expansions of this bone **(all genera not in other CS)**

— CS–1: Posterior portion of autopalatine significantly expanded dorsoventrally due to presence of ventral and dorsal laminar expansions of this bone **(*Amphilius, Paramphilius, Phractura, Doumea, Trachyglanis, Belonoglanis, Andersonia, Zaireichthys, Leptoglanis*)**

290. *Dorsoventral enlargement of centre of ossification of posterior portion of autopalatine (ordered multistate character) (character inspired from de Pinna, 1996).* Contrary to other catfish examined [State 0: e.g. Fig. 3.63], in siluriforms of CS–1 [State 1: e.g. Fig. 3.25], and especially of CS–2 [State 2: e.g. Fig. 3.120], the posterior tip of the bony portion of the autopalatine is considerably broader dorsoventrally than the rest of the bone seemingly due to marked enlargement of the centre of ossification of the autopalatine (see de Pinna, 1996, and character above).
   — CS–0: Posterior portion of autopalatine not significantly expanded dorsoventrally due to enlargement of its centre of ossification **(all genera not in other CS)**
   — CS–1: Posterior portion of autopalatine expanded dorsoventrally due to enlargement of its centre of ossification **(*Erethistes, Hara, Bunocephalus, Aspredo, Xyliphius, Liobagrus, Akysis, Parakysis, Amblyceps, Chaca, Pseudopimelodus, Microglanis, Synodontis, Mochokus, Franciscodoras, Anadoras, Acanthodoras, Doras, Centromochlus*)**
   — CS–2: Dorsoventral expansion of posterior portion of the autopalatine due to marked enlargement of centre of ossification of this bone even more pronounced than in CS–1 **(*Plotosus, Paraplotosus, Cnidoglanis, Neosilurus, Glyptothorax, Glyptosternon, Bagarius, Gagata*)**

291. *Transversal expansion of posterior portion of autopalatine* (*character inspired from de Pinna and Vari, 1996*). Contrary to all the other catfish examined [State 0: e.g. Fig. 3.63], in *Cetopsis* the posterior tip of the bony portion of the autopalatine is markedly expanded transversally [State 1].
   — CS–0: Posterior portion of autopalatine not significantly expanded transversally **(all genera not in other CS)**
   — CS–1: Posterior portion of autopalatine significantly expanded transversally **(*Cetopsis*)**

292. *Transversal expansion of anterior portion of autopalatine*. Contrary to all the other catfish examined [State 0: e.g. Fig. 3.63], in *Glyptosternon* the anterior tip of the bony portion of the autopalatine is markedly expanded transversally [State 1].
   — CS–0: Anterior portion of autopalatine not significantly expanded transversally **(all genera not in other CS)**
   — CS–1: Anterior portion of autopalatine significantly expanded transversally **(*Glyptosternon*)**

293. *Lateral encurvation of posterior portion of autopalatine (ordered multistate character) (character inspired from Brown and Ferraris, 1988; de Pinna, 1996).*

Contrary to other catfish examined [State 0: e.g. Fig. 3.68], in siluriforms of CS–1 [State 1: e.g. Fig. 3.25], and especially of CS–2 [State 2: e.g. Fig. 3.120], the posterior tip of the bony portion of the autopalatine exhibits a marked lateral incurvation.

- CS–0: Posterior portion of autopalatine not laterally curved **(all genera not in other CS)**
- CS–1: Posterior portion of autopalatine presenting lateral incurvation **(*Trichomycterus, Hatcheria, Nematogenys, Callichthys, Corydoras, Astroblepus*)**
- CS–2: Posterior portion of autopalatine markedly curved laterally **(*Parakysis, Chaca*)**
- ?: Since some specimens examined presented a lateral incurvation of the posterior tip of the autopalatine and others not **(*Akysis*)**
- Inapplicable: Since in specimens of these genera the portion of the autopalatine situated posterior to the articulatory surface of this bone for the neurocranium is remarkably reduced in size **(*Loricaria, Hypoptopoma, Lithoxus, Scoloplax*)**

294. *Type of palatine-maxillary system (character inspired from Gosline, 1975).* Plesiomorphic situation for Siluroidea seems to be that found in most catfish of this group, in which there is a 'rocking' palatine-maxillary system, that is, one in which the abduction of the maxillary barbel is essentially associated with a mesial displacement of the posterior extremity of the autopalatine (see Diogo et al., 2003a) [State 0: e.g. Fig. 4.21B]. Catfish of CS–1 present a 'sliding' palatine-maxillary system, i.e., a system in which the abduction of the maxillary barbel is essentially associated with posterior displacement of the posterior tip of the autopalatine [State 1: e.g. Fig. 4.22A]. This character is discussed in detail in Chapters 4 and 5.
   - CS–0: 'Rocking' palatine-maxillary system **(all genera not in other CS)**
   - CS–1: 'Sliding' palatine-maxillary system **(*Bagrus, Bagrichthys, Hemibagrus, Pimelodus, Calophysus, Hypophthalmus, Pseudoplatystoma, Heptapterus, Goeldiella, Rhamdia*)**
   - ?: Since it was very difficult to appraise this character in the specimens examined **(*Silurus, Wallago, Rita*)**
   - Inapplicable: Since the palatine-maxillary system of diplomystids differs markedly from that in Siluroidea (see Diogo et al., 2003a) **(*Diplomystes*)**

295. *Anteriorly directed mesial articulatory surface of autopalatine for neurocranium (character inspired from de Pinna, 1993).* Among catfish presenting a mesial articulatory surface of the autopalatine for the neurocranium (see below) the plesiomorphic condition seems to be that in which this articulatory surface is mainly directed posteromesially [State 0: e.g. Fig. 3.33], but in siluriforms of CS–1 this articulatory surface is essentially directed anteromesially [State 1: e.g. Fig. 3.85].
   - CS–0: Articulatory surface of autopalatine for neurocranium mainly directed posteromesially **(all genera not in other CS)**

— CS–1: Articulatory surface of autopalatine for neurocranium essentially directed anteromesially ***(Clarias, Uegitglanis, Heterobranchus, Heteropneustes, Plotosus, Paraplotosus, Neosilurus, Cnidoglanis, Chaca)***
— Inapplicable: Since these catfish present an articulatory surface of the autopalatine for the neurocranium markedly directed dorsally (see below) ***(Diplomystes, Nematogenys, Trichomycterus, Hatcheria, Callichthys, Corydoras, Loricaria, Hypoptopoma, Lithoxus, Scoloplax, Astroblepus)***

296. *Type of articulation between autopalatine and neurocranium*. As noted above, in diplomystid and loricarioid catfish examined the articulatory surface of the autopalatine for the neurocranium is markedly directed dorsally. In primitive cypriniforms and characiforms the articulatory surface of the autopalatine for the neurocranium is situated on the dorsomedial and mesial surface of the autopalatine respectively while such an 'articulation—either direct or indirect via cartilage—is lacking in gymnotoids, even in those forms with a chondroidal autopalatine' (Arratia, 1992: 89). Thus, the comparison with other ostariophysans does not clarify whether the markedly dorsally directed articulatory surface of the autopalatine for the neurocranium represents or not a plesiomorphic situation for the Siluriformes. However, since this is the situation found in diplomystids, the codification of this character is based on the hypothesis that this situation represents the plesiomorphic state [State 0: e.g. Figs. 3.39, 3.67], and hence an essentially mesially directed articulatory surface (see above) represents the derived condition [State 1: e.g. Fig. 3.3], with a detailed discussion of this character given ulteriorly in chapters 4 and 5.
— CS–0: Articulatory surface of autopalatine for neurocranium markedly directed dorsally ***(Diplomystes, Nematogenys, Trichomycterus, Hatcheria, Callichthys, Corydoras, Loricaria, Hypoptopoma, Lithoxus, Scoloplax, Astroblepus)***
— CS–1 Articulatory surface of autopalatine for neurocranium essentially directed mesially **(all genera not in other CS)**

297. *Dorsoventral elongation of articulatory surface of autopalatine for neurocranium*. Contrary to all other catfish examined [State 0: e.g. Fig. 3.68], in *Heteropneustes* the articulatory surface of the autopalatine for the neurocranium is remarkably, uniquely elongated dorsoventrally [State 1: e.g. Fig. 3.86].
— CS–0: Articulatory surface of autopalatine for neurocranium not remarkably elongated dorsoventrally **(all genera not in other CS)**
— CS–1: Articulatory surface of autopalatine for neurocranium remarkably elongated dorsoventrally ***(Heteropneustes)***

298. *Marked incurvation of articulatory surface of autopalatine for neurocranium*. Contrary to all other catfish examined [State 0: e.g. Fig. 3.68], in *Mochokus* and *Synodontis* [State 1] the articulatory surface of the autopalatine for

the neurocranium is remarkably curved, with the anterior margin of this surface markedly projected mesially, projecting well beyond to the main body of the autopalatine.

— CS–0: Articulatory surface of autopalatine for neurocranium not remarkably curved **(all genera not in other CS)**

— CS–1: Articulatory surface of autopalatine for neurocranium markedly curved **(*Synodontis, Mochokus*)**

299. *Presence of long, strong ligament connecting dorsal surface of entoectopterygoid and posterior tip of autopalatine (character inspired from Gosline, 1975; Mo, 1991).* Peculiarly in catfish of CS–1 [State 1: e.g. Fig. 3.39], and contrarily to all other siluriforms examined [State 0: e.g. Fig. 3.64], a long, strong ligament connects the dorsal surface of the entoectopterygoid and the posterior extremity of the autopalatine.

— CS–0: Absence of long, strong ligament between dorsal surface of entoectopterygoid and posterior extremity of autopalatine **(all genera not in other CS)**

— CS–1: Presence of long, strong ligament between dorsal surface of entoectopterygoid and posterior extremity of autopalatine **(*Bagrus, Bagrichthys, Hemibagrus*)**

— Inapplicable: Since there is no entoectopterygoid **(*Aspredo*)**

300. *Presence of long, thin ligament connecting anterolateral surface of lateral ethmoid and dorsomesial surface of autopalatine (character inspired from Schaefer, 1990).* Contrary to other catfish examined [State 0: e.g. Fig. 3.64], in specimens of genus *Astroblepus* examined [State 1] a long and thin ligament connects the anterolateral surface of the lateral ethmoid and the dorsomesial surface of the autopalatine.

— CS–0: Absence of long, thin ligament between anterolateral surface of lateral ethmoid and dorsomesial margin of autopalatine **(all genera not in other CS)**

— CS–1: Presence of long, thin ligament between anterolateral surface of lateral ethmoid and dorsomesial margin of autopalatine **(*Astroblepus*)**

301. *Presence of long, strong ligament connecting anterior surface of entoectopterygoid and posterior margin of maxilla (character inspired from Gosline, 1975; Mo, 1991).* Peculiarly in catfish of CS–1 [State 1: e.g. Fig. 3.39], and contrarily to all other siluriforms examined [State 0: e.g. Fig. 3.64], a long, strong ligament connects the anterior surface of the entoectopterygoid and the posterior margin of the maxilla.

— CS–0: Absence of long, strong ligament between anterior surface of entoectopterygoid and posterior margin of maxilla **(all genera not in other CS)**

— CS–1: Presence of long, strong ligament between anterior surface of entoectopterygoid and posterior margin of maxilla **(*Bagrus, Bagrichthys, Hemibagrus, Rita, Pseudopimelodus, Microglanis, Pimelodus, Calophysus, Pseudoplatystoma*)**

— Inapplicable: Since the entoectopterygoid is absent **(*Aspredo*)**

302. *Presence of long, thin ligament connecting dorsal surface of sesamoid bone 1 of suspensorium and posteroventral surface of autopalatine.* Contrary to other siluriforms examined [State 0: e.g. Fig. 3.64], in catfish of CS–1 a long, thin ligament connects the dorsal surface of the sesamoid bone 1 of the suspensorium and the posteroventral surface of autopalatine [State 1: e.g. Fig. 3.28].
   — CS–0: Absence of long, thin ligament between dorsal surface of sesamoid bone 1 of suspensorium and posteroventral margin of autopalatine **(all genera not in other CS)**
   — CS–1: Presence of long, thin ligament between dorsal surface of sesamoid bone 1 of suspensorium and posteroventral margin of autopalatine ***(Clarias, Heterobranchus, Heteropneustes, Glyptothorax, Bagarius, Erethistes, Hara, Bunocephalus, Aspredo, Xyliphius, Franciscodoras, Doras, Anadoras, Acanthodoras)***
   — ?: Since it was not possible to properly observe this character in the specimens examined ***(Glyptosternon, Gagata, Centromochlus, Ageneiosus, Auchenipterus, Uegitglanis)***
   — Inapplicable: Since there is no sesamoid bone 1 of the suspensorium ***(Rita, Synodontis, Mochokus, Trichomycterus, Hatcheria, Callichthys, Corydoras, Loricaria, Hypoptopoma, Lithoxus, Scoloplax, Astroblepus)***

303. *Presence of ligament between anteroventral surface of autopalatine and dorsolateral surface of premaxilla.* Plesiomorphically in catfish there is no well-defined ligament between the anteroventral surface of autopalatine and the dorsolateral surface of premaxilla [State 0: e.g. Fig. 3.64]. In catfish of CS–1 the anteroventral surface of the autopalatine is connected by means of a short, thick ligament to the dorsolateral surface of the premaxilla [State 1].
   — CS–0: Absence of short, thick ligament connecting dorsolateral surface of premaxilla and anteroventral surface of autopalatine **(all genera not in other CS)**
   — CS–1: Presence of short, thick ligament connecting dorsolateral surface of premaxilla and anteroventral surface of autopalatine ***(Phractura, Doumea, Belonoglanis, Trachyglanis, Andersonia)***

304. *Presence of ligament between posterolateral surface of premaxilla and anteromesial surface of autopalatine (character inspired from Oliveira, 2001).* Contrary to other siluriforms examined [State 0: e.g. Fig. 3.64], in catfish of CS–1 a well-defined ligament connects the anteromesial surface of autopalatine and the posterolateral surface of premaxilla [State 1].
   — CS–0: Absence of well-defined ligament connecting anteromesial surface of autopalatine and the posterolateral surface of premaxilla **(all genera not in other CS)**
   — CS–1: Presence of well-defined ligament connecting anteromesial surface of autopalatine and posterolateral surface of premaxilla ***(Plotosus, Neosilurus, Cnidoglanis, Paraplotosus)***

305. *Presence of well-defined, strong ligament between anterior surface of autopalatine and mesethmoid.* Contrary to other siluriforms examined [State 0: e.g. Fig. 3.64], in catfish of CS–1 there is well-defined, strong ligament connecting the anterior surface of autopalatine and mesethmoid [State 1].
    — CS–0: Absence of well-defined, strong ligament connecting anterior surface of autopalatine and mesethmoid **(all genera not in other CS)**
    — CS–1: Presence of well-defined, strong ligament connecting anterior surface of autopalatine and mesethmoid ***(Phractura, Doumea, Belonoglanis, Trachyglanis, Andersonia)***
306. *Presence of accessory tooth-plate on ethmoid region (character inspired from Gosline, 1975; Mo, 1991; Arratia, 1992).* Plesiomorphically catfish seemingly lack accessory tooth-plates in the ethmoid region [State 0: e.g. Fig. 3.73]. Such accessory tooth-plates are present in siluriforms of CS–1, however [State 1: e.g. Fig. 4.27a].
    — CS–0: Absence of accessory tooth-plates in ethmoid region **(all genera not in other CS)**
    — CS–1: Presence of accessory tooth-plates in ethmoid region ***(Pseudoplatystoma, Chrysichthys, Schilbe, Pangasius, Clarotes, Genidens)***
307. *Presence of ligamentous connection between anterior margin of suspensorium and ethmoid region (character inspired from Schaefer, 1990).* Although the plesiomorphic condition for siluriforms, present in most of these fishes, is to present a strong ligamentous connection between the anterior margin of the suspensorium and the ethmoid region [State 0: e.g. Fig. 3.66], such a ligamentous connection is absent in catfish of CS–1 [State 1].
    — CS–0: Presence of ligamentous connection between anterior margin of suspensorium and ethmoid region **(all genera not in other CS)**
    — CS–1: Absence of ligamentous connection between anterior margin of suspensorium and ethmoid region ***(Chaca, Loricaria, Hypoptopoma, Lithoxus, Scoloplax, Astroblepus)***
308. *Sesamoid bone 2 of suspensorium.* The plesiomorphic condition for siluriforms is seemingly that in which the sesamoid bone 2 of the suspensorium (see terminology of Diogo and Chardon, 2001a) is present (see, e.g., Alexander, 1965; Gosline, 1975) [State 0: e.g. Fig. 3.73]. In catfish of CS–2 this bone is absent [State 1: e.g. Fig. 3.3].
    — CS–0: Presence of sesamoid bone 2 of the suspensorium ***(Diplomystes, Bagrichthys, Bagrus, Hemibagrus, Chrysichthys, Clarotes, Genidens, Arius, Auchenoglanis, Pimelodus, Calophysus, Austroglanis)***
    — CS–1: Absence of sesamoid bone 2 of the suspensorium **(all genera not in other CS)**
    — ?: Since it was not possible to properly appraise this character in the specimens examined ***(Malapterurus)***
309. *Sesamoid bone 1 of suspensorium.* The plesiomorphic condition for siluriforms is seemingly that in which the sesamoid bone 1 of the

suspensorium (sensu Diogo and Chardon, 2001a) is present (see, e.g., Alexander, 1965; Gosline, 1975) [: e.g. Fig. 3.73]. In specimens of genera of CS–1 examined this bone is absent [State 1: e.g. Fig. 3.3].

— CS–0: Presence of sesamoid bone 1 of the suspensorium **(all genera not in other CS)**

— CS–1: Absence of sesamoid bone 1 of the suspensorium **(*Rita, Synodontis, Mochokus, Trichomycterus, Hatcheria, Callichthys, Corydoras, Loricaria, Hypoptopoma, Lithoxus, Scoloplax, Astroblepus*)**

310. *Presence of anterolateral projection of sesamoid bone 1 of suspensorium (character inspired from Lundberg and McDade, 1986)*. Contrary to other catfish examined [State 0: e.g. Fig. 3.73], in the pseudopimelodins analysed the sesamoid bone 1 of the suspensorium exhibits a prominent anterolateral projection, which confers a very peculiar, characteristic shape to this bone [State 1].

— CS–0: Absence of prominent anterolateral projection of sesamoid bone 1 of the suspensorium **(all genera not in other CS)**

— CS–1: Presence of prominent anterolateral projection of sesamoid bone 1 of the suspensorium **(*Pseudopimelodus, Microglanis*)**

— Inapplicable: Since there is no sesamoid bone 1 of the suspensorium **(*Rita, Synodontis, Mochokus, Trichomycterus, Hatcheria, Callichthys, Corydoras, Loricaria, Hypoptopoma, Lithoxus, Scoloplax, Astroblepus*)**

311. *Elongation of sesamoid bone 1 of the suspensorium (character inspired from Gosline, 1975; Ghiot, 1978)*. Contrary to other catfish examined [State 0: e.g. Fig. 3.73], in specimens of the three genera of CS–1 analysed the sesamoid bone 1 of the suspensorium is a peculiarly thin, elongated the structure [State 1].

— CS–0: Sesamoid bone 1 of the suspensorium not peculiarly thin, elongated **(all genera not in other CS)**

— CS–1: Sesamoid bone 1 of the suspensorium a peculiarly thin, elongated structure **(*Pimelodus, Calophysus, Hypophthalmus*)**

— Inapplicable: Since there is no sesamoid bone 1 of the suspensorium **(*Rita, Synodontis, Mochokus, Trichomycterus, Hatcheria, Callichthys, Corydoras, Loricaria, Hypoptopoma, Lithoxus, Scoloplax, Astroblepus*)**

312. *Sesamoid bone 1 of suspensorium presenting peculiar L-shape (ordered character state) (character inspired from Chen and Lundberg, 1994; de Pinna, 1996)*. Contrary to other catfish examined [State 0: e.g. Fig. 3.73], in specimens of *Akysis and Parakysis* examined [State 1], and especially of *Amblyceps* and *Liobagrus* [State 2: e.g. Fig. 3.3], the sesamoid bone 1 of the suspensorium exhibits a very characteristic, peculiar L-shape.

— CS–0: Sesamoid bone 1 of suspensorium not presenting peculiar L-shape **(all genera not in other CS)**

— CS–1: Sesamoid bone 1 of suspensorium presenting peculiar L-shape (***Akysis, Parakysis***)
— CS–2: Sesamoid bone 1 of suspensorium a markedly L-shaped structure (***Amblyceps, Liobagrus***)
— Inapplicable: Since sesamoid bone 1 of suspensorium absent (***Rita, Synodontis, Mochokus, Trichomycterus, Hatcheria, Callichthys, Corydoras, Loricaria, Hypoptopoma, Lithoxus, Scoloplax, Astroblepus***)

313. *Ligamentous connection between anterior margin of suspensorium and ethmoid region (unordered multistate character) (character inspired from Arratia, 1992).* As described above, the vast majority of siluriforms present a strong ligamentous connection between the anterior margin of the suspensorium and the ethmoid region. The plesiomorphic situation for these fishes seems to be that in which the ligament runs from the suspensorium to attach on the prevomer, as is the case in diplomystids and the great majority of catfish [State 0: e.g. Fig. 3.66], while attachment occurs on the ventromesial [State 1: e.g. Fig. 3.3] or the ventrolateral [State 2] surfaces of the lateral ethmoid, representing thereby apomorphic states.
— CS–0: Ligament connecting anterior margin of suspensorium and ethmoid region attaching on prevomer **(all genera not in other CS)**
— CS–1: Ligament connecting anterior margin of suspensorium and ethmoid region attaching on ventromesial surface of lateral ethmoid (***Amphilius, Paramphilius, Bagrus, Hemibagrus, Liobagrus, Amblyceps, Pimelodus, Calophysus, Hypophthalmus, Pseudoplatystoma, Heptapterus, Goeldiella, Rhamdia, Doras, Acanthodoras, Anadoras, Centromochlus, Ageneiosus, Auchenipterus***)
— CS–2: Ligament connecting anterior margin of suspensorium and ethmoid region attaching on ventrolateral surface of lateral ethmoid (***Leptoglanis, Zaireichthys, Malapterurus***)
— ?: Since it was very difficult to discern this character in the specimens examined (***Franciscodoras***)
— Inapplicable: Since the prevomer is absent or profoundly modified (***Bunocephalus, Aspredo, Xyliphius, Microglanis Pseudopimelodus***), since the sesamoid bone 1 of the suspensorium is absent (***Rita, Synodontis, Mochokus, Trichomycterus, Hatcheria, Callichthys, Corydoras, Loricaria, Hypoptopoma, Lithoxus, Scoloplax, Astroblepus***), or since, as noted above, there is no ligamentous connection between the anterior margin of the suspensorium and the ethmoid region (***Chaca***)

314. *Sesamoid bone 1 presenting a markedly well-defined, rectangular shape (character inspired from Howes and Fumihito, 1991).* Plesiomorphically in siluriforms the sesamoid bone 1 of the suspensorium does not present a

markedly well-defined, rectangular shape [State 0: e.g. Fig. 3.73], but such is the case in specimens of genus *Silurus* analysed [State 1].

— CS–0: Sesamoid bone 1 of the suspensorium not presenting markedly well-defined, rectangular shape **(all genera not in other CS)**

— CS–1: Sesamoid bone 1 of the suspensorium presenting markedly well-defined, rectangular shape **(*Silurus*)**

— Inapplicable: Since sesamoid bone 1 of the suspensorium is absent **(*Rita, Synodontis, Mochokus, Trichomycterus, Hatcheria, Callichthys, Corydoras, Loricaria, Hypoptopoma, Lithoxus, Scoloplax, Astroblepus*)**

315. *Marked posterior bifurcation of sesamoid bone 1 of suspensorium (character inspired from Mo, 1991)*. As stated by Mo (1991), the posteriorly bifurcate, roughly ʌ-shaped sesamoid bone 1 of the suspensorium of *Austroglanis* [State 1: e.g. Fig. 3.34] is peculiarly shaped and differs from that of other catfishes [State 0: e.g. Fig. 3.73].

— CS–0: Sesamoid bone 1 of suspensorium not a posteriorly bifurcated, roughly ʌ-shaped structure **(all genera not in other CS)**

— CS–1: Sesamoid bone 1 of the suspensorium a posteriorly bifurcate, roughly ʌ-shaped structure **(*Austroglanis*)**

— Inapplicable: Since the sesamoid bone 1 of the suspensorium is lacking **(*Rita, Synodontis, Mochokus, Trichomycterus, Hatcheria, Callichthys, Corydoras, Loricaria, Hypoptopoma, Lithoxus, Scoloplax, Astroblepus*)**

316. *Anterior extension of ligament connecting anterior margin of suspensorium and ethmoid region*. In *Nematogenys* [State 1: e.g. Fig. 3.89] the attachment of the ligament connecting the anterior margin of the suspensorium to the ethmoid region lies more anteriorly than in other siluriforms examined [State 0: e.g. Fig. 3.73], with this attachment lying significantly anterior to the anterior margin of the lateral ethmoid.

— CS–0: Ligament connecting anterior margin of suspensorium and ethmoid region not attaching significantly anterior to anterior margin of lateral ethmoid **(all genera not in other CS)**

— CS–1: Ligament connecting anterior margin of suspensorium and ethmoid region attaching significantly anterior to anterior margin of lateral ethmoid **(*Nematogenys*)**

— Inapplicable: Since there is no ligamentous connection between anterior margin of suspensorium and ethmoid region **(*Chaca, Loricaria, Hypoptopoma, Lithoxus, Scoloplax, Astroblepus*)**

317. *Presence of sesamoid bone 3 of suspensorium*. The plesiomorphic condition for siluriforms seems to be that in which the sesamoid bone 3 of the suspensorium (see terminology of Diogo et al., 2001a) is absent [State 0: e.g. Fig. 3.89], with this bone being uniquely present in specimens of *Diplomystes* examined [State 1: e.g. Fig. 3.73] (see, e.g., Arratia, 1987; Diogo et al., 2001a).

— CS–0: Absence of sesamoid bone 3 of suspensorium **(all genera not in other CS)**

— CS–1: Presence of sesamoid bone 3 of suspensorium **(*Diplomystes*)**

318. *Double articulation between anterior part of suspensorium and ethmoid region (character inspired from Oliveira et al., 2001).* As described by Oliveira et al. (2001), in plotosids, but also in *Malapterurus*, contrary to other catfish [State 0: e.g. Fig. 3.66], a double articulation occurs between the anterior part of the suspensorium and the ethmoid region [State 1: e.g. Fig. 3.112].
    - — CS–0: Absence of double articulation between anterior part of suspensorium and ethmoid region **(all genera not in other CS)**
    - — CS–1: Presence of double articulation between anterior part of suspensorium and ethmoid region ***(Plotosus, Neosilurus, Paraplotosus, Cnidoglanis, Malapterurus)***
319. *Presence of entoectopterygoid (character inspired from Friel, 1994).* Contrary to other catfish examined [State 0: e.g. Fig. 3.66], in *Aspredo* the entoectopterygoid is absent [State 1].
    - — CS–0: Presence of entoectopterygoid **(all genera not in other CS)**
    - — CS–1: Absence of entoectopterygoid ***(Aspredo)***
320. *Arrow-shaped entoectopterygoid.* The peculiarly, somewhat arrow-shaped entoectopterygoid present in the specimens examined of *Austroglanis* [State 1: e.g. Fig. 3.34] differs markedly from that of other teleosts [State 0: e.g. Fig. 3.63].
    - — CS–0: Entoectopterygoid not a peculiarly, arrow-shaped structure **(all genera not in other CS)**
    - — CS–1: Entoectopterygoid a peculiarly, arrow-shaped structure ***(Austroglanis)***
    - — Inapplicable: Since the entoectopterygoid is absent ***(Aspredo)***
321. *Presence of firm, long suture between dorsal surface of entoectopterygoid and neurocranium (character inspired from Arratia, 1990).* Contrary to all other catfish examined [State 0: e.g. Fig. 3.63], in the loricariids analysed there is a marked, long, firm suture between the dorsal surface of the entoectopterygoid and the neurocranium [State 1].
    - — CS–0: Absence of long, firm suture between dorsal surface of entoectopterygoid and neurocranium **(all genera not in other CS)**
    - — CS–1: Presence of long, firm suture between dorsal surface of entoectopterygoid and neurocranium ***(Loricaria, Hypoptopoma, Lithoxus)***
    - — Inapplicable: Since the entoectopterygoid is absent ***(Aspredo)***
322. *Presence of voluminous, globular structure on anteroventral surface of entoectopterygoid.* Contrary to all other catfish examined [State 0: e.g. Fig. 3.66], in *Astroblepus* there is a voluminous, globular structure on the anteroventral surface of the entoectopterygoid, in which attaches a strong ligament connecting this bone to the premaxilla [State 1].
    - — CS–0: Absence of voluminous, globular structure on anteroventral surface of entoectopterygoid **(all genera not in other CS)**
    - — CS–1: Presence of voluminous, globular structure on anteroventral surface of entoectopterygoid ***(Astroblepus)***
    - — Inapplicable: Since the entoectopterygoid is absent ***(Aspredo)***

323. *Presence of well-defined, deep, anteroposteriorly elongated concavity formed by frontal and lateral ethmoid to receive anteromedial surface of entoectopterygoid.* Uniquely in *Cranoglanis* [State 1: e.g. Fig. 3.62], a well-defined, deep, anteroposteriorly elongated concavity is formed by the frontal and lateral ethmoid to receive the anteromesial surface of the entoectopterygoid, with the attachment between this latter bone and the neurocranium lying further dorsally (almost at the level of the cranial roof) than in other siluriforms [State 0: e.g. Fig. 3.63].
   - CS–0: Absence of well-defined, deep, anteroposteriorly elongated concavity on frontal and lateral ethmoid to receive entoectopterygoid **(all genera not in other CS)**
   - CS–1: Presence of well-defined, deep, anteroposteriorly elongated concavity on frontal and lateral ethmoid to receive entoectopterygoid **(*Cranoglanis*)**
   - Inapplicable: Since the entoectopterygoid is absent **(*Aspredo*)**
324. *Presence of teeth on entoectopterygoid (character inspired from Arratia, 1992).* Contrary to all other catfish examined [State 0: e.g. Fig. 3.66], specimens of the two genera of CS–1 analysed present teeth on the anteroventral surface of the entoectopterygoid [State 1: e.g. Fig. 4.27A].
   - CS–0: Absence of teeth on anteroventral surface of entoectopterygoid **(all genera not in other CS)**
   - CS–1: Presence of teeth on anteroventral surface of entoectopterygoid **(*Chrysichthys, Pseudoplatystoma*)**
   - Inapplicable: Since the entoectopterygoid is absent **(*Aspredo*)**
325. *Shape of entoectopterygoid (ordered multistate character) (character inspired from He et al., 1999).* In most plesiomorphic catfish the entoectopterygoid is a well-developed, large bone [State 0: e.g. Fig. 3.66]. In siluriforms of CS–1 this bone is relatively reduced in size but markedly larger, however, than sesamoid bone 1 of the suspensorium [State 1]. A different configuration is found in *Leptoglanis* in which the entoectopterygoid is significantly more reduced in size than in catfish of CS–1, with a significant part of its medial surface surrounded by the lateral margin of sesamoid bone 1 of the suspensorium [State 2]. In siluriforms of CS–3 the entoectopterygoid is notably reduced in size, with its medial surface completely surrounded by lateral margin of this sesamoid bone [State 3: e.g. Fig. 3.3], while reduction of the entoectopterygoid is especially marked in specimens of genus *Andersonia* [State 4].
   - CS–0: Entoectopterygoid well developed **(all genera not in other CS)**
   - CS–1: Entoectopterygoid relatively reduced in size **(*Zaireichthys, Akysis, Auchenoglanis*)**
   - CS–2: Entoectopterygoid significantly more reduced than in CS–1, with significant part of medial surface surrounded by lateral margin of sesamoid bone 1 of suspensorium **(*Leptoglanis*)**
   - CS–3: Entoectopterygoid notably reduced in size, with medial surface completely surrounded by lateral surface of sesamoid bone 1 of

suspensorium **(*Phractura, Doumea, Belonoglanis, Trachyglanis, Glyptothorax, Bagarius, Liobagrus, Amblyceps, Erethistes, Hara, Parakysis*)**
- CS–4: Entoectopterygoid especially reduced to a small, somewhat oval structure **(*Andersonia*)**
- ?: Since it was very difficult to appraise this character in the specimens examined **(*Bunocephalus, Xyliphius*)**
- Inapplicable: Since the entoectopterygoid is absent **(*Aspredo*)**

326. *Entoectopterygoid elongate, roughly rectangular in shape (character inspired from de Pinna and Vari, 1995)*. As stated by de Pinna and Vari (1995), in cetopsids, contrary to other catfish [State 0: e.g. Fig. 3.66], the entoectopterygoid is a markedly elongate, roughly rectangular structure [State 1: e.g. Fig. 3.46].
- CS–0: Entoectopterygoid not a particularly elongate, roughly rectangular structure **(all genera not in other CS)**
- CS–1: Entoectopterygoid a markedly elongate, roughly rectangular structure **(*Cetopsis, Hemicetopsis, Helogenes*)**
- Inapplicable: Since the entoectopterygoid is absent **(*Aspredo*)**

327. *Contact between hyomandibulo-metapterygoid and entoectopterygoid (character inspired from Mo, 1991; Arratia, 1992)*. Plesiomorphically in catfish the hyomandibulo-metapterygoid comes into contact with the entoectopterygoid [State 0: e.g. Fig. 3.66], but in specimens of genera of CS–1 examined these structures do not come into contact [State 1: e.g. Fig. 3.30].
- CS–0: Hyomandibulo-metapterygoid and entoectopterygoid in contact **(all genera not in other CS)**
- CS–1: Hyomandibulo-metapterygoid and entoectopterygoid not in contact **(*Clarias, Uegitglanis, Heterobranchus, Heteropneustes, Rita, Chrysichthys, Clarotes, Pseudoplatystoma, Heptapterus, Goeldiella, Rhamdia, Doras, Centromochlus, Ageneiosus*)**
- ?: Since it was not possible to properly observe this character in the specimens examined **(*Malapterurus*)**
- Inapplicable: Since the entoectopterygoid is absent **(*Aspredo*)**

328. *Presence of prominent posterodorsal crest of entoectopterygoid (character inspired from Vidthayanon, 1992)*. Contrary to other catfish examined [State 0: e.g. Fig. 3.63], in *Pangasius* and *Helicophagus* the entoectopterygoid exhibits a prominent posterodorsal crest which, together with an also prominent anterodorsal crest of the hyomandibulo-metapterygoid, constitutes a particularly elongate crest for attachment of the levator arcus palatini [State 1: e.g. Fig. 3.99].
- CS–0: Absence of prominent posterodorsal crest of entoectopterygoid **(all genera not in other CS)**
- CS–1: Presence of prominent posterodorsal crest of entoectopterygoid **(*Pangasius, Helicophagus*)**
- Inapplicable: Since the entoectopterygoid is absent **(*Aspredo*)**

329. *Entoectopterygoid a boomerang-shaped structure (character inspired from Schaefer, 1990)*. Contrary to all other catfish examined [State 0: e.g. Fig. 3.66], in specimens of *Scoloplax* analysed the entoectopterygoid exhibits a unique, boomerang-shaped structure [State 1].
   — CS–0: Entoectopterygoid not a unique, boomerang-shaped structure **(all genera not in other CS)**
   — CS–1: Entoectopterygoid a unique, boomerang-shaped structure **(*Scoloplax*)**
   — Inapplicable: Since the entoectopterygoid is absent **(*Aspredo*)**
330. *Quadrato-symplectic pierced (CI=0.5, RI=0)*. Contrary to other siluriforms examined [State 0: e.g. Fig. 3.66], in specimens examined of *Doumea* and *Phractura* the quadrato-symplectic (see terminology of Diogo et al., 2001a) is pierced by several small, circular foramina [State 1: e.g. Fig. 3.16].
   — CS–0: Quadrato-symplectic not pierced by several small, circular foramina **(all genera not in other CS)**
   — CS–1: Quadrato-symplectic pierced by several small, circular foramina **(*Phractura, Doumea*)**
331. *Mesial expansion of articulatory surface of quadrato-symplectic for mandible (ordered multistate character)*. Plesiomorphically in catfish the articulatory surface of the quadrato-symplectic for the mandible is basically directed anteroventrally [State 0: e.g. Fig. 3.63], but in *Cetopsis* and *Bunochepalus* [State 1], and especially in *Hemicetopsis* [State 2], this articulatory surface is markedly expanded mesially.
   — CS–0: Articulatory surface of quadrato-symplectic not presenting mesial expansion **(all genera not in other CS)**
   — CS–1: Articulatory surface of quadrato-symplectic presenting mesial expansion **(*Cetopsis, Bunocephalus*)**
   — CS–2: Mesial expansion of articulatory surface of quadrato-symplectic more pronounced than in CS–1 **(*Hemicetopsis*)**
332. *Presence of fossa or foramen on anteroventromesial surface of quadrato-symplectic (ordered multistate character) (character inspired from Mo, 1991)*. Plesiomorphically catfish lack major fossas or foramens on the anteroventromesial surface of the quadrato-symplectic [State 0: e.g. Fig. 3.66]. In siluriforms of CS–1, however, a deep fossa occurs on the anteroventromesial surface of this bone [State 1: e.g. Fig. 3.34]. In catfish of CS–2 the anteroventral surface of the quadrato-symplectic is completely pierced by a well-developed foramen [State 2: e.g. Fig. 3.116].
   — CS–0: Quadrato-symplectic lacking major fossas or foramens on anteroventromesial surface **(all genera not in other CS)**
   — CS–1: Quadrato-symplectic presenting deep fossa on anteroventromesial surface **(*Bagrus, Hemibagrus, Paramphilius, Heptapterus, Austroglanis*)**
   — CS–2: Anteroventral surface of quadrato-symplectic completely pierced by well-developed foramen **(*Schilbe, Goeldiella, Rhamdia, Pseudopimelodus, Microglanis, Ageneiosus*)**

— ?: Since it was not possible to properly appraise this character in the specimens examined **(*Rita*)**

333. *Suture between hyomandibulo-metapterygoid and neurocranium (character inspired from Schaefer, 1987).* Uniquely in *Lithoxus* [State 1] among the catfish examined [State 0: e.g. Fig. 3.63], a long, firm suture lies between the posterior margin of the hyomandibulo-metapterygoid and the neurocranium.
   — CS–0: Absence of suture between hyomandibulo-metapterygoid and neurocranium **(all genera not in other CS)**
   — CS–1: Presence of suture between hyomandibulo-metapterygoid and neurocranium **(*Lithoxus*)**

334. *Development of anterodorsal spine of hyomandibulo-metapterygoid (ordered multistate character).* In catfish the anterodorsal spine of the hyomandibulo-metapterygoid, if present, is characteristically a thin, sharply-pointed structure [State 0: e.g. Fig. 3.63], but in siluriforms of CS–1 [State 1], and especially of CS–2 [State 2: e.g. Fig. 3.29], this spine is a particularly developed, enlarged structure.
   — CS–0: Anterodorsal spine of hyomandibulo-metapterygoid not a particularly developed, enlarged structure **(all genera not in other CS)**
   — CS–1: Anterodorsal spine of hyomandibulo-metapterygoid a particularly developed, enlarged structure **(*Ictalurus, Amiurus, Franciscodoras, Anadoras, Acanthodoras, Ageneiosus, Auchenipterus*)**
   — CS–2: Anterodorsal spine of hyomandibulo-metapterygoid more developed and enlarged than in CS–1 **(*Bunocephalus, Xyliphius, Centromochlus*)**

335. *Orientation of crest of hyomandibulo-metapterygoid for levator arcus palatini (character inspired from Howes and Fumihito, 1991).* Contrary to other catfish examined [State 0: e.g. Fig. 3.99], in the silurids analysed the lateral crest of the hyomandibulo-metapterygoid for attachment of the muscle levator arcus palatini is not essentially oriented anteroposteriorly or obliquely, but rather dorsoventrally [State 1].
   — CS–0: Crest of hyomandibulo-metapterygoid for levator arcus palatini essentially oriented anteroposteriorly or obliquely **(all genera not in other CS)**
   — CS–1: Crest of hyomandibulo-metapterygoid for levator arcus palatini essentially oriented dorsoventrally **(*Silurus, Wallago*)**

336. *Deep dorsal concavity of hyomandibulo-metapterygoid to receive ventral process of sphenotic (character inspired from Adriaens and Verraes, 1998).* Contrary to all other catfish examined [State 0: e.g. Fig. 3.68], in siluriforms of CS–1 the hyomandibulo-metapterygoid exhibits a deep dorsal concavity in which a prominent ventral process of the sphenotic is lodged [State 1].

— CS–1: Presence of deep dorsal concavity of hyomandibulo-metapterygoid to receive ventral process of sphenotic **(*Clarias, Heterobranchus*)**
— ?: Since it was not possible to properly examine this character in the specimens studied (*Uegitglanis*)

337. *Shape of neurocranial concavity for hyomandibulo-metapterygoid (unordered multistate character)*. Plesiomorphically in siluriforms the neurocranium presents a somewhat deep, thin, elongated concavity to articulate with the dorsal surface of the hyomandibulo-metapterygoid [State 0: e.g. Fig. 3.66]. However, in specimens examined of genus *Arius* the neurocranium exhibits a markedly developed, volcano-shaped fossa to receive the dorsal surface of the hyomandibulo-metapterygoid [State 1]. A different configuration is found in *Chaca*, in which there is a remarkably deep, large, roundish neurocranial concavity to receive the dorsal surface of this bone [State 2].
— CS–0: Somewhat deep, thin, elongated neurocranial concavity to articulate with dorsal surface of hyomandibulo-metapterygoid **(all genera not in other CS)**
— CS–1: Markedly developed, volcano-shaped fossa to receive dorsal surface of hyomandibulo-metapterygoid **(*Arius*)**
— CS–2: Remarkably deep, large, roundish neurocranial concavity to receive dorsal surface of hyomandibulo-metapterygoid **(*Chaca*)**

338. *Presence of prominent posterodorsal projection of hyomandibulo-metapterygoid firmly attached to neurocranium by strong connective tissue (ordered multistate character)*. Contrary to other catfish [State 0: e.g. Fig. 3.63], in specimens of *Trichomycterus* and *Hatcheria* examined [State 1], and especially of *Nematogenys* [State 2: e.g. Fig. 3.88], the hyomandibulo-metapterygoid exhibits a markedly developed, elongated posterodorsal projection, which is firmly attached to the neurocranium by means of massive, strong connective tissue.
— CS–0: Absence of markedly developed, elongated posterodorsal projection of hyomandibulo-metapterygoid firmly attaching to neurocranium by massive, strong connective tissue **(all genera not in other CS)**
— CS–1: Presence of markedly developed, elongated posterodorsal projection of hyomandibulo-metapterygoid firmly attaching to neurocranium by massive, strong connective tissue **(*Trichomycterus, Hatcheria*)**
— CS–2: Presence of particularly well-developed, elongated posterodorsal projection of hyomandibulo-metapterygoid firmly attaching to neurocranium by massive, strong connective tissue **(*Nematogenys*)**

339. *Development of opercular and interopercular bones (character inspired from de Pinna, 1998)*. Peculiarly in those catfish of CS–1 [State 1], and contrarily

to the other siluriforms examined [State 0: e.g. Fig. 3.63], the opercle, as well as the interopercle (when present), are particularly compact, voluminous, thick structures.

— CS–0: Opercle and interopercle (when present) not particularly compact, voluminous, thick structures **(all genera not in other CS)**
— CS–1: Opercle and interopercle (when present) particularly compact, voluminous, thick structures **(*Trichomycterus, Hatcheria*)**

340. *Presence of prominent, posteriorly pointed posterodorsal projection of opercle (ordered multistate character)*. Contrary to other catfish examined [State 0: e.g. Fig. 3.63], in specimens of genera of CS–1 examined the opercle exhibits a markedly developed, elongated, posteriorly pointed posterodorsal projection [State 1: e.g. Fig. 3.82].

— CS–0: Absence of markedly developed, elongated, posteriorly pointed posterodorsal projection of opercle **(all genera not in other CS)**
— CS–1: Presence of markedly developed, elongated, posteriorly pointed posterodorsal projection of opercle **(*Clarias, Uegitglanis, Heterobranchus, Heteropneustes*)**

341. *Ossification of opercle*. Peculiarly in catfish of CS–1 [State 1: e.g. Fig. 3.17], and contrarily in the other siluriforms examined [State 0: e.g. Fig. 3.63], the opercle remains a notably thin, poorly ossified structure in the adults.

— CS–0: Well-ossified opercle in adult specimens **(all genera not in other CS)**
— CS–1: Notably poorly ossified opercle in adult specimens **(*Amphilius, Paramphilius, Leptoglanis, Cetopsis, Helogenes, Hemicetopsis*)**
— ?: Since it was not possible to properly appraise this character in the specimens examined **(*Zaireichthys*)**

342. *Opercle presenting a characteristic L-shape (ordered multistate character) (character inspired from Friel, 1994)*. As described by Friel (1994), contrary to other catfish analysed [State 0: e.g. Fig. 3.63], in specimens of *Bunocephalus* and *Xyliphius* examined [State 1: e.g. Fig. 3.25], and especially of *Aspredo* [State 2], the opercle is characteristically compressed anteroposteriorly into a roughly L-shaped structure.

— CS–0: Opercle not an L-shaped structure **(all genera not in other CS)**
— CS–1: Opercle an L-shaped structure **(*Bunocephalus, Xyliphius*)**
— CS–2: Anteroposterior compression and characteristic L-shaped structure of opercle even more pronounced than in CS–1 **(*Aspredo*)**

343. *Distance from anteroventral surface of opercle and posterior margin of suspensorium*. In the plesiomorphic condition, the anteroventral surface of the opercle is not significantly posterior to the posterior margin of the suspensorium [State 0: e.g. Fig. 3.63], while in catfish of CS–1 the anteroventral surface of the opercle is considerably distant from the posterior margin of the suspensorium [State 1] (see Diogo, 2003b).

— CS–0: Anteroventral surface of opercle not considerably distant from posterior margin of suspensorium **(all genera not in other CS)**

— CS–1: Anteroventral surface of opercle considerably distant from posterior margin of suspensorium **(*Phractura, Doumea, Belonoglanis, Andersonia, Trachyglanis, Zaireichthys*)**

344. *Reduction of size of opercle*. Plesiomorphically in catfish the opercle is a well-developed, broad bone considerably larger than the interopercle [State 0: e.g. Fig. 3.63], but in *Zaireichthys* the opercle is markedly reduced in size, inclusively slightly narrower than the interopercle [State 1].
   — CS–0: Opercle not markedly reduced in overall size **(all genera not in other CS)**
   — CS–1: Opercle markedly reduced in overall size **(*Zaireichthys*)**
345. *Lap joint present between opercle and interopercle (character inspired from de Pinna and Vari, 1995)*. Contrary to other catfish examined [State 0: e.g. Fig. 3.63], in cetopsids the opercle and interopercle partly overlap each other along modified laminar surfaces [State 1: e.g. Fig. 3.43].
   — CS–0: Opercle and interopercle not partly overlapping each other along modified laminar surfaces **(all genera not in other CS)**
   — CS–1: Opercle and interopercle partly overlapping each other along modified laminar surfaces **(*Cetopsis, Hemicetopsis, Helogenes*)**
   — Inapplicable: Since in the specimens examined of these genera the interopercle is missing **(*Loricaria, Hypoptopoma, Lithoxus, Scoloplax, Astroblepus, Aspredo*)**
346. *Orientation of articulation between opercle and hyomandibulo-metapterygoid (character inspired from de Pinna, 1993)*. Contrary to other siluriforms examined [State 0: e.g. Fig. 3.63], in catfish of CS–1 the opercle articulation with the hyomandibulo-metapterygoid is not essentially oriented anteriorly, but rather dorsally [State 1: e.g. Fig. 3.82].
   — CS–0: Opercular articulation with hyomandibulo-metapterygoid essentially oriented anteriorly **(all genera not in other CS)**
   — CS–1: Opercular articulation with hyomandibulo-metapterygoid essentially oriented dorsally **(*Clarias, Uegitglanis, Heterobranchus, Heteropneustes, Loricaria, Hypoptopoma, Lithoxus, Astroblepus*)**
347. *Presence of well-developed, large, roughly roundish dorsal laminar projection of opercle*. Contrary to other catfish examined [State 0: e.g. Fig. 3.63], in *Neosilurus* the opercle exhibits a markedly developed, large, roundish dorsal laminar projection [State 1: e.g. Fig. 3.112].
   — CS–0: Absence of markedly developed, large, roundish dorsal laminar projection of opercle **(all genera not in other CS)**
   — CS–1: Presence of markedly developed, large, roundish dorsal laminar projection of opercle **(*Neosilurus*)**
348. *Elongation of articulation between opercle and hyomandibulo-metapterygoid*. In catfish of CS–1 [State 1: e.g. Fig. 3.62] the articulatory surface of the hyomandibulo-metapterygoid for the opercular bone is markedly more elongated than in other siluriforms [State 0: e.g. Fig. 3.29].
   — CS–0: Articulatory surface of opercle for hyomandibulo-metapterygoid not markedly elongated **(all genera not in other CS)**

— CS–1: Markedly elongated articulatory surface of opercle for hyomandibulo-metapterygoid (***Cranoglanis, Centromochlus, Ictalurus, Liobagrus, Cetopsis, Hemicetopsis, Arius, Ancharius, Zaireichthys***)

349. *Double articulation between opercle and hyomandibulo-metapterygoid*. Contrary to the other catfish examined, in which there is a single articulatory facet of the opercle for the hyomandibulo-metapterygoid [State 0: e.g. Fig. 3.66], in *Phractura* the opercle exhibits two well-developed articulatory facets for this latter bone [State 1: e.g. Fig. 3.18].
   — CS–0: Absence of double articulation between opercle and hyomandibulo-metapterygoid **(all genera not in other CS)**
   — CS–1: Presence of double articulation between opercle and hyomandibulo-metapterygoid (***Phractura***)

350. *Presence of prominent, sharply pointed, dorsally oriented projection of opercle (ordered multistate character)*. Contrary to other catfish examined [State 0: e.g. Fig. 3.63], in siluriforms of CS–1 [State 1], and especially of CS–2 [State 2: e.g. Fig. 3.89] the opercular bone exhibits a particularly prominent, sharply pointed, dorsally oriented projection.
   — CS–0: Opercular bone not presenting particularly prominent, sharply pointed, dorsally oriented projection **(all genera not in other CS)**
   — CS–1: Opercular bone presenting particularly prominent, sharply pointed, dorsally oriented projection (*Corydoras*)
   — CS–2: Sharply-pointed, dorsally oriented projection of opercle more developed than in CS–1 (***Nematogenys, Trichomycterus, Hatcheria***)

351. *Size of interopercle (ordered multistate character) (character inspired from Schaefer, 1988)*. Plesiomorphically in catfish the interopercle is a well-developed, relatively large bone [State 0: e.g. Fig. 3.66]. In siluriforms of CS–1 the interopercle is markedly reduced in size [State 1], with this bone completely absent in siluriforms of CS–2 [State 2].
   — CS–0: Interopercle present, well-developed **(all genera not in other CS)**
   — CS–1: Interopercle markedly reduced in size (***Xyliphius, Bagrichthys***)
   — CS–2: Interopercle absent (***Aspredo, Loricaria, Hypoptopoma, Lithoxus***)

352. *Absence of ligamentous connection between interopercle and mandible (character inspired from Schaefer and Lauder, 1986; de Pinna, 1998)*. Contrary to all other catfish analysed [State 0: e.g. Fig. 3.69], in specimens examined of *Astroblepus* there is no ligamentous connection between the interopercle and the mandible [State 1].
   — CS–0: Presence of ligamentous connection between interopercle and mandible **(all genera not in other CS)**
   — CS–1: Absence of ligamentous connection between interopercle and mandible (***Astroblepus***)
   — Inapplicable: Since the interopercle is absent (***Aspredo, Loricaria, Hypoptopoma, Lithoxus***)

353. *Presence of deep anterior concavity of interopercle for articulation with preopercle.* Contrary to all other catfish analysed [State 0: e.g. Fig. 3.63], in specimens of *Chaca* examined the interopercular bone exhibits a deep anterior concavity for articulation with preopercle [State 1: e.g. Fig. 3.49].
   — CS–0: Interopercle not presenting deep anterior concavity for articulation with preopercle **(all genera not in other CS)**
   — CS–1: Interopercle presenting deep anterior concavity for articulation with preopercle **(*Chaca*)**
   — Inapplicable: Since the interopercle is absent **(*Aspredo, Loricaria, Hypoptopoma, Lithoxus*)**
354. *Presence of large foramen on posteromesial surface of interopercle.* Contrary to other catfish [State 0: e.g. Fig. 3.47], in *Austroglanis* the interopercle exhibits a large, deep fossa on its posteromesial surface [State 1: e.g. Fig. 3.37].
   — CS–0: Interopercle not presenting large foramen on posteromesial surface **(all genera not in other CS)**
   — CS–1: Interopercle presenting large foramen on posteromesial surface **(*Austroglanis*)**
   — Inapplicable: Since the interopercle is lacking **(*Aspredo, Loricaria, Hypoptopoma, Lithoxus*)**
355. *Presence of well-defined, long, strong ligament running from anterodorsal surface of interopercle and posterodorsal surface of preopercle.* Contrary to other catfish [State 0: e.g. Fig. 3.63], in specimens examined of the four genera of CS–1, the anterodorsal surface of the interopercle is connected by a well-defined, long, strong ligament to the posterodorsal surface of the preopercle [State 1: e.g. Fig. 3.49].
   — CS–0: Anterodorsal surface of interopercle not connected by well-defined, long, strong ligament to posterodorsal surface of preopercle **(all genera not in other CS)**
   — CS–1: I Anterodorsal surface of interopercle connected by well-defined, long, strong ligament to posterodorsal surface of preopercle **(*Chaca, Nematogenys, Trichomycterus, Hatcheria*)**
   — Inapplicable: Since the interopercle is lacking **(*Aspredo, Loricaria, Hypoptopoma, Lithoxus*)**
356. *Presence of prominent anterodorsal expansion of interopercle.* Contrary to other catfish [State 0: e.g. Fig. 3.63], in specimens of *Trichomycterus* and *Hatcheria* examined the interopercle exhibits a large, prominent anterodorsal expansion, which extends markedly anterior to the posterior margin of the preopercle [State 1].
   — CS–0: Interopercle not presenting large, prominent anterodorsal expansion **(all genera not in other CS)**
   — CS–1: Interopercle presenting large, prominent anterodorsal expansion **(*Trichomycterus, Hatcheria*)**
   — Inapplicable: Since the interopercle is absent **(*Aspredo, Loricaria, Hypoptopoma, Lithoxus*)**

357. *Presence of prominent, anteriorly pointed anteroventral projection of interopercle.* Contrary to other catfish [State 0: e.g. Fig. 3.63], in specimens of the two genera of CS–1 analysed the interopercle exhibits a prominent, anteriorly pointed anteroventral projection [State 1].
   — CS–0: Interopercle not presenting prominent, anteriorly pointed anteroventral projection **(all genera not in other CS)**
   — CS–1: Interopercle presenting prominent, anteriorly pointed anteroventral projection ***(Pseudoplatystoma, Hypophthalmus)***
   — Inapplicable: Since the interopercle is lacking ***(Aspredo, Loricaria, Hypoptopoma, Lithoxus)***
358. *Articulatory facet of interopercle for posterior ceratohyal (unordered multistate character).* As explained in Diogo (2003b) plesiomorphically in siluriforms the interopercle lacks major well-developed, prominent medial articulatory facets for the posterior ceratohyal [State 0]. However, in catfish of CS–1 the interopercle exhibits a well-developed, prominent, circular, volcano-shaped anteromesial facet for the posterior ceratohyal [State 1: e.g. Fig. 3.18]. A different configuration is found in catfish of CS–2, in which there is a large, deep, roundish fossa on the mesial surface of the interopercle to receive the posterior ceratohyal [State 2: e.g. Fig. 3.37].
   — CS–0: Interopercle lacking major well-developed, prominent mesial articulatory facets for posterior ceratohyal **(all genera not in other CS)**
   — CS–1: Interopercle presenting well-developed, prominent, circular, volcano-shaped anteromesial facet for posterior ceratohyal ***(Phractura, Doumea, Belonoglanis, Andersonia, Trachyglanis)***
   — CS–2: Large, deep, roundish fossa on mesial surface of interopercle to receive posterior ceratohyal ***(Glyptothorax, Glyptosternon, Liobagrus, Akysis, Parakysis, Amblyceps, Synodontis, Mochokus, Austroglanis)***
   — Inapplicable: Since the interopercle is missing ***(Aspredo, Loricaria, Hypoptopoma, Lithoxus)***
359. *Double ligamentous connection between interopercle and mandible.* Peculiarly among the catfish examined [State 0: e.g. Fig. 3.69], in *Trichomycterus* and *Hatcheria* the interopercle and the mandible are connected not by a single strong ligament, but by two long, strong ligaments [State 1].
   — CS–0: Absence of double ligamentous connection between interopercle and mandible **(all genera not in other CS)**
   — CS–1: Presence of double ligamentous connection between interopercle and mandible ***(Trichomycterus, Hatcheria)***
   — Inapplicable: Since the interopercle is lacking ***(Aspredo, Loricaria, Hypoptopoma, Lithoxus)*** or since there is no connection between this bone and the mandible ***(Astroblepus)***
360. *Interopercle expanded along dorsoventral axis (ordered multistate character) (character inspired from de Pinna and Vari, 1995).* Contrary to other catfish

[State 0: e.g. Fig. 3.63], in *Hemicetopsis* and *Cetopsis* [State 1: e.g. Fig. 3.43], and especially in *Helogenes* [State 2], the interopercle is markedly expanded dorsoventrally.

— CS–0: Interopercle not markedly expanded dorsoventrally **(all genera not in other CS)**

— CS–1: Interopercle markedly expanded dorsoventrally **(*Hemicetopsis, Cetopsis*)**

— CS–2: Interopercle even more expanded dorsoventrally than in CS–1 **(*Helogenes*)**

— Inapplicable: Since the interopercle is absent **(*Aspredo, Loricaria, Hypoptopoma, Lithoxus*)**

361. *Length of ligament connecting interopercle and mandible.* Contrary to other catfish [e.g. Fig. 3.64], in *Belonoglanis* the ligament connecting the interopercle and the mandible is peculiarly, remarkably elongated anteroposteriorly [State 1].

— CS–0: Ligament connecting interopercle and mandible not peculiarly, remarkably elongated anteroposteriorly **(all genera not in other CS)**

— CS–1: Ligament connecting interopercle and mandible peculiarly, remarkably elongated anteroposteriorly **(*Andersonia*)**

— Inapplicable: Since the interopercle is lacking **(*Aspredo, Loricaria, Hypoptopoma, Lithoxus*)** or since there is no connection between this bone and the mandible **(*Astroblepus*)**

362. *Ligament connecting interopercle and mandible attaching on anterodorsal surface of interopercle (character inspired from de Pinna and Vari, 1995).* Contrary to all other siluriforms examined, in which the ligament connecting the interopercle and the mandible is essentially associated with the anteroventral margin of the interopercle [State 0: e.g. Fig. 3.64], in cetopsids this ligament is associated with the anterodorsal surface of the interopercle [State 1: e.g. Fig. 3.43].

— CS–0: Ligament connecting interopercle and mandible essentially associated with anteroventral margin of interopercle **(all genera not in other CS)**

— CS–1: Ligament connecting interopercle and mandible essentially associated with anterodorsal margin of interopercle **(*Cetopsis, Hemicetopsis, Helogenes*)**

— Inapplicable: Since the interopercle is missing **(*Aspredo, Loricaria, Hypoptopoma, Lithoxus*)** or since there is no connection between this bone and the mandible **(*Astroblepus*)**

363. *Presence of prominent posteroventral laminar projection of interopercle.* Contrary to other catfish [State 0: e.g. Fig. 3.63], in specimens of *Laides* analysed the interopercle exhibits a prominent posteroventral projection of laminar bone [State 1].

— CS–0: Interopercle not presenting prominent posteroventral projection of laminar bone **(all genera not in other CS)**

— CS–1: Interopercle presenting prominent posteroventral projection of laminar bone **(*Laides*)**
— Inapplicable: Since the interopercle is lacking **(*Aspredo, Loricaria, Hypoptopoma, Lithoxus*)**

364. *Presence of well-developed, quadrangular bone associating dorsally with interopercle (character inspired from Schaefer, 1988).* Contrary to all other catfish examined [State 0: e.g. Fig. 3.63], *Astroblepus* exhibits a well-developed, quadrangular bone situated ventral to the interopercle and associated dorsally to this latter bone [State 1] (this quadrangular bone was considered a modified interhyal by Schaefer (1988), but its position, its ligamentous connection with the opercular bone, and its association with the interopercle are completely atypical for an interhyal).
— CS–0: No well-developed, quadrangular bone situated ventral to interopercle and associated dorsally to this latter bone **(all genera not in other CS)**
— CS–1: Well-developed, quadrangular bone situated ventral to interopercle and associated dorsally to this latter bone **(*Astroblepus*)**
— Inapplicable: Since the interopercle is missing **(*Aspredo, Loricaria, Hypoptopoma, Lithoxus*)**

365. *Presence of anterodorsal process of preopercle.* Plesiomorphically in catfish the preopercle lacks major anterodorsal processes [State 0: e.g. Fig. 3.63], but in *Amphilius* and *Paramphilius* this bone exhibits a well-developed, triangular anterodorsal process pointed anteriorly [State 1: e.g. Fig. 3.6].
— CS–0: Preopercle not presenting well-developed, triangular anterodorsal process pointed anteriorly **(all genera not in other CS)**
— CS–1: Preopercle presenting well-developed, triangular anterodorsal process pointed anteriorly **(*Amphilius, Paramphilius*)**

366. *Presence of prominent anteroventral lamina of preopercle (ordered multistate character).* Contrary to other siluriforms [State 0: e.g. Fig. 3.63], in catfish of CS–1 [State 1], and especially of CS–2 [State 2: e.g. Fig. 3.25], the preopercle exhibits a well-developed anteroventral lamina, which covers part of the lateral fibres of the muscle adductor mandibulae.
— CS–0: Absence of well-developed anteroventral lamina of preopercle **(all genera not in other CS)**
— CS–1: Presence of well-developed anteroventral lamina of preopercle **(*Aspredo, Xyliphius, Franciscodoras, Anadoras, Acanthodoras, Doras*)**
— CS–2: Anteroventral lamina of preopercle more developed than in CS–1 **(*Bunocephalus*)**

367. *Shape of interhyal (character inspired from Mo, 1991).* Contrary to all other catfish examined [State 0: e.g. Fig. 3.47], in *Auchenoglanis* the interhyal is markedly developed, presenting unique alate laminae [State 1].
— CS–0: Interhyal, when present, a relatively small bone **(all genera not in other CS)**
— CS–1: Interhyal markedly developed, presenting unique alate laminae **(*Auchenoglanis*)**

368. *Overall shape of hyoid arch*. In the specimens of *Hypophthalmus* analysed [State 1], the hyoid arch is markedly thinner and more elongated than in other catfish examined [State 0: e.g. Fig. 3.69].
    — CS–0: Hyoid arch not markedly thinner and elongated **(all genera not in other CS)**
    — CS–1: Hyoid arch markedly thinner and elongated **(*Hypophthalmus*)**
369. *Size of posterior ceratohyal (ordered multistate character)*. Contrary to other catfish analysed, in which the posterior ceratohyal is a large, essentially triangular bone [State 0: e.g. Fig. 3.69], in *Doumea* and *Phractura* [State 1: e.g. Fig. 3.9], and especially in *Andersonia, Belonoglanis* and *Trachyglanis* [State 2: e.g. Fig. 3.14], the posterior ceratohyal is a highly modified structure considerably reduced in size.
    — CS–0: Posterior ceratohyal a large, essentially triangular bone **(all genera not in other CS)**
    — CS–1: Posterior ceratohyal essentially a quadrangular, stout, and relatively small bone **(*Doumea, Phractura*)**
    — CS–2: Posterior ceratohyal essentially a quadrangular, stout bone markedly reduced in size **(*Belonoglanis, Andersonia, Trachyglanis*)**
370. *Presence of prominent posterolateral projection of posterior ceratohyal*. Plesiomorphically in siluriforms the posterior ceratohyal lacks major projections or processes [State 0: e.g. Fig. 3.69], but in *Bunocephalus* this bone exhibits a prominent posterolateral projection [State 1: e.g. Fig. 3.27].
    — CS–0: Absence of prominent posterolateral projection of posterior ceratohyal **(all genera not in other CS)**
    — CS–1: Presence of prominent posterolateral projection of posterior ceratohyal **(*Bunocephalus*)**
371. *Contact between posterior ceratohyal and mesial surface of interopercle*. Plesiomorphically in siluriforms the posterior ceratohyal comes into contact with the mesial surface of the interopercle [State 0: e.g. Fig. 3.69]. In *Helogenes* these bones are not in contact [State 1].
    — CS–0: Posterior ceratohyal contacting mesial surface of interopercular **(all genera not in other CS)**
    — CS–1: Posterior ceratohyal not contacting mesial surface of interopercular **(*Helogenes*)**
372. *Orientation of posterior ceratohyal*. Contrary to all the other catfish examined [State 0: e.g. Fig. 3.69], *Chaca* exhibits a particularly peculiar, unusual configuration of the hyoid arch, with orientation of the posterior ceratohyal somewhat perpendicular to that of both the ventral hypohyal and the anterior ceratohyal in a tridimensional axis [State 1].
    — CS–0: Orientation of posterior ceratohyal not perpendicular to that of both ventral hypohyal and anterior ceratohyal **(all genera not in other CS)**
    — CS–1: Orientation of posterior ceratohyal essentially perpendicular to that of both ventral hypohyal and anterior ceratohyal **(*Chaca*)**

373. *Presence of prominent posterodorsal projection of posterior ceratohyal.* Plesiomorphically in siluriforms the posterior ceratohyal lacks major projections or processes [State 0: e.g. Fig. 3.69], but in *Bunocephalus* this bone exhibits a prominent posterodorsal projection, with the whole bone thus markedly elongated anteroposteriorly [State 1: e.g. Fig. 3.27].
   - CS–0: Absence of prominent posterodorsal projection of posterior ceratohyal **(all genera not in other CS)**
   - CS–1: Presence of prominent posterodorsal projection of posterior ceratohyal **(*Bunocephalus*)**
374. *Presence of additional ligament between posterior ceratohyal and mesial surface of suspensorium (character inspired from Mo, 1991).* Plesiomorphically in siluriforms the posterior ceratohyal is connected by a ligament to the mesial surface of the suspensorium [State 0: e.g. Fig. 3.66]. However, specimens of genus *Bagrichthys* examined present an additional ligament connecting the dorsal surface of the posterior ceratohyal to the mesial surface of the suspensorium, namely to the quadrato-symplectic [State 1: e.g. Fig. 3.27].
   - CS–0: Absence of additional ligament connecting posterior ceratohyal and suspensorium **(all genera not in other CS)**
   - CS–1: Presence of additional ligament connecting posterior ceratohyal and suspensorium **(*Bagrichthys*)**
375. *Presence of prominent ventrolateral crest of anterior ceratohyal.* Plesiomorphically in siluriforms the anterior ceratohyal lacks major projections or processes [State 0: e.g. Fig. 3.69], but in specimens of the four genera of CS–1 examined, this bone exhibits a prominent ventrolateral crest for insertion of the hyohyoideus inferior [State 1: e.g. Fig. 3.87].
   - CS–0: Absence of prominent ventrolateral crest on anterior ceratohyal **(all genera not in other CS)**
   - CS–1: Presence of prominent ventrolateral crest on anterior ceratohyal **(*Clarias, Heteropneustes, Heterobranchus, Uegitglanis*)**
   - ?: Since it was not possible to properly appraise this character in the specimens examined **(*Chaca*)**
376. *Presence of well-developed ventrolateral laminar expansion of anterior ceratohyal.* Contrary to other catfish [State 0: e.g. Fig. 3.69], in the specimens examined of *Nematogenys* the anterior ceratohyal exhibits a well-developed ventrolateral laminar expansion [State 1].
   - CS–0: Absence of well-developed ventrolateral laminar expansion of anterior ceratohyal **(all genera not in other CS)**
   - CS–1: Presence of well-developed ventrolateral laminar expansion of anterior ceratohyal **(*Nematogenys*)**
377. *Presence of prominent anterolateral process on anteroventrolateral margin of anterior ceratohyal.* Contrary to other catfish [State 0: e.g. Fig. 3.69], in *Glyptosternon* there is a prominent anterolateral process on the anteroventrolateral margin of the anterior ceratohyal [State 1].
   - CS–0: Absence of prominent anterolateral process of anterior ceratohyal **(all genera not in other CS)**

— CS–1: Presence of prominent anterolateral process of anterior ceratohyal **(*Glyptosternon*)**

378. *Enlargement of posterior margin of anterior ceratohyal and anterior margin of posterior ceratohyal.* Contrary to other catfish [State 0: e.g. Fig. 3.69], in siluriforms of CS–1 the margins of contact between the anterior and the posterior ceratohyal are peculiarly, remarkably enlarged [State 1].

— CS–0: Margins of contact between anterior and posterior ceratohyal not remarkably enlarged **(all genera not in other CS)**

— CS–1: Margins of contact between anterior and posterior ceratohyal remarkably enlarged **(*Zaireichthys, Siluranodon*)**

379. *Presence of laminar expansions on posterior margin of anterior ceratohyal (character inspired from de Pinna, 1996).* Contrary to other catfish [State 0: e.g. Fig. 3.69], in *Erethistes* and *Hara* the posterior margin of the anterior ceratohyal exhibits well-developed laminar expansions, which form a prominent process directed laterally [State 1].

— CS–0: Absence of laminar projections on posterior margin of anterior ceratohyal **(all genera not in other CS)**

— CS–1: Presence of laminar projections on posterior margin of anterior ceratohyal **(*Erethistes, Hara*)**

380. *Presence of well-developed, long, sharply pointed posterodorsal process of anterior ceratohyal.* Contrary to other catfish [State 0: e.g. Fig. 3.69], in specimens of *Nematogenys* examined the anterior ceratohyal exhibits a well-developed, long, sharply pointed posterodorsal process surrounding a significant part of the dorsal surface of the posterior ceratohyal [State 1: e.g. Fig. 3.92].

— CS–0: Absence of well developed, long, sharply pointed posterodorsal process of anterior ceratohyal **(all genera not in other CS)**

— CS–1: Presence of well-developed, long, sharply pointed posterodorsal process of anterior ceratohyal **(*Nematogenys*)**

381. *Presence of anterolateral projection of anterior ceratohyal (character inspired from Mo, 1991).* Contrary to other catfish, in which the anterior ceratohyal is essentially truncated anteriorly [State 0: e.g. Fig. 3.69], in specimens examined of siluriforms of CS–1 the anterior ceratohyal exhibits a marked anterolateral projection of laminar bone [State 1: e.g. Fig. 3.14].

— CS–0: Absence of marked anterolateral projection of anterior ceratohyal **(all genera not in other CS)**

— CS–1: Presence of marked anterolateral projection of anterior ceratohyal **(*Chrysichthys, Siluranodon, Pseudeutropius, Phractura, Andersonia, Leptoglanis, Doumea, Belonoglanis, Trachyglanis, Cranoglanis*)**

382. *Presence of prominent, broad anteroventral lamina of anterior ceratohyal.* Contrary to other catfish [State 0: e.g. Fig. 3.69], in *Nematogenys* the anterior ceratohyal exhibits a well-developed, prominent, broad anteroventral lamina of anterior ceratohyal [State 1].

— CS–0: Absence of prominent, broad anteroventral lamina of anterior ceratohyal **(all genera not in other CS)**

— CS–1: Presence of prominent, broad anteroventral lamina of anterior ceratohyal **(*Nematogenys*)**

383. *Presence of well-developed anteroventromesial process of ventral hypohyal.* Plesiomorphically in siluriforms the anterior ceratohyal lacks major projections or processes on its ventral surface [State 0: e.g. Fig. 3.69], but in *Auchenoglanis* this bone exhibits a well-developed anteroventromesial process [State 1].
   — CS–0: Absence of well-developed anteroventromesial process of ventral hypohyal **(all genera not in other CS)**
   — CS–1: Presence of well-developed anteroventromesial process of ventral hypohyal **(*Auchenoglanis*)**
384. *Presence of prominent posteroventral crest of ventral hypohyal.* Contrary to other catfish [State 0: e.g. Fig. 3.69], *Chaca* exhibits a prominent crest on the posteroventral surface of the ventral hypohyal, in which inserts a strong anterior tendon of the muscle sternohyoideus [State 1].
   — CS–0: Absence of prominent posteroventral crest of ventral hypohyal **(all genera not in other CS)**
   — CS–1: Presence of prominent posteroventral crest of ventral hypohyal **(*Chaca*)**
385. *Anterior margin of parurohyal.* Characteristically in catfish the anterior surface of the parurohyal is not pronouncedly concave [State 0: e.g. Fig. 3.69], but in *Leptoglanis* and *Zaireichthys* the anterior margin of this bone has a markedly concave shape in ventral view, with the anteromesial portion of this bone lying considerably posterior to its anterolateral edges [State 1] (see Diogo, 2003b).
   — CS–0: Anterior margin of parurohyal not markedly concave in ventral view **(all genera not in other CS)**
   — CS–1: Anterior margin of parurohyal markedly concave in ventral view **(*Leptoglanis, Zaireichthys*)**
386. *Posterior margin of parurohyal.* Characteristically in catfish the posterior surface of the parurohyal is not markedly truncated in ventral view [State 0: e.g. Fig. 3.69], but this is the case in those siluriforms of CS–1, with only a very small posteromesian process being eventually in such a view [State 1: e.g. Fig. 3.92].
   — CS–0: Posterior margin of parurohyal not markedly truncated in ventral view **(all genera not in other CS)**
   — CS–1: Posterior margin of parurohyal markedly truncated in ventral view **(*Plotosus, Paraplotosus, Cnidoglanis, Neosilurus, Helogenes, Nematogenys, Loricaria, Hypoptopoma, Lithoxus, Scoloplax, Astroblepus, Cranoglanis, Bagrichthys, Heptapterus, Goeldiella, Rhamdia, Mochokus, Ictalurus, Amiurus, Austroglanis*)**
387. *Markedly rectangular, thin parurohyal (character inspired from de Pinna, 1996).* Contrary to other catfish examined [State 0: e.g. Fig. 3.69], in specimens

of genus *Glyptothorax* analysed the parurohyal is a markedly unique, thin, rectangular structure in ventral view [State 1].

— CS–0: Parurohyal not a markedly rectangular, thin structure **(all genera not in other CS)**

— CS–1: Parurohyal a markedly rectangular, thin structure **(*Glyptothorax*)**

388. *Reduction in size of parurohyal.* Contrary to other catfish examined [State 0: e.g. Fig. 3.69], in *Synodontis* the parurohyal is markedly reduced in size to a very small, somewhat triangular structure [State 1].

— CS–0: Parurohyal not markedly reduced in size **(all genera not in other CS)**

— CS–1: Parurohyal markedly reduced in size **(*Synodontis*)**

389. *Transversal expansion of parurohyal (ordered multistate character).* In *Chaca* [State 1], and especially in *Heteropneustes* [State 2: e.g. Fig. 3.87], the parurohyal is markedly more expanded transversally than in all the other catfish examined [State 0: e.g. Fig. 3.69].

— CS–0: Parurohyal not markedly expanded transversally **(all genera not in other CS)**

— CS–1: Parurohyal markedly expanded transversally **(*Chaca*)**

— CS–2: Parurohyal even more expanded transversally than in CS–1 **(*Heteropneustes*)**

390. *Markedly enlarged hypobranchial foramen of parurohyal.* In specimens examined of *Amblyceps* and *Liobagrus* [State 1: e.g. Fig. 3.4] the hypobranchial foramen of the parurohyal (sensu Arratia and Schultze, 1990) is markedly more enlarged than in other catfish analysed [State 0: e.g. Fig. 3.69], with length of the foramen superior to half the length of the main body of the parurohyal.

— CS–0: Hypobranchial foramen of parurohyal not markedly enlarged **(all genera not in other CS)**

— CS–1: Hypobranchial foramen of parurohyal markedly enlarged **(*Liobagrus, Amblyceps*)**

391. *Presence of prominent, sharply pointed anterolateral processes of parurohyal.* Contrary to other catfish [State 0: e.g. Fig. 3.69], in *Cetopsis* and *Hemicetopsis* the parurohyal exhibits two prominent, sharply pointed processes [State 1].

— CS–0: Parurohyal not presenting two prominent, sharply pointed processes **(all genera not in other CS)**

— CS–1: Parurohyal presenting two prominent, sharply pointed processes **(*Cetopsis, Hemicetopsis*)**

392. *Enlargement of inner branchiostegal rays (ordered multistate character).* Plesiomorphically in catfish the inner branchiostegal rays are considerably thinner and shorter than the lateral ones [State 0: e.g. Fig. 3.26], but in siluriforms of CS–1 [State 1], and especially of CS–2 [State

2: e.g. Fig. 3.44], the inner branchiostegal rays are peculiarly developed, enlarged structures.

— CS–0: Inner branchiostegal rays not considerably thinner and shorter than outer ones **(all genera not in other CS)**

— CS–1: Inner branchiostegal rays markedly developed **(*Helogenes*)**

— CS–2: Inner branchiostegal rays even larger and more developed than in CS–1 **(*Cetopsis, Hemicetopsis*)**

393. *Branchiostegal rays forming a closed circular arrangement* (*character inspired from de Pinna, 1996*). As described by de Pinna (1996), the branchiostegal rays of the amblycipitids examined [State 1] are oriented in a manner different from that of other siluriforms [State 0: e.g. Fig. 3.69], presenting a highly unique, closed circular arrangement.

— CS–0: Branchiostegal rays not presenting a highly unique, closed, circular arrangement **(all genera not in other CS)**

— CS–1: Branchiostegal rays presenting a highly unique, closed, circular arrangement **(*Liobagrus, Amblyceps*)**

394. *Branchiostegal rays presenting long, thin cartilage for articulation with hyoid arch (character inspired from Reis, 1998a). In Callichthys* [State 1], contrary to other catfish [State 0: e.g. Fig. 3.69], the branchiostegal rays present a long, thin cartilage for articulation with the hyoid arch.

— CS–0: Branchiostegal rays not presenting long, thin cartilage for articulation with hyoid arch **(all genera not in other CS)**

— CS–1: Branchiostegal rays presenting long, thin cartilage for articulation with hyoid arch **(*Callichthys*)**

395. *Presence of 'branchiostegal cartilage'* (*character inspired from Reis, 1998a*). As described by Reis (1998a), the specimens examined of *Callichthys* and *Corydoras*, contrary to other catfish [State 0: e.g. Fig. 3.69], present a well-developed, singular 'branchiostegal cartilage' (see terminology of Reis, 1998a) on the hyoid region.

— CS–0: Absence of 'branchiostegal cartilage' **(all genera not in other CS)**

— CS–1: Presence of 'branchiostegal cartilage' **(*Callichthys, Corydoras*)**

396. *Tooth-bearing area of lower jaw situated in deep concave area of dentary bone* (*character inspired from de Pinna, 1993*). Contrary to other siluriforms [State 0: e.g. Fig. 3.88], in specimens of the five genera of CS–1 the tooth-bearing area of the mandible is situated in a deep concave area of the dentary bone [State 1].

— CS–0: Tooth-bearing area of lower jaw not situated in deep concave area of dentary **(all genera not in other CS)**

— CS–1: Tooth-bearing area of lower jaw situated in deep concave area of dentary **(*Synodontis, Loricaria, Hypoptopoma, Lithoxus, Astroblepus*)**

397. *Pronounced anteroposterior compression of mandible*. Plesiomorphically in catfish the mandible is not significantly compressed anteroposteriorly [State 0: e.g. Fig. 3.65], but in the doumeins examined the mandible is a

notably curved, anteroposteriorly compressed structure [State 1: e.g. Fig. 3.19].

— CS–0: Mandible not remarkably curved and compressed anteroposteriorly **(all genera not in other CS)**

— CS–1: Mandible remarkably curved, compressed anteroposteriorly **(*Phractura, Belonoglanis, Trachyglanis, Doumea, Andersonia*)**

398. *Number of mandibular teeth.* Plesiomorphically catfish have a lower jaw with many teeth [State 0: e.g. Fig. 3.65]. In specimens of genera of CS–1 examined, very few, or even no teeth occur on the lower jaw [State 1: e.g. Fig. 3.19].

— CS–0: Numerous teeth on lower jaw **(all genera not in other CS)**

— CS–1: Very few, or even no teeth on lower jaw **(*Scoloplax, Phractura, Doumea, Andersonia, Belonoglanis, Trachyglanis, Siluranodon, Bagrichthys, Hypophthalmus, Corydoras*)**

399. *Outer row of teeth on dentary enlarged and wide set (character inspired from de Pinna and Vari, 1995).* As stated by de Pinna and Vari (1995) plesiomorphically in catfish the individual teeth of the dentary increase in size gradually towards the outer margin of the bone [State 0: e.g. Fig. 3.65]. In *Helogenes*, in contrast, the lateral teeth of the anterior portion of the dentary are markedly larger than all the other teeth on the bone [State 1].

— CS–0: Outer row of teeth on dentary not enlarged and wide set **(all genera not in other CS)**

— CS–1: Outer row of teeth on dentary markedly enlarged and wide set **(*Helogenes*)**

— Inapplicable: Due to the significant reduction in number, or complete absence of mandibular teeth (see above) **(*Scoloplax, Phractura, Doumea, Andersonia, Belonoglanis, Trachyglanis, Siluranodon, Bagrichthys, Hypophthalmus, Corydoras*)**

400. *Completely undifferentiated coronoid process of mandible (character inspired from de Pinna, 1993).* In the mochokids examined [State 1], contrary to other catfish analysed [State 0: e.g. Fig. 3.65], the coronoid process of the mandible is completely undifferentiated.

— CS–0: Presence of distinct coronoid process of mandible **(all genera not in other CS)**

— CS–1: No distinct coronoid process of mandible **(*Synodontis, Mochokus*)**

401. *Coronoid process essentially constituted by posterodorsal margin of dentary.* Contrary to other catfish, in which the coronoid process of the mandible is essentially constituted by both the anterodorsal and posterodorsal surfaces of the angulo-articular and the dentary [State 0: e.g. Fig. 3.63], in siluriforms of CS–1 this process is essentially formed by the posterodorsal surface of the dentary [State 1: e.g. Fig. 3.81].

— CS–0: Coronoid process of mandible essentially constituted by anterodorsal and posterodorsal surfaces of angulo-articular and dentary **(all genera not in other CS)**

— CS–1: Coronoid process of mandible essentially constituted by posterodorsal surface of dentary (***Glyptothorax, Akysis, Parakysis, Glyptosternon, Gagata, Erethistes, Hara***)
— Inapplicable: Since the coronoid process is markedly reduced in size, not allowing proper appraisal of this character (***Bagarius, Liobagrus, Amblyceps***) or since there is no distinct coronoid process (***Synodontis, Mochokus***)

402. *Coronoid process markedly expanded anteroposteriorly (character inspired from Mo, 1991)*. Contrary to other catfish [State 0: e.g. Fig. 3.65], in siluriforms of CS–1 the coronoid process of the mandible is markedly expanded anteroposteriorly [State 1: e.g. Fig. 3.55].
— CS–0: Coronoid process of mandible not markedly expanded anteroposteriorly **(all genera not in other CS)**
— CS–1: Coronoid process of mandible markedly expanded anteroposteriorly (***Chrysichthys, Pangasius, Helicophagus, Clarotes***)
— Inapplicable: Since there is no distinct coronoid process of the mandible (***Synodontis, Mochokus***)

403. *Right and left halves of lower jaw independently movable (character inspired from Howes, 1983a; Schaefer and Lauder, 1986)*. Contrary to all other catfish examined [State 0: e.g. Fig. 3.69], in loricariids and astroblepids the right and left halves of the lower jaw are independently movable, loosely attached to each other on the symphysis [State 1].
— CS–0: Right and left halves of lower jaw firmly attached to each other on mesial symphysis **(all genera not in other CS)**
— CS–1: Right and left halves of lower jaw not firmly attached to each other on mesial symphysis (***Loricaria, Hypoptopoma, Lithoxus, Astroblepus***)

404. *Presence of well-developed, large 'cartilage plug' associated with mesial surfaces of right and left halves of lower jaw (character inspired from Howes, 1983a; Schaefer and Lauder, 1986)*. Contrary to all other catfish examined [State 0: e.g. Fig. 3.69], in loricariids and astroblepids the mesial surfaces of the right and left halves of the lower jaw are associated with a well-developed, large 'cartilage plug' (see terminology of Howes, 1983a) [State 1].
— CS–0: Absence of 'cartilage plug' associated with lower jaw **(all genera not in other CS)**
— CS–1: Presence of 'cartilage plug' associated with lower jaw (***Loricaria, Hypoptopoma, Lithoxus, Astroblepus***)

405. *Dorsal tip of coronoid process markedly curved mesially*. Peculiarly in the nematogenyids and trichomycterids examined, and contrary to other catfish [State 0: e.g. Fig. 3.65], the coronoid process of the mandible is markedly curved medially, with its dorsal tip projecting medially beyond the main body of the mandible [State 1: e.g. Fig. 3.91].
— CS–0: Dorsal tip of coronoid process not markedly curved medially **(all genera not in other CS)**

— CS–1: Dorsal tip of coronoid process markedly curved medially **(*Nematogenys, Trichomycterus, Hatcheria*)**

— Inapplicable: Since there is no distinct coronoid process of the mandible **(*Synodontis, Mochokus*)**

406. *Anterodorsal margin of angulo-articular markedly dorsal to posterodorsal margin of dentary (character inspired from Mo, 1991)*. As described by Mo (1991), in the auchenoglanidin specimens examined, contrary to other siluriforms [State 0: e.g. Fig. 3.65], the anterodorsal margin of the angulo-articular, constituting the posterior surface of the coronoid process of the mandible, is markedly dorsal to the posterodorsal margin of the dentary, which constitutes the anterior surface of this process [State 1].

— CS–0: Anterodorsal margin of angulo-articular not markedly dorsal to posterodorsal margin of dentary **(all genera not in other CS)**

— CS–1: Anterodorsal margin of angulo-articular markedly dorsal to posterodorsal margin of dentary **(*Auchenoglanis*)**

— ?: Since in some specimens of these genera the situation seems to be similar to that of CS–1, while in other specimens it seems to be closer to that of CS–0 **(*Austroglanis, Malapterurus*)** or since it was not possible to properly observe this character in the specimens examined **(*Ancharius*)**

407. *Mandibular bony tunnel to enclose the ramus mandibularis*. In the plesiomorphic condition the nerve ramus mandibularis is differentiated into two branches, of which the main branch passes between the dentary and Meckel's cartilage (see Diogo and Chardon, 2000a) [State 0: e.g. Fig 3.22A]. In catfish of CS–1 the main branch of the ramus mandibularis is completely enclosed in a well-developed mandibular bony tunnel, distinctly visible in a ventral view of the mandible [State 1: e.g. Fig. 3.19D].

— CS–0: Main branch of ramus mandibularis not enclosed in well-developed mandibular bony tunnel **(all genera not in other CS)**

— CS–1: Main branch of ramus mandibularis enclosed in well-developed mandibular bony tunnel **(*Doumea, Phractura, Andersonia, Belonoglanis, Trachyglanis, Synodontis, Mochokus, Loricaria, Lithoxus, Scoloplax, Astroblepus, Hypoptopoma*)**

408. *Posterolateral projection of dentary covering significant part of lateral surface of dentary (character inspired from Mo, 1991)*. The plesiomorphic condition for catfish seems to be that in which the angulo-articular occupies a significant part of the lateral surface of the mandible [State 0: e.g. Fig. 3.63]. In specimens of genera of CS–1 examined, a great part of the angulo-articular is laterally covered by a posterolateral laminar projection of the dentary, with only a small part of the angulo-articular visible in a lateral view of the mandible [State 1: e.g. Fig. 3.17].

— CS–0: Significant part of angulo-articular visible in lateral view of mandible **(all genera not in other CS)**

— CS–1: Only a small part of angulo-articular visible in lateral view of mandible **(*Nematogenys, Trichomycterus, Hatcheria, Callichthys,***

***Corydoras, Loricaria, Hypoptopoma, Lithoxus, Scoloplax, Astroblepus, Amphilius, Paramphilius, Glyptothorax, Glyptosternon, Erethistes, Chaca, Hara, Bunocephalus, Aspredo, Xyliphius, Heptapterus, Bagarius, Liobagrus, Akysis, Parakysis, Amblyceps, Doumea, Belonoglanis, Andersonia, Leptoglanis, Trachyglanis*)**

— ?: Since it was not possible to properly appraise this character in the specimens examined **(*Zaireichthys, Gagata*)**

409. *Prominent mesial process on posteromesial surface of angulo-articular.* Characteristically in catfish there is no prominent medial process of the angulo-articular [State 0: e.g. Fig. 3.65], but in *Amphilius* there is a prominent, roughly triangular, anteromesially oriented process on the posteromesial surface of this bone [State 1].

— CS–0: Absence of prominent, roughly triangular, anteromesially oriented process of angulo-articular **(all genera not in other CS)**

— CS–1: Presence of prominent, roughly triangular, anteromesially oriented process of angulo-articular **(*Amphilius*)**

410. *Posterolateral laminar projection of angulo-articular (character inspired from de Pinna, 1993).* Contrary to other catfish [State 0: e.g. Fig. 3.63], in the mochokids examined the angulo-articular exhibits a prominent posterolateral laminar projection, which covers, in lateral view, the articulation between this bone and the quadrato-symplectic [State 1].

— CS–0: Absence of prominent posterolateral projection of angulo-articular **(all genera not in other CS)**

— CS–1: Presence of prominent posterolateral projection of angulo-articular **(*Synodontis, Mochokus*)**

411. *Posterodorsal crest of angulo-articular.* Contrary to other catfish [State 0: e.g. Fig. 3.63], in the specimens examined of *Genidens* [State 1] the angulo-articular exhibits a prominent posterodorsal crest to receive the adductor mandibulae A1-OST [State 1].

— CS–0: Absence of prominent posterodorsal crest of angulo-articular **(all genera not in other CS)**

— CS–1: Presence of prominent posterodorsal crest of angulo-articular **(*Genidens*)**

412. *Presence of smooth, circular surface on posteroventromesial surface of angulo-articular (ordered multistate character).* Contrary to other catfish [State 0: e.g. Fig. 3.63], in specimens of *Cetopsis* examined [State 1: e.g. Fig. 3.48], and particularly of *Helogenes* [State 2], the angulo-articular exhibits a well-developed, smooth, circular surface on its posteroventromesial surface.

— CS–0: Absence of smooth, circular surface on posteroventromesial surface of angulo-articular **(all genera not in other CS)**

— CS–1: Presence of smooth, circular surface on posteroventromesial surface of angulo-articular **(*Cetopsis*)**

— CS–2: Smooth, circular surface on posteroventromesial surface of angulo-articular more developed than in CS–1 **(*Helogenes*)**

413. *Anteromesial process of dentary bone*. Contrary to other catfish [State 0: e.g. Fig. 3.69], in *Auchenoglanis* the dentary bone exhibits a well-developed anteromesial process, which is markedly curved anteriorly [State 1].
    — CS–0: Absence of prominent anteromesial process of dentary bone **(all genera not in other CS)**
    — CS–1: Presence of prominent anteromesial process of dentary bone **(*Auchenoglanis*)**
414. *Posteroventromesial process of dentary bone (character inspired from Mo, 1991)*. Contrary to other siluriforms [State 0: e.g. Fig. 3.69], in catfish of CS–1 the dentary bone exhibits a prominent, posteromesially oriented, posteroventromesial process near the lower jaw symphysis [State 1: e.g. Fig. 3.51].
    — CS–0: Absence of prominent posteroventromesial process of dentary bone **(all genera not in other CS)**
    — CS–1: Presence of prominent posteroventromesial process of dentary bone **(*Erethistes, Arius, Chrysichthys, Clarotes, Callichthys, Corydoras, Chaca, Malapterurus, Plotosus, Cnidoglanis, Neosilurus, Paraplotosus*)**
    — ?: Since it was not possible to properly appraise this character appropriately in the specimens examined **(*Auchenoglanis*)**
415. *Anteroventrolateral extension of dentary bone*. Contrary to all other siluriforms studied [State 0: e.g. Fig. 3.69], in *Auchenoglanis* a large, prominent anteroventrolateral extension of the dentary bone occurs, in which originates a significant part of the fibres of the intermandibularis [State 1].
    — CS–0: Absence of prominent anteroventrolateral extension of dentary bone **(all genera not in other CS)**
    — CS–1: Presence of prominent anteroventrolateral extension of dentary bone **(*Auchenoglanis*)**
416. *Presence of dorsolateral lamina of dentary bone (ordered multistate character)*. Contrary to other catfish [State 0: e.g. Fig. 3.63], in the specimens of *Cetopsis* examined [State 1: e.g. Fig. 3.48], and especially of *Hemicetopsis* [State 2], the dentary bone exhibits a well-developed, broad dorsolateral lamina, which covers a significant part of the lateral surface of the mandibular teeth in lateral view.
    — CS–0: Absence of dorsolateral lamina of dentary bone **(all genera not in other CS)**
    — CS–1: Presence of well-developed dorsolateral lamina of dentary bone **(*Cetopsis*)**
    — CS–2: Dorsolateral lamina of dentary bone even more developed than in CS–1 **(*Hemicetopsis*)**
417. *Presence of anterodorsomesial laminar projection of dentary bone (ordered multistate character) (character inspired from Oliveira et al., 2001)*. Contrary to other catfish [State 0: e.g. Fig. 3.69], in siluriforms of CS–1 [State 1], and especially of CS–2 [State 2], the dentary bone exhibits a well-developed, laminar anterodorsomesial projection.
    — CS–0: Absence of laminar anterodorsomesial projection of dentary bone **(all genera not in other CS)**

— CS–1: Presence of well-developed, laminar anterodorsomesial projection of dentary bone (***Plotosus, Paraplotosus, Chaca***)
— CS–2: Laminar anterodorsomesial projection of dentary bone more developed than in CS–1 (***Neosilurus***)

418. *Anteroventromesial extension of dentary bone.* Contrary to all other siluriforms studied [State 0: e.g. Fig. 3.69], in *Pangasius* there is a highly developed, broad anteroventromesial extension of the dentary bone, which lies ventral to the intermandibularis and thus covers a great part of this muscle in ventral view [State 1: e.g. Fig. 3.98].
— CS–0: Absence of prominent anteroventromesial extension of dentary bone **(all genera not in other CS)**
— CS–1: Presence of prominent anteroventromesial extension of dentary bone (***Pangasius***)

419. *Absence of ascending portion of Meckel's cartilage (character inspired from Mo, 1991).* Plesiomorphically in siluriforms Meckel's cartilage presents a well-developed ascending portion and a well-developed horizontal portion [State 0: e.g. Fig. 3.65]. In specimens of CS–1 examined, the ascending portion of Meckel's cartilage is lacking [State 1: e.g. Fig. 3.19].
— CS–0: Presence of ascending portion of Meckel's cartilage **(all genera not in other CS)**
— CS–1: Absence of ascending portion of Meckel's cartilage (***Clarias, Heteropneustes, Amphilius, Paramphilius, Leptoglanis, Zaireichthys, Phractura, Doumea, Belonoglanis, Trachyglanis, Silurus, Wallago, Andersonia, Synodontis, Mochokus, Callichthys, Corydoras, Loricaria, Hypoptopoma, Lithoxus, Scoloplax, Astroblepus***)

420. *Marked posterior extension of horizontal portion of Meckel's cartilage (character inspired from de Pinna, 1993).* As described by de Pinna (1993), contrary to other catfish examined [State 0: e.g. Fig. 3.65], in schilbids the horizontal portion of Meckel's cartilage is markedly extended posteriorly, with its posterior margin markedly posterior to the posterior margin of the coronoid process of the mandible [State 1: e.g. Fig. 3.119].
— CS–0: Meckel's cartilage not markedly extended posteriorly **(all genera not in other CS)**
— CS–1: Meckel's cartilage markedly extended posteriorly (***Schilbe, Laides, Pseudeutropius, Siluranodon, Ailia***)

421. *Dorsal extension of ascending portion of Meckel's cartilage.* Plesiomorphically in siluriforms the dorsal margin of the ascending portion of Meckel's cartilage lies somewhat at the same level as the dorsal margin of the coronoid process of the mandible [State 0: e.g. Fig. 3.65]. In specimens of CS–1 examined, the ascending portion of this cartilage is markedly extended dorsally, with its dorsal margin markedly dorsal to that of the coronoid process [State 1: e.g. Fig. 3.25].
— CS–0: Ascending portion of Meckel's cartilage lies at about the same level as dorsal margin of coronoid process **(all genera not in othern CS)**

— CS–1: Ascending portion of Meckel's cartilage markedly extended dorsally, its dorsal margin dorsal to that of coronoid process **(*Bunocephalus, Xyliphius, Franciscodoras, Anadoras, Acanthodoras, Doras, Ageneiosus, Auchenipterus*)**

— ?: Since it was not possible to properly observe this character in the specimens examined **(*Centromochlus*)**

— Inapplicable: Since there is no ascending portion of Meckel's cartilage **(*Clarias, Heteropneustes, Amphilius, Paramphilius, Leptoglanis, Zaireichthys, Phractura, Doumea, Belonoglanis, Trachyglanis, Silurus, Wallago, Andersonia, Synodontis, Mochokus, Callichthys, Corydoras, Loricaria, Hypoptopoma, Lithoxus, Scoloplax, Astroblepus*)**

422. *Dorsal extension of ascending portion of Meckel's cartilage.* Contrary to other catfish examined [State 0: e.g. Fig. 3.65], in specimens of siluriforms of CS–1 analysed the ascending and horizontal portions of Meckel's cartilage are markedly distant from each other [State 1: e.g. Fig. 3.90].

— CS–0: Ascending and horizontal portions of Meckel's cartilage not markedly distant from each other **(all genera not in other CS)**

— CS–1: Ascending and horizontal portions of Meckel's cartilage markedly distant from each other **(*Nematogenys, Trichomycterus, Hatcheria, Malapterurus*)**

— Inapplicable: Since there is no ascending portion of Meckel's cartilage **(*Clarias, Heteropneustes, Amphilius, Paramphilius, Leptoglanis, Zaireichthys, Phractura, Doumea, Belonoglanis, Trachyglanis, Silurus, Wallago, Andersonia, Synodontis, Mochokus, Callichthys, Corydoras, Loricaria, Hypoptopoma, Lithoxus, Scoloplax, Astroblepus*)**

423. *Development of horizontal portion of Meckel's cartilage (ordered multistate character) (character inspired from Mo, 1991; de Pinna, 1993).* Plesiomorphically in catfish the horizontal portion of Meckel's cartilage is considerably elongated anteroposteriorly [State 0: e.g. Fig. 3.65]. In specimens of genera of CS–1 analysed [State 1], and especially of CS–2 [State 2: e.g. Fig. 3.19], the horizontal portion of Meckel's cartilage is markedly compressed anteroposteriorly.

— CS–0: Horizontal portion of Meckel's cartilage not markedly compressed anteroposteriorly **(all genera not in other CS)**

— CS–1: Horizontal portion of Meckel's cartilage significantly compressed anteroposteriorly **(*Amphilius, Uegitglanis, Leptoglanis, Zaireichthys, Paramphilius, Synodontis, Mochokus, Corydoras*)**

— CS–2: Horizontal portion of Meckel's cartilage even more compressed anteroposteriorly than in CS–1 **(*Andersonia, Phractura, Belonoglanis, Trachyglanis, Doumea, Loricaria, Hypoptopoma, Lithoxus, Astroblepus*)**

— ?: Since it was not possible to properly appraise this character in the specimens examined **(*Trichomycterus, Hatcheria*)**

424. *Absence of coronomeckelian bone (character inspired from Mo, 1991; de Pinna, 1993).* Plesiomorphically catfish present a coronomeckelian bone on the mesial face of the mandible [State 0: e.g. Fig. 3.65]. In specimens examined of siluriforms of CS–1 this bone is lacking [State 1].
   — CS–0: Presence of coronomeckelian bone **(all genera not in other CS)**
   — CS–1: Absence of coronomeckelian bone **(*Synodontis, Mochokus, Callichthys, Corydoras, Loricaria, Hypoptopoma, Lithoxus, Scoloplax, Astroblepus*)**
425. *Development of coronomeckelian bone.* The plesiomorphic condition for siluriforms seems to be that in which the coronomeckelian bone is a well-developed, broad structure [State 0: e.g. Fig. 3.65] (see, e.g., Diogo, 2003b). In catfish of CS–1, the coronomeckelian bone is considerably smaller than in catfish of CS–0 [State 1: e.g. Fig. 3.22].
   — CS–0: Well-developed coronomeckelian bone **(*Diplomystes, Nematogenys, Trichomycterus, Hatcheria*)**
   — CS–1: Coronomeckelian bone considerably smaller than in CS–0 **(all genera not in other CS)**
   — Inapplicable: Since the coronomeckelian bone is lacking **(*Synodontis, Mochokus, Callichthys, Corydoras, Loricaria, Hypoptopoma, Lithoxus, Scoloplax, Astroblepus*)**
426. *Remarkable anteroposterior elongation of coronomeckelian bone.* Contrary to all other catfish examined [State 0: e.g. Fig. 3.65], in *Franciscodoras* the coronomeckelian bone is a peculiarly, remarkably elongated, thin structure [State 1].
   — CS–0: Coronomeckelian bone not a remarkably elongated, thin structure **(all genera not in other CS)**
   — CS–1: Coronomeckelian bone a remarkably elongated, thin structure **(*Franciscodoras*)**
   — Inapplicable: Since the coronomeckelian bone is missing **(*Synodontis, Mochokus, Callichthys, Corydoras, Loricaria, Hypoptopoma, Lithoxus, Scoloplax, Astroblepus*)**
427. *Position of coronomeckelian bone (ordered character state) (character inspired from de Pinna, 1993).* In specimens examined of *Glyptothorax* [State 1], and especially of *Leptoglanis* [State 2], the coronomeckelian bone lies in a markedly more dorsal position than in other siluriforms analysed [State 0: e.g. Fig. 3.65], with its dorsal margin situating dorsal to the dorsal margin of the coronoid process.
   — CS–0: Dorsal margin of coronomeckelian bone not situating dorsally to dorsal margin of coronoid process **(all genera not in other CS)**
   — CS–1: Dorsal margin of coronomeckelian bone situating dorsal to dorsal margin of coronoid process **(*Glyptothorax*)**
   — CS–2: Dorsal margin of coronomeckelian bone situating markedly dorsal to that of coronoid process **(*Leptoglanis*)**
   — ?: Since it was not possible to properly appraise this character in the specimens examined **(*Zaireichthys, Bagarius*)**

— Inapplicable: Since the coronomeckelian bone is lacking (***Synodontis, Mochokus, Callichthys, Corydoras, Loricaria, Hypoptopoma, Lithoxus, Scoloplax, Astroblepus***)

Miscellaneous

428. *Nasal with expansions beyond canal-bearing portion (character inspired from de Pinna, 1996).* As described by de Pinna (1996), *Akysis* [State 1], contrary to the other catfish examined [State 0], exhibits a nasal bone with well-developed expansions beyond the canal-bearing portion.
 — CS–0: Nasal not presenting expansions beyond canal-bearing portion **(all genera not in other CS)**
 — CS–1: Nasal presenting expansions beyond canal-bearing portion (***Akysis***)
429. Presence of 'thoracic sucker' (*character inspired from de Pinna, 1996*). Contrary to other catfishes [State 0], specimens of genus *Glyptothorax* analysed present a well-developed 'thoracic sucker' (see terminology of de Pinna, 1996)[State 1].
 — CS–0: Absence of 'thoracic sucker' **(all genera not in other CS)**
 — CS–1: Presence of 'thoracic sucker' (***Glyptothorax***)
430. *Anterior portion of lateral line running closely parallel to lateral margin of Weberian lamina* (character inspired from de Pinna, 1996). As described by de Pinna (1996), the lateral-line sensory canal and associated ossicles in siluriforms primitively follow a path along the body of the fish with no close associations with internal skeletal structures [State 0]. However, in specimens of the five genera of CS–1 examined, the anterior portion of the canals runs closely parallel to the lateral margin of the Weberian lamina [State 1].
 — CS–0: Anterior portion of lateral line not running closely parallel to lateral margin of Weberian lamina **(all genera not in other CS)**
 — CS–1: Anterior portion of lateral line running closely parallel to lateral margin of Weberian lamina (***Erethistes, Hara, Bunocephalus, Aspredo, Xyliphius***)
431. *First lateral-line ossicle enlarged, adjoined to posterior portion of posttemporo-supracleithrum (character inspired from de Pinna, 1996).* Contrary to other catfish [State 0], in siluriforms of CS–1 the first lateral-line ossicle is markedly enlarged, adjoined to posterior portion of posttemporo-supracleithrum [State 1].
 — CS–0: First lateral-line ossicle not markedly enlarged **(all genera not in other CS)**
 — CS–1: First lateral-line ossicle markedly enlarged, adjoined to posterior portion of posttemporo-supracleithrum (***Akysis, Parakysis, Bunocephalus, Aspredo, Xyliphius***)

432. *Development of subpreopercle* (*character inspired from de Pinna, 1996*). As described by de Pinna (1996), in *Parakysis* [State 1] the subpreopercle is considerably more developed than in the other siluriforms examined [State 0].
   — CS–0: Subpreopercle not markedly developed **(all genera not in other CS)**
   — CS–1: Subpreopercle markedly developed **(*Parakysis*)**
433. *First infraorbital with marked inner expansion* (*character inspired from Reis, 1998a*). Contrary to other siluriforms examined [State 0], in the callichthyids analysed the first infraorbital exhibits a markedly developed inner expansion, which situates ventral to the eye and partly supports it [State 1].
   — CS–0: First infraorbital not presenting markedly developed inner expansion **(all genera not in other CS)**
   — CS–1: First infraorbital presenting markedly developed inner expansion **(*Callichthys, Corydoras*)**
434. *Tympanic area* (*character inspired from Higuchi, 1992; de Pinna, 1998*). As described by Higuchi (1992) and de Pinna (1998), peculiarly in doradids [State 1], and contrary to other catfish [State 0], the tympanic area is bordered by the posttemporo-supracleithrum, the postoccipital region, the infraneural scutes and the humeral process of the pectoral girdle.
   — CS–0: Tympanic area not peculiarly bordered by posttemporo-supracleithrum, postoccipital region, infraneural scutes and humeral process of pectoral girdle **(all genera not in other CS)**
   — CS–1: Tympanic area peculiarly bordered by posttemporo-supracleithrum, postoccipital region, infraneural scutes and humeral process of pectoral girdle **(*Franciscodoras, Anadoras, Acanthodoras, Doras*)**
435. *Abduction of maxillary barbel* (*character inspired from Ferraris, 1988b; de Pinna, 1998*). Contrary to other siluriforms, in which the maxillary barbel, when abducted, is essentially directed laterally or ventrolaterally [State 0], in the auchenipterids examined this barbel, when abducted, is essentially directed dorsolaterally [State 1].
   — CS–0: Maxillary barbel, when abducted, essentially directed laterally or ventrolaterally **(all genera not in other CS)**
   — CS–1: Maxillary barbel, when abducted, essentially directed dorsolaterally **(*Centromochlus, Ageneiosus, Auchenipterus*)**
436. *Adduction of maxillary barbel* (*character inspired from Ferraris, 1988b; de Pinna, 1998*). Peculiarly in auchenipterids [State 1] among the catfish examined [State 0], the maxillary barbel, when adducted, lies in a well-developed, deep concavity on the lateral surface of the cheek.
   — CS–0: Maxillary barbel, when adducted, not lying in a well-developed, deep concavity on lateral surface of cheek **(all genera not in other CS)**

— CS–1: Maxillary barbel, when adducted, lying in a well-developed, deep concavity on lateral surface of cheek **(*Centromochlus, Ageneiosus, Auchenipterus*)**

437. *Presence of onodontes (ordered multistate character) (character inspired from Howes, 1983a; Schaefer, 1990; de Pinna, 1998).* As described in de Pinna (1998), plesiomorphically catfish lack onodontes [State 0], but those specimens examined of siluriforms of CS–1 present onodontes, mainly on the pectoral region [State 1], with these onodontes being inclusively spread to the opercular series in catfish of CS–2 [State 2].
   — CS–0: Absence of onodontes **(all genera not in other CS)**
   — CS–1: Presence of onodontes, mainly on pectoral region **(*Nematogenys*)**
   — CS–2: Onodontes inclusively spread to opercular series **(*Trichomycterus, Hatcheria, Callichthys, Corydoras, Loricaria, Hypoptopoma, Lithoxus, Scoloplax*)**

438. *Additional 'mandibular barbel' (character inspired from Baskin, 1972; de Pinna, 1998).* As described by Baskin (1972) and de Pinna (1998), the situation found in callichthyids [State 1] is unique among the catfish examined [State 0], since the fishes of this family present a well-developed additional 'mandibular barbel' which, in fact, is clearly not homologous with true mandibular barbels, due to its association with the lateral and anteroventral surfaces of the mandible and its lack of connection to the protractor hyoideus muscle (see above).
   — CS–0: Absence of well-developed additional 'mandibular barbel' **(all genera not in other CS)**
   — CS–1: Presence of well-developed additional 'mandibular barbel' **(*Callichthys, Corydoras*)**

439. *Nasal organ encapsulated on lateral extension of lateral ethmoid (character inspired from Howes, 1983a; Reis, 1998a).* Peculiarly in catfish of CS–1 [State 1] among the Siluriformes examined [State 0], the nasal organ is encapsulated in a marked lateral extension of the lateral ethmoid.
   — CS–0: Nasal organ not encapsulated in marked lateral extension of lateral ethmoid **(all genera not in other CS)**
   — CS–1: Nasal organ encapsulated in marked lateral extension of lateral ethmoid **(*Callichthys, Corydoras, Loricaria, Hypoptopoma, Lithoxus*)**

440. *Presence of well-developed, broad 'rostral plate' (character inspired from Schaefer, 1990; de Pinna 1998).* As described by Schaefer (1990) and de Pinna (1998), the scoloplacid catfish examined [State 1] differ from other siluriforms [State 0] in having a well-developed, broad 'rostral plate' on the anterior region of the skull.
   — CS–0: Absence of well-developed, broad 'rostral plate' **(all genera not in other CS)**
   — CS–1: Presence of well-developed, broad 'rostral plate' **(*Scoloplax*)**

3.2 CLADISTIC ANALYSIS, DIAGNOSIS FOR CLADES, AND COMPARISON WITH PREVIOUS HYPOTHESES

The 440 characters listed above were coded for each of the 87 genera analysed, resulting in the data matrix shown in Tables 3.1 to 3.4. The phylogenetic analysis of these 440 characters, using the implicit enumeration algorithm (ie*) of the Hennig86 computer program (Farris, 1988), resulted in 12 equally parsimonious trees with a length of 898 steps, CI = 0.52, and RI = 0.78. The strict consensus of these 12 equally parsimonious trees resulted in the almost completely resolved cladogram illustrated in Fig. 3.123, with a length of 902 steps, CI = 0.52, and RI = 0.78. As can be seen in Fig. 3.123, the resultant cladogram presents only three trichotomies. Two of them (the trichotomy inside the Schilbidae leading to *Laides*, to *Ailia*, and to *Siluranodon* and the trichotomy inside the Sisoridae leading to *Gagata*, to *Bagarius*, and to the clade formed by *Glyptosternon* and *Glyptothorax*) refer to intrafamilial relationships. Thus, only one (the trichotomy leading to Akysidae, to Amblycipitidae, and the clade including Sisoridae, Erethistidae and Aspredinidae) directly concerns interfamilial relationships within the Siluriformes which, as explained in Chapters 1 and 2, is the actual major purpose of this phylogenetic analysis. The relationships among the catfish families examined, derived from the phylogenetic hypothesis presented in the consensus tree of Fig. 3.123, are illustrated in Fig. 3.124.

As mentioned above, the cladistic analysis of the present work includes more characters, as well as more terminal taxa, than the two analyses available on catfish higher level phylogeny, i.e., Mo's 1991 (126 characters, 40 terminal taxa) and de Pinna's 1993 (239 characters, 80 terminal taxa). However, the consistency and retention indexes of the consensus cladogram obtained, CI = 0.52 and RI = 0.78, are significantly superior to that of these two studies (Mo's 1991 cladogram I: CI = 0.34, RI = 0.64; Mo's 1991 cladogram II: CI = 0.36, RI = 0.72; de Pinna's 1993 cladogram: CI = 0.41, RI = 0.72). Notably, the CI of 0.52 is significantly superior to that expected for an analysis of 87 terminal taxa (see Sanderson and Donoghue, 1989).

The information presented in the cladogram of Fig. 3.123 is summarised below in a synapomorphy list that includes for each node a commentary with a comparison with previous hypotheses and in which the numbering for diagnostic characters follows that in the 'Character Description and Comparison' Section above.

The character state changes mentioned in this synapomorphy list are restricted to those unambiguous character state changes occurring in the different nodes, and can be included in two main categories: 1) state changes that occurred only once within the Siluriformes (in bold); 2) state changes subsequently reversed in a more terminal node and/or independently acquired in another node within the Siluriformes (non-bold). The synapomorphy list refers to *all* clades numbered in Fig. 3.123, thus referring to the nodes supporting the relationships between the different catfishes families analysed and

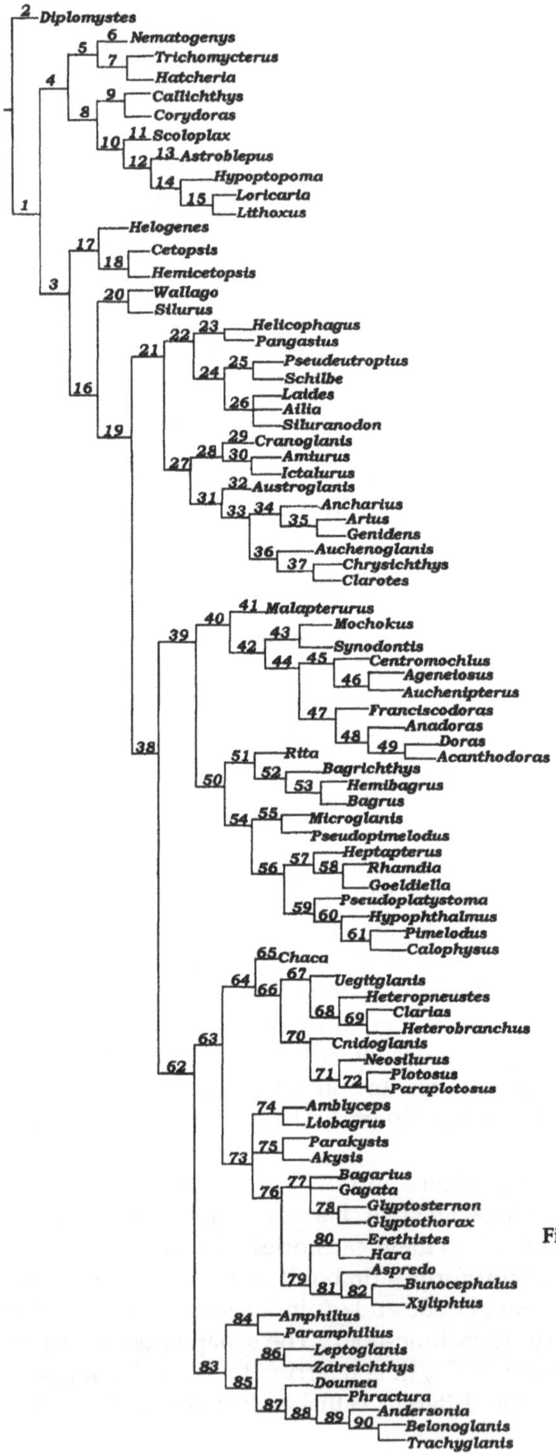

Fig. 3.123 Strict consensus tree of 12 equally parsimonious trees obtained in the cladistic analysis of the present work (CI = 0.52; RI = 0.78), all the terminal taxa examined are represented [for more details, see text].

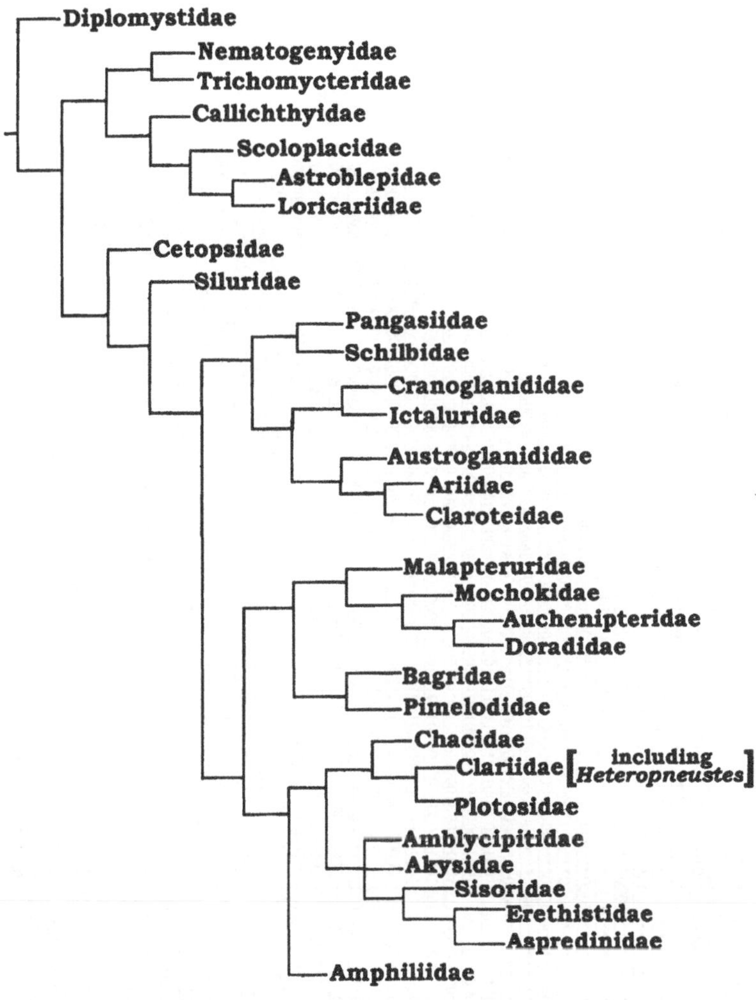

Fig. 3.124 Relationships among extant catfish families, derived from the phylogenetic hypothesis illustrated on the consensus tree obtained in the cladistic analysis of the present work (see Fig. 3.123) [for more details, see text].

the nodes leading to each of these families. Moreover, since, as explained in Section 2.1, this is the first analysis at the higher level phylogeny of Siluriformes in which all terminal taxa are individual genera, this synapomorphy list also provides, in some cases, important information on internal phylogenetic groups inside the families. With respect to the specific character states of each of the various individual genera examined, these are annexed in a short summary at the end of this Section, except for individual genera representing monogeneric families such as Diplomystidae, Nematogenyidae, Scoloplacidae, Astroblepidae, Cranoglanididae, Austroglanididae, Malapteruridae, Chacidae and Heteropneustidae (see above).

Table 3.1: Data matrix of 440 characters for each genus analysed; order of characters follows that presented and numbered in the text; inapplicable and missing character states for a certain genus are indicated with an asterisk (*) and a small square (#) respectively (Characters 001 to 110)

	00000000000000000000000000000000000000000000000000000000000000000000000000000000000000000000000000001111111111
	00000000011111111112222222222333333333344444444445555555555666666666677777777778888888888999999999900000000001
	12345678901234567890123456789012345678901234567890123456789012345678901234567890123456789012345678901234567890
Diplomystes	0************************00000***000000000000000000****00000000000000000000000000000000000000000000000000001000*
Austroglanis	21000000001000000000000000100101000000000000011001000000000000000000000000000000101000000000000000000000000011
Scoloplax	0************************01000***001000001000000000****000000000000000001000000000000000000001**00000010000000*
Loricaria	0************************01000***101100001010000000****0000011100000000010000000010100000000000000000010000000*
Hypoptopoma	0************************01000***101000001010000000****0000011100000000010000000010000000000000000000010000000*
Lithoxus	0************************01000***101100001010000000****0000011100000000010000000010100000000000000000010000000*
Astroblepus	0************************01100***101000001000000000****0000011000000000000000000010100000000000000000010000000*
Callichthys	0************************01000***000200000000000000****000000000000000002000000011000000000000000000000000000*
Corydoras	0************************01000***000200000000000000****000000000000000002000000011000000000000000000000000000*
Nematogenys	100***00#000#0#*0*******#00000***0000#000100000000****00000000000000000000010100000100000000020000000000000010
Trichomycterus	0************************00000***0000#000100001000****0000000000000000000001000000000000000001000000000010010
Hatcheria	0************************00000***0000#000100001000****000000000000000000000100000000000000000000000000000#0010
Chaca	200001000000000*0*******00100100100000001****011011010000000000000000300200010000000010000011**00000000000000*
Plotosus	21010000000000000000000000100100100000#10000000010010200000000000000000000001000000000000000000000210000000111 0
Neosilurus	210100000000000000000000001001001000001000000001001020000000000000000000000010000000000000000001000000000000110
Cnidoglanis	21010000000000000000000000100100100000000000000100102000000000000000000000001000000000000000000100100000000010*
Paraplotosus	2101000000000000000000000010010010000000000000010010200000000000000000000000100000000000000000010100000000001110
Malapterurus	21000000000000000100000001###01000010000000000001001000300000000000000000000000000000000000000010000000000000020
Auchenoglanis	21000000000000000100000020100100001000000100001100100000000000000000100100010000000000000000000000000000000020
Chrysichthys	21000000000000000110000000100100000000000100001100100000000000000000100000000000000000010000000000000000000010
Clarotes	21000000000000000110000000100100000000000100001100100000000000000000100000000000000000010000000000000000000010
Ageneiosus	01***********************010010000001000010000110010001000010000000010000001000000000100000000000000000000000*
Auchenipterus	2100000000000000000#000000001001000000100000000011001000100001000000000000000000000000000#0000000000100000000000*
Centromochlus	2100000000000000000#000000001001000000100001001011001000100001000000000000000000000000000000000000000000000000*
Franciscodoras	21000000001000000100000000100100001010000100001100100010000000000000000000000000000000001000000000000000000000000*
Doras	210#0000000000000#00000000100100001010000100001100100010000100000000000000000000000001000000000000000000000000*
Anadoras	21010000000000000#00000000100100001010000100001100100010000100000000000000000000000001000000000000000000000000*
Acanthodoras	210#0000000000000#00000000100100000010000100001100100010000100000000000000000000000001000000000000000000000000*
Amiurus	21010000000000000000000000100100000000000000001100100000000000000000100000000000000001010000000000000000000010
Ictalurus	2101000000000000000000000010010000000000000000110010000000000000000010000000000000000101000000000000000000000*
Cranoglanis	2100000000100010000000000010010110000000000000110010000000000000000000000000000000000010200000000000000000000*
Mochokus	211#000000000000000000000010010000101000010000110010001000000000000000000000000000000000000000000000000000000*
Synodontis	2111000100000100000000000010110000201000010000110010001000000000000000000000000000000001000000000000000000000000*
Microglanis	2*0*000000010000000001000000100000000000000000110010000000000000000001001000000000001000000001**00000000000000*
Pseudopimelodus	2*0*00000001000000000100000010000000000000000011001000000000000000000100100000000000100000000000000000000000000*
Rhamdia	2*0*000000020000000000100010010000000000000000110010000001000000000000010000000000000000000000000000000000000010
Goeldiella	2*0*00000002000000000010001001000000000000000011001000000100000000000001000000000000010000000000000000000000000*
Heptapterus	2*0*0000000200000000001000100100000000000000000100100000010000000000000100000000000000000000000000000000000000*
Pimelodus	2*0#00000002000000010010001001000000000000000011001000000000000000000002000000000000100000000000100000000000010
Calophysus	2*0#00000002000000010010001001000000000000000011001000000000000000000002000000000000100000000000100000000000010
Hypophthalmus	2*0#0000000200000001001000100100000000000000001100102000000000000000000200000000100010000000000010000000000000*

(Contd.)

Pseudoplatystoma	2*0#000000002000000001001000100100000000000000001100100000000000000000000002000000000000010000000000000000000000000*
Bagrichthys	21010000000000000000000000000#010000200000000000011001000000000000000000000011000000000000000000000000000000000000*
Hemibagrus	21010000000000000000000000000001000000000000000011001100000000000000000000001000000000000000000000000000000000000*
Bagrus	21010000000000001000000000000001000000000000000011001100000000000000000000001000000000000000000000000000000000000*
Rita	110***00#000#0##0#*00000#010010000000000000010000110010000000000000000000001000000000000000000000000000000000000*
Wallago	100**000#000#0#*0*******#000010000000000000000000100100000000000000000001010000000000000000100100000000000000000000*
Silurus	100**000#000#0#*0*******#000010000000000000000000100100000000000000000001000000000000000001001000000001000000000000*
Laides	210000010#000000000100000001001000000000000000010001100102000000000000000000000000000000002000000000000000000000000*
Pseudeutropius	210000010#000000000100000001001000000000000000000001100100000000000000000000000000000000002000000000000000000000000*
Ailia	210000010#000000000100000001001000000000000000000001100102000000000000000000000000000000002000000000000000001000000*
Siluranodon	210000010#000000000100000001001000000000001001000110010200000000000000000000000001000020000000000000000000000000*
Shilbe	210000000100000000010000000100100000000000000000011001000000000000000000010000000000000020000000000000000000000000*
Helicophagus	110***00#100#0##0#*00000#010010000000000000000000011001000000000000000000000000100000000002000000000000000000000000*
Pangasisus	110***00#100#0##0#*00000#01001000000000000100000110010000000010000000000000001000000000020000000000000000000000000010
Parakysis	210#00000000#00000000000000100100000000000100000010010000000000000000000000000000000000000000000000000000000000000*
Akysis	210000000000100000000000000100100000000000100001100100000000000000000000000000000000000000000000000000000000000*
Amblyceps	210100002000100000000000000100100000000000000000110010000000000000000000000001000000000000000010000000000000000000000*
Liobagrus	210100000000100000000000000100100000000000000000010010000000000000000000000001000000000000000010000000000000000000000*
Erethistes	210000000000000000000000000100100000000000100001100100000000000000000000000000000000000000000000000000000000000*
Hara	210000000000000000000000000100100000000000100001100100000000000000000000000000000000000000000000000000000000000*
Aspredo	200000200000000*0*******001001000000200001000011001021001001000000000000000000100000000000001**0000000000000000*
Bunocephalus	200000200000000*0*******0010010000002000010000110010210010010000000000000000000000000000000001**0000000000000000*
Xyliphius	200000200000000*0*******00100100000020000100001100102100100100000000000000000100000000000001**0000000000000000*
Gagata	2100100010000000000000000001001000000000001000011001000000000000000000000000000000000010100000000#0000000000000000100000*
Bagarius	210000000000000000000000000100100000000000100001100100000000000000000000000000000010000000000001000000000000000000000010
Glyptosternon	2101000020000#00000000000000100100000000000100001200100000000000000000000000000000000000001000000110000000000000000000000*
Glyptothorax	21010000200010000000000000010010000000000010000110010000000000000000000000000000000000100010000001100000000000000000000*
Cetopsis	2000001000000000#0000010000010000000#000000000#000****1000000000100000000000000000000000000100000000000000101001000*
Hemicetopsis	2000001000000000#0000010000010000000#000000000#000****100000000010000000000000000000000000000000000000000100001000*
Helogenes	2100001000000000#0000010000010000000#000000000#000****0000000000010100000000000000000000000000000010000001000#0000*
Amphilius	2100000000000000*021001000010010000000000010001101010003000000000000000020000000000000000100000000000000000000000000*
Paramphilius	21000000000000000210000000###01000000000000100001010100000000000000000000000000000000000100000000000000000000000000000*
Andersonia	21000000000000000110000001**01000020000000100001020100030000000000000000000000100000000000001000000000000001000000000000*
Belonoglanis	21000000000000000110000001**010000200000001000010201000300000000000000100002000000000000010000000000000010000000000000*
Doumea	21000000000000000110000001**0100002000000010000102010002000000000000000000000010000000000000100000000000001000000000000*
Phractura	21000000000000000110000001**01000020000000100001020100030000000000000000000001000000000000010000000000000010000000000000*
Trachyglanis	21000000000000000110000001**010000200000001000010201000300000000000000100002000000000000010000000000000010000000000000*
Leptoglanis	21000000000000000110000000###01000020000000100001010100010000000001000000000100000000000001000001000000001000000000000000*
Zaireichthys	2100000000000000##00000000###0100002000000010000101010002000000000000000000000000000000000100000100000000#00001000000000*
Arius	21000000000000000110000000000100000000000000000011001000000000000000000000000001000000000002010000000000000000000000000*
Ancharius	2100000000000000##0000000100100000000000000000011001000000000#0000000001000001000000000002010000000000000000000000000*
Genidens	21000000000000000110000000000100000000000000000011001000000000000000000000000001000000000002010000000000000000000000000*
Clarias	21010000000000001100100000100111000001000000000110210200000000000000000000000000000000000000000000000000110000001000000*
Uegitglanis	2101000000000000001#00#000000100110000000000000000110210#00000000000000000000001000000000000000000000000100000010000000*
Heterobranchus	21010000000000001100100000100111000001000000000110210200000000000000000000000000000000000000000000000000110000001000000*
Heteropneustes	21010000000000000110000000001001100000000000000001102102000000000000000000000000100000000000000000000000110000001000000*

Table 3.2: Data matrix of 440 characters for each genus analysed; order of characters follows that presented and numbered in the text; inapplicable and missing character states for a certain genus are indicated with an asterisk (*) and a small square (#) respectively (Characters 111 to 220)

	11111111111111111111111111111111111111111111111111111111111111111111111111111111111111111222222222222222222222
	11111111122222222223333333333444444444455555555556666666666777777777788888888889999999999000000000011111111112
	12345678901234567890123456789012345678901234567890123456789012345678901234567890123456789012345678901234567890
Diplomystes	00000000000000000000000000000000000000000000000000000000000000000000000000000000000000000000000000000000*****0
Austroglanis	00000000000000100000100000000000000000000000000100000000000000001000000100000000000100000000000000000000200000 1
Scoloplax	00000010000000*0000000*00100000000000102*1*****1100**0*00000001101000120000000000000000000000000000010*00#####*
Loricaria	00000000000000*0000000*00100000000000002*1*****1100**0*00000000100000120000000000010000000000000000 10*00#####*
Hypoptopoma	00000000000000*0000000*00100000000000002*1*****1100**0*00000000100001120000000000000000000000000000 10*00#####*
Lithoxus	00000000000000*0000000*00100000000000002*1*****1100**0*00000000100000120000000000000000000000000000 10*00#####*
Astroblepus	00000000000000*0000000*00100000000000002*0*****1100**0*00000000100000000*0000200000***0000000000000 10*00#####*
Callichthys	10000010000000*0000000*00100000000000002*00****1000100100000000110000120000000000000000000000000000 1100000200*
Corydoras	00000010000000*0000000*00100000000000002*00****1000100100000000110000130000000000000000000000000000 1000000200*
Nematogenys	00000000000000*0000000*000000000000000001000000000000*000000000#00000000*0000200010*****01000000000110000000 1*
Trichomycterus	00100010000000*0000020*00100000000000002*00****000000*000000000#0000000000000000000000**010000000001100000000*
Hatcheria	00100010000000*0000020*00100000000000002*00****000000*000000000#0000000000000000000000**010000000001100000000*
Chaca	00000000000000*1000020*00000010010*020100100000000011000000000010000001000000020000000010000000000000 10020001*
Plotosus	00010000000000000000000000000000000010010000000000000000000000001000000100000000000100000000000000000000000001 0
Neosilurus	00010000000000000000000000000000000010010000000000000000000000001000000100000000000100000000000000000001000001 0
Cnidoglanis	00010000000000000000000000000000000010010000000000000000000000001000000100000000000100000000000000000000000001 0
Paraplotosus	00010000000000000000000000000000000010010000000000000000000000001000000100000000000100000000000000000001000001 0
Malapterurus	0000000001 0000*00000200000000000000000100000000100****000000001000001 00*0000300110*****00000001 00000##0####1*
Auchenoglanis	00000000000000100000100000000000000000000000000001000000000000010000010000010000000100000000000000000020*****1
Chrysichthys	00000000000000100000100000000000000000000001000010000000000000100000100000000000001000000000000000000000000000
Clarotes	00000000000000100000100000000000000000000001000010000000000000100000100000000000001000000000000000000000000000
Ageneiosus	01000001000000*0000020*0000000000000000000100000000 0**00100000010000001000 02**01**00000020000000000100000000 1*
Auchenipterus	00000001000000*0000020*000000000100000000010000000000000100000010000001 30002**01**00100020000000000100000000 1#
Centromochlus	00000001000000*0000020*000000000100000000010000000000000100000010000001 20002**01**00100020000000000100000000 1#
Franciscodoras	00000001000010*0000020*0000000001000000010001000000000001000000100000012 0002**01**001000200000000001000000000*
Doras	00000001000010*0000020*00000000010000000000010001 00**0*00000000100000013 0002**01**0010002000000000010000#####*
Anadoras	00000001000010*0000020*0000000001000000010001000100**0*000000001000000120002**01**00100020000000000 1000000000*
Acanthodoras	00000001000010*0000020*000000000100000000000 1000100**0*000000001000000130002**01**00100020000000000 1000000000*
Amiurus	00000000000010010000000000000000000000000000001000000000100000010000001000000000000000100000000000000000*****1
Ictalurus	00000000000010010000010000000000000000000000001000000000100000010000001000000000000000100000000000000000002001
Cranoglanis	00000000100000000000010000000000000000000000101000000010000000010000001000000000000000010000000000000 100002101
Mochokus	00000000000000000000 20*000000000100000000000100000000100100000010000001000000000000000020000000000010000******
Synodontis	00000000000000000000 20*000000000100000000011000000000100100000010000001000000000000000020000000000010000******
Microglanis	00000000000000000000000000000000000000010000000000000000100000010000001000000000000000020000000000000000000010
Pseudopimelodus	00000000000000000000000000000000000000010000000000000000100000010000001000000000000100020000000000000000000010
Rhamdia	00000000000000000000000000000000000000000000000000000000100000010000001100000000000000010000000000000000000000
Goeldiella	00000000000000000000000000000000000000000000000000000000100000010000001000000000000000010000000000000000*****0
Heptapterus	0000000000000000000000000000000000000000000000000000**0010000001000000100000000000000 0**0000000000000000000000
Pimelodus	00000000000000000000100000000000000000000000000000000100100000010000001000000000000100000000000000000000#####0
Calophysus	0000000000000000000000000000000000000000000000000000**0010000001000000100000000000000 0**0000000000000000#####0
Hypophthalmus	0000000000000000000000*00000000000000000000#00001000**000000000100000010000000000000000**00000000000000000000 10

(Contd.)

Pseudoplatystoma	0000000000000000000000000000000000000000000000000000000000000100000100000100000000000000000000000000000000000000000000
Bagrichthys	0000000001000000000010000000000000000000000000000000000000000000100000110000000000001000200000000000000000000000
Hemibagrus	0000000001000000000000000000000000000000000000000000000000000000100000120000000000000000200000000000000010001000
Bagrus	0000000001000000000000000000000000000000000000000000000000000000100000120000000000000000200000000000000010001000
Rita	000000000100000000000000000000000000000000000000000000000000000010000010000 2**00**0010002000000000000000#0000000
Wallago	0000000000000*000000 0*00000000000000001000000000000**000000000000000000000000000000000000000000000000000000000000
Silurus	0000000000000*000000 0*000000000000000100000000000000000000000000000000000000000100000000000000000000000000000
Laides	0000000000000100000100000000000000000000000000100000000000000001000001000000000001000000000000000100000000000
Pseudeutropius	0000000000000100000100000000000000000000000000100000000000000001000001000000000000000020000000000100000000000
Ailia	0000000000000*0000000*000000000000000002*00****000000**00000000100000100*000000010***00000000000100000000000
Siluranodon	0000000000000100000100000000000000000000000000100000000000000001000001000*000000010***00000000000100000000000
Shilbe	0000000000000100000100000000000000000000000000100000000000000001000001000000000000000000000000000100000000000
Helicophagus	00000000000001000001010000000000000000000000001010000000000000000100000100000000000000000000000000000000*****0
Pangasisus	00000000000001000001010000000000000000000000001010000000000000000100000100000000000000000000000000000000*****0
Parakysis	000000000000010*0000000*0000100000101010100101000000001*01000000010000010000 2**00**000000000000000100010000#####*
Akysis	000000000000010*0000000*000010000001010100101000000002001000000100000100000000000000000000000000000000010000******
Amblyceps	0000000000000*0000000*000010000000000100100000000000100100000010000010000000000000000000000000101010000******
Liobagrus	0000000000000*0000000*000010000000000100100000000000100100000010000010000000000000000000000000101010000******
Erethistes	000000000000010*1000010*00001110000000301001000000000003001000000100000130000010000010000000001000000010000******
Hara	000000000000010*1000010*00001110000000301001000000000003001000000100000130000010000010000000001000000010000******
Aspredo	000000000000010*1000000*0000110100 10*0201001000000100**3*00000000100000130002**00**000001200001000000010000******
Bunocephalus	000000000000010*1000020*0000110100 10*020100100000000103000000000100000120002**00**0000002000010000000 10000******
Xyliphius	000000000000010*1000000*0000110100 10*020100100000000030120000000100000120002**00**00000120000100000001 0000******
Gagata	0000100000000020*0000000*0000110000 0102010010#0000000030010000000100000100000000000000000000000000000010000******
Bagarius	0000100000000020*0000000*00001100000 0102010010#000000003001000000010000011000000000000000000000000000001000000000*
Glyptosternon	000020000000010*0000000*00001000000010002*01****000000**010000000100010100*0000001010*****0000000000000010000*****0
Glyptothorax	000020000000010*0000000*00001100000 0102000010#000000003001000000010000012000000000100000000000000000000010000*****0
Cetopsis	0100000000000000000100000000000000000000000000000000100100****000101000000000000010000400#00000**0100000000101000000000020
Hemicetopsis	0100000000000000000100000000000000000000000000000000100100****000101000000000000010000400#00000**0100000000101000000000020
Helogenes	01000000000001*0001000*000000000000000000000100000000100****00010000000000000000010000400#00000**010000000#010000*****0
Amphilius	0000000000000*0000000*0000000000000000000010010000010 1****00000001100000100000000201000000**0000000011010001******
Paramphilius	00000000001000*0000000*0000000000100000001001000001 01****000000011000001 00*0000201010*****000000001101000 1******
Andersonia	0000000000000*0100011*0001000000 000000100100000000000100000001100000102000003010000 00**000000000000000000******
Belonoglanis	0000000000000*0000012*000000000000000001001000000000001000000011000001022000030100000 0**002200000000000000******
Doumea	0000000000000*0000010*00000000000000000100100000000000100000001100000100010003010000 00**001100000000000000******
Phractura	0000000000000*0000010*00000000000000000100100000000000100000001100000100000003010000 00**000000000000000000******
Trachyglanis	0000000000000*000001 1*00000000000000000100100000000000100000001100100101100003010000 00**002200000000000000******
Leptoglanis	0000000000000*0000000*00010000000110000001010000000000100000001100000100000001010000000000000001000000000 0******
Zaireichthys	0000000000000*0000000*000100000002100002*01****00000001000000011000001000000010100000000000000010000000000******
Arius	000201000000002001001000100000000000000100000101000000000000000001000001000010000000010000000000000000000020010001
Ancharius	0000000#000000100000000000000000000000#000001010000000000100000100000110002**00**0000000000000000000000020000001
Genidens	000201000000002001001000100000000000000100000101000000000000000001000001000010000000010000000000000000000020010001
Clarias	0000000000000*0000000*2000000000 000202*00****00000000100010001000001000000000000000100000000000000020010001*
Uegitglanis	0000000000000*0000000*1000000000 000102*00****00000000000100010000001000000000#0000#00000000000000020010001*
Heterobranchus	0000000000000*0000000*2000000000 000202*00****00000000100010001000001000000000000000100000000000000020010001*
Heteropneustes	0000000000000*0000000*2000000000 000202*00****000000000001010100000010000000001#####0000000000000002001000#*

Table 3.3: Data matrix of 440 characters for each genus analysed; order of characters follows that presented and numbered in the text; inapplicable and missing character states for a certain genus are indicated with an asterisk (*) and a small square (#) respectively (Characters 221 to 330)

	22222222222222222222222222222222222222222222222222222222222222222222222222222222333333333333333333333333333333 22222222223333333333444444444455555555556666666666777777777788888888889999999999000000000011111111112222222223 12345678901234567890123456789012345678901234567890123456789012345678901234567890123456789012345678901234567890
Diplomystes	0001000****00000000000000000000000*0000000000000000000000000100000*0000000**00000000000000000000001000000000000
Austroglanis	1000000****0000000000000001001000000001000100010 1#000100100000000000000000001000000000000000000100001000000000000
Scoloplax	100105*00011000000000000000010000000***0100010101000000100000000100*010000*0*000000*0000111*******00000000000010
Loricaria	100106*00011100001***01000011**000***0100010001100000100000000100*000000*0*000000*0000111*******0000100000000
Hypoptopoma	100106*000111000000000000001000000***0100010001100000100000000100*000000*0*000000*0000111*******0000100000000
Lithoxus	100106*00011100000***00000001**000***0100010001100000100000000100*000000*0*000000*0000111*******0000100000000
Astroblepus	100105*0001100000000000000010000000***0100010101000000100000000100*0100001 0*000010*0000111*******00000100000000
Callichthys	100105*****000000000000000010011000***010000000100000010000000100*0100001 0*000000*000001 1******000000000000000
Corydoras	100105*****00000100000000010011000***01000#000100000010000000100*0100001 0*000000*000001 1******000000000000000
Nematogenys	0000000****00000100000000000#011000***01000000010000000000000000*0100001 0*00000000000001000000010000000000000
Trichomycterus	100100*****0000000000100200#001100***01000#000100000000100000000100*0100001 0*000000*000001 1******000000000000000
Hatcheria	100100*****0000000000100200#001100***01000#000100000000100000000100*0100001 0*000000*000001 1******000000000000000
Chaca	1000100****00101000000000000010000000**001000111010000001000000000000#0201002011000000000001 10000*00*000000000000000
Plotosus	1000100****00001000011000000001000**001000100010001101000000000001000200001100000000100010000000000100000000000
Neosilurus	1000100****000010000000000000001000**001000100010001101000000000001000200001100000000100010000000000100000000000
Cnidoglanis	1000100****000010000000000000001000**001000100010001101000000000001000200001100000000100010000000000100000000000
Paraplotosus	1000100****00001000010000000001000**001000110010001101000000000001000200001100000000100010000000000100000000000
Malapterurus	1000030100000000000000000001000000010000100010001 00#0001000000000000000000000001000000000000#0000200001000000000#000
Auchenoglanis	1000031****0000000000000001100000000**001000100010010001001000000010000000000010000000000000000000000000000100000
Chrysichthys	1010001****000000000000000010000000**001000100010010001001100000010000000000010000000000100000000000000000001001000
Clarotes	1010001****000000000000000010000000**00100010001001000100110000001000000000001000000000010000000000000000000001000
Ageneiosus	1100020****000000000000000001000001000010001010100000010000001100010000000000100000#00000100001000000000000001000
Auchenipterus	1100020****000000000000000001000001000010001010100000010000001000010000000000100000#00000100000000000000000000000
Centromochlus	1000020000100000000000000000100000110001000101010000001000000000001000100000100000#00000100001000000000000001000
Franciscodoras	1000020100000000000000000000100000100001000100010000001000000000001000100000100000100000010000#00000000000000000
Doras	1000020****000000000000000001000001000010001000100000010000000000010001000001000001000001000010000000000000001000
Anadoras	1000020****000000000000000001000001000010001000100000010000000000010001000001000001000001000010000000000000000000
Acanthodoras	1000021****000000000000000001000001000010001000100000010000000000010001000001000001000001000010000000000000000000
Amiurus	1000000000000000000000000000100010000**001000100010 0#00010000000000000000000000010000000000000100000000000000000000
Ictalurus	1000000****000000000000000001000100 0**001000100010 0#00010000000000000000000000010000000000000100000000000000000000
Cranoglanis	1010001****00000000000000002100010010000100010001 00#0001000000000000000000000001000000000000010000000000000010000000
Mochokus	100002*000000000000000000000100000101001000100010000001000000000000000001000001 01000*0000011******000000000000000
Synodontis	100002*000000000000000000000100000101001000100010000001000000000000000001000001 01000*0000011******000000000000000
Microglanis	1000010****0000000000000000100010010000100010001000000100000000000#0001000000100001000000000100*0000000000000000
Pseudopimelodus	1000010****0000000000000000100010010000100010001000000100000000000#0001000000100001000000000100*0000000000000000
Rhamdia	10000100000000000000000000001000100 0**00100010001000000100000000000020000000101000000000001000010000000000000001000
Goeldiella	10000100000000000000000000001000100 0**00100010001000000100000000000020000000101000000000001000010000000000000001000
Heptapterus	10000100000000000000000000001001100 0**00100010001000000100000000000020000000101000000000001000010000000000000001000
Pimelodus	1000011****000000000000000001000100 0**11100010001000000100000000000020000000101000010000000001010000000000000000000
Calophysus	1000011****000000000000000001000100 0****1000100010000001000000000000200000001010000100000000010100000000000000000000
Hypophthalmus	10000110000000000000000000001000100 0****1000100010000001000000000000200000001010000000000000100001000000000000000000

(Contd.)

Pseudoplatystoma	10000100000000000000000000100010 00**11100010001000000100000000000200000001010000100001010000100000000001001000
Bagrichthys	1000010****00000000000000000000010**00100010001000000100000000000200000001010010100000000000000000000000000000
Hemibagrus	1000010****00000000000000000000010**00100010001000000100000000000200000001010010100000000000001000000000000000000
Bagrus	1000010****00000000000000000000010**00100010001000000100000000000200000001010010100000000000001000000000000000000
Rita	10000#*00000000*00000000000000000010**00100010001000000100000000000200000000#0100001*0000011******00000000001000
Wallago	100104000100000000000000001000100**0010001011100000001000000000#1##0000000#010000000000000100000000000000000000000
Silurus	100104000100000000000000001000100**00100010011100#00010000000000#1##0000000#010000000000001000001000000000000000000
Laides	10000##****000000000000000100010 00**0010001001100100010010000000#000000000000100000000000001000000000000000000000
Pseudeutropius	10000#1****000000000000000100010 00**00100010011001000101100000000100000000000010000000000000100000000000000000000000
Ailia	1000000000000000000000000001000100**0010001001100100010000000000000000000000001000000000000010000000000000000000000
Siluranodon	1000000000000000000000000001000100**001000100110010001001000000000000000000000010000000000000100000000000000000000000
Shilbe	1010001****000000000000000100010 00**00100010011001000100100000001000000000000100000000010100000000000000000000000
Helicophagus	1000000****00000000200000010000000**001000100010 0#000101000000000000000000000100000000000001000000000000000000000100
Pangasisus	1000000****00000000200000010000001000010001010100#000100000000000000000000000100000000000101000000000000000000000100
Parakysis	10000#1000000000000000000##00000 0***0100010001000000100000000000100010020010000000000010001000000000000300000
Akysis	10000#00000000010000000000##00000 0***0100010001000000100000000000100010 0#0010000000000010001000000000000100000
Amblyceps	100002000000000100000000001000000 0***01000100010000001000000000001000100000100000000000100021000000000000300000
Liobagrus	100002000000000100000000001000000 0***01000101010000001000000000001000100000100000000000100021000000000000300000
Erethistes	100002000000000100000000000100000 0***010001000100000010000000000010001000001000001000001000000000000000000300000
Hara	100002000000000100000000000100000 0***010001000100000010000000000010001000001000001000001000000000000000000300000
Aspredo	100002011000000*00*000010010000 0**00100010101000001100000000000100010000100*0*10000010000*0000010********00
Bunocephalus	1000020110000000000000010010000 0**0010001000100000010000000000010001000001000001000001000*00000000000#00000
Xyliphius	1000020110000000000000010010000 0**0010001000100000010000000000010001000001000001000001000*00000000000#00000
Gagata	100002000000000100000000000100000 0***01000102010000001000000000001000200000100000#000001000000000000000000000000
Bagarius	100002000000000100000000000100000 0***010001020100000010000000000010002000001000001000001000000000000000000300000
Glyptosternon	100002000000000100000000000100000 0***010001010100000010000000000010002010001000000#000001000000000000000000000000
Glyptothorax	100002000000000100000000000100000 0***010001000100000010000000000010002000001000001000001000000000000000000300000
Cetopsis	1000000****00000001220010000000000***011000001100000#001110000000030010000100000000000100000000000000000010000
Hemicetopsis	1000000****000000012200000000000 0**0011000001100000#0011100000000300000001000000000001000000000000000000010000
Helogenes	1000000****000000012000000000000 0**001100#0011000000#00110000000000000000001000000000000100000000000000000010000
Amphilius	100002000000000000100000000000000***0100010001000000100000000100010100000010000000000010000100000000000000000
Paramphilius	100002000000000000100000000000000***0100110001000000100000000100010100000010000000000010000100000000000000000
Andersonia	1000020000000001000000000001000000***0100010001000000100000000100010100000010000001010010000000000000000400000
Belonoglanis	1000020000000001000000000001000000***0100010001000000100000000100010100000010000001010010000000000000000300000
Doumea	1000020000000001000000000001000000***0100010001000000100000000100010100000010000001010010000000000000000300001
Phractura	1000020000000001000000000001000000***0100010001000000100000000100010100000010000001010010000000000000000300001
Trachyglanis	1000020000000000000000000001000000***0100010001000000100000000100010100000010000001010010000000000000000300000
Leptoglanis	1000020000000000000000000000000000***0101010101000000100000000100010100000010000000000010000200000000000200000
Zaireichthys	1000020000000000000000000000000000***0100010101000000100000000100010100000010000000000010000200000000000100000
Arius	1010001****000001000000000100000010000100010001001000100100000000200000000010000000000000000000000000000000000000
Ancharius	10#0001****0000000000000001000000100001000100010010001001000000000000000000010000000000010000000000000000000000000
Genidens	1010001****000001000000000100000010000100010001001000100100000000200000000010000000000010000000000000000000000000
Clarias	100001000000000100010000020000000***0100010011000000100000000000100000000110000010000010000000000000000001000
Uegitglanis	100001000000000100010000020000000***01000100110000001000000000001000000001100000#00000100000000000000000001000
Heterobranchus	100001000000000100010000020000000***0100010011000000100000000000100000000110000010000010000000000000000001000
Heteropneustes	100001000000000100010000020000000***0100010011000000100000000000100000000111000010000010000000000000000001000

Table 3.4: Data matrix of 440 characters for each genus analysed; order of characters follows that presented and numbered in the text; inapplicable and missing character states for a certain genus are indicated with an asterisk (*) and a small square (#) respectively (Characters 331 to 440)

	33333333333333333333333333333333333333333333333333333333333333333333344444444444444444444444444444444444444444 33333333344444444445555555555666666666677777777778888888888999999999900000000001111111111222222222233333333334 12345678901234567890123456789012345678901234567890123456789012345678901234567890123456789012345678901234567890
Diplomystes	00000000000000000000000000000000000000000000000000000000000000000000000000000000000000000000000000000000000000
Austroglanis	0100000000000000000000010001000000000000000000000000000001000000000000000000000#0000000000000010000000000000000
Scoloplax	00000000000000*00000000000000000000000000000000000000000100000000010000000000011000000000010**01***0000000002001
Loricaria	00000000000000*100002*************000000000000000000000100000000010000000110011000000000010**21***0000000002010
Hypoptopoma	00000000000000*100002*************000000000000000000000100000000010000000110011000000000010**21***0000000002010
Lithoxus	00100000000000*100002*************000000000000000000000100000000010000000110011000000000010**21***0000000002010
Astroblepus	00000000000000*1000001000000*0**010000000000000000000000000000000010000000110011000000000010**21***0000000001000
Callichthys	0000000000000000000000000000000000000000000000000000000000000001100000000000001000010000100**01***0010002110
Corydoras	0000000000000000100000000000000000000000000000000000000000000001001*00000000001000010000100**11***0010002110
Nematogenys	00000002000000000002000010000000000000000000010001010001000000000000000000010010000000000001000000000000001000
Trichomycterus	000000011000000000020000110010000000000000000000000000000000000000000000000100100000000000001#00000000000002000
Hatcheria	000000011000000000020000110010000000000000000000000000000000000000000000000100100000000000001#00000000000002000
Chaca	00000020000000000000001010000000000000000100000000000100001000000000000000000010000100100000000100000000000000
Plotosus	00000000000000000000000000000000000000000000000000000100000000000000000000000000000100100000000100000000000000
Neosilurus	00000000000000001000000000000000000000000000000000000100000000000000000000000000000100200000000100000000000000
Cnidoglanis	00000000000000000000000000000000000000000000000000000100000000000000000000000000000100000000000100000000000000
Paraplotosus	00000000000000000000000000000000000000000000000000000100000000000000000000000000000100100000000100000000000000
Malapterurus	000000000000000000000000000000000000000000000000000000000000000000000000000#000000010000000000100000000000000
Auchenoglanis	0000000000000000000000000000000000001000000000000000001000000000000000000000000100001#10000000000100000000000000
Chrysichthys	00000000000000000000000000000000000000000000000000000010000000000000000001000000000001000000000010000000000000
Clarotes	00000000000000000000000000000000000000000000000000000000000000000000000001000000000001000000000010000000000000
Ageneiosus	01010000000000000000000000000000000000000000000000000000000000000000000000000000000000000010001000000000110000
Auchenipterus	00010000000000000000000000000000000000000000000000000000000000000000000000000000000000000010001000000000110000
Centromochlus	000200000000000001000000000000000000000000000000000000000000000000000000000000000000000000#0001000000000110000
Franciscodoras	00010000000000000000000000000000000010000000000000000000000000000000000000000000000000000010001100000001000000
Doras	00000000000000000000000000000000000010000000000000000000000000000000000000000000000000000010001000000001000000
Anadoras	00010000000000000000000000000000000010000000000000000000000000000000000000000000000000000010001000000001000000
Acanthodoras	00010000000000000000000000000000000010000000000000000000000000000000000000000000000000000010001000000001000000
Amiurus	00010000000000000000000000000000000000000000000000000100000000000000000000000000000000000000001000000000000000
Ictalurus	00010000000000000100000000000000000000000000000000000100000000000000000000000000000000000000001000000000000000
Cranoglanis	00000000000000000100000000000000000000000000000000100001000000000000000000000000000000000000001000000000000000
Mochokus	000000000000000000000000000200000000000000000000000000010000000000000001**000010010000000010**11*0*0000000000000
Synodontis	000000000000000000000000000200000000000000000000000000000100000000100001**000010010000000010**11*0*0000000000000
Microglanis	02000000000000000000000000000000000000000000000000000000000000000000000000000000000000000000001000000000000000
Pseudopimelodus	02000000000000000000000000000000000000000000000000000000000000000000000000000000000000000000001000000000000000
Rhamdia	02000000000000000000000000000000000000000000000000000100000000000000000000000000000000000000001000000000000000
Goeldiella	02000000000000000000000000000000000000000000000000000100000000000000000000000000000000000000001000000000000000
Heptapterus	01000000000000000000000000000000000000000000000000000100000000000000000000000100000000000000001000000000000000
Pimelodus	00000000000000000000000000000000000000000000000000000000000000000000000000000000000000000000001000000000000000
Calophysus	00000000000000000000000000000000000000000000000000000000000000000000000000000000000000000000001000000000000000
Hypophthalmus	00000000000000000000000000100000000000100000000000000000000000000000000000000001*000000000000001000000000000000

Pseudoplatystoma	000000000000000000000000001000000000000000000000000000000000000000000000000000000000000000000000001000000000000000
Bagrichthys	000000000000000000001000000000000000000000001000000000000010000000000001*00000000000000000000000001000000000000000
Hemibagrus	010000000000000000000000000000000000000000000000000000000000000000000000000000000000000000000000001000000000000000
Bagrus	010000000000000000000000000000000000000000000000000000000000000000000000000000000000000000000000001000000000000000
Rita	0#0000000000000000000000000000000000000000000000000000000000000000000000000000000000000000000000001000000000000000
Wallago	000010000000000000000000000000000000000000000000000000000000000000000000000000000000000000010**001000000000000000
Silurus	000010000000000000000000000000000000000000000000000000000000000000000000000000000000000000010**001000000000000000
Laides	000000000000000000000000000000001000000000000000000000000000000000000000000000000000000000000100001000000000000000
Pseudeutropius	000000000000000000000000000000000000000000000000001000000000000000000000000000000000000000000100001000000000000000
Ailia	00000000000000000000000000000000000000000000000000000000000000000000001*0000000000000000000001000010000000000000000
Siluranodon	000000000000000000000000000000000000000000000001001000000000000000001*0000000000000000000001000010000000000000000
Shilbe	020000000000000000000000000000000000000000000000000000000000000000000000000000000000000000000100001000000000000000
Helicophagus	000000000000000000000000000000000000000000000000000000000000000000000000001000000000000000000000001000000000000000
Pangasisus	000000000000000000000000000000000000000000000000000000000000000000000000001000000000000000010000001000000000000000
Parakysis	000000000000000000000000000000000000000000000000000000000000000000000000010000001000000000000000001000001100000000
Akysis	000000000000000000000000000000000000000000000000000000000000000000000000010000001000000000000000001001001000000000
Amblyceps	0000000000000000000000000000000000000000000000000000000000000100100000000*000000100000000000000000100000000000000
Liobagrus	000000000000000000100000000000000000000000000000000000000000000100100000000*00000010000000000000000010000000000000
Erethistes	000000000000000000000000000000000000000000000000010000000000000000000000010000001000000100000000001000010000000000
Hara	000000000000000000000000000000000000000000000000010000000000000000000000010000001000000000000000001000010000000000
Aspredo	000000000000200*00002************0010000000000000000000000000000000000000000000100000000000000000010000110000000000
Bunocephalus	100200000010000000000000000000000002000100100000000000000000000000000000000000001000000000000010001000011000000000
Xyliphius	000200000010000000001000000000000001000000000000000000000000000000000000000000001000000000000010001000011000000000
Gagata	00000000000000000000000000000000000000000000000000000000000000000000000001000000#00000000000000000100000000000000
Bagarius	0000000000000000000000000000000000000000000000000000000000000000000000000*000000100000000000000000010#0000000000000
Glyptosternon	000000000000000000000000000020000000000000000001000000000000000000000000010000001000000000000000001000000000000000
Glyptothorax	000000000000000000000000000020000000000000000000000000000010000000000000010000001000000000000000001010100000000000
Cetopsis	100000000010001001000000000000201000000000000000000000000000000120000000000000000000100010000000001000000000000000
Hemicetopsis	200000000010001001000000000000101000000000000000000000000000000120000000000000000000000200000000001000000000000000
Helogenes	000000000010001000000000000000301000000100000000000000000100000010000001000000000000200000000000001000000000000000
Amphilius	000000000010000000000000000000000001000000000000000000000000000000000000000000110000000000001 0**101000000000000000
Paramphilius	010000000010000000000000000000000001000000000000000000000000000000000000000000010000000000001 0**101000000000000000
Andersonia	000000000000100000000000000010000000000200000000000100000000000000000011*00000011000000000000 10**201000000000000000
Belonoglanis	000000000000100000000000000010010000000200000000000100000000000000000011*00000011000000000000 10**201000000000000000
Doumea	000000000000100000000000000010000000000100000000000100000000000000000011*00000011000000000000 10**201000000000000000
Phractura	000000000000100000100000000010000000000100000000000100000000000000000011*00000011000000000000 10**201000000000000000
Trachyglanis	000000000000100000000000000010000000000200000000000100000000000000000011*00000011000000000000 10**201000000000000000
Leptoglanis	000000000010000000000000000000000000000000000000000100010000000000000000000000010000000000000 10**1012000000000000000
Zaireichthys	0000000000#0110001000000000000000000000000000000000001000001000000000000000000000000#00000000000 10**1010#0000000000000
Arius	000000100000000000100000000000000000000000000000000000000000000000000000000000000000001000000000001000000000000000
Ancharius	000000000000000000100000000000000000000000000000000000000000000000000000000000#0000000000000000001000000000000000
Genidens	000000000000000000000000000000000000000000000000000000000000000000000000000000000002000000000000001000000000000000
Clarias	000001000100000010000000000000000000000000000100000000000000000000000000000000000000000000000010**001000000000000000
Uegitglanis	000000000100000010000000000000000000000000000100000000000000000000000000000000000000000000000010**101000000000000000
Heterobranchus	000001000100000010000000000000000000000000000100000000000000000000000000000000000000000000000010**001000000000000000
Heteropneustes	000000000100000010000000000000000000000000000100000000000000200000000000000000000000000000000010**001000000000000000

Clade 1 (All Genera Excluding *Diplomystes*)

Diagnosis. [221:0→1], **[259:0→1]**, **[267:0→1]**, [308:0→1]

Comments. As explained in Chapters 1 and 2, the sister-group relationship between *Diplomystes* and all other catfishes clearly constitutes the most consensual subject concerning siluriform higher level phylogeny (see, e.g., Regan, 1911b; Alexander, 1965; Chardon, 1968; Gosline, 1975; Arratia, 1987; Mo, 1991; de Pinna, 1993). Two of the synapomorphies given above are homoplasy-free within the siluriform genera analysed (259: absence of maxillary teeth; 267: absence of firm ligamentous connection between mesial surface of mandible). One (221: separation of extensor tentaculi from adductor arcus palatini) is almost homoplasy free (it only changes subsequently to CS0 in *Nematogenys*). The other one (308: absence of sesamoid bone 2 of the suspensorium) is highly homoplasic. It should be noted, however, that besides the diplomystids, the presence of maxillary teeth was also documented in the fossil catfish †*Hypsidoris*, which led some authors to hypothesise a rather plesiomorphic position of this latter genus within the Siluriformes (see Section 1.3). According to these authors, the absence of maxillary teeth would separate not only the Diplomystidae, but also the Hypsidoridae, from the remaining catfishes (see Section 1.3).

Clade 2 (*Diplomystes*)

Diagnosis. [106:0→1], **[280:0→1]**, **[317:0→1]**

Comments. Of the three characters unambiguously assigned to *Diplomystes*, only the latter two seemingly constitute autapomorphic features (character 106, presence of a prominent anteroventrolateral process of the pterosphenoid, is also present in *Trichomycterus*): 1) anterior portion of autopalatine with a donut-like aspect (character 280); 2) presence of a somewhat developed sesamoid bone 3 of the suspensorium (character 317) (for a detailed insight into diplomystid apomorphies, see Arratia, 1987).

Clade 3 (All Genera Excluding *Diplomystes, Nematogenys, Trichomycterus, Hatcheria, Callichthys, Corydoras, Scoloplax, Astroblepus, Hypoptopoma, Loricaria, Lithoxus*)

Diagnosis. [1:0→2], **[30:0→1]**, **[296:0→1]**, **[425:0→1]**

Comments. The monophyly of this clade is strongly supported by these four unambiguous characters, of which three are homoplasy-free. The differentiation of the protractor hyoideus in a well-differentiated pars ventralis, pars lateralis and pars dorsalis (char. 30), and the articulatory surface of the autopalatine for the neurocranium essentially directed mesially (char. 296), are present in all catfishes of this clade and absent in all siluriforms outside it. The coronomeckelian reduced in size (char. 425) is also homoplasy free, but inapplicable in some genera inside the clade (*Synodontis* and *Mochokus*) and outside it (*Callichthys, Corydoras, Scoloplax, Astroblepus, Hypoptopoma, Loricaria, Lithoxus*). The transition from no mandibular barbels to two pairs of mandibular barbels (char. 1) is not homoplasy free but stills constitutes a strong argument to support the monophyly of the clade since, according to the present analysis, only a genus inside it, *Ageneiosus*, reverted from CS–2 to CS–0 (some other genera of the clade passed from CS–2 to CS–1, but only *Ageneiosus* reverted from CS–2 to CS–0; in the same way, 1 of the 11 genera outside the clade, *Nematogenys*, passed from CS–0 to CS–1, but not from CS–0 to CS–2).

That this clade encompasses three homoplasy-free characters clearly constitutes a remarkably strong support for its monophyly. As noted by de Pinna (1996, 1998), homoplasy-free features are particularly rare among such a vast and complex group as the Siluriformes. Therefore, the presence of not one, but three such features in a clade clearly constitutes a very strong argument to support its monophyly. The other unambiguous synapomorphic character, although homoplasic, should also not be neglected.

It is also important to note here that besides these four unambiguous synapomorphies, the present cladistic analysis also points out another feature that, although with an ambiguous distribution on the cladogram, could eventually constitute a putative synapomorphy to support this clade. This feature concerns the passage from a well-developed, essentially transversal to a thin, essentially anteroposterior arrector ventralis (char. 48). This character does not appear as an unambiguous synapomorphy since it was not possible to discern the configuration of the arrector ventralis in the cetopids examined due to the peculiar configuration of the pectoral girdle of these catfishes (see description of the character). However, whether this feature would prove to be a synapomorphy of all extant catfishes excluding the Diplomystidae and the Loricarioidea or excluding these two latter groups plus the Cetopsidae, it would nonetheless constitute strong evidence to support a rather basal position of the Loricarioidea within the Siluriformes (the presence of state 0 of this character in the Amphiliidae would anyway be assigned to a secondary, homoplasic reversion in both cases).

Positioning of the loricarioids as the sister-group of all remaining extant non-diplomystid catfishes constitutes an important outcome of the present phylogenetic analysis. In fact, according to the phylogenetic analysis of Mo (1991) and de Pinna (1993), the loricarioids would be closely related with the amphiliid catfishes (see Section 1.3). However, it should be noted that the evidence provided by these authors to support such a close relationship is relatively sparse. Mo (1991) provided a single feature, the absence of the posterior cartilage of the autopalatine, to support such a relationship. But this feature is also present in the malapterurids and absent in the loricarioid *Nematogenys* and hence is not homoplasy free. With respect to de Pinna (1993), this author provided five features to support the clade Amphiliidae + Loricarioidea, but all these are highly homoplasic, with the latter two referring inclusively to plesiomorphic, general features that were supposedly acquired by secondary reversion: 1) Meckel cartilage without an ascending portion; 2) markedly truncated lateral ethmoid; 3) anterior face of lateral ethmoid for autopalatine directed anteriorly or anterolaterally; 4) reversion of posterior process of basipterygium broadly fused basally; 5) reversion of flange extending along ventral surface of basipterygium.

So, not even a single unique, homoplasy-free feature has been presented to date to support the sister-group relationship between Amphiliidae and the Loricarioidea. The evidence presented in this work to group amphiliids with all the remaining non-diplomystid and non-loricarioid catfishes (three unambiguous homoplasy-free features, one unambiguous homoplasic feature, and one ambiguously distributed character) thus seems to be considerably stronger than the evidence presented so far to group Amphiliidae with Loricarioidea. The difference between the hypothesis presented in this study and that of Mo (1991) and de Pinna (1993) could be related with the inclusion, in this work, of characters not usually included in the investigation of siluriform higher level phylogeny, such as, e.g., those concerning myological features (of the five characters listed above to support a markedly basal position of the Loricarioidea, two refer to myological features: differentiation of the protractor hyoideus in three bundles and the thin, essentially anteroposteriorly oriented arrector ventralis).

It is also important to note that although the cladistic analysis of Arratia (1992) included only a few catfish families, its results somehow supported, like the present study, a markedly basal position of the loricarioids within the siluriforms (see Section 1.3 above). Also, the outstanding work of Chardon (1968), albeit not based on an explicit cladistic analysis, suggested an also markedly basal position of the Loricarioidea. In the 'evolutionary scheme' provided on page 241 of Chardon's work there is a 'generalised stock I' at the very base of the Siluriformes, from which evolve the Diplomystidae and a 'generalised stock II'. Subsequently, from this 'generalised stock II' evolve five different groups: 1) the malapterurids; 2) the cetopsins; 3) a group including the silurids, amblycepitids and the helogenins; 4) a group including the Loricarioidea and the Aspredinididae; and 5) a group including all the remaining catfishes. Therefore, apart from the Aspredinidae, the Loricarioidea appear to have evolved clearly separate from all other non-diplomystid families in Chardon's work. Notably, the Loricarioidea appear remarkably distantly related to the Amphiliidae (see discussion above).

It is also noteworthy that several other authors who have studied in detail the osteological structures of the loricarioid catfishes have pointed out that some of these fishes, such as nematogenyids and trichomycterids, present certain plesiomorphic characters that, in some cases, are only comparable with those present in the Diplomystidae (see, e.g., Arratia and Chang, 1975; Schaefer, 1990; Schaefer and Lauder, 1996; Azpelicueta and Rubilar, 1998). Even those authors placing the nematogenyid and trichomycterid loricarioids in a rather derived position within the siluriforms (Mo, 1991; de Pinna, 1993) agree with the morphological primitiveness of certain nematogenyid features (e.g., the thin medial arm of the scapulo-coracoid not suturing medially with its counterpart). They explain, however, this primitive morphological configuration as being due to a homoplasic, secondary reversion (see above).

Another author suggesting a rather plesiomorphic position of the Loricarioidea, this one based on palaeontological data, was Reis (1998b). Referring to a fossil callichthyid catfish, †*Corydoras revelatus*, from the late Palaeocene 'some 58.5 million years ago', he noted that 'the relatively derived position of *Corydoras* in loricarioid phylogeny makes †*Corydoras revelatus* extremely old'. This led him to ask whether this might not mean that 'loricarioids undergo an earlier differentiation than other catfishes and modern teleosts in South America?' (Reis, 1998b: 359-361).

Taking into account all the points mentioned above, there are thus some good arguments to support the markedly basal position of the Loricarioidea within the order Siluriformes.

Clade 4 (*Nematogenys, Trichomycterus, Hatcheria, Callichthys, Corydoras, Scoloplax, Astroblepus, Hypoptopoma, Loricaria, Lithoxus*)

Diagnosis. [150:0→2], [288:0→1], [293:0→1], [408:0→1], [437:0→1]

Comments. The grouping of the nematogenyids, trichomycterids, callicthyids, scoloplacids, astroblepids and loricariids in the superfamily Loricarioidea constitutes, as explained above, a clearly consensual issue nowadays (see, e.g., Howes, 1983a; Schaefer, 1990; Mo, 1991, de Pinna, 1992, 1993, 1998). However, as de Pinna (1998) noted, this concensually recognised group relies on only a few characters, with not even one of them being homoplasy free within the Siluriformes. The few characters listed by de Pinna (1998: Fig. 6) were: the presence of onodontes; mesial processes of exoccipitals not meeting at midline; arms of basipterygium lacking anterior cartilages in adults; jaw teeth with bifid cups.

The present cladistic analysis pointed out four additional synapomorphic features to characterise the loricarioids, but these, too, are not homoplasy free: 1) mesial limb of posttemporo-supracleithrum undifferentiated (also present in clariids, *Zaireichthys, Glyptosternon* and *Ailia*, and absent in *Nematogenys*: see char. 150); 2) autopalatine markedly compressed dorsoventrally, with dorsal salience to form the articulatory surface for the neurocranium (absent in loricariids: see char. 288); 3) posterior portion of autopalatine presenting a moderately pronounced lateral encurvation (inapplicable in loricariids and scoloplacids: see char. 293); 4) only small part of angulo-articular visible in lateral view of mandible (also present in chacids, aspredinids, *Heptapterus*, and the vast majority of amphiliids and sisoroids: see char. 408). [Note: character 437 of the present analysis refers to the presence of onodontes, which is precisely one of the four characters listed in de Pinna's 1998 overview, as explained above.]

Clade 5 (*Nematogenys, Trichomycterus, Hatcheria*)

Diagnosis. **[78:0→1]**, [109:0→1], [200:0→1], [211:0→1], **[338:0→1]**, [350:0→2], [355:0→1], **[405:0→1]**

Comments. These eight characters strongly support a hypothesis already concensually accepted, i.e., the sister-group relationship between nematogenyids and trichomycterids. In fact, although Howes (1983a) suggested that the nematogenyids were probably the sister-group of all other loricarioids, he emphasised that this hypothesis was weakly supported and even contradicted by some derived features present in both Trichomycteridae and Nematogenyidae (see Section 1.3). Schaefer (1990) was fully aware of this problem and hence preferred to consider the relationships of the Trichomycteridae and Nematogenyidae as unresolved a priori. Mo (1991) analysed this problem and concluded that the nematogenyids and trichomycterids were probably sister-groups, with the clade formed by these two groups being the sister-group of all other loricarioids. This hypothesis was subsequently strongly supported by de Pinna (1992), who included a new, undescribed group of trichomycterids in his phylogenetic analysis, the Copionodontinae. This group proved to be the most basal taxon within the Trichomycteridae and pointed out that some characters commonly used to place the members of this family as the sister-group of all non-nematogenyid loricarioids were, in reality, plesiomorphically absent in trichomycterids. A sister-group relationship between the Nematogenyidae and Trichomycteridae was also corroborated by de Pinna (1998), who provided some new data to support it (see de Pinna, 1998: 296-297). The grouping of nematogenyid and trichomycterid catfishes clearly appears, therefore, to be a particularly well-supported hypothesis within the higher level phylogeny of the Siluriformes.

Clade 6 (*Nematogenys*)

Diagnosis. [1:0→1], [76:0→1], [84:0/1→2], [150:0→1], [188:0→2], [192:0→1], [219:0→1], [221:1→0], [237:0→1], **[316:0→1]**, **[338:1→2]**, **[376:0→1]**, **[380:0→1]**, **[382:0→1]**, [386:0→1]

Comments. It is interesting to note that, although *Nematogenys* was consensually included in a separate family, the Nematogenyidae, no clear, unique autapomorphic features have been provided so far in the literature to differentiate nematogenyids from all the other siluriforms. The single 'autapomorphic' feature listed in de Pinna's 1998 extensive and detailed overview, the lateral insertion of the levator operculi on the opercle, is also found, albeit less pronounced, as noted by de Pinna, in other taxa such as *Heptapterus* or *Austroglanis* (see char. 250 of the present work).

The cladistic analysis in the present work pointed out four autapomorphic features present in *Nematogenys* and in no other catfishes examined: 1) ligament connecting anterior margin of suspensorium and ethmoid region attaching significantly anterior to anterior margin of lateral ethmoid (char. 316); 2) presence of prominent ventrolateral crest of anterior ceratohyal (char. 376); 3) presence of well-developed, long, sharply pointed posterodorsal process of anterior ceratohyal (char. 380); 4) presence of prominent, broad anteroventral lamina of anterior ceratohyal (char. 382). [Note: character 338 was not included in this list since the presence of a highly developed, elongated posterodorsal projection of the hyomandibulo-metapterygoid firmly attaching to the neurocranium by massive, strong connective tissue, is also present, although in a less pronounced way, in the trichomycterid catfishes examined.]

Clade 7 (*Trichomycterus, Hatcheria*)

Diagnosis. [41:0→1], **[113:0→1]**, [131:0→2], [242:0→1], **[245:0→2]**, **[252:0→1]**, **[275:0→1]**, **[339:0→1]**, **[356:0→1]**, **[359:0→1]**

Comments. As noted in de Pinna's 1998 overview, the family Trichomycteridae, as presently defined, is a clearly monophyletic group. However, the nine synapomorphies listed above to support the monophyly of the two trichomycterid genera examined should not in any way be regarded as diagnostic features of the Trichomycteridae. In fact, as explained in Section 2.2, among all catfish families included in the present cladistic analysis, the Trichomycteridae is probably the most poorly represented. This is due to the fact that the only two genera that could be included in the analysis, *Trichomycterus* and *Hatcheria*, are both from the subfamily Trichomycterinae, which is just one of the eight trichomycterid subfamilies (see Section 2. 2).

Clade 8 (*Callichthys, Corydoras, Scoloplax, Astroblepus, Hypoptopoma, Loricaria, Lithoxus*)

Diagnosis. [27:0→1], **[158:0→1]**, [180:0→1], [181:0→2], [226:0→5], [247:0→1], [419:0→1], [424:0→1]

Comments. The grouping of callicthyids, scoloplacids, astroblepids and loricariids in a monophyletic clade is consensually accepted nowadays, with several characters adduced as support (see, e.g., Howes, 1983a; Schaefer, 1990; Mo, 1991; de Pinna, 1992, 1993, 1998). The present cladistic analysis pointed out some new, but rather homoplasic, synapomorphic features to support this clade: intermandibularis reduced in size (char. 27); presence of interdigitation between well-developed scapulo-coracoids (char. 180); markedly developed posterior process of scapulo-coracoid (char. 181); extensor tentaculi with two markedly elongated ventral bundles attaching on posteroventrolateral surface of autopalatine and a more dorsal bundle essentially oriented dorsoventrally and attaching on posterodorsal surface of this bone (char. 226); presence of well-developed adductor hyomandibularis (char. 247). [Note: characters 158, i.e., posttemporo-supracleithrum and pterotic fused into a single element 419, absence of ascending portion of Meckel's cartilage; and 423, absence of coronomeckelian bone, were based on previous descriptions of other authors: for description of these characters, see Section 3.1.]

Clade 9 (*Callichthys, Corydoras*)

Diagnosis. [37:0→2], [73:0→2], [81:0→1], **[175:0→1]**, [250:0→1], **[395:0→1]**, [414:0→1], **[433:0→1]**, **[438:0→1]**, [439:0→1]

Comments. As stated in Section 1.2, the monophyly of the Callichthyidae is well supported in the literature. The present phylogenetic analysis also supported it, but failed to detect any new, undescribed callichthyid autapomorphies (characters 175, i.e., presence of a firm association between the posterolateral processes of the scapulo-coracoid and cleithrum; 395, presence of 'branchiostegal cartilage', 433; first infraorbital presenting markedly developed inner expansion; and 438, presence of well-developed additional 'mandibular barbel', were based on previous descriptions of other authors: for description of these characters, see Section 3.1).

Clade 10 (*Scoloplax, Astroblepus, Hypoptopoma, Loricaria, Lithoxus*)

Diagnosis. **[36:0→1]**, [102:0→1], [159:0→1], **[232:0→1]**, [263:0→1], [307:0→1], [407:0→1]

Comments. The close relationship between scoloplacids, astroblepids and loricariids is also well-supported in the literature (see, e.g., Schaefer, 1990; Mo, 1991; de Pinna, 1993, 1998). Of the seven characters listed above, four provide new evidence to support this close relationship, namely: 1) absence of anterior fontanel (char. 102); 2) presence of one, single, well-developed dorsal process of cleithrum for articulation with posttemporo-supracleithrum (char. 159); 3) posterodistal margin of maxilla not markedly concave (char. 263); 4) main branch of ramus mandibularis enclosed in well-developed mandibular bony tunnel (char. 407). [Note: characters 36, i.e., hyohyoideus inferior with bilateral bifurcation; 232, presence of a retractor premaxillae; and 307, absence of ligamentous connection between anterior margin of suspensorium and ethmoid region, were based on previous descriptions of other authors: for description of these characters, see Section 3.1.]

Clade 11 (*Scoloplax*)

Diagnosis. **[70:0→1]**, [93:0→1], [148:0→1], [173:0→1], **[176:0→1]**, **[329:0→1]**, [398:0→1], **[440:0→1]**

Comments. Inclusion of genus *Scoloplax* in a separate catfish family is also consensually accepted nowadays, with several characters having been provided to characterise the scoloplacids (see, e.g., Schaefer, 1990; Mo, 1991; de Pinna, 1993, 1998). The present cladistic analysis pointed out no new, undescribed scoloplacid autapomorphies, since the mesethmoid presenting a prominent anteromesial process (char. 70), the entoectopterygoid being a peculiarly boomerang-shaped structure (char. 329), and the presence of a well-developed, broad 'rostral plate' (char. 440) were based on previous descriptions of other authors (for description of these characters, see Section 3.1). It is important to note that the other homoplasy-free character listed above, the presence of a large dorsolateral foramen of the cleithrum (char. 176), does not constitute a scoloplacid autapomorphy. Instead, it seems to constitute an autapomorphy of a clade inside the Scoloplacidae including two of the four species of this family, *Scoloplax distolothrix* and *Scoloplax empousa* (for more details on this subject, see Schaefer, 1990).

Clade 12 (*Astroblepus, Hypoptopoma, Loricaria, Lithoxus*)

Diagnosis. **[34:0→1]**, **[61:0→1]**, **[62:0→1]**, [346:0→1], [396:0→1], **[403:0→1]**, **[404:0→1]**, [423:0→2]

Comments. The sister-group relationship between the astroblepid and loricariid catfishes constitutes a well-grounded hypothesis supported by several synapomorphic features (see, e.g., Howes, 1983a; Schaefer, 1990; Mo, 1991; de Pinna, 1993, 1998). The present cladistic analysis confirmed these synapomorphies (see descriptions of the seven characters listed above in Section 3.1), but revealed no additional characters for diagnosing this clade.

Clade 13 (*Astroblepus*)

Diagnosis. **[28:0→1]**, [180:1→0], [181:2→0], [188:0→2], **[300:0→1]**, **[322:0→1]**, [437:0→1]

Comments. Inclusion of genus *Astroblepus* in a separate catfish family has also been supported by a series of studies (see, e.g., Schaefer, 1990; Mo, 1991; de Pinna, 1993, 1998). The present cladistic analysis revealed an autapomorphic character present in the astroblepids examined and in no other catfishes: presence of a voluminous, globular structure on the anteroventral surface of the entoectopterygoid, in which attaches a strong ligament connecting this bone to the premaxilla (char. 322) [characters 28, i.e. muscle intermandibularis not continuous in midline and 300, presence of a long and thin ligament between the anterolateral surface of the lateral ethmoid and the dorsomesial margin of the autopalatine, were based on previous descriptions of other authors: for description of these characters, see Section 3.1].

Clade 14 (*Hypoptopoma, Loricaria, Lithoxus*)

Diagnosis. **[44:0→1]**, **[63:0→1]**, [73:0→1], [226:5→6], **[233:0→1]**, [247:1→0], [248:0→1], **[268:0→1]**, [288:0→1], **[321:0→1]**, [351:0→2], [439:0→1]

Comments. The loricariids are consensually considered a specialised, monophyletic catfish group diagnosed by several synapomorphic features (see, e.g., Alexander, 1965; Gosline, 1975; Chardon, 1968; Howes, 1983a; Schaefer, 1990; Mo, 1991; de Pinna, 1993, 1998). The present work detected two new autapomorphic characters present in the loricariids examined found in no other siluriforms: 1) abductor superficialis 1 situated on posterior surface of pectoral girdle (char. 44); 2) presence of smooth, roundish surface on proximoventral of maxilla (char. 268). [Note: characters 63, i.e. presence of two well-defined ligaments between ventral surface of mesethmoid and premaxilla; 233, presence of a retractor palatini and 321, presence of a firm, long suture between dorsal surface of the entoectopterygoid and neurocranium, were based on previous descriptions of other authors: for description of these characters, see Section 3.1.]

Clade 15 (*Loricaria, Lithoxus*)

Diagnosis. [37:0→1]

Comments. Although a study on the intrarelationships within family Loricariidae is clearly beyond the main scope of this work, the present cladistic analysis did support a closer relationship between *Loricaria* and *Lithoxus* than between either of these genera and genus *Hypoptopoma*. This accords with the commonly accepted view that within Loricariidae, Loricariinae (including *Loricaria*) and Ancistrinae (including *Lithoxus*) are more closely related to each other than to Hypoptopomatinae (including *Hypoptopoma*) (see de Pinna, 1998).

Clade 16 (All Genera Excluding *Diplomystes, Nematogenys, Trichomycterus, Hatcheria, Callichthys, Corydoras, Scoloplax, Astroblepus, Hypoptopoma, Loricaria, Lithoxus, Helogenes, Cetopsis, Hemicetopsis*)

Diagnosis. **[51:0→1]**, [263:0→1]

Comments. The results of the present cladistic analysis support a markedly basal position of the cetopsids within the order Siluriformes, with these catfishes appearing as the most basal extant group after the diplomystids and the loricarioids (see Fig. 3.123). This essentially agrees with the two other published cladistic papers dealing with the phylogenetic position of the cetopsids among Siluriformes, Mo's (1991) and de Pinna's (1998). In Mo's 1991 paper the cetopsids appear as the most basal non-diplomystid catfishes. Cetopsids and diplomystids are separated from the remaining catfishes by a 'computer node' in Mo's cladogram 1 (see Mo, 1991: Fig. 4) and by the fact that in these two groups the 'ramus mandibularis nerve (does not run) inside hyomandibular for a distance' in Mo's cladogram 2. [Note: although Mo considered the Cetopsidae of the present study a non-monophyletic group, the genus *Hemicetopsis*, the remaining cetopsines and the helogenines appear in a more basal position than all other non-diplomystid catfishes, including the fossil hypsidorids, in Mo's 1991 cladograms]. De Pinna (1998: 292) stated that the cetopsids, together with the fossil hypsidorids, are the most basal non-diplomystid catfishes since 'they lack some synapomorphies of all other catfishes except for diplomystids and in some instances also hypsidorids', without specifying, however, which synapomorphies. It is also important to cite here the studies of Lundberg and Baskin (1969) and Chardon (1968), which, although not based on an explicit cladistic analysis, also suggested a rather plesiomorphic position of the cetopsids within the Siluriformes. Lundberg and Baskin (1969: 42) clearly stated that the Cetopsidae (which at that time did not include genus *Helogenes*) present some plesiomorphic characters found only, in the Neotropics, in the Diplomystidae. With respect to Chardon (1968), his 'evolutionary scheme' (p. 241) also suggests a markedly basal position of cetopsids (the cetopsin and the helogenin cetopsids, although not forming a natural group, appear in a rather basal position in this scheme).

The present study detected two derived features that separate cetopsids from all the remaining non-diplomystid and non-loricarioid Siluriformes, with the first of these features being homoplasy free within the order: 1) horizontal lamina of the scapulo-coracoid present and similar to that of the cleithrum (char. 51); and 2) posterodistal margin of maxilla not markedly concave (char. 263).

Taking into account the points mentioned above, the markedly basal position of the cetopsids within catfishes thus appears to be a well-grounded, relatively robust hypothesis.

Clade 17 (*Helogenes, Cetopsis, Hemicetopsis*)

Diagnosis. **[7:0→1]**, **[24:0→1]**, [112:0→1], [159:0→1], **[169:0→1]**, [183:1→0], [188:0→4], [200:0→1], [239:0→1], [240:0→2], **[260:0→1]**, [277:0→1], [278:0→1], **[326:0→1]**, [341:0→1], **[345:0→1]**, [360:0→1], **[362:0→1]**, **[392:0→1]**

Comments. In 1995, de Pinna and Vari published an excellent work supporting the union in an expanded Cetopsidae of catfishes previously included in this family and genus *Helogenes* formerly included in family Helogenidae. This latter genus was assigned to the new cetopsid subfamily Helogeninae, with the remaining cetopsids being assigned to subfamily Cetopsinae (see de Pinna and Vari, 1995 for more details).

The present cladistic analysis clearly corroborates de Pinna and Vari's 1995 conclusions, with several characters supporting the sister-group relationship between *Helogenes* and the remaining cetopsids. In particular, the present analysis detected four additional, undescribed autapomorphic features present in Cetopsidae sensu Pinna and Vari (1995) and in no other catfishes examined: 1) cartilages of mandibular barbels not differentiated into supporting and moving part, with mandibular barbels situated near the posterior end of these cartilages (char. 7); 2) presence of muscle 6 of the mandibular barbels (char. 24); 3) presence of prominent posteroventral projection of cleithrum (char. 169); 4) marked development of inner branchiostegal rays (char. 392). [Note: characters 260, i.e., maxilla with one, single, articulatory facet for autopalatine; 326, entoectopterygoid a peculiarly elongate, roughly rectangular structure; 345, opercle and interopercle partly overlapping each other along modified laminar surfaces and 362, ligaments connecting interopercles and mandibles essentially associated with anterodorsal margin of interopercles, were based on previous descriptions of other authors: for descriptions of these characters, see Section 3.1.]

Clade 18 (*Cetopsis, Hemicetopsis*)

Diagnosis. **[56:0→1]**, **[65:0→1]**, **[100:0→1]**, **[130:0→1]**, **[156:0→1]**, **[171:0→1]**, **[241:0→2]**, **[279:0→1]**, [288:0→3], [331:0→1], [348:0→1], **[391:0→1]**, **[392:1→2]**, **[416:0→1]**

Comments. As expected, the present analysis supported a closer relationship between the cetopsid genera *Cetopsis* and *Hemicetopsis* of subfamily Cetopsinae than between either of these genera and genus *Helogenes* of subfamily Helogeninae (see above). Notably, this analysis pointed out several new autapomorphic features found only in cetopsins and in no other catfishes examined, which could, thus, eventually constitute potential Cetopsinae autapomorphies: 1) mesethmoid cornua with well-developed, ventral, vertical projection of laminar bone (char. 56); 2) frontal markedly compressed transversally (char. 100); 3) presence of well-developed lateral projection of parieto-supraoccipital (char. 130); 4) presence of posteroventral projection of cleithrum (char. 171); 5) parurohyal presenting two prominent, sharply pointed processes (char. 391); 6) particularly well-developed inner branchiostegal rays (char. 392); 7) presence of well-developed dorsolateral lamina of dentary bone (char. 416) [Note: characters 65, i.e., presence of a 'layer of cartilage-like tissue' protecting the olfactory capsule; 156, mesial limb of posttemporo-supracleithrum not mainly associated with the exoccipital; 241, levator arcus palatini originates on dorsal surface of neurocranium at level of ethmoid region and 279, marked enlargement of anterior cartilage of autopalatine, were based on previous descriptions of other authors: for descriptions of these characters, see Section 3.1.]

Clade 19 (All Genera Excluding *Diplomystes, Nematogenys, Trichomycterus, Hatcheria, Callichthys, Corydoras, Scoloplax, Astroblepus, Hypoptopoma, Loricaria, Lithoxus, Helogenes, Cetopsis, Hemicetopsis, Wallago, Silurus*)

Diagnosis. [2:0→1], [27:0→1], [47:0→1], [174:0→1], [180:0→1]

Comments. The five synapomorphies listed above, although not completely homoplasy free, strongly support a clade formed by all extant Siluriformes except diplomystids, loricarioids, cetopsids and silurids, and, thus, a rather plesiomorphic position of the silurid catfishes within the order. Character 2 concerns the firm association between the cartilages of the mandibular barbels and the mandible. It was, according to the present cladistic analysis, only independently acquired, outside the clade, in genus *Helogenes*, and secondarily lost, within the clade, in genus *Chaca* and in the node leading to the three aspredinid genera examined (see below). Character 27 concerns reduction in size of the muscle intermandibularis. It was only independently acquired, outside the clade, in the node leading to callichthyids + scoloplacids + astroblepids + loricariids (see above), and secondarily reverted, within the clade, in the nodes leading to *Bagrichthys* + *Bagrus* + *Hemibagrus*, to *Pseudopimedodus* + *Microglanis*, and to *Arius* + *Genidens* (see below). Character 47 concerns the origin of the abductor profundus in, or near the mesial symphysis of the pectoral girdle. According to the present analysis, it was homoplasically acquired outside the clade, in the node leading to *Trichomycterus* + *Hatcheria* (see above), and homoplasically reverted within the clade, in four different genera, *Liobagrus, Parakysis, Malapterurus, Heptapterus,* and in the node leading to the plotosid genera examined (see below). Character 174 refers to the presence of a pronounced ankylosis between the cleithrum and the scapulo-coracoid. It is almost homoplasy free, with a single homoplasic occurrence inside the order, namely its independent acquisition, outside the clade, in the node leading to callichthyids + scoloplacids + astroblepids + loricariids (see above). This is also the case of character 180, i.e., the presence of an interdigitation between the well-developed scapulo-coracoids, which was only independently acquired in the node leading to callichthyids + scoloplacids + astroblepids + loricariids and subsequently reverted in genus *Astroblepus*.

These results supporting a rather plesiomorphic position of the silurids among the order Siluriformes are not really surprising. As mentioned in the Introduction, of the two papers available on the higher level phylogeny of this order, Mo (1991) and de Pinna (1998), that of Mo had also suggested a rather plesiomorphic position of the Siluridae. In Mo's cladogram II, admittedly that preferred by him, the Siluridae appear as the sister-group of a clade formed by all catfishes except diplomystids and cetopsids (see Fig. 1.6). The works of Chardon (1968) and Lundberg and Baskin (1969), although not based on an explicit cladistic analysis, also suggested a somewhat primitive status of the silurids within the siluriforms. Lundberg and Baskin (1969: 37) stated that the silurids exhibited a combination of some rather primitive features and some particular, rare characters, as is to be expected for a rather plesiomorphic group presenting a long history independent of most other catfishes. With respect to Chardon (1968), the 'evolutionary scheme' presented on his page 241 also suggests, as explained above, a markedly basal position of the silurids. In fact, with careful reading of the excellent voluminous work of Chardon (1968), one can distinctly discern throughout the text his marked conviction of an important, special position of the silurids within the Siluriformes, comparable only, perhaps, to that of the Diplomystidae. Apart from these major studies encompassing the vast majority of the siluriform families, a few other studies also suggest a rather plesiomorphic position of silurids, such as those of Tilak (1963a,b; 1967a,b) on Euro-Asiatic catfishes.

Clade 20 (*Wallago, Silurus*)

Diagnosis. [1:0→1], [69:0→1], **[89:0→1]**, [150:0→1], [226:0→4], **[230:0→1]**, [247:0→1], [251:0→1], **[284:0→1]**, **[335:0→1]**, [419:0→1]

Comments. As noted in the Chapters 1 and 2, Siluridae is consensually considered a monophyletic group diagnosed by several synapomorphic features (see, e.g., Bornbusch, 1991b, Howes and Fumihito, 1991). The present work revealed no additional, undescribed autapomorphic character exclusively present in silurids, since the four homoplasy-free characters listed above were all based on previous descriptions of other authors (char. 89: presence of a thin elongated lateral laminar projection of the lateral ethmoid contacting the sphenotic; char. 230: differentiation of the retractor tentaculi into two well-developed divisions; char. 284: autopalatine reduced to a very short, nodular, irregularly shaped structure; char. 335: crest of the hyomandibulo-metapterygoid for the levator arcus palatini essentially oriented dorsoventrally: for descriptions of these characters, see Section 3.1).

Clade 21 (*Helicophagus, Pangasius, Pseudeutropius, Schilbe, Laides, Ailia, Siluranodon, Cranoglanis, Amiurus, Ictalurus, Austroglanis, Ancharius, Arius, Genidens, Auchenoglanis, Chrysichthys, Clarotes*)

Diagnosis. **[125:0→1]**, [131:0→1], [157:0→1], **[270:0→1]**

Comments. Grouping of pangasiid, schilbid, cranoglanidid, ictalurid, austroglanidid, ariid and claroteid catfishes examined in a monophyletic clade appears to be a strongly supported, robust result of the present cladistic analysis. Character 125 concerns the presence of a well-developed, deep fossa between the dorsomesial limb of the posttemporo-supracleithrum, extrascapular and pterotic. It is homoplasy free among the specimens examined, neither found in any catfish outside this clade nor secondarily lost within the clade (the character is inapplicable in *Ailia*, since in the specimens of this genus the extrascapular is missing: see the description of this character in Section 3.1). Character 131 (state 1) refers to the presence of a particularly conspicuous posterior process of the parieto-supraoccipital. Although homoplasic it was, according to the present cladistic analysis, only independently acquired outside the clade in *Bagrichthys*, *Pimelodus*, in the node leading to doumeins and in the node leading to erethistids, and secondarily lost within the clade in *Ailia*, *Amiurus*, *Cranoglanis* and *Ancharius* (see below). Character 157, the presence of a well-developed posterior laminar projection of the mesial limb of the posttemporo-supracleithrum, is a very interesting, often neglected feature that is not found in any catfish outside of the clade and that, inside of it, was only secondarily lost in *Auchenoglanis* (see below). With respect to character 270, the massive, somewhat cartilaginous or cord-like tissue connecting the coronoid process and the maxilla, it appears to be completely homoplasy free within the siluriforms examined.

In such a complex and diverse group as the Siluriformes, having two unambiguous and relatively poorly homoplasic characters and particularly two unambiguous and completely homoplasy-free features to diagnose such a large clade including seven different families, clearly confers a strong argument to support the monophyly of this clade.

The group including these seven families constitutes a new clade not previously described and/or suggested in the literature. However, it partially agrees, in some points, with the phylogenetic results of de Pinna's 1993 unpublished thesis. De Pinna (1993) refers to a clade including austroglanidids, pangasiids, schilbids, claroteins and ariids, but also some bagrids and some pimelodids. The major differences between the present study and de Pinna's results concerning this specific point, are essentially his inclusion of some bagrids and some pimelodids in the clade and exclusion of ictalurids, claroteins and cranoglanidids from it (see below). A different situation is found in the particularly confusing results of Mo's 1991 phylogenetic analysis. In Mo's cladogram I (see Fig. 1.6.A) the vast majority of the seven families included in clade 21 of the present analysis appear in a large, unresolved polytomy. In Mo's cladogram II (see Fig. 1.6.B) these seven families appear as somewhat closely related, in a monophyletic clade that includes all of them, but also the Plotosidae, †Hypsidoridae, Bagridae, Doradoidea and Pimelodidae.

Interestingly, in Chardon's 1968 'evolutionary scheme' the pangasiids, schilbids, ariids and ictalurids (the cranoglanidids were not figured in this 'evolutionary scheme') appear near each other in a pile supposedly evolved from the 'Pimelodidae/Bagridae stock' (which, at that moment, included the claroteids and austroglanidids).

The partial phylogenetic analysis on catfish higher level phylogeny based on mitochondrial DNA currently conducted by Peng is also worthy of mention here, since it supports a rather close relationship between ariids, claroteids, ictalurids, cranoglanidids and pangasiids (Peng, pers. comm.).

Clade 22 (*Helicophagus, Pangasius, Pseudeutropius, Schilbe, Laides, Ailia, Siluranodon*)

Diagnosis. **[10:0→1]**, [86:1→2]

Comments. Contrary to the previously undescribed, but rather robust clade 21 discussed above, a group including the pangasiid and schilbid catfishes does not constitute a new hypothesis on siluriform phylogeny. Although the Bagridae (e.g., Regan, 1911b; Rastogi, 1963, 1964; Chardon, 1968), Siluridae (e.g., Howes, 1985a; Rastogi, 1963, 1964), or even Plotosidae (e.g., Mo, 1991) have been proposed as potential sister-groups of the Schilbidae, most authors agree nowadays that the schilbid catfishes are very likely closely related to the pangasiids (see, e.g., de Pinna, 1998; Teugels, 1996, 2003; Pouyaud et al., 2000). The present phylogenetic analysis supported this view, pointing out two additional characters supporting a sister-group relationship between pangasiids and schilbids, with the first of these characters being inclusively homoplasy-free within Siluriformes: 1) cartilages of mandibular barbels bifurcated anteriorly (char. 10); 2) presence of complete foramen between the dorsal surfaces of lateral ethmoid and frontal (char. 86).

Clade 23 (*Helicophagus, Pangasius*)

Diagnosis. [1:0→1], [75:0→1], **[133:0→1]**, [155:0→1], [240:0→2], **[328:0→1]**, [402:0→1]

Comments. The pangasiids are consensually considered a monophyletic group (see, e.g., Mo, 1991; Vidthayanon, 1992; de Pinna, 1993, 1998; Teugels, 1996, 2003; Pouyaud et al., 2000). The present cladistic analysis revealed one new, undescribed autapomorphic feature found in the pangasiids and in no other catfishes examined, namely, the presence of a well-developed foramen between the extrascapular, pterotic and parieto-supraoccipital (char. 133). [Note: character 328, i.e. presence of a prominent posterodorsal crest of the entoectopterygoid, was based on previous descriptions of other authors: for a description of this character, see Section 3.1].

Clade 24 (*Pseudeutropius, Schilbe, Laides, Ailia, Siluranodon*)

Diagnosis. **[209:0→1]**, [251:0→1], [266:0→1], [277:0→1], **[420:0→1]**

Comments. As discussed in Chapters 1 and 2, although the name 'Schilbidae' is commonly cited as a valid siluriform family, 'most authors agree that this family is a non-monophyletic assemblage' (Teugels, 1996: 15). Mo (1991: 194-195), for example, considered Schilbidae as 'obviously a non-monophyletic assemblage'. According to him, there is a '*Schilbe* group' representing real schilbids, one phylogenetically distinct '*Ailia* group' closer to the Clariidae and Heteropneustidae and a third, also phylogenetically distinct '*Pseudeutropius* group' closer to the Bagridae and/or Pangasiidae. Mo's (1991) view was contested two years later, however, in de Pinna's 1993 unpublished thesis. De Pinna included three different groups of schilbids in his analysis, namely the '*Laides*', '*remaining Schilbinae*' and '*Ailiinae*' groups. He proposed that the schilbids constitute, in fact, a monophyletic group, which presented a peculiar, unique feature: 'Meckel's cartilage extending posteriorly much further beyond limit of dentary-anguloarticular in coronoid process' (de Pinna, 1993: 151). However, as these results were not published, Teugels (2003), in the most recent overview of the taxonomy, phylogeny and biogeography of the catfishes, continued to state that there were no published autapomorphies to support the monophyly of the Schilbidae, and that, in fact, this family was probably a non-monophyletic assemblage. The five groups of schilbids analysed in the present study were thus, as noted above, carefully chosen to test appropriately the monophyly of family Schilbidae. In fact, these five groups embrace and hence concomitantly represent all the different schilbid divisions considered in Regan's 1911b work (which first divided the family into subfamilies Schilbinae, Siluranodontinae and Ailiinae), Mo's 1991 (*Schilbe* group, *Pseudeutropius* group, *Ailia* group) and de Pinna's 1993 (*Laides*, *Ailia* group and *Schilbe* group). So, genus *Schilbe* represents both the Schilbinae of

Regan (1911b) and the *Schilbe* groups of Mo (1991) and de Pinna (1993). *Siluranodon* represents the subfamily Siluranodontinae of Regan (1911b). *Ailia* represents both the Ailiinae of Regan (1911b) and the *Ailia* groups of Mo (1991) and de Pinna (1993). *Pseudeutropius* represents the *Pseudeutropius* group of Mo (1991). Lastly, *Laides* represents the *Laides* group of de Pinna (1993).

The present cladistic analysis strongly supports the Schilbidae as a monophyletic group, characterised by three unambiguous homoplasic synapomorphies (char. 251: levator operculi originating on both the neurocranium and dorsolateral surface of the hyomandibulo-metapterygoid; char. 266: maxilla reduced in size; char. 277: anterior cartilage of the autopalatine markedly elongated anteroposteriorly), and in particular by two unique, autapomorphic characters (char. 209: adductor mandibulae A2 essentially lateral to A1-OST; char. 420: Meckel's cartilage markedly extended posteriorly). As explained above, in such a vast, diverse and complex group as the Siluriformes, the presence of unique, distinct, non-homoplasic features in a certain group is quite rare and clearly constitutes a very strong argument to support the monophyly of that group. The presence of not one, but two such features in the schilbids, as well as the three additional homoplasic synapomorphies listed above, clearly constitutes a very strong argument on behalf of the monophyly of these catfishes.

Clade 25 (*Pseudeutropius, Schilbe*)

Diagnosis. [227:0→1], [285:0→1]

Comments. Although a study on the intrarelationships within family Schilbidae is beyond the main scope of this work, the present analysis seems to suggest two main phylogenetic units within the family. One is constituted by the *Schilbe* and *Pseudeutropius* groups, the other by the *Siluranodon, Ailia* and *Laides* groups (see definitions of these groups above).

The two derived features supporting the first of these units (it is important to mention here that both these features are absent in the sister-group Pangasiidae) are: 1) muscle extensor tentaculi inserting not only on autopalatine, but also on sesamoid bone 1 of the suspensorium (char. 227: also found in *Cranoglanis*, claroteids, ariids, *Parakysis*, *Acanthorodas*, and some pimelodids); 2) autopalatine presenting a prominent anteroventrolateral process (char. 285: only found elsewhere in claroteids).

With respect to the second unit, there are also two derived features supporting it (which are also absent in the sister-group Pangasiidae): 1) vast majority or totality of ventral division of arrector dorsalis lying on dorsal surface of pectoral girdle (char. 53: also found in plotosids, aspredinids, *Heteropneustes*, *Clarias*, *Heterobranchus* and *Hypophthalmus*); 2) no distinct coracoid bridge (char. 192: only found elsewhere in a few individual genera, such as *Paramphilius*, *Nematogenys*, *Glyptosternon*, *Malapterurus* or *Astroblepus*).

Although these are clearly preliminary results, and a much more complete study is clearly needed to test them, they would have an interesting direct implication. This would be that the African and the Asian schilbids do not constitute separate groups, since the African *Siluranodon* would be more closely related to certain Asian genera, such as *Laides* or *Ailia,* than to the also African genus *Schilbe.*

Clade 26 (*Laides, Ailia, Siluranodon*)

Diagnosis. [53:0→2], [192:0→1]

Comments. See comments of clade 25 above.

Clade 27 (*Cranoglanis, Amiurus, Ictalurus, Austroglanis, Ancharius, Arius, Genidens, Auchenoglanis, Chrysichthys, Clarotes*)

Diagnosis. [88:0→1], [194:0→1]

Comments. This clade including cranoglanidids, ictalurids, austroglanidids, claroteids and ariids is supported by an almost homoplasy-free synapomorphy, the lateral ethmoid presenting a well-developed, broad dorsolateral projection of laminar bone surrounding the anterodorsolateral surface of the eye (char. 88: absent in all catfishes outside the clade, and only secondarily lost, within it, in *Auchenoglanis*). Also, it is supported by the markedly enlarged coracoid bridge (char. 194: also present, outside the clade, in doradids and auchenipterids, as well as in genera *Rita, Bagrichthys, Pseudopimelodus* and *Pimelodus,* and only secondarily lost, within the clade, in genus *Ancharius*). As explained above, in one of the clades of de Pinna (1993) austroglanidids, claroteids and ariids do appear somewhat related. But the inclusion of this clade in an unresolved pentatomy including precisely cranoglanidids and ictalurids, as well as two other clades composed by several families, does not allow an appraisal of the phylogenetic relationships among austroglanidis, claroteids, ariids, cranoglanidids and ictalurids (see Section 1.3). With respect to Mo's 1991 phylogenetic analysis, the results, as noted above, are particularly confusing concerning this subject, with most of these five groups appearing in a large polytomy in Mo's cladogram I and in a clade also including some other catfish families in Mo's cladogram II. As for Chardon's 1968 'evolutionary scheme', these five groups appear somewhat closer to each other but do not constitute a natural group (see above). Therefore, as was the case concerning the clade including Pangasiidae, Schilbidae, Austroglanididae, Claroteidae, Ariidae, Cranoglanididae and Ictaluridae (see above), this is the first time that a cladistic analysis has provided evidence for a monophyletic clade including austroglanidids, claroteids, ariids, cranoglanidids and ictalurids.

Clade 28 (*Cranoglanis, Amiurus, Ictalurus*)

Diagnosis. [217:0→2], [251:0→1]

Comments. The phylogenetic relationships between cranoglanidids and other catfishes have long posed one of the most puzzling questions regarding catfish interrelationships. The first author to deal with this problem was Jáyaram (1956). He suggested that Cranoglanididae could be closely related to Pangasiidae, Ictaluridae and/or Bagridae. However, as stated by some authors, such as Chardon (1968) or Burgess (1989), the arguments given by Jayaram (1956) to support his hypotheses were rather fragile. Chardon (1968), based on characters of the Weberian apparatus, suggested, also without convincing arguments (that is, without giving a list of derived features simultaneously present in both taxa, thus supporting a close relationship between them), that the Cranoglanididae are probably closely related to the Bagridae. Lundberg and Baskin (1969: 37), after an analysis of the caudal skeleton of numerous catfishes, stated that 'the caudal skeleton yields no information on the systematic position of the monotypic Cranoglanididae'. With respect to Mo's 1991 explicit cladistic analysis, in his cladogram I the relationships of Cranoglanididae were unresolved, while in cladogram II the family was grouped with Austroglanididae (see Fig. 1.4). But the author gave no convincing argument for this latter hypothesis (the grouping of the two families in cladogram II is based on a synapomorphy [synapomorphy no. 49], not subsequently described by the author). The recent, detailed phylogenetic studies carried out by de Pinna also did not completely succeed in solving the phylogenetic position of the Cranoglanididae, with the latter being included in a 'large pentatomy' that comprises several catfish taxa (see Fig. 1.11).

The present phylogenetic analysis supports a sister-group relationship between the ictalurid and the cranoglanidid catfishes examined with two synapomorphic characters. The first is a very interesting, not previously described character that is very rare and almost homoplasy-free among the Siluriformes: adductor mandibulae A3'' presenting a large anterior tendon inserting on the mandible and on the posterior portion of the primordial ligament (char. 217: only independently acquired in callichthyids). The other character, the levator operculi originating on both the neurocranium and dorsolateral surface of the hyomandibulo-metapterygoid (char. 251), is a rather homoplasic character also found outside the clade Ictaluridae + Cranoglanididae, in the nematogenyid, trichomycterid, callichthyid, silurid, schilbid, plotosid and pimelodid catfishes.

The close relationship between cranoglanidids and ictalurids suggested in the present phylogenetic analysis is not a completely new hypothesis. As mentioned above, a close relationship between ictalurids and cranoglanidids was first hypothesised by Jayaram (1956). Also, in private talks with Carl Ferraris (pers. comm.), this author, based on his own studies, as well as in the biogeographical distribution of these two groups (ictalurids in North America, cranoglanidids in China), also suggested a close relationship between them.

This view likewise seems supported by the discovery of a seemingly 'ictalurid' fossil from the Eocene of Inner Mongolia (Stucky, 1982), as well as by the still partial phylogenetic analysis on catfish higher level phylogeny based on mitochondrial DNA currently conducted by Peng (Peng, pers. comm.).

Clade 29 (*Cranoglanis*)

Diagnosis. [11:0→1], **[15:0→1]**, [32:0→1], [33:0→1], [88:1→2], **[119:0→1]**, [155:1→0], [165:0→1], [201:0→1], [212:0→1], **[218:0→1]**, [223:0→1], [227:0→1], [246:0→2], [254:0→1], **[323:0→1]**, [381:0→1]

Comments. Interestingly, although *Cranoglanis* is consensually included in a separate catfish family, no unique, autapomorphic features were published by other authors to differentiate cranoglanidids from other siluriforms. The present cladistic analysis pointed out four features present in cranoglanidids found in no other catfishes examined: 1) cartilages of mandibular barbels in deep contact with each other on the midline (char. 15); 2) presence of prominent, stout, roundish posterodorsal process of epioccipital (char. 119); 3) adductor mandibulae A3'' subdivided into two well-developed divisions (char. 218); 4) presence of well-defined, deep, anteroposteriorly elongated concavity on frontal and lateral ethmoid to receive entoectopterygoid (char. 323).

Clade 30 (*Amiurus, Ictalurus*)

Diagnosis. [4:0→1], [69:0→1], **[122:0→1]**, [168:0→1], [334:0→1]

Comments. As noted in Chapters 1 and 2, ictalurids are consensually considered a monophyletic group (see, e.g., Lundberg, 1975a, 1982, 1992; Grande and Lundberg, 1988; Mo, 1991; de Pinna, 1993). The present work detected no undescribed autapomorphic character exclusively present in the ictalurid catfishes examined. [Note: the single homoplasy-free character listed above, i.e., the presence of a prominent ventrolateral process of the pterotic receiving a subpterotic process of the post-temporo-supracleithrum (char. 122), was based on previous descriptions of other authors: for description of this character, see Section 3.1].

Clade 31 (*Austroglanis, Ancharius, Arius, Genidens, Auchenoglanis, Chrysichthys, Clarotes*)

Diagnosis. [213:0→2], [277:0→1], [308:1→0]

Comments. The phylogenetic relationships between *Austroglanis* and the other catfishes have been a particularly puzzling issue. Skelton et al.'s (1984) formal definition of the genus suggested it could be related not only to the 'remaining bagrids' (which then included both Claroteidae and Bagridae), but also to pimelodids and/or ictalurids. Skelton et al. (1984), however, did not formally provide a list of synapomorphic characters to support this suggestion.

The phylogenetic results of Mo (1991) were somewhat confusing: in his cladogram I Austroglanididae was placed in a large, unresolved polytomy, while in his cladogram II it was placed as the sister-group of Cranoglanididae. Moreover, the author provided no synapomorphic characters to support this latter hypothesis, with grouping of the two families in his cladogram II based on a synapomorphy (synapomorphy no. 49) he failed to describe subsequently.

A very different hypothesis was formulated by de Pinna (1993), in which Austroglanidae was placed as the sister-group of a large clade including genus *Horabagrus*, the schilbids, pangasiids, claroteins, *Ancharius*, remaining ariids, some bagrids and some pimelodids. De Pinna provided three characters to support this phylogenetic hypothesis, namely: 1) frontal and lateral ethmoid connected by a lateral bridge of bone, mesially delimiting a space or foramen; 2) (reversion of) first pharyngobranchial absent; 3) transverse process of 4th vertebra with a well-defined posterior arm.

With respect to the present analysis, it partially supports and partially contradicts each of these three studies. In fact, the austroglanidids appear closely related to ictalurids (as suggested by Skelton et al., 1984) and cranoglanidids (as suggested by Mo, 1991), but particularly to ariids (as suggested by de Pinna, 1993) and claroteids (as suggested by Skelton et al., 1984 and de Pinna, 1993). Three phylogenetically unambiguous characters support, in this analysis, the monophyletic clade constituted by austroglanidids, ariids and claroteids, the first two of which are relatively homoplasy-free among the siluriforms. These are: 1) adductor mandibulae A3′ differentiated into lateral and mesial divisions attached respectively, to posterodorsolateral and posterodorsomesial surfaces of the coronomeckelian (char. 213: not found in other catfishes outside the clade and within the clade, only secondarily lost in claroteins); 2) anterior cartilage of autopalatine markedly elongated anteroposteriorly (char. 277: present in all members of the clade and found outside it only in cetopsids and schilbids); 3) presence of sesamoid bone 2 of the suspensorium (char. 308: acquired in the node leading to non-diplomystid catfishes and subsequently secondarily reverted in the nodes leading to the clade formed by bagrids excluding *Rita*, to the clade *Pimelodus* + *Calophysus*, clade *Pseudopimelodus* + *Microglanis*, and clade austroglanidids + ariids + claroteids, with a subsequent transition to state 1 inside this latter clade in genus *Ancharius*).

Clade 32 (*Austroglanis*)

Diagnosis. [11:0→1], [32:0→1], **[110:0→1]**, [131:1→0], [250:0→1], **[269:0→1]**, **[315:0→1]**, **[320:0→1]**, [332:0→1], **[354:0→1]**, [358:0→1], [425:1→0]

Comments. Skelton (1981) and Skelton et al. (1984) provided the background for exclusion of austroglanidids from the claroteid genus *Gephyroglanis* (which at that time was included, together with the other claroteid genera, in 'Bagridae'), and thus the formal definition of genus *Austroglanis* Skelton et al., 1984. A step forward was undertaken in Mo's 1991 study, in which the

'Bagridae' was, as explained in Chapter 1, divided into Bagridae sensu Mo (1991), Claroteidae and Austroglanididae, with *Austroglanis* thus assigned its own family. Placement of genus *Austroglanis* in a separate family was subsequently supported in de Pinna's 1993 unpublished thesis.

The present analysis pointed out three new, undescribed features exclusively found in the austroglanidids examined: 1) presence of marked bifurcation of anterodorsolateral salience of sphenotic (char. 110); 2) entoectopterygoid a peculiarly, arrow-shaped structure (char. 320); 3) interopercle presenting a large foramen on its posteromesial surface (char. 354). [Note: characters 269, i.e., presence of a thin posterior laminar flange along the posterior margin of the maxilla, and 315, sesamoid bone 1 of the suspensorium being a posteriorly bifurcated, roughly ∧-shaped structure, were based on previous descriptions of other authors: for description of these characters, see Section 3.1.] However, one should be particularly careful in considering these three features as potential Austroglanididae autapomorphies, since it was only possible to include in the present analysis one of the three austroglanidid species, *Austroglanis gilli*. For example, attending to the descriptions of Skelton (1981) and Skelton et al. (1984), it seems that in the other two austroglanidid species, *A. sclateri* and *A. barnardi*, the entoectopterygoid is not peculiarly arrow-shaped and that the interopercle does not present a large foramen on its posteromesial surface.

Clade 33 (*Ancharius, Arius, Genidens, Auchenoglanis, Chrysichthys, Clarotes*)

Diagnosis. [18:0→1], [227:0→1]

Comments. The above two characters support the grouping of ariids and claroteids. Character 18, i.e., presence of a muscle depressor of the internal mandibular barbels, is a rather homoplasic feature also found in *Malapterurus*, *Franciscodoras*, amphiliids, and in the node leading to *Heteropneustes* + *Clarias* + *Heterobranchus*. With respect to character 227, i.e., insertion of a significant part of the fibres of the extensor tentaculi on the mesial and/or dorsal surface of sesamoid bone 1 of the suspensorium, this, too, is a rather homoplasic character. It is not reverted in any member of the clade but was seemingly independently acquired outside it in *Parakysis*, *Acanthodoras*, *Cranoglanis*, in the node leading to *Calophysus* + *Pimelodus* + *Hypophthalmus*, and in the node leading to *Schilbe* + *Pseudeutropius*.

The cosmopolitan ariids are commonly placed near the African mochokid and the Neotropical doradid and auchenipterid catfishes (see, e.g., Mo, 1991; Lundberg, 1993; de Pinna, 1998) mainly because they possess a well-developed 'elastic spring apparatus' and other features on the posterior region of the cranium (see Section 1.3). According to Chardon (1968), the taxonomic distribution of the elastic spring apparatus, as well as its rather different configuration in various catfish groups, clearly seems to indicate that it evolved several times within the order Siluriformes. Interestingly, in his comments on the relationships of Ariidae (1968: 69), Chardon suggested a close relationship between these fishes and Bagridae, which at that time included Bagridae sensu Mo, Claroteidae, and Austroglanididae. More specifically, Chardon supported this point citing the remarkable tolerance to high salinity levels of some claroteid fishes such as *Chrysichthys* (the tolerance to high levels of salinity of the fishes of this genus is, according to Laleye, 'completely astonishing': Laleye, pers. comm.).

It is also important to note here that the close relationship of ariids and doradoids suggested by de Pinna's 1998 cladogram (see Fig. 1.11) is based on the studies of Mo (1991) and Lundberg (1993), and not on to phylogenetic research done by him. In fact, in the cladistic analysis of de Pinna's 1993 unpublished thesis, Ariidae appear closely related to clarotein catfishes (but not to the auchenoglanidins), as suggested by Chardon's 1968 study and by the present work.

Clade 34 (*Ancharius, Arius, Genidens*)

Diagnosis. [75:0→1], [86:1→2], [155:0→1], [185:0→1], [254:0→1]

Comments. As observed in Chapters 1 and 2, the freshwater genus *Ancharius* from Madagascar is traditionally included in family Ariidae. However, some authors, such as Glaw and Vences (1994), have suggested that *Ancharius* should have its own family, Anchariidae, while other authors, such as Mo (1991), have argued that the genus should be included in the African family Mochokidae (see Section 1.3). The results of the present work accord with those of de Pinna's unpublished thesis (1993), according to which *Ancharius* is indeed the sister-group of the remaining ariids. This sister-group relationship is supported by the five unambiguous synapomorphies above: 1) marked lateral bifurcation of premaxilla (char. 75: only present outside the clade in the node leading to pangasiids and in *Ageneiosus*); 2) complete foramen between dorsal surfaces of lateral ethmoid and frontal (char. 86, state 2: only present outside the clade in the node leading to pangasiids + schilbids); 3) presence of suture between mesial limb of posttemporo-supracleithrum and neurocranium (char. 155: present in several groups outside the clade, namely in pangasiids, *Cranoglanis*, akysids, *Leptoglanis*, and in the node leading to mochokids + doradids + auchenipterids); 4) mesocoracoid arch reduced to thin structure fused with main body of scapulo-coracoid (char. 185: present outside the clade in aspredinids, *Parakysis*, *Rita*, and in the node leading to the clade doradids + auchenipterids); 5) presence of protractor of Müllerian process (char. 254: present outside the clade in *Pangasius*, *Cranoglanis*, in the node leading to malapterurids + mochokids + doradids + auchenipterids pangasiids, and in the node leading to pseudopimelodins).

Although the five synapomorphies listed above are not homoplasy free within the Siluriformes, it should be noted that the first two have only been independently acquired, within the whole order, in one or two other nodes outside family Ariidae. Moreover, it is also very important to note that besides these five characters there are several other derived features, including two autapomorphic ones, supporting a close relationship between *Ancharius*, the remaining ariids and the claroteids, between these groups and the cranoglanidids + ictalurids, and between all these five taxa and the clade formed by schilbids + pangasiids (see above). Therefore, the results of the present study, like those of de Pinna's unpublished thesis (1993), strongly contradict a close relationship between genus *Ancharius* and the mochokids or the remaining doradoids.

With respect to the proposition of removing *Ancharius* from Ariidae and placing in its own family Anchariidae, in view of the results of de Pinna (1993) and this work supporting the sister-group relationship *Ancharius* + remaining ariids and the absence of unique features characterising genus *Ancharius* (see below), I see no major reasons for doing so. Perhaps an alternative option, if future studies support the phylogenetic results of de Pinna (1993) and the present study, would be to include genus *Ancharius* in an ariid subfamily Anchariinae, with the remaining ariids placed, for example, in a subfamily Ariinae.

Clade 35 (*Arius, Genidens*)

Diagnosis. [27:1→0], **[114:0→2], [116:0→1], [128:0→1], [135:0→1], [216:0→1]**, [286:0→2]

Comments. As expected (see above), within the Ariidae, genera *Arius* and *Genidens* appear more related to each other than to genus *Ancharius*. This is supported by seven synapomorphies, of which four are completely homoplasy free within the siluriforms examined, namely: 1) presence of prominent posterior process of exoccipital (char. 116); 2) presence of a well-developed ventral process of the basioccipital (char. 128); 3) presence of a ventral superficial ossification of the complex centrum completely covering the aortic groove (char. 135); 4) adductor mandibulae A3'' covering a great part of the dorsolateral surface of the hyomandibulo-metapterygoid (char. 216).

Clade 36 (*Auchenoglanis, Chrysichthys, Clarotes*)

Diagnosis. [42:0→1], [86:1→0], [162:0→1], [285:0→1]

Comments. As explained in Chapter 1, family Claroteidae was formally described by Mo (1991). This author provided strong evidence to support the monophyly of the family, as well as each of its two subfamilies, Claroteinae and Auchenoglanidinae (see Mo, 1991). De Pinna (1998), however, surprisingly stated that Claroteidae sensu Mo (1991) is polyphyletic, since each of its two subfamilies is more closely related to other siluriform groups than to each other. De Pinna's view was not accepted by most catfish specialists, however, who continued to refer to Claroteidae sensu Mo as a monophyletic group (see, e.g., Teugels' 2003 detailed and up-to-date overview on catfish systematics). The present analysis reinforces the validity of Mo's Claroteidae, with the auchenoglanidin (*Auchenoglanis*) and clarotein (*Clarotes, Chrysichthys*) examined appearing more closely related to each other than to any of the other 84 catfish genera examined (see Fig. 3.123). The four characters supporting this hypothesis are: 1) abductor superficialis 1 does not reach anteroventral surface of cleithrum (char. 42: highly homoplasic, also found in several other catfishes); 2) absence of well-defined dorsal concavity between lateral ethmoid and frontal (char. 86: reversion to state 0, highly homoplasic); 3) presence of well-developed cartilage between dorsal processes of cleithrum (char. 162: relatively homoplasy-free, also only found in callichthyids, *Bunocephalus* and *Chaca*); 4) presence of prominent anteroventrolateral projection of autopalatine (char. 285: relatively homoplasy-free, also only found in the node leading to *Pseudeutropius* + *Schilbe*).

Clade 37 (*Chrysichthys, Clarotes*)

Diagnosis. [213:2→0], [278:0→1], [306:0→1], [327:0→1], [402:0→1]

Comments. As expected, the present analysis supported a closer relationship between genera *Clarotes* and *Chrysichthys* of subfamily Claroteinae than between these genera and genus *Auchenoglanis* of subfamily Auchenoglanidinae (see above). However, the present analysis revealed no new potential Claroteinae autapomorphy.

Clade 38 (*Malapterurus, Mochokus, Synodontis, Centromochlus, Ageneiosus, Auchenipterus, Franciscodoras, Anadoras, Doras, Acanthodoras, Rita, Bagrichthys, Hemibagrus, Bagrus, Microglanis, Pseudopimelodus, Heptapterus, Rhamdia, Goeldiella, Pseudoplatystoma, Hypophthalmus, Pimelodus, Calophysus, Amphilius, Paramphilius, Leptoglanis, Zaireichthys, Doumea, Phractura, Andersonia, Belonoglanis, Trachyglanis, Chaca, Uegitglanis, Heteropneustes, Clarias, Heterobranchus, Cnidoglanis, Neosilurus, Plotosus, Paraplotosus, Parakysis, Akysis, Amblyceps, Liobagrus, Gagata, Bagarius, Glyptosternon, Glyptothorax, Erethistes, Hara, Aspredo, Bunocephalus, Xyliphius*)

Diagnosis. [226:0→2]

Comments. This is clearly the most fragile clade of the cladogram presented in Figure 3.123, as it is only characterised by a single and highly homoplasic feature that is often subsequently modified to other states in members within the clade: the most posterior bundles of the extensor tentaculi inserting on the posterodorsal and posteroventral extremities of the autopalatine (char. 226, state 2: for description of this character, see Section 3.1). The phylogenetic meaning of this is that although in the most parsimonious, consensus-strict cladogram of Fig. 3.123 clade 21 appears as the sister-group of a group including clades 39 and 62, the relationships among these three major catfish clades are not clearly established. This is not so unexpected, since in Mo's cladogram I (Fig. 1.6A) and in de Pinna's 1998 cladogram (Fig. 1.11), a great part of the catfish taxa, excluding the most basal ones, appear grouped in large, unresolved polytomies.

Clade 39 (*Malapterurus, Mochokus, Synodontis, Centromochlus, Ageneiosus, Auchenipterus, Franciscodoras, Anadoras, Doras, Acanthodoras, Rita, Bagrichthys, Hemibagrus, Bagrus, Microglanis, Pseudopimelodus, Heptapterus, Rhamdia, Goeldiella, Pseudoplatystoma, Hypophthalmus, Pimelodus, Calophysus*)

Diagnosis. [198:0→2], [313:0→1]

Comments. Grouping the malapterurid, mochokid, auchenipterid, doradid, bagrid and pimelodid catfishes examined in a large monophyletic clade is supported by the two unambiguous synapomorphic characters cited above. The presence of a highly developed anterior process of the dorsal condyle of the pectoral spine (char. 198) is a very interesting feature that is relatively homoplasy free within the whole order Siluriformes. Such a process is only completely lost, within this large clade, in pimelodins, and is only independently acquired, outside the clade, in genus *Pseudeutropius* and in aspredinids (see below). The other synapomorphy concerns the attachment of the ligament between the suspensorium and the ethmoid region on the ventromesial surface of the lateral ethmoid (char. 313). This, too, is a relatively homoplasy-free feature that within the clade, was only modified in *Malapterurus* and secondarily lost in *Auchenipterus* and *Bagrichthys* and outside it, was only independently acquired in the amphiliin and amblycipitid catfishes.

Although the grouping of mochokids, doradids and auchenipterids is not new (see, e.g., Chardon, 1968; Mo, 1991; Lundberg, 1993; de Pinna, 1993, 1998), grouping these three groups with malapterurids, bagrids and pimelodids constitutes an essentially new phylogenetic hypothesis. However, it is interesting to note that in Mo's cladogram II (Fig. 1.6B), bagrids, mochokids, doradids, auchenipterids and most pimelodids are closely united in a clade only including some other few catfishes such as pangasiids and ariids (see Section 1.3). However, Mo (1991: 208) provided no synapomorphic feature to support such a clade, stating that its definition had been due to a 'computer generated node'.

Clade 40 (*Malapterurus, Mochokus, Synodontis, Centromochlus, Ageneiosus, Auchenipterus, Franciscodoras, Anadoras, Doras, Acanthodoras*)

Diagnosis. [35:0→1], [131:0→2], [254:0→1]

Comments. The phylogenetic relationships between the electric catfishes, or Malapteruridae, and the other siluriforms have long been a puzzling issue. Günther (1864) was the first to discuss the systematic position of *Malapterurus* and assigned the genus to his subfamily Stenobranchiae, which also included taxa now recognised as the families Mochokidae, Doradidae and Auchenipteridae. This view was essentially followed by Bridge and Haddon (1894). Regan (1911b), however, proposed a close relationship between malapterurids and bagrids. Chardon (1968) listed the bagrids, but also the silurids and the pangasiids, as potential sister-groups of the electric catfishes. A close relationship between silurids and malapterurids was also suggested by Howes (1985a). However, Bornbusch (1991b) contradicted such a close relationship and considered the hypotheses of both Chardon (1968) and Howes (1985a) as inconclusive. Mo's (1991) results were confusing, with the Malapteruridae appearing in a large polytomy in his cladogram I, and as the sister-group of a clade including families such as Ictaluridae, Claroteidae, Plotosidae, Schilbidae, Cranoglanididae, Austroglanididae, Bagridae, Pangasiidae, Pimelodidae, Ariidae, Mochokidae, Doradidae and Auchenipteridae, in his cladogram II (see Fig. 1.6). De Pinna (1993, 1998) considered the malapterurids, as explained above, to be closely related to the auchenoglanidin claroteids (see Fig. 1.11).

The results of the present cladistic analysis interestingly come back to the original hypothesis of Günther (1964), with malapterurids appearing closely related to the Mochokidae, Doradidae and Auchenipteridae. The synapomorphies supporting this close relationship are: 1) hypertrophication of muscle hyohyoideus inferior (char. 35: secondarily lost, within the clade, in auchenipterids, and independently acquired outside it, in *Auchenoglanis*, *Bagrichthys*, and in the node leading to leptoglanidins + doumeins); 2) posterior truncation of parieto-supraoccipital (char. 131: present in all members of the clade and only found outside it in trichomycterids, *Chaca* and *Bunocephalus*); 3) presence of a muscle protractor of Müllerian process (char. 254: present in all members of the clade and found outside it in pseudopimelodins, ariids, *Cranoglanis* and *Pangasius*).

Clade 41 (*Malapterurus*)

Diagnosis. [25:0→1], [47:1→0], [97:0→1], [109:0→2], [120:0→1], [150:0→1], [159:0→1], [188:0→3], [191:0→1], [192:0→1], [206:0→1], [226:2→3], [228:0→1], [246:0→1], [313:1→2], [318:0→1], [414:0→1]

Comments. Electric catfishes of genus *Malapterurus* are consensually placed in a separate family, Malapteruridae (see, e.g., Regan, 1911b; Chardon, 1968; Howes, 1985a; Mo, 1991; de Pinna, 1993, 1998; Teugels, 1996, 2003). The present cladistic analysis revealed no new, undescribed autapomorphic feature characterising the members of this family.

Clade 42 (*Mochokus, Synodontis, Centromochlus, Ageneiosus, Auchenipterus, Franciscodoras, Anadoras, Doras, Acanthodoras*)

Diagnosis. [37:0→1], [42:0→1], [142:0→1], [155:0→1], [168:0→1], [210:0→1], [248:0→1], [290:0→1]

Comments. As explained by de Pinna (1998), several studies—Chardon (1968), Mo (1991), Lundberg (1993) or de Pinna (1993, 1998)—have provided strong evidence to support a close relationship between the African Mochokidae and the Neotropical Auchenipteridae and Doradidae. The present analysis corroborates this close relationship, pointing out five additional synapomorphic features to support the monophyly of the clade constituted by these three families: 1) hyohyoideus abductor hypertrophied, with median aponeurosis firmly attached to pectoral girdle (char. 37); 2) abductor superficialis 1 not reaching anteroventral surface of cleithrum (char. 42); 3) presence of markedly developed anterolateral laminar projection of cleithrum (char. 168); 4) adductor mandibulae A2 directly inserted on mesial surface of mandible (char. 210); 5) adductor operculi or part differentiated from it not contacting hyomandibulo-metapterygoid (char. 248). [Note: characters 142, i.e., nuchal plates forming a markedly enlarged nuchal shield in which the median X-shaped nuchal plate encloses almost completely, or completely, the anterior nuchal plate; 155, presence of suture between mesial limb of posttemporo-supracleithrum and neurocranium; and 290, posterior portion of autopalatine expanded dorsoventrally due to enlargement of its centre of ossification, were based on previous descriptions of other authors: for description of these characters, see Section 3.1.]

Clade 43 (*Mochokus, Synodontis*)

Diagnosis. **[3:0→1]**, [165:0→1], **[256:0→1]**, **[298:0→1]**, [309:0→1], [358:0→2], **[400:0→1]**, [407:0→1], **[410:0→1]**, [419:0→1], [423:0→1], [424:0→1]

Comments. The mochokids are consensually considered a monophyletic group (see, e.g., Regan, 1911b; Chardon, 1968; Mo, 1991; de Pinna, 1993, 1998; Teugels, 1996, 2003). The present analysis pointed out one new, homoplasy-free feature found in the mochokids and in no other catfishes examined, which thus could eventually constitute a Mochokidae autapomorphy: articulatory surface of autopalatine for neurocranium remarkably curved (char. 298). [Note: characters 3, i.e., marked ramification of mandibular barbels; 256, protractor of Müllerian process originating on region of dorsal fin and its support; 400, coronoid process of mandible completely undifferentiated and 410, presence of a prominent posterolateral projection of angulo-articular, are based on previous descriptions of other authors: for description of these characters, see Section 3.1.]

Clade 44 (*Centromochlus, Ageneiosus, Auchenipterus, Franciscodoras, Anadoras, Doras, Acanthodoras*)

Diagnosis. **[118:0→1]**, [181:0→2], [185:0→2], **[189:0→1]**, [194:0→1], [286:0→1], [334:0→1]

Comments. The close relationship between doradid and auchenipterid catfishes has long been recognised and supported with strong evidence by several authors (e.g., Regan, 1911b; Chardon, 1968; Curran, 1989; Mo, 1991; Lundberg, 1993; de Pinna, 1993, 1998). This work corroborates the grouping of auchenipterids and doradids in a monophyletic clade, adducing three additional synapomorphic features to support this clade: 1) articulatory facet of scapulo-coracoid for pectoral spine a markedly elongated, thin structure (char. 189); 2) highly enlarged coracoid bridge (char. 194); 3) anterodorsal spine of hyomandibulo-metapterygoid a particularly developed, enlarged structure (char. 334). [Note: characters 118, i.e., epioccipital constituting a significant part of the cranial roof; 181, presence of a well-developed posterior process of the scapulo-coracoid; 185, mesocoracoid arch and main body of scapulo-coracoid undistinguished from each other and 286, autopalatine roughly tubular structure with markedly salient mesial articulatory surface for the neurocranium, were based on previous descriptions of other authors: for description of these characters, see Section 3.1.]

Clade 45 (*Centromochlus, Ageneiosus, Auchenipterus*)

Diagnosis. [35:1→0], [265:0→1], [435:0→1], [436:0→1]

Comments. The auchenipterids are consensually considered a monophyletic group (see, e.g., Regan, 1911b; Chardon, 1968; Curran, 1989; de Pinna, 1998; Teugels, 1996, 2003; Soares-Porto, 1998). The present analysis revealed no new, homoplasy-free autapomorphic feature exclusively found in the auchenipterid catfishes examined [Note: characters 435, i.e., essentially dorsolaterally orientation of the adducted maxillary barbel, and 436, adducted maxillary barbel lying in a well-developed, deep concavity on the lateral surface of the cheek, were based on previous descriptions of other authors: for description of these characters, see Section 3.1.]

Clade 46 (*Ageneiosus, Auchenipterus*)

Diagnosis. **[222:0→1]**, **[281:0→1]**, [290:1→0]

Comments. As expected, the present analysis supported a closer relationship between genera *Ageneiosus* and *Auchenipterus* of auchenipterid subfamily Auchenipterinae than between either of these genera and genus *Centromochlus* of subfamily Centromochlinae (see Chapters 1 and 2). Notably, this analysis detected two autapomorphic features found in auchenipterins but in no other catfishes examined, which could eventually constitute potential Auchenipterinae autapomorphies: 1) extensor tentaculi markedly hypertrophied (char. 222); 2) presence of prominent dorsolateral crest of autopalatine (char. 281).

Clade 47 (*Franciscodoras, Anadoras, Doras, Acanthodoras*)

Diagnosis. [123:0→1], [366:0→1], [434:0→1]

Comments. Like the auchenipterids, the doradids are also consensually considered a monophyletic group (see, e.g., Regan, 1911b; Chardon, 1968; Higuchi, 1992; de Pinna, 1993, 1998; Teugels, 1996, 2003). The present analysis revealed no new homoplasy-free autapomorphic features exclusively found in doradid catfishes. [Note: char. 434, i.e., tympanic area peculiarly bordered by the posttemporo-supracleithrum, postoccipital region, infraneural scutes and humeral process of pectoral girdle, was based on previous descriptions of other authors: for description of this character, see Section 3.1.]

Clade 48 (*Anadoras, Doras, Acanthodoras*)

Diagnosis. [159:0→1], [168:1→0]

Comments. As expected, doradid genera *Anadoras* of subfamily Astrodoradinae, *Acanthodoras* of subfamily Platydoradinae, and *Doras* of subfamily Doradinae, appear united in a clade having the rather plesiomorphic doradid genus *Franciscodoras* as sister-group (see Higuchi, 1992; de Pinna, 1998). Two synapomorphies support this clade: 1) presence of one, single, well-developed dorsal process of the cleithrum for articulation with posttemporo-supracleithrum (char. 159); 2) secondary loss of markedly developed anterodorsal laminar projection of cleithrum (char. 168).

Clade 49 (*Doras, Acanthodoras*)

Diagnosis. [181:2→3]

Comments. In de Pinna's 1998 cladogram of doradid interrelationships (de Pinna, 1998: Fig. 14), which is based on the unpublished thesis of Higuchi (1992), subfamilies Astrodoradinae, Platydoradinae and Doradinae appear in an unresolved trichotomy. The character above, i.e., remarkable development of the posterior process of the scapulo-coracoid (char. 181), however, supports a closer relationship between genus *Acanthodoras* of subfamily Platydoradinae and genus *Doras* of subfamily Doradinae, than between either of these genera and genus *Anadoras* of subfamily Astrodoradinae.

Clade 50 (*Rita, Bagrichthys, Hemibagrus, Bagrus, Microglanis, Pseudopimelodus, Heptapterus, Rhamdia, Goeldiella, Pseudoplatystoma, Hypophthalmus, Pimelodus, Calophysus*)

Diagnosis. [72:0→1], [226:2→1], [286:1→2], [301:0→1]

Comments. The four characters listed above strongly support grouping the bagrid and pimelodid catfishes examined in a monophyletic clade. A well-developed dorsolateral process of the premaxilla for insertion of the premaxillo-maxillary ligament (char. 72) is a very interesting feature that is almost homoplasy free within the Siluriformes, found in all catfishes of this clade and only present outside it, in genus *Chaca* (see below). With respect to differentiation of the extensor tentaculi in at least four bundles promoting respectively adduction, abduction, dorsal rotation and ventral rotation of the maxillary barbel (char. 226), it is found in all members of the clade and, outside it, is present only in the node leading to *Clarias* + *Uegitglanis* + *Heterobranchus* + *Heteropneustes*. The autopalatine being a markedly tubular, simple structure lacking major mesial projections or saliences (char. 286) is also a relatively homoplasy-free feature present in all bagrids and pimelodids examined and, apart from these catfishes, is found only in the node leading to *Arius* + *Genidens*. Lastly, character 301 concerns the presence of a long and strong ligament between the anterior surface of the entoectopterygoid and the posterior margin of the maxilla. This, too, is an almost homoplasy-free feature not found in any other catfish and, among the clade including bagrids and pimelodids, only secondarily absent in heptapterins and genus *Hypophthalmus*.

The grouping of bagrids and pimelodids, supported by four synapomorphic that are relatively homoplasy free within such a large, complex and diverse group as the order Siluriformes, thus clearly appears as a particularly well-grounded, robust hypothesis. Therefore, the present analysis corroborates the suggestion by authors such as Eigenmann (1890), Regan (1911b), Chardon (1968), de Pinna (1993, 1998) and Bockmann (1998), of a close relationship between Pimelodidae (or part of it: see below) and Bagridae.

Clade 51 (*Rita, Bagrichthys, Hemibagrus, Bagrus*)

Diagnosis. [120:0→1], **[253:0→1]**

Comments. As explained in the Introduction, Mo (1991) provided strong evidence to support the monophyly of family Bagridae, as defined by him, as well as of its two subfamilies, Ritinae and Bagrinae (see Mo, 1991). De Pinna (1998), however, stated that Bagridae sensu Mo (1991) was polyphyletic, since some bagrid groups were more closely related to other siluriform taxa than to each other. As with de Pinna's statement on the polyphyly of the Claroteidae (see above), this statement was not followed by most catfish specialists, who continued to refer to Bagridae sensu Mo as a monophyletic group (Teugels, 2003). The present analysis reinforces the validity of Mo's Bagridae. The two characters supporting the monophyly of this group, likewise described by Mo (1991), are: 1) presence of well-developed posterodorsolateral projection of pterotic (char. 120: almost completely homoplasy-free, found in all members of Bagridae and only found, among non-bagrids, in *Malapterurus*); 2) presence of protractor posttemporalis (char. 253: homoplasy-free within the Siluriformes, found in all bagrids and in no other catfishes).

Clade 52 (*Bagrichthys, Hemibagrus, Bagrus*)

Diagnosis. [27:1→0], [181:0→1], **[299:0→1]**, [308:1→0]

Comments. As expected, the bagrid genera *Bagrichthys*, *Hemibagrus* and *Bagrus* from subfamily Bagrinae appear as more closely related to each other than to genus *Rita* of subfamily Ritinae (see Chapter I). The present analysis revealed no new, undescribed autapomorphic feature to characterise the bagrin catfishes examined. [Note: character 299, i.e., presence of a long and strong ligament between the dorsal surface of the entoectopterygoid and the posterior extremity of the autopalatine, was based on previous descriptions of other authors: for description of this character, see Section 3.1.]

Clade 53 (*Hemibagrus, Bagrus*)

Diagnosis. **[52:1→0]**, [181:1→2], [213:0→1], [217:0→1]

Comments. Although a study on the interrelationships within Bagridae is beyond the scope of the present work, it should be noted that the bagrin specimens examined of genera *Hemibagrus* and *Bagrus* appear more closely related to each other than to those of genus *Bagrichthys*, as would be expected according to Mo's 1991 study. Notably, the present analysis revealed one new, homoplasy-free character to support this close relationship: ventral division of the arrector dorsalis considerably reduced in size (char. 52).

Clade 54 (*Microglanis, Pseudopimelodus, Heptapterus, Rhamdia, Goeldiella, Pseudoplatystoma, Hypophthalmus, Pimelodus, Calophysus*)

Diagnosis. **[12:0→1]**, **[23:0→1]**, [168:0→1], [247:0→1], [251:0→1]

Comments. The monophyly of the clade formed by all pseudopimelodin, pimelodin and heptapterin catfishes examined, and thus of family Pimelodidae as a whole (see Chapters 1 and 2), supported by the five characters above, clearly constitutes one of the most important results of the present cladistic analysis. In particular, the first two of these five characters concern the presence, in all pimelodids examined and in no other catfish, of noticeable, distinct, easily recognised features and thus clearly constitute very strong evidence to support the monophyly of the family. These characters are: 1) presence of cartilaginous plates carrying the mandibular barbels (char. 12); 2) presence of muscle 4 of the mandibular barbels (char. 23). As explained above, the uniform, exclusive presence of a well-defined, distinct homoplasy-free feature in a certain catfish group is extremely rare in a taxon as large, complex and diverse as the Siluriformes. Moreover, the other three synapomorphic characters, although homoplasic, clearly provide strong support for Pimelodidae monophyly. Character 168 (presence of markedly developed anterolateral laminar projection of cleithrum) is found only among non-pimelodids, in mochokids, auchenipterids, ictalurids and *Franciscodoras*, and is secondarily lost, among pimelodids, in the very peculiar and morphologically derived genus *Hypophthalmus*. Character 247 (presence of well-developed adductor hyomandibularis) is also found in amblycipitids, silurids, callichthyids, *Chaca, Astroblepus, Scoloplax,* and in those catfishes of clade 21, but is not secondarily lost anywhere within the Pimelodidae. Character 251 (levator operculi originating on both the neurocranium and the dorsolateral surface of the hyomandibulo-metapterygoid) is present in all pimelodids and, among non-pimelodids, in trichomycterids + nematogenyids, in callicthyids, silurids, schilbids, plotosids, and in the node leading to cranoglanidids + ictalurids.

The present cladistic analysis thus contradicts the nowadays often-accepted view that Pimelodidae constitute a polyphyletic assemblage (see Sections 1.2 and 1.3). The strong evidence provided here to support Pimelodidae monophyly is probably associated with the inclusion, in this work, of characters not usually included in studies of catfish phylogeny, such as those concerning the pectoral girdle, cranial muscles, or structures associated with the mandibular barbels. The five characters listed above do indeed refer to such characters.

Clade 55 (*Microglanis, Pseudopimelodus*)

Diagnosis. [27:1→0], [69:0→1], [84:0→1], [150:0→1], [219:0→1], [254:0→1], [290:0→1], [308:1→0], [310:0→1], [332:0/1→2]

Comments. As expected, genera *Microglanis* and *Pseudopimelodus* from Pseudopimelodidae appear more closely related to each other than to genera *Heptapterus, Rhamdia* and *Goeldiella* of Heptapterinae or to genera *Pseudoplatystoma, Hypophthalmus, Pimelodus,* and *Calophysus* of Pimelodidae (see, e.g., Lundberg et al., 1991a; de Pinna, 1998; Teugels, 2003). The present analysis revealed no new, undescribed potential pseudopimelodin autapomorphy, since character 310, i.e., presence of a prominent anterolateral projection of sesamoid bone 1 of the suspensorium, was based on previous descriptions of other authors (for description of this character, see Section 3.1).

Clade 56 (*Heptapterus, Rhamdia, Goeldiella, Pseudoplatystoma, Hypophthalmus, Pimelodus, Calophysus*)

Diagnosis. [12:1→2], [198:2→1]

Comments: As seen above, the present analysis provided strong evidence to support the monophyly of family Pimelodidae as a whole. Therefore, one question is raised: if subfamilies Heptapterinae, Pimelodinae and Pseudopimelodinae do thereby constitute a monophyletic assemblage, which two of these three subfamilies are more closely related to each other? This analysis suggests a closer relationship between the heptapterins and the pimelodins than between either of these two groups and the pseudopimelodins. Two features support this: 1) the more developed cartilaginous plates carrying the mandibular barbels (char. 12, state 2); 2) the presence of well-developed anterior process of the dorsal condyle of the pectoral spine (char. 198, reverted from state 2 to state 1 in the node leading to this clade and subsequently to state 0 in pimelodins: see below). The close relationship between heptapterins and pimelodins is also supported by recently obtained, and still unpublished, molecular sequence data from 12S and 16S mitochondrial rDNA genes, which, according to Lundberg (1998: 59), 'contain a clear phylogenetic signal for Heptapterinae and Pimelodinae, plus monophyly of these two groups together'.

Clade 57 (*Heptapterus, Rhamdia, Goeldiella*)

Diagnosis. **[58:0→1]**, [301:1→0], [386:0→1]

Comments. As expected, genera *Heptapterus, Rhamdia* and *Goeldiella* of subfamily Heptapterinae appear more closely related to each other than to the other pimelodid genera examined (see, e.g., Lundberg and McDade, 1986; Ferraris, 1988a; Lundberg et al., 1988, 1991a; Bockmann, 1994, 1998; de Pinna, 1998). The present analysis detected no new potential Heptapterinae autapomorphy, since the character 58, i.e., mesethmoid exhibiting two well-developed ventrolateral processes mainly directed laterally as well as anteriorly, was based on previous descriptions of other authors (for description of this character, see Section 3.1).

Clade 58 (*Rhamdia, Goeldiella*)

Diagnosis. [332:1→2]

Comments. Although a study of the interrelationships within Heptapterinae is beyond the main scope of this work, the specimens examined of genera *Rhamdia* and *Goeldiella* appear more closely related to each other than to specimens of genus *Heptapterus*, as expected (see Lundberg et al., 1991a). The character supporting this concerns the presence of a well-developed foramen of the anteroventral surface of the quadrato-symplectic (char. 332).

Clade 59 (*Pseudoplatystoma, Hypophthalmus, Pimelodus, Calophysus*)

Diagnosis. **[20:0→1]**, [72:1→2], **[85:0→1]**, [198:1→0], **[257:0→1]**, **[258:0→1]**

Comments. The four genera of subfamily Pimelodinae appear more closely related to each other than to the other pimelodid genera examined, as expected (see, e.g., Lundberg et al., 1988, 1991b; de Pinna, 1998). The present analysis revealed no potential new Pimelodinae autapomorphy. [Note: character 20, i.e., presence of muscle 1 of the mandibular barbels; 85, articulatory facet of the neurocranium for the autopalatine remarkably elongated anteroposteriorly; 257, presence of a muscle tensor tripodis and 258, presence of a drumming muscle of the swim bladder, were based on previous descriptions of other authors: for descriptions of these characters, see Section 3.1.]

Clade 60 (*Hypophthalmus, Pimelodus, Calophysus*)

Diagnosis. [97:0→1], [227:0→1]

Comments. Although a study of Pimelodinae interrelationships is clearly outside the main purpose of this work, it is interesting to note that the specimens examined of *Hypophthalmus*, *Pimelodus* and *Calophysus* appear more closely related to each other than to specimens of *Pseudoplatystoma*, as expected (see Lundberg et al., 1991b; de Pinna, 1998). The characters supporting this hypothesis are: 1) dorsal margins of frontal and pterotic in contact with each other in dorsal view (char. 97); 2) significant part of fibres of extensor tentaculi inserting on mesial and/or dorsal surface of sesamoid bone 1 (char. 227).

Clade 61 (*Pimelodus, Calophysus*)

Diagnosis. [109:0→1], [308:1→0], **[311:0→1]**

Comments. Although this is also clearly outside the main aim of the present work, it should be noted that the specimens examined of genera *Pimelodus* and *Calophysus* appear more closely related to each other than to specimens of genus *Hypophthalmus*. Three characters support this hypothesis, with the third character inclusively homoplasy-free among all the siluriforms examined: 1) presence of well-developed anterodorsolateral salience of sphenotic (char. 109); 2) presence, due to secondary reversion, of sesamoid bone 2 of the suspensorium (char. 308); 3) sesamoid bone 1 of the suspensorium a peculiarly thin, elongated structure (char. 311).

These results contradict those of de Pinna (1993, 1998), according to which genus *Hypophthalmus*, included in the '*Pimelodus* group', would be more related to genus *Pimelodus*, also from this group, than to genus *Calophysus* from the '*Calophysus* group' (see de Pinna, 1998: Fig. 16). However, it should be noted that de Pinna clearly emphasised, as have other authors such as Howes (1983b) or Lundberg et al. (1991b), the highly difficult task of establishing a solid, robust hypothesis on the phylogenetic position of *Hypophthalmus* within Pimelodinae (see de Pinna, 1998: 321-322).

Clade 62 (*Amphilius, Paramphilius, Leptoglanis, Zaireichthys, Doumea, Phractura, Andersonia, Belonoglanis, Trachyglanis, Chaca, Uegitglanis, Heteropneustes, Clarias, Heterobranchus, Cnidoglanis, Neosilurus, Plotosus, Paraplotosus, Parakysis, Akysis, Amblyceps, Liobagrus, Gagata, Bagarius, Glyptosternon, Glyptothorax, Erethistes, Hara, Aspredo, Bunocephalus, Xyliphius*)

Diagnosis. [150:0→1], [153:0→1], [408:0→1]

Comments. This clade is supported by the three characters listed above. The anteroposteriorly compressed mesial limb of the post-temporo-supracleithrum (char. 150), is a rather homoplasic feature. It is also found outside clade 62 in

Nematogenys, Helogenes, Malapterurus, silurids and pseudopimelodins, and is modified within the clade leptoglanidins, glyptosternins, and in the node leading to clariids plus *Heteropneustes* (see below). However, character 153, i.e., marked bifurcation of the posttemporo-supracleithrum to receive the dorsal surface of the cleithrum, is almost homoplasy free. It was not found elsewhere among the siluriforms examined and within clade 62, is only absent in the node leading to clariids, plotosids and *Heteropneustes*. Character 408 concerns the covering of a great part of the lateral surface of the angulo-articular by a posterolateral laminar projection of the dentary. It is only found outside clade 62 in loricarioids and *Heptapterus,* and is only secondarily lost, within this clade, in the node leading to *Heteropneustes,* clariids and plotosids.

Interestingly, all three of these features are absent in the group including Clariidae and *Heteropneustes* and, except for the first, rather homoplasic feature, also in Plotosidae. Inclusion of these three taxa in clade 62 has thus more to do with indirect evidence supporting the grouping of these taxa with Chacidae and Sisoroidea (see below) on the one hand, and of Sisoroidea with Amphiliidae on the other, than with direct evidence supporting the grouping of clariiids + plotosids + heteropneustids with Amphiliidae.

Although the clade including Amphiliidae, Chacidae, Clariidae, Heteropneustidae, Plotosidae and Sisoroidea constitutes an essentially new one, some of its groups have already been associated in previous studies. For example, the close relationship between amphiliids and sisoroids has been defended by many authors (see, e.g., Regan, 1911b; Harry, 1953; Chardon, 1968; Mo, 1991; de Pinna, 1993, 1998; He et al., 1999), constituting nowadays a relatively well-grounded and consensual hypothesis (see Diogo, 2003a). Also, some authors, e.g., Lundberg and Baskin (1969), Tilak (1971) or Brown and Ferraris (1988), have proposed a close relationship between chacids and plotosids. A close relationship between chacids and sisoroids has also been proposed by some authors, such as Gauba (1970) or Tilak (1971). Interestingly, de Pinna (1993, 1998) suggested a somewhat close relationship between chacids, clariids, *Heteropneustes* and plotosids, while Mo (1991) suggested a somewhat close relationship between chacids, clariids, *Heteropneustes,* amphiliids and sisoroids (see Section 1.3).

Clade 63 (*Chaca, Uegitglanis, Heteropneustes, Clarias, Heterobranchus, Cnidoglanis, Neosilurus, Plotosus, Paraplotosus, Parakysis, Akysis, Amblyceps, Liobagrus, Gagata, Bagarius, Glyptosternon, Glyptothorax, Erethistes, Hara, Aspredo, Bunocephalus, Xyliphius*)

Diagnosis. [148:0→1], [236:0→1], [286:0→1], [290:0→1]

Comments. Four characters support this clade. Character 148, concerning the presence of a transcapular process of the posttemporo-supracleithrum, is only secondarily lost within the clade in plotosids, in amblycipitids and in *Glyptosternon*, and found outside it in *Scoloplax*. Character 236, i.e., attachment of a significant part of the fibres of the adductor arcus palatini on sesamoid bone 1 of the suspensorium, is an interesting, almost homoplasy-free, feature. It is not found elsewhere in the siluriforms and is only secondarily lost inside the clade in *Parakysis* and in the highly peculiar aspredinids (see below). The autopalatine a roughly tubular structure with a salient mesial articulatory surface for the neurocranium (char. 286, state 1) is also an almost homoplasy-free feature. It is only found outside the clade in the node leading to doradids + auchenipterids, and is not reverted within the clade. With respect to the dorsoventral expansion of the back of the autopalatine due to an enlargement of its centre of ossification (char. 290), this was found outside the clade in the pimelodins analysed, as well as in the doradoids examined, except for *Ageneiosus* and *Auchenipterus*. Inside the clade, it was absent only in the group constituted by clariids and *Heteropneustes*.

A clade including Chacidae, Clariidae, Heteropneustidae, Plotosidae and Sisoroidea has not been previously described in the literature, but some authors had already suggested a close relationship between certain members of this clade (see comments on clade 62 above).

Clade 64 (*Chaca, Uegitglanis, Heteropneustes, Clarias, Heterobranchus, Cnidoglanis, Neosilurus, Plotosus, Paraplotosus*)

Diagnosis. [53:0→1], [212:0→1], [219:0→1], [226:2→0], [295:0→1]

Comments. A particularly close relationship between clariids, *Heteropneustes*, chacids and plotosids was suggested by de Pinna (1993, 1998), who included these four groups, together with the silurids, malapterurids and auchenoglanidins, in an unresolved polytomy (see Section 1.3). Previous studies had suggested either a close relationship between chacids and plotosids (see, e.g., Day, 1877; Jayaram and Majumdar, 1964; Tilak, 1971; Lundberg and Baskin, 1979; Brown and Ferraris, 1988) or a somewhat close relationship between chacids and clariid/heteropneustid catfishes (see, e.g., Mo, 1991), but not among all the four groups.

The evidence provided in the present analysis to support grouping in a monophyletic clade of the chacid, heteropneustid, clariid and plotosid catfishes examined is rather strong. Placement of at least part of the ventral division of the arrector ventralis on the dorsal side of the pectoral girdle (char. 53) is a feature found only, outside the clade, in aspredinids, *Hypophthalmus* and the node leading to *Laides* + Ailia + *Siluranodon* (see above). It is not reverted anywhere within the clade. The adductor A3' lying at least partly mesial to the levator arcus palatini (char. 212) is found only, outside the clade, in *Cranoglanis*, and within clade is secondarily lost only in *Plotosus* and *Cnidoglanis*. Character 219 concerns the origin of the adductor mandibulae A3″ on the suspensorium and the neurocranium, near the region of its articulation with the hyomandibulo-metapterygoid. It is found in several groups outside the clade, such as *Nematogenys, Malapterurus, Hypophthalmus,* auchenipterids and pseudopimelodins, but is not secondarily lost in members of the clade. Character 226, concerning the differentiation of the extensor tentaculi, is a rather homoplasic character subsequently modified in the clariids and found outside the clade in several other catfishes (see description of character in Section 3.1). But the most important synapomorphy supporting this clade concerns a completely homoplasy-free, autapomorphic feature: the essentially anteromesially oriented articulatory surface of the autopalatine for the neurocranium (char. 295).

Interestingly, long ago some authors, such as Hamilton (1822) or Chardon (1968), pointed out the morphological similitude between chacid and aspredinid catfishes. In fact, besides the features listed above to support clades 62 and 63 and, consequently, the somewhat close relationship between sisoroid aspredinids and chacids (see above), these two groups share other rather peculiar and rare morphological characters. In particular, the configuration of the posterodorsal region of the skull in the members of these groups is, as underscored by Chardon (1968), remarkably similar. For example, both groups present a well-developed, deep fossa between the posttemporo-supracleithrum, parieto-supraoccipital and, eventually, the epioccipital (char. 126). Such a fossa was only found, aside from these groups, in the erethistid catfishes examined (for description of this character, see Section 3.1). Also, both Aspredinidae and Chacidae present a well-developed, dorsal lamina of the Weberian apparatus in contact with the dorsal surface of the body (char. 141), a feature not found elsewhere in the Siluriformes (for description of this character, see Section 3.1). A peculiar, and also rather rare feature present in the chacid and aspredinidid siluriforms is the markedly thin and mesially extended dorsomesial limb of the posttemporo-supracleithrum (char. 144). This was only found, aside from members of these two groups, in specimens examined of genus *Parakysis* (for description of this character, see Section 3.1). But the morphological similarity between chacids and aspredinids is certainly not restricted just to the configuration of the structures of the posterodorsal region of the cranium. For instance, in both these groups the prevomer is missing (char. 93). The absence of the prevomer is a highly peculiar and rare character in order Siluriformes, being also found only, among the other catfishes examined, in *Microglanis* and *Scoloplax*.

However, the four homoplasic and the one non-homoplasic characters supporting the grouping of chacids, clariids, heteropneustids and plotosids, and in particular the 10 homoplasic and the 5 non-homoplasic characters supporting the grouping of aspredinids and erethistids, clearly seem to indicate that the 3 homoplasic and the non-homoplasic peculiar characters shared by chacids and aspredinids (see above) are due to convergent/parallel evolution. This convergent/parallel evolution could probably be associated, as explained by Chardon, to convergent/parallel adaptation to the peculiar 'burrowing' behaviour exhibited by these fishes (see Chardon, 1968:161, for more details on this subject).

Clade 65 (*Chaca*)

Diagnosis. [2:1→0], **[6:0→1]**, **[41:0→1]**, [69:0→3], [72:0→2], **[77:0→1]**, [86:0→1], **[92:0→1]**, [93:0→1], [126:0→1], [131:0→2], [141:0→1], [144:0→1], [148:0→2], [162:0→1], **[153:1→0]**, [188:1→2], **[199:0→1]**, [215:0/1→2], **[234:0→1]**, [247:0→1], [264:0→1], [265:0→1], [288:0→2], [293:0→2], [307:0→1], **[337:0→2]**, [353:0→1], [355:0→1], **[372:0→1]**, **[384:0→1]**, [389:0→1], [417:0→1]

Comments. Inclusion of genus *Chaca* in a separate catfish family has been commonly accepted for a long time (see, e.g., Regan, 1991; Chardon, 1968; Lundberg and Baskin, 1969; Gosline, 1975; Mo, 1991; de Pinna, 1993, 1998). Indeed, the fishes of this genus differ remarkably morphologically from other catfishes, as revealed by the numerous characters listed above. Notably, the present cladistic analysis revealed 8 new autapomorphic characters in chacids, not found in other catfishes examined, which could thus constitute potential Chacidae autapomorphies: 1) marked separation of cartilages of mandibular barbels (char. 6); 2) abductor superficialis present (char. 41); 3) presence of thin anterior projection of ethmoid cartilage (char. 92); 4) second dorsal process of the cleithrum markedly developed (char. 153); 5) highly developed, dorsally expanded adductor arcus palatini (char. 234); 6) remarkably deep, large, roundish neurocranial concavity to receive dorsal surface of hyomandibulo-metapterygoid (char. 337); 7) orientation of posterior ceratohyal perpendicular to that of both ventral hypohyal and anterior ceratohyal (char. 372); 8) presence of prominent posteroventral crest of ventral hypohyal (char. 384). [Note: characters 77, i.e., presence of a prominent posterolateral process of the premaxilla; 199, presence of a thin anterior projection of the ethmoid cartilage and 353, interopercle presenting a deep anterior concavity for articulation with the preopercle, were based on previous descriptions of other authors: for description of these characters, see Section 3.1.]

Clade 66 (*Uegitglanis, Heteropneustes, Clarias, Heterobranchus, Cnidoglanis, Neosilurus, Plotosus, Paraplotosus*)

Diagnosis. [4:0→1], [53:1→2], [76:1→0], [153:0→1], [408:1→0]

Comments. The sister-group relationship between plotosids and clariids plus *Heteropneustes* appears as a somewhat surprising but well-supported outcome of the present cladistic analysis. In fact, as mentioned above, the only phylogenetic studies proposing a somewhat close relationship between these groups were those of de Pinna (1993, 1998), in which Plotosidae, Heteropneustidae and Clariidae were included (see Fig. 1.11), together with Siluridae, Malapteruridae, Chacidae and Auchenoglanidinae, in an unresolved polytomy (see Section 1.3). Besides all the characters presented above referring to clades 62, 63 and 64, the five synapomorphic characters here support a close relationship between plotosids and clariids plus *Heteropneustes*. Characters 153, i.e., bifurcation of the posttemporo-supracleithrum for the cleithrum not remarkably developed, and 408, lateral surface of the angulo-articular widely visible in a lateral view of the mandible refer, as explained above, to secondary reversions to the plesiomorphic state within clade 62. Character 4, contact between the anterior parts of the mandibular barbels, is a rather homoplasic character found in all members of the clade but also found in glyptosternins, amblycepitids, bagrids, *Anadoras*, *Synodontis* and ictalurids. However, character 53, the vast majority or totality of the ventral division of arrector dorsalis lying on dorsal surface of pectoral girdle, is an interesting feature not secondarily lost within the clade, and outside it only present in aspredinids, *Hypophthalmus*, and *Laides* + *Ailia* + *Siluranodon*. The presence of a prominent posteroventral extension of the premaxilla (char. 76) is a feature secondarily lost within the clade, in the node leading to *Clarias* + *Heterobranchus*, but found outside it, only in genus *Nematogenys*.

The relatively strong evidence provided here to support the monophyly of the clade including clariids + *Heteropneustes* + plotosids seems once again to be associated with the inclusion in this work of features not usually included in studies on catfish phylogeny, such as those concerning the structures associated with the mandibular barbels or the pectoral girdle musculature. Of the five synapomorphic characters listed above, indeed two refer to such features.

Clade 67 (*Uegitglanis, Heteropneustes, Clarias, Heterobranchus*)

Diagnosis. **[17:0→1]**, **[31:0→1]**, [50:0/1→2], [97:0→1], [103:0→1], **[134:0→1]**, [150:1→2], **[170:0→1]**, [212:1→2], [226:0→1], [240:0→1], [246:0→2], [266:0→1], [290:1→0], [327:0→1], **[340:0→1]**, [346:0→1], **[375:0→1]**, [419:0→1]

Comments. The cladistic analysis of the present work strongly supports the grouping of the clariid and *Heteropneustes* catfishes examined in a monophyletic clade. This is not surprising since a close relationship between clariids and *Heteropneustes* has been suggested by several authors, e.g., Srinivasachar (1958, 1959), Tilak (1967), Chardon (1968), Lundberg and Baskin (1969), Datta et al. (1975), Mo (1991), de Pinna (1993, 1998) or Teugels (1996, 2003). Notably, the present analysis revealed 5 new autapomorphic characters to diagnose this clade, which could constitute potential autapomorphies of family Clariidae (including *Heteropneustes*), as explained below: 1) presence of well-developed, mesial cartilaginous complex connecting the internal mandibular barbels midline (char. 17); 2) pars lateralis of protractor hyoideus lying completely dorsal to the pars ventralis of this muscle (char. 31); 3) dorsolateral portion of cleithrum markedly elongated anteroposteriorly (char. 170); 4) presence of markedly developed, elongated, posteriorly pointed posterodorsal projection of opercle (char. 340); 5) presence of prominent ventrolateral crest of anterior ceratohyal. [Note: character 134, referring to shape of the infraorbital 4 and suprapreopercle, was already pinpointed in previous studies as a feature uniting clariids and *Heteropneustes*, see, e.g., Chardon, 1968.]

Clade 68 (*Heteropneustes, Clarias, Heterobranchus*)

Diagnosis. **[134:1→2]**, [148:1→2]

Comments. Although, as noted above, there is a consensus in most available studies concerning the close relationship between *Heteropneustes* and clariids, there is some controversy in those studies concerning the phylogenetic status of Heteropneustidae. In fact, in most of these studies the genus *Heteropneustes* is included in a separate, monogeneric family, Heteropneustidae, the sister-group of a clade including all clariid genera, i.e., family Clariidae as it is commonly recognised (see Sections 1.2 and 1.3).

However, this view was contradicted by Chardon (1968), who suggested that the phylogenetic distance between *Heteropneustes* and some clariid genera such as *Clarias, Heterobranchus, Channallabes, Dolichallabes* or *Clariallabes* is not greater than that between these latter genera and two genera commonly included in family Clariidae, *Horaglanis* and *Uegitglanis*. This suggestion was recently corroborated in de Pinna's 1993 unpublished thesis. According to him, genera *Uegitglanis* and *Horaglanis* are respectively the successive sister-groups of a monophyletic clade composed by the genus *Heteropneustes* plus all the remaining clariid genera. Thus, according to de Pinna (1993), *Heteropneustes* should be included in Clariidae, and consequently family Heteropneustidae should be synonymised with Clariidae, the latter comprising *Horaglanis* (its more basal genus), *Uegitglanis* and a clade including *Heteropneustes* and the remaining clariid genera. This view was maintained in de Pinna's 1998 overview, in which *Heteropneustes* is included in family Clariidae and *Horaglanis* considered the most plesiomorphic clariid genus.

The present work supports de Pinna's 1993 and 1998 studies. Although it was not possible to include *Horaglanis* in the cladistic analysis given here, this analysis detected two derived features supporting a closer relationship between *Heteropneustes* and ordinary clariid genera (i.e., those genera commonly assigned to Clariidae excluding *Horaglanis* and *Uegitglanis*) such as *Clarias* or *Heterobranchus*, than between the latter genera and clariid genus *Uegitglanis*. The first of these features concerns development of the infraorbital 4 and suprapreopercle. Plesiomorphically in catfishes these bones are 'simple tube-like structures' (de Pinna, 1993: 184-185). In *Uegitglanis* these two bones are somewhat more developed than in most other catfishes, with their connection to the neurocranium, however, relatively feeble (char. 134, state 1). In *Heteropneustes*, as well as specimens of the two ordinary clariid genera examined, both the infraorbital 4 and the suprapreopercle are highly developed, markedly enlarged bones firmly ankylosed to the neurocranium (char. 134, state 2) (for description of this character, see Section 3.1). The other feature concerns development of the transcapular process. As explained above, plesiomorphically in catfishes the posttemporo-supracleithrum does not present a prominent posterodorsal process attaching to the parapophysis of vertebra 4. In *Uegitglanis* the posttemporo-supracleithrum bears a poorly developed posterodorsal process, which contacts with, but is not firmly attached to, the anterior margin of this parapophysis (char. 148, state 1). In *Heteropneustes,* as well as in specimens of the two ordinary clariid genera examined, *Clarias* and *Heterobranchus,* the posttemporo-supracleithrum has a prominent posterodorsal projection that is firmly ankylosed with the anterolateral margin of the parapophysis of the fourth vertebra (char. 148, state 2). Such a development of the transcapular process is also found only in catfishes of the sisoroid families Sisoridae, Erethistidae and Aspredinidae (see description of this character in Section 3.1) and, according to the present cladistic analysis, this morphological similarity is due to homoplasy (see below).

As first noted by Chardon (1968), to consider the ordinary clariid genera more closely related to *Heteropneustes* than to the clariid genera *Horaglanis* and *Uegitglanis* might seem unreasonable. This is because it is commonly mentioned in the literature that *Heteropneustes* and clariid catfishes have two different, albeit probably homologous, accessory respiratory organs: *Heteropneustes* has pulmonary sacs, while clariids have a suprabranchial organ. In actuality, this feature is very complex. In fact, some authors, e.g. Gianferrari (1923) and David (1936), refer to a vestigial suprabranchial organ in *Uegitglanis,* while others such as Chardon (1968) and Burgess (1989) state that *Uegitglanis* does not present a suprabranchial organ. The bad state of conservation of the branchial region in the single specimen of *Uegitglanis zammoronoi* analysable in the present study did not allow clarification of whether a suprabranchial organ was/was not present. With respect to *Horaglanis,* Menon's 1951 original description of the genus does refer to a rudimentary suprabranchial organ. However, the somewhat superficial and even sometimes erroneous description of *Horaglanis* given by Menon raised serious doubts for Chardon (1968) as to whether this genus does indeed present a suprabranchial organ or not. Moreover, even among ordinary clariid genera the suprabranchial organ may be reduced (e.g. *Channallabes, Dolichallabes, Gymnallabes*) or even completely absent (e.g. *Platyclarias, Xenoclarias, Platyallabes*) (see, e.g. Burgess, 1989). Therefore, until more conclusive studies become available on whether a suprabranchial organ is present or not in *Horaglanis* and *Uegitglanis,* as well as on the phylogenetic relationships among clariids, it is unreasonable to refer to the presence of a suprabranchial organ as an unambiguous synapomorphy of all clariids excluding *Heteropneustes.*

Clade 69 (*Clarias, Heterobranchus*)

Diagnosis. [21:0→1], [32:0→1], **[38:0→1]**, [76:1→0], [166:0→1], [336:0→1]

Comments. As expected, specimens of the two ordinary clariid genera examined appeared more closely related to each other than to genus *Heteropneustes* (see above). Of the six characters listed above, one was notably found only in these ordinary clariids and in no other catfishes examined: sternohyoideus covering a significant part of the anteroventral surface of the pectoral girdle (char. 38).

Clade 70 (*Cnidoglanis, Neosilurus, Plotosus, Paraplotosus*)

Diagnosis. [47:1→0], **[108:0→1]**, [114:0→1], **[147:0→1]**, [148:1→0], [251:0→1], **[271:0→1]**, **[272:0→1]**, [290:1→2], **[304:0→1]**, [318:0→1], [486:0→1]

Comments. As pointed out in the Introduction, Plotosidae is consensually considered a monophyletic group. The present analysis revealed two autapomorphic characters present in plotosids and in no other siluriforms examined: 1) posterior portion of parasphenoid markedly compressed transversally (char. 108); 2) presence of anterodorsal foramen of posttemporo-supracleithrum (char. 147). [Note: characters 271, i.e. presence of two well-developed ligaments between the maxilla and the coronoid process; 272, enlarged base of the maxillary barbels and 304, presence of a well-defined ligament connecting the anteromesial surface of the autopalatine and the posterolateral surface of the premaxilla, were based on previous descriptions of other authors: for description of these characters, see Section 3.1.]

Clade 71 (*Neosilurus, Plotosus, Paraplotosus*)

Diagnosis. [109:0→1], [417:0→1]

Comments. Although, as explained above, the phylogenetic analysis of this work is mainly focused on the interrelationships between the various catfish families, it provides the first partial, explicit cladistic insight into relationships within Plotosidae. The two characters above, i.e., the well-developed anterodorsolateral salience of the sphenotic (char. 109) and the anterodorsomesial laminar projection of the dentary (char. 417), suggest a closer relationship between genera *Neosilurus, Plotosus* and *Paraplotosus* than between any of these genera and the other plotosid genus examined, *Cnidoglanis*.

Clade 72 (*Plotosus, Paraplotosus*)

Diagnosis. **[98:0→1]**, **[107:0→1]**, **[241:0→1]**

Comments. Within genera *Neosilurus, Plotosus* and *Paraplotosus*, the following three homoplasy-free characters strongly indicate a closer relationship between *Plotosus* and *Paraplotosus* than between any of these genera and *Neosilurus*: 1) dorsomesial process of the frontal remarkably developed (char. 98); 2) presence of a lateral process of the pterosphenoid (char. 107); 3) origin of the levator arcus palatini on the dorsal surface of the neurocranium (char. 241).

Clade 73 (*Parakysis, Akysis, Amblyceps, Liobagrus, Gagata, Bagarius, Glyptosternon, Glyptothorax, Erethistes, Hara, Aspredo, Bunocephalus, Xyliphius*)

Diagnosis. **[138:0→1], [164:0→1]**, [167:0→1], [210:0→1], [325:0→3], [401:0→1]

Comments. The cladistic analysis of the present work supports de Pinna's (1996) phylogenetic hypothesis, according to which the Amblycipitidae, Akysidae, Sisoridae, Erethistidae and Aspredinidae form a monophyletic clade, the superfamily Sisoroidea (see Section 1.3). In fact, the present analysis not only confirmed the synapomorphies given by de Pinna to support Sisoroidea monophyly (which includes characters 138 and 164 of the present analysis, see De Pinna, 1996: 59-60), but also pointed out four additional synapomorphies to support this superfamily: 1) presence of well-developed anteromesial process of cleithrum (char. 167); 2) A2 directly inserted on mesial surface of mandible (char. 210); 3) entoectopterygoid remarkably reduced in size, with medial surface completely surrounded by lateral surface of sesamoid bone 1 of suspensorium (char. 325); 4) coronoid process of mandible essentially constituted by posterodorsal surface of dentary (char. 401).

According to de Pinna (1996), within superfamily Sisoroidea, Amblycipitidae occupy the most plesiomorphic position, with the intra-relationships among the superfamily being (Amblycipitidae + (Akysidae + (Sisoridae + (Aspredinidae + Erethistidae)))). However, as recognised by de Pinna (1996: 76), contrary to the grouping Sisoridae + Aspredinidae + Erethistidae, which is very well supported (and was later corroborated by Diogo et al., 2001c, 2002), the proposal of a sister-group relationship between Akysidae and these three families is based on scarce evidence. In fact, this proposal relied 'on three synapomorphies, one of these (supratemporal fossae) shows reversals, and the other two (supracleithrum strongly attached to skull; posterior nuchal plate with anterior facet for articulation with anterior nuchal plate) have a number of putatively homoplasic occurrences elsewhere in siluriforms' (De Pinna, 1996: 76). The evidence presented by de Pinna (1996) to support a sister-group relationship between Akysidae and the clade (Sisoridae + (Aspredinidae + Erethistidae)) was not significantly stronger than the evidence supporting the alternative, most traditional hypothesis defending a sister-group relationship between Akysidae and Amblycipitidae (see, e.g., Reagan, 1911b; Chen and Lundberg, 1994).

The results obtained in the present cladistic analysis confirmed the difficulty to conclusively solve the phylogenetic position of the amblycipitid and akysid catfishes within the superfamily Sisoroidea. In fact, as explained above, the node leading to these two groups and the remaining sisoroids is the single unresolved interfamilial node of the strict-consensus cladogram presented on Fig. 3.123. The present study corroborated the presence, in the Akysidae, Sisoridae, Aspredinidae and Erethistidae, of the three derived features provided by de Pinna (1996) to support the grouping of these families (see above). However, it also pointed out some important, major conflicting characters with this hypothesis.

One of these characters is the peculiar 7-shape of sesamoid bone 1 of the suspensorium. De Pinna (1996: 70) mentioned this peculiar feature, noting that it was present in the amblycipitids and in *Parakysis*. However, this peculiar feature is also present in the other akysid genus, *Akysis*, as noted by Chen and Lundberg (1994: 795) and confirmed in the present study (see char. 312). The presence of this peculiar, derived feature in Amblycipitidae and Akysidae genera examined conflicts with the grouping of the Akysidae with families Sisoridae, Aspredinidae and Erethistidae.

Another conflicting character is the bifurcation of the basal cartilages of the external mandibular barbels. Contrary to most catfishes, in some akysids and in the amblycipitids, as well as in some glyptosternin Sisoridae, the basal cartilages of the outer mandibular barbels are posteriorly bifurcate. Attending to the well-supported grouping of Sisoridae, Aspredinidae and Erethistidae (see below), the taxonomic distribution of this character supports a sister-group relationship between Akysidae and the Amblycipitidae, with an independent, homoplasic acquisition in the glyptosternin Sisoridae.

In addition to the two characters above, it is also important to take cognizance of the three characters discussed by de Pinna (1996: 69), namely, the 'morphology of the first proximal pectoral radial', 'presence of a spurlike process on the quadrate' and 'state of the humerovertebral ligament'. These refer to derived features present in the amblycipitids and *Parakysis*, but not *Akysis*, and could thus either be interpreted as independently acquired in amblycipitids and *Parakysis*, or acquired in amblycipitids plus akysids and subsequently reverted in *Akysis*. In the latter case, these three characters would support a sister-group relationship between the families Amblycipitidae and Akysidae.

Clade 74 (*Amblyceps, Liobagrus*)

Diagnosis. [4:0→1], **[74:0→1]**, [90:0→1], [148:1→0], [206:0→1], [208:0→1], [247:0→1], [312:0/1→2], [313:0→1], **[390:0→1]**, **[393:0→1]**

Comments. As explained in Chapters 1 and 2, Amblycipitidae is consensually considered a monophyletic group, diagnosed by several synapomorphic features (see, e.g., Mo, 1991; Chen, 1994; Chen and Lundberg, 1994; de Pinna, 1996). The present analysis traced one new character present in the amblycipitids and in no other catfishes examined, which could eventually constitute a potential amblycipitid autapomorphy: hypobranchial foramen of parurohyal markedly enlarged (char. 390). [Note: characters 74, i.e., presence of a prominent posterolateral process of the premaxilla and 393, branchiostegal rays presenting a highly peculiar, closed round arrangement, were both based on previous descriptions of other authors: for description of these characters, see Section 3.1.]

Clade 75 (*Parakysis, Akysis*)

Diagnosis. [146:0→1], [155:0→1], [431:0→1]

Comments. The akysids are also considered a monophyletic group by most authors (see, e.g., Mo, 1991; de Pinna, 1996, 1998; Teugels, 1996, 2003). The present work supported this monophyly, but detected no additional, undescribed autapomorphic characters in the akysids examined.

Clade 76 (*Gagata, Bagarius, Glyptosternon, Glyptothorax, Erethistes, Hara, Aspredo, Bunocephalus, Xyliphius*)

Diagnosis. [139:0→1], [148:1→2], **[164:1→3]**, [248:0→1], [302:0→3]

Comments. The relationships of the Neotropical Aspredinidae, included in this clade together with the Asiatic Sisoridae and Erethistidae, have been a particularly puzzling issue for a long time. The first study dealing in some detail with this subject was that of Günther (1864), who placed the aspredinids together with loricarioids and sisorids in his 'sixth subfamily Siluridae Proteropodes'. Chardon (1968), based on characters of the Weberian apparatus, also suggested a close relationship between loricarioids and aspredinids. However, subsequent studies (Baskin, 1973; Howes, 1983a) showed that Chardon's arguments were not convincing, that is, they were supported by no putative derived synapomorphy. Ferraris (1989) was the first to suggest that aspredinids were probably related to Asiatic taxa, namely to Akysidae. Mo (1991), in the first explicitly phylogenetic analysis of siluriform phylogeny, placed the aspredinids as either basal to or in a polytomy with a clade containing clariids, heteropneustids, amblycipitids, akysids, sisorids, amphiliids and loricarioids (see Fig. 1.6). Somewhat similar hypotheses were suggested subsequently by de Pinna (1993), who also placed Aspredinidae in a polytomy including Amblycipitidae, Akysidae, Sisoridae, Amphiliidae and Loricarioidea, and by Chen (1994), who placed Aspredinidae as the sister-group of a clade composed of Amblycipitidae, Sisoridae and Akysidae. However, these hypotheses were challenged by Friel (1994), principally concerned with the interrelationships of aspredinids, but also the relationships among these fishes and other Siluriformes. He proposed doradoids (see above) as the closest relatives of aspredinids. This hypothesis in turn was subsequently challenged by de Pinna (1996), who considered the Sisoridae of previous authors to be a paraphyletic assemblage, with a subunit (which he subsequently named Erethistidae) more closely related to Aspredinidae than to the remaining sisorid taxa (see Section 1.3). The relationships of Aspredinidae have thus long been problematical, as pointed out in de Pinna's recent and detailed overview concerning the phylogenetic relationships of Neotropical catfishes (1998). In that overview de Pinna stated that 'clearly, the position of aspredinids within siluriforms is a complex issue, plagued by some striking morphological homoplasies' (de Pinna, 1998: 319).

The cladistic analysis of the present work strongly supports the hypothesis of de Pinna (1996), with the Neotropical Aspredinidae being grouped in a strongly supported clade together with the Asian Sisoridae and Erethistidae. In fact, this analysis not only corroborates the strong evidence provided in de Pinna's 1996 work to group these three families, but also provides two additional derived features to support this grouping: 1) neither adductor operculi nor part differentiated from it in contact with hyomandibulo-metapterygoid (char. 248: also found only in doumeins, loricariids and doradoids); 2) presence of long and thin ligament between dorsal surface of sesamoid bone 1 of suspensorium and posteroventral margin of autopalatine (char. 302: also found only in doradids and clariids).

It is interesting to note here that despite the clearly strong evidence provided in de Pinna's 1996 and the present analyses supporting the close relationship between aspredinids and the sisorids, and in particular the erethistids, the aspredinids share, as stressed by de Pinna, some 'striking' anatomical similarities with other catfish groups (see above). Some remarkable, highly peculiar features found in both the Neotropical aspredinids and the Asiatic chacids were already listed in the discussion of clade 64 concerning the phylogenetic position of *Chaca* (see above). But the rather 'mosaic', complex combination of peculiar characters present in Aspredinidae means these fishes also share some remarkable peculiar features with other groups that are morphologically very different from chacids, for example, the doradoids, and more particularly the doradids. Some of these characters concern, for example, the prominent dorsolateral projections of the laminar bone of the mesethmoid (char. 60), the mesocoracoid arch and the main body of the scapulo-coracoid being undistinguished from each other (char. 186), the presence of the highly developed anterior process of the dorsal condyle of the pectoral spine (char. 198) or the presence of the well-developed anteroventral lamina of the preopercle (char. 366). These characters, and other features, inclusively led, as noted above, to Friel's proposal of a sister-group relationship between aspredinid and doradoid catfishes (see Friel, 1994). However, within clade 38 (see Fig. 3.123), the 5 synapomorphies supporting the group doradoids + malapterurids (2 in clade 39 + 3 in clade 40) and in particular the 21 synapomorphies, 5 of which are completely homoplasy-free, supporting the group aspredinids + erethistids (3 in clade 62 + 4 in clade 63 + 6 in clade 73 + 5 in clade 76 + 3 in clade 79), clearly seem to indicate that the peculiar characters found in both doradoids and aspredinids are due to a convergent/ parallel evolution. This interesting issue will be discussed in more detail in Chapters 4 and 5.

Clade 77 (*Gagata, Bagarius, Glyptosternon, Glyptothorax*)

Diagnosis. **[115:0→1]**, [146:0→1], [265:0→1/2], [290:1→2]

Comments. The phylogeny and systematics of the Sisoridae were recently revised by de Pinna (1996), who concluded that six genera previously included in this family—*Conta, Erethistes, Erethistoides, Hara, Laguvia* and *Pseudolaguvia*—were more closely related to the Neotropical Aspredinidae than to the remaining 14 sisorid genera. Therefore, these six genera were assigned to family Erethistidae, which, according to de Pinna (1996), is the sister-group of the Neotropical Aspredinidae, with the clade formed by these two families in turn the sister-group of Sisoridae sensu stricto. Still according to de Pinna (1996), Sisoridae (sensu stricto) is divided into subfamilies Sisorinae and Glyptosterninae, with the former comprising tribes Sisorini (including *Sisor, Gagata, Nangra*) and Bagariini (*Bagarius*), and the latter comprising tribes Glyptothoracini (*Glyptothorax*) and Glyptosternini (*Glyptosternon, Glaridoglanis, Oreoglanis, Exostoma, Myersglanis, Coraglanis, Euchiloglanis, Pseudexostoma, Pseudecheneis*).

The phylogenetic analysis of this work corroborated the monophyly of Sisoridae sensu de Pinna (1996), and pointed out one additional, homoplasy-free autapomorphic feature to diagnose this family: the presence of a well-developed fossa between the ventral surfaces of the pterotic and the exoccipital (char. 115). However, it is noteworthy that in the present analysis, within the Sisoridae, representatives of tribes Sisorini—*Gagata*—and Bagariini—*Bagarius*—appeared in an unresolved trichotomy that also included the group formed by the glyptosternin representatives of tribes Glyptothoracini (*Glyptothorax*) and Glyptosternini (*Glyptosternon*) (Fig. 3.123). Thus, although the analysis supported de Pinna's 1996 Sisoridae and, within this group, subfamily Glyptosterninae, it did not corroborate the sisorid subfamily Sisorinae as defined by him (see above). This is not wholly surprising considering that some studies have suggested a far more plesiomorphic position of *Bagarius* within Sisoridae than proposed by de Pinna (1996), with this genus inclusively seen as a sisorid 'living fossil' by some authors (see, e.g., Hora, 1939; Gauba, 1962; Tilak, 1963c; He, 1997).

Clade 78 (*Glyptosternon, Glyptothorax*)

Diagnosis. [4:0→1], [9:0/1→2], [84:0→1], [90:0→1], **[115:1→2]**, [190:0→1], [358:0→2]

Comments. As explained above, the sisorid subfamily Glyptosterninae sensu de Pinna (1996) was supported by the present phylogenetic analysis, with this analysis revealing a new autapomorphic feature found in the glyptosternin specimens examined and in no other catfishes. This feature concerns the remarkable development of the fossa lying between the ventral surfaces of the pterotic and the exoccipital (char. 115, state 2).

Clade 79 (*Erethistes, Hara, Aspredo, Bunocephalus, Xyliphius*)

Diagnosis. [126:0→1], **[203:0→1]**, **[430:0→1]**

Comments. The cladistic analysis of the present work strongly supports de Pinna's 1996 grouping of Aspredinidae and Erethistidae. It not only corroborated de Pinna's synapomorphies (which include the characters 203 and 430 of the present work), but also revealed one additional synapomorphic feature shared by aspredinids and erethistids: the presence of a well-developed, deep fossa between the posttemporo-supracleithrum, the parieto-supraoccipital and, eventually, the epioccipital (char. 126: also found only in *Chaca*).

Clade 80 (*Erethistes, Hara*)

Diagnosis. [131:0→1], **[140:0→1]**, [148:2→3], [193:0→1], **[379:0→1]**

Comments. The present analysis supported de Pinna's 1996 Erethistidae, but did not find any new potential Erethistidae autapomorphies, since the homoplasy-free characters 193 (presence of lateral ramification of parapophysis 5) and 379 (laminar projections on posterior margin of anterior ceratohyal) refer to features described by de Pinna (1996).

Clade 81 (*Aspredo, Bunocephalus, Xyliphius*)

Diagnosis. [2:1→0], **[7:0→2]**, [37:0→2], [53:0→2], **[54:0→1]**, **[57:0→1]**, [60:0→1], [93:0→1], [141:0→1], [144:0→1], [185:0→2], [198:0→2], [228:0→1], **[229:0→1]**, [236:1→0], **[245:0→1]**, **[342:0→1]**, [366:0→1], [401:1→0], [431:0→1]

Comments. The aspredinids have long been concensually recognised as an independent, highly peculiar group of catfishes presenting several derived features (see, e.g. Cope, 1871; Gill, 1872; Regan, 1911b; Chardon, 1968; Lundberg and Baskin, 1969; Friel, 1994; de Pinna, 1996, 1998). The present analysis detected four additional autapomorphic features characterising the aspredinids examined, which refer to structures not often included in studies on catfish phylogeny, such as the structures associated with the mandibular barbels or the cranial and pectoral muscles. These four features are: 1) cartilages of the mandibular barbels not differentiated in supporting and moving part, very reduced in size (char. 7); 2) dorsal division of arrector dorsalis considerably reduced in size (char. 54); 3) retractor tentaculi lateral to levator arcus palatini (char. 229); 4) dilatator operculi with posterior bundle originating on posterodorsal surface of hyomandibulo-metapterygoid and anterior bundle originating on neurocranium (char. 245). [Note: characters 57, i.e., mesethmoid presenting a small Y-shaped cartilage on the posteromesial surface and 342, opercle a markedly thin, characteristically L-shaped structure, were based on previous descriptions of other authors: for description of these characters, see Section 3.1.]

Clade 82 (*Bunocephalus, Xyliphius*)

Diagnosis. [334:0→2], [421:0→1]

Comments. Although a study of the interrelationships within Aspredinidae is outside the main objective of the present work, it is interesting to note that genus *Bunocephalus* of the Bunocephalinae and *Xyliphius* of the Hoplomyzontinae appear more closely related to each other than to genus *Aspredo* of Aspredininae. This contradicts the sister-group relationship of the two latter subfamilies proposed by Friel (1994). The two characters listed above concern the remarkable development and enlargement of the anterodorsal spine of the hyomandibulo-metapterygoid (char. 334) and the marked dorsal extension of Meckel's cartilage (char. 421).

Clade 83 (*Amphilius, Paramphilius, Leptoglanis, Zaireichthys, Doumea, Phractura, Andersonia, Belonoglanis, Trachyglanis*)

Diagnosis. [18:0→1], [19:0→1], [48:1→0], **[49:0→1]**, [84:0→1], [173:0→1], [190:0→1], [283:0→1], **[287:0→1]**, **[289:0→1]**, [419:0→1], [423:0→1]

Comments. This last major clade to be discussed concerns family Amphiliidae. As described in Chapters 1 and 2, the monophyly, limits, and interrelationships of Amphiliidae have been the subject of endless controversies and so this taxon was particularly carefully analysed in the present work. Family Amphiliidae was erected by Regan (1911b), who divided it into two major groups. He placed genera *Amphilius* and *Paramphilius* in subfamily Amphiliinae and genera *Doumea, Phractura, Paraphractura, Andersonia, Trachyglanis* and *Belonoglanis* in subfamily Doumeinae. However, in 1937, David and Poll suggested that genus *Leptoglanis*, usually considered a bagrid, should also be included in family Amphiliidae. Barnard (1942) suggested that the genus *Paramphilius* should be synonymized with the genus *Amphilius* and, consequently, that Amphiliinae would be a monogeneric subfamily. Both the suggestions of David and Poll (1937) and Barnard (1942) were subsequently refuted by Harry (1953), who agreed with the taxonomic arrangement of family Amphiliidae originally proposed by Regan (1911b), except with respect to the validity of genus *Paraphractura*, considered a synonym of *Phractura* by Harry (1953). Bailey and Stewart (1984) suggested again that genus *Leptoglanis* should be transferred from Bagridae to Amphiliidae. In addition, Bailey and Stewart (1984) suggested that *Zaireichthys*, a genus originally described by Roberts (1967) and considered closely related to *Leptoglanis*, should also be transferred from Bagridae to Amphiliidae. This suggestion was strongly supported by Mo (1991), who, based on a phylogenetic analysis of the interrelationships of the various catfish families, pointed out that genera *Leptoglanis* and *Zaireichthys* should be included in Amphiliidae. However, the most important taxonomic change proposed since the first recognition of the family Amphiliidae by Regan (1911b) is that proposed by He et al. (1999). These authors provided a phylogenetic analysis in which *Leptoglanis* was surprisingly considered a non-amphiliid catfish a priori and genus *Zaireichthys* was not included. The phylogenetic results of He et al. (1999) suggested that family Amphiliidae is not monophyletic since Doumeinae are more closely related to genus *Leptoglanis* and sisorid genera *Euchiloglanis* and *Glyptothorax* than to Amphiliinae. Consequently, family Amphiliidae should be 'restricted to the genera *Amphilius* and *Paramphilius*' and subfamily Doumeinae plus *Leptoglanis* should 'be recognised as a separate family, the Doumeidae' (He et al., 1999: 144). In addition, He et al. (1999: 141) concluded that 'there are no synapomorphies to define the two genera of subfamily Amphiliinae, the reason for placing these two genera together is the lack of synapomorphies shared by the other genera only'.

Due to the amazing help and support of the late Dr. G.G. Teugels, it was possible to include all the Amphiliidae genera in the present cladistic analysis. Thus, it was possible to provide a careful phylogenetic analysis of the relationships among all these genera, including genus *Zaireichthys*, and thus to test the monophyly and limits of this polemic family. The results of this work strongly support the monophyly of all nine amphiliid genera, including *Zaireichthys* and *Leptoglanis* (see above), with Amphiliidae diagnosed by the several unambiguous synapomorphic characters listed above. Notably, three of these characters concern derived, homoplasy-free autapomorphic features that are uniformly present in all amphiliids and in no other catfishes examined: 1) arrector ventralis markedly bifurcate mesially (char. 49); 2) autopalatine markedly bifurcate posteriorly (char. 287); 3) posterior portion of autopalatine significantly expanded dorsoventrally due to presence of ventral and dorsal laminar expansions of this bone (char. 289). As explained above, the uniform, exclusive presence of well-defined, distinct, homoplasy-free features in a certain major catfish group is extremely rare in a taxon as large, complex and diverse as order Siluriformes. Thus, the presence of not only one, but three such features in Amphiliidae, together with the several other characters listed above, clearly constitutes a very strong argument on behalf of the monophyly of this family.

Hence, the present work contradicts the results of He et al. (1999). As explained in Chapters 1 and 2, in what concerns the interfamilial relationships within Siluriformes, the cladogram of He et al. (1999) is rather confusing, simultaneously contradicting several other points commonly accepted in those studies dealing with higher level phylogeny of Siluriformes. For example, in the cladogram of He et al. (1999), Amblycipitidae is placed in a more basal position within the siluriforms than the fossil catfish family †Hypsidoridae, thus contradicting the studies of Grande (1987), Mo (1991) and de Pinna (1993, 1998) (see Fig. 1.12). Also, in He et al.'s (1999) cladogram, Amblycipitidae is phylogenetically separated from Sisoridae by the †Hypsidoridae, Bagridae and Amphiliinae, while it is commonly accepted (de Pinna, 1993, 1996, 1998; Chen, 1994; Chen and Lundberg, 1994; Friel, 1994; Diogo et al., 2003b; this book) that Amblycipitidae is more closely related to Sisoridae than to any of these three groups. Lastly, He et al. (1999)

suggested that Sisoridae sensu de Pinna (1996) is not a monophyletic group since *Glyptothorax* (of sisorid tribe Glyptothoracini) is more closely related to the clade doumeins + *Leptoglanis* than to *Euchiloglanis* (of sisorid tribe Glyptosternini), while recent studies (de Pinna, 1996, 1998; Diogo et al., 2002b; this book) strongly support the monophyly of Sisoridae.

Clade 84 (*Amphilius, Paramphilius*)

Diagnosis. [18:1→2], [159:0→1], **[161:0→1]**, **[207:0→1]**, [208:0→1], [210:0→1], **[214:0→1]**, [239:0→1], [313:0→1], **[365:0→1]**

Comments. With respect to the amphiliid subfamily Amphiliinae—which appeared as a monophyletic group in He et al.'s 1999 work, although surprisingly these authors were not able to adduce any synapomorphy to support its monophyly, this, too, appears as a clearly monophyletic taxon in the present analysis. Amphiliins are diagnosed by several synapomorphic characters, four of which seemingly constitute Amphiliinae autapomorphies: 1) dorsal process of cleithrum essentially oriented anterodorsally (char. 161); 2) adductor mandibulae A1-OST differentiated into diverse well-developed sections (char. 207); 3) adductor mandibulae A3'-v differentiated into small and large divisions attached respectively to posterodorsal and posterior surfaces of coronomeckelian (char. 214); 4) preopercle presenting well-developed, triangular anterodorsal process pointed anteriorly (char. 365).

Clade 85 (*Leptoglanis, Zaireichthys, Doumea, Phractura, Andersonia, Belonoglanis, Trachyglanis*)

Diagnosis. [35:0→2], [97:0→1], [325:0→1/2/3]

Comments. As mentioned above, family Amphiliidae was erected by Regan (1911b), who divided it into two major groups, Amphiliinae and Doumeinae. However, in 1937, David and Poll included, without comment, genus *Leptoglanis*, usually recognised as a bagrid, in family Amphiliidae. This was contested by Harry (1953: 180), who viewed this genus as 'a bagrid, possessing a normal, large, free air bladder'; likewise by Jayaram (1966: 1106) since 'besides this primary character (the large and free air bladder), *Leptoglanis* has a strong pectoral spine (versus absent in the Amphiliidae) and the dorsal fin more anteriorly placed than in the Amphiliidae'. However, Bailey and Stewart (1984) emphasised that *Leptoglanis*, as well as *Zaireichthys*, a genus originally described by Roberts (1967) and considered closely related to *Leptoglanis*, should be included in family Amphiliidae. This because the swim bladder of these two genera consists in paired vesicles largely encapsulated', similar to that of amphiliids (Bailey and Stewart, 1984: 9). This suggestion was strongly corroborated by Mo (1991: 69), who provided a list of derived features shared by these two genera and 'all or some amphiliids'.

The present study also supports the inclusion of *Leptoglanis* and *Zaireichthys* in Amphiliidae and in particular in an amphiliid clade including these two genera and the doumeins (see Fig. 3.123). In fact, besides the 12 synapomorphies uniting *Leptoglanis* and *Zaireichthys* to all amphiliids, three of which are unique within siluriforms (see above), there are three additional characters associating these two genera to the doumein amphiliins: 1) hyohyoideus inferior even more hypertrophied, with median aponeurosis almost indistinguishable (char. 35); 2) frontal and pterotic contacting in dorsal view (char. 97); 3) reduction of entoectopterygoid (char. 325). The inclusion of *Leptoglanis* and *Zaireichthys* in Bagridae, suggested by some authors (see above), is contradicted. This is not really surprising. The two principal characters usually employed to justify the placement of *Leptoglanis* and *Zaireichthys* in Bagridae, i.e., 'large, free air bladder' and the 'strong pectoral spine', of which only the latter is indeed present in these two genera (see above), are plesiomorphic characters within Siluriformes. Hence they do not constitute evidence for the inclusion of these two genera in family Bagridae.

Clade 86 (*Leptoglanis, Zaireichthys*)

Diagnosis. [90:0→1], [137:0→1], **[145:0→1]**, [146:0→1], **[204:0→1]**, [265:0→1], [313:0→2], **[385:0→1]**

Comments. *Leptoglanis* and *Zaireichthys* form a monophyletic, morphological distinct amphiliid subfamily, Leptoglanidinae, which is diagnosed by the unambiguous distribution of several characters, and especially by three potential Leptoglanidinae autapomorphies, namely: 1) presence of anteroventrolateral process of posttemporo-supracleithrum (char. 145); 2) proximal radials of pectoral girdle markedly completely fused along their length (char. 204); 3) anterior margin of parurohyal markedly concave in ventral view (char. 385).

Clade 87 (*Doumea, Phractura, Andersonia, Belonoglanis, Trachyglanis*)

Diagnosis. **[49:1→2]**, [131:0→1], [235:0→1], [248:0→1], **[303:0→1]**, **[305:0→1]**, [325:0/1/2→3], [358:0→1], **[369:0→1]**, **[397:0→1]**, [398:0→1], [407:0→1], [423:1→2]

Comments. The present analysis strongly corroborated the monophyly of amphiliin subfamily Doumeinae, as suggested by authors such as Regan (1911), Harry (1953) or He et al. (1999). Notably, the Doumeinae appear defined by five unique, homoplasy-free, autapomorphic features: 1) differentiation of arrector ventralis markedly pronounced, with differentiation in additional muscle (char. 49); 2) presence of thick, short ligament connecting dorsolateral surface of premaxilla and anteroventral surface of autopalatine (char. 303); 3) presence of well-defined, strong ligament connecting anterior surface of autopalatine and mesethmoid (char. 305); 4) posterior ceratohyal essentially a quadrangular, stout, but relatively small bone (char. 369); 5) mandible remarkably curved, compressed anteroposteriorly (char. 397).

Clade 88 (*Phractura, Andersonia, Belonoglanis, Trachyglanis*)

Diagnosis. [55:2→3]

Comments. Both the work of He et al. (1999) and the present study agree in the sister-group relationship between *Belonoglanis* and *Trachyglanis*, but differ in the phylogenetic position of *Phractura, Doumea* and *Andersonia*. In fact, two hypotheses concerning the phylogenetic position of these three genera are proposed by He et al. (1999). Their most parsimonious cladogram (see Fig. 1.12) suggests that *Andersonia* is the sister-group of the other four doumein genera, with *Belonoglanis* the sister-group of *Trachyglanis* and *Doumea* the sister-group of *Phractura*. The bootstrap analysis of this cladogram (He et al., 1999: Fig. 1.12), however, suggests that *Phractura* and *Doumea* constitute a monophyletic clade, which is the sister-group of a clade having *Andersonia* as the most basal taxon and *Belonoglanis* and *Trachyglanis* as sister-groups. This latter hypothesis is similar to that of the present study (see Fig. 3.123), the only difference is that in the present study (see Fig. 3.123), *Phractura* appears as the sister-group of the clade constituted by *Andersonia, Belonoglanis* and *Trachyglanis*, and not as the sister-group of *Doumea*. Grouping *Phractura, Andersonia, Belonoglanis* and *Trachyglanis* is supported by the fact that in the specimens of these genera examined, the totality of the dorsal division of the arrector dorsalis lies ventral to the pectoral girdle. Such a configuration is only found, apart from these four genera, in *Amphilius* and *Malapterurus* (for description of this character, see Section 3.1).

Clade 89 (*Andersonia, Belonoglanis, Trachyglanis*)

Diagnosis. **[132:0→1]**, [182:0→1/2], **[369:1→2]**

Comments. This clade is supported by the three characters listed above, and in particular by the homoplasy-free, autapomorphic characters 132 and 369, which concern respectively the presence of well-developed lateral laminae of the parieto-supraoccipital (char. 132), and the quadrangular, stout, but markedly reduced in size, posterior ceratohyal (char. 369, state 2).

Clade 90 (*Belonoglanis, Trachyglanis*)

Diagnosis. **[68:0→1]**, [73:1→2], [183:0→1], [201:0→2], [202:0→2]

Comments. Lastly, grouping of genera *Belonoglanis* and *Trachyglanis* is supported by the five characters listed above, of which the first one, i.e., posterodorsal surface of the mesethmoid peculiarly Y-shape (char. 68), is notably present in the specimens of these two genera and in no other catfishes examined.

3.3 CHARACTER STATE CHANGES FOR INDIVIDUAL GENERA

As explained above, this summary concerns the specific character state changes characterising the individual genera of those non-monogeneric families included in the present cladistic analysis. These character state changes hence refer neither to the siluriform higher level phylogeny nor to the monophyly and/or intrarelationships within the various siluriform families, and, thus, are presented here as a short annex to the main scope of this work.

The list below, like the synapomorphy list above, is restricted to unambiguous character state changes, which are differentiated into two main categories, namely: 1) state changes that occurred only once within the Siluriformes (in bold); 2) homoplasic state changes independently acquired in another node within the Siluriformes and/or referring to secondary reversions to a more plesiomorphic state (not-bold). The list below also follows the order of the synapomorphy list presented above and thus of the cladogram presented in Figure 3.123.

Trichomycterus: No character state changes

Hatcheria: No character state changes

Callichthys: **[111:0→1]**, [211:0→1], **[394:0→1]**

Corydoras: [181:2→3], [237:0→1], [350:0→1], [398:0→1], [423:0→1]

Hypoptopoma: [179:0→1]

Loricaria: [193:0→1], **[238:0→1]**, **[243:0→1]**, **[249:0→1]**

Lithoxus: [333:0→1]

Helogenes: [2:0→1], **[67:0→1]**, [69:0→1], **[64:0→1]**, **[101:0→1]**, **[124:0→1]**, **[129:0→1]**, [150:0→1], [360:0/1/2→3], **[371:0→1]**, [386:0→1], **[399:0→1]**, [412:0/1→2]

Cetopsis: **[87:0→1]**, [102:0→1], **[244:0→1]**, **[291:0→1]**

Hemicetopsis: [331:1→2], **[416:1→2]**

Wallago: **[71:0→1]**, [265:0→1]

Silurus: [97:0→1], **[195:0→1]**, **[314:0→1]**

Helicophagus: [276:0→1]

Pangasius: [42:0→1], **[59:0→1]**, [109:0→1], [254:0→1], [265:0→1], [306:0→1], **[418:0→1]**

Pseudeutropius: [191:0→2], [276:0→1], [381:0→1]

Schilbe: [69:0→1], [223:0→1], [306:0→1], [332:0→2]

Laides: [363:0→1]

Ailia: [103:0→1], [131:1→0], [150:0→2], [277:1→0]

Siluranodon: **[40:0→1]**, [81:0→1], [378:0→1], [381:0→1]

Amiurus: [109:0→1], [131:0→1]

Ictalurus: No character state changes

Ancharius: [131:1→0], [168:0→1], [181:0→1], [185:1→2], [194:1→0]

Arius: **[337:0→1]**, [414:0→1]

Genidens: [306:0→1], [411:0/1→2]

Auchenoglanis: [25:0→2], [35:0→1], **[66:0→1]**, [73:0→1], [88:1→0], [109:1→2], **[186:0→1]**, [226:0→3], [246:0→1], [325:0→1], **[367:0→1]**, **[383:0→1]**, **[406:0→1]**, **[413:0→1]**, **[415:0→1]**

Chrysichthys: [324:0→1], [381:0→1]

Clarotes: No character state changes

Mochokus: [386:0→1]

Synodontis: [8:0→1], **[14:0→1]**, **[29:0→1]**, [35:1→2], **[154:0→1]**, **[388:0→1]**, [396:0→1]

Centromochlus: **[45:0→1]**, **[255:0→1]**, [334:1→2], [348:0→1]

Ageneiosus: [1:2→0], [69:0→1], [75:0→1], [112:0→1], [142:1→0], [181:2→0], [194:1→0], **[282:0→1]**, [332:0→1]

Auchenipterus: [42:1→0], [97:1→0], [181:1→3], [314:1→0]

Franciscodoras: [11:0→1], **[426:0→1]**

Anadoras: No character state changes

Doras: [327:0→1], [334:1→0]

Acanthodoras: [35:1→0], [227:0→1]

Rita: [1:2→1], [42:0→1], [185:0→2], [309:0→1], [327:0→1]

Bagrichthys: [35:0→2], [73:0→1], [131:0→2], [313:1→0], [351:0→1], **[374:0→1]**, [386:0→1], [398:0→1]

Hemibagrus: No character state changes

Bagrus: [16:0→1]

Microglanis: [93:0→1]

Pseudopimelodus: [194:0→1]

Heptapterus: [47:1→0], [250:0→1], [408:0→1]

Rhamdia: [109:0→1], [181:0→1]

Goeldiella: [86:0→1]

Pseudoplatystoma: [306:0→1], [324:0→1]

Hypophthalmus: [53:0→2], [81:0→1], [160:0→1], [168:1→0], [301:1→0], **[368:0→1]**, [398:0→1]

Pimelodus: [131:0→1], [194:0→1]

Calophysus: No character state changes

Uegitglanis: [423:0→1]

Heteropneustes: **[172:0→1]**, **[297:0→1]**, [389:0→2]

Clarias: No character state changes

Heterobranchus: No character state changes

Cnidoglanis: [92:0→1], [212:1→0]

Neosilurus: **[347:0→1]**, [417:1→2]

Plotosus: [96:1→0], **[98:1→2]**, [99:0→1], [212:1→0], [242:0→1]
Paraplotosus: [264:0→1]
Amblyceps: [9:0→2]
Liobagrus: [47:1→0], [265:0→1], [348:0→1]
Parakysis: [47:1→0], [144:0→1], [185:0→2], [206:0→1], [227:0→1], [236:1→0], **[432:0→1]**
Akysis: **[164:1→2]**, [325:0→1], **[428:0→1]**
Bagarius: [80:0→1], [109:0→1]
Gagata: **[5:0→1]**, [81:0→1], **[83:0→1]**, **[104:0→1]**, [325:3→0]
Glyptosternon: [48:1→2], [139:1→0], [148:2→0], **[178:0→1]**, [192:0→1], **[292:0→1]**, [325:3→0], **[377:0→1]**
Glyptothorax: [80:0→1], [181:0/1→2], [265:1→0], **[387:0→1]**, [427:0→1], **[429:0→1]**
Erethistes: [419:0→1]
Hara: No character state changes
Aspredo: **[79:0→1]**, [159:0→1], [265:0→1], **[273:0→1]**, **[319:0→1]**, **[342:1→2]**, [351:0/1→2]
Bunocephalus: [131:0→2], [162:0→1], [331:0→1], [366:1→2], **[370:0→1]**, **[373:0→1]**
Xyliphius: [81:0→1], [166:0→1], [167:0/1→2]
Amphilius: **[22:0→1]**, **[46:0→1]**, [55:0→3], [69:0→2], **[429:0→1]**
Paramphilius: **[121:0→1]**, **[143:0→1]**, [192:0→1], **[262:0→1]**, [332:0→1]
Leptoglanis: **[55:0→1]**, [64:0→1], [261:0→1]
Zaireichthys: [102:0→1], **[145:1→2]**, **[344:0→1]**, [348:0→1], [378:0→1]
Doumea: **[184:0→1]**, [201:0→1], [202:0→1]
Phractura: [349:0→1]
Andersonia: **[127:0→1]**, [137:0→1], [325:3→4]
Belonoglanis: **[132:1→2]**, [183:1→2], **[361:0→1]**
Trachyglanis: **[177:0→1]**, [235:1→0]

3.4 RESULTS OF PHYLOGENETIC ANALYSIS: MAJOR OUTLINES

After presenting in detail each node of the cladogram resulting from the strict consensus of the 12 equally parsimonious trees obtained in the phylogenetic analysis of this Chapter, it is important, before passing to the next one, to emphasise here in broad outline the most significant points regarding the results of this analysis.

The phylogenetic comparison of 440 characters in 87 genera representing all the 32 extant catfish families resulted in the almost completely resolved strict consensus cladogram illustrated in Fig. 3.123, with a length of 902 steps, Consistency Index (CI) = 0.52 and Retention Index (RI) = 0.78. The indexes of this consensus cladogram are thus significantly superior to those of Mo (1991: cladogram I: CI = 0.34, RI = 0.64; cladogram II: CI = 0.36, RI = 0.72) and de Pinna (1993: CI = 0.41, RI = 0.72). The consensus cladogram of this work is more resolved than those of de Pinna's excellent 1993 work and of Mo's 1991 study: it presents only three trichotomies, of which only one directly concerns interfamilial relationships within Siluriformes (see below).

The most significant original outcome of the phylogenetic analysis of the present work clearly concerns the positioning of the Loricarioidea as the sister-group of all the remaining non-diplomystid catfishes. Another important aspect concerns the monophyly of Schilbidae, questioned by some authors, e.g. Mo (1991), but supported in de Pinna's 1993 unpublished thesis. Also worthy of mention is the sister-group relationship between *Ancharius* and the remaining ariids, likewise questioned by authors such as Mo (1991) but, again, supported by de Pinna (1993). Grouping family Ariidae, including *Ancharius*, and families Schilbidae, Pangasiidae, Ictaluridae, Cranoglanididae, Austroglanididae and Claroteidae in a strong, well-supported clade also constitutes an important and original outcome of the present work. Within this well-supported clade, it is especially interesting to note the position of the ariids, which are often associated with the mochokid, doradid and auchenipterid catfishes, but that appear, as in de Pinna's 1993 thesis, close to the claroteids. However, it should be noted that de Pinna's Claroteidae excluded the claroteid subfamily Auchenoglanidinae sensu Mo (1991), while the present analysis corroborates the monophyly of family Claroteidae as defined by Mo. With respect to the phylogenetic position of *Malapterurus*, it was interesting to verify that this genus essentially comes back to the original position to which Günther had assigned it in 1864, i.e., close to the consensually accepted clade formed by the African Mochokidae and the Neotropical Doradidae and Auchenipteridae.

Like the Claroteidae, the Bagridae appear as a monophyletic clade, as defined by Mo (1991), and contrary to de Pinna (1993). Pimelodidae also clearly appear as a natural group, which surely also constitutes one of the most significant outcomes of the present work. Within the Pimelodidae, it is interesting to note the sister-group relationship between Heptapterinae and Pimelodinae, but it should be kept in mind that the evidence provided to support this sister-group relationship is considerably weaker that that supporting the monophyly of the family as a whole. The monophyly of Amphiliidae also constitutes an important point, since it was seriously questioned in He's 1997 unpublished thesis, as well as in He et al.'s 1999 controversial paper. Within Amphiliidae, the sister-group relationship between the subfamilies including genera *Zaireichthys* and *Leptoglanis*, i.e., Leptoglanidinae and Doumeinae is noteworthy. With respect to the phylogenetic position of *Heteropneustes*, the present work corroborated the suggestions of Chardon (1968) and de Pinna (1993), according to which this genus would lie at the very core of family Clariidae.

The phylogenetic position of Amblycipitidae and Akysidae is not fully resolved in the present work, contrary to de Pinna's 1996 study, in which the amblycipitids appeared as the sister-group of all the remaining sisoroids. However, it should be noted that de Pinna himself recognised, in that study, that the evidence provided to support the sister-group relationship between the Amblycipitidae and the remaining Sisoroidea was not much stronger than that indicating a sister-group relationship between this former family and the Akysidae. Besides the unresolved node 74 (Fig. 3.123) leading to the Akysidae, Amblycipitidae and remaining Sisoroidea, another interfamilial node appears as somewhat fragile. This concerns the node leading to clade 38, that is, to the clade including on the one hand the malapterurids, doradoids, bagrids and pimelodids, and on the other, the amphiliids, chacids, clariids, *Heteropneustes*, plotosids and sisoroids. In fact, although this clade appears in all the most parsimonious trees obtained in the present analysis, and hence in the strict-consensus cladogram illustrated in Fig. 3.123, the evidence to support it is relatively weak. In other words, although the relationships between clades 62 and 21 appear to be fully resolved, with clades 39 and 62 ranked as sister-groups (Fig. 3.123), one should keep in mind that this sister-group relationship relies on somewhat weak evidence. With respect to the other major nodes referring to the interfamilial relationships of the catfishes, they are relatively well supported.

Although the main aim of the cladistic analysis of the present work mainly concerns the higher level phylogeny of Siluriformes, it is worth underscoring here some points regarding the interrelationships within certain families. For example, although very limited in this respect, the present analysis essentially corroborated the expected intrafamilial relationships within Loricariidae, Cetopsidae, Ariidae, Claroteidae, Auchenipteridae, Doradidae, Bagridae or subfamily Heptapterinae. However, within the pimelodid subfamily Pimelodinae, genera *Calophysus* and *Pimelodus* appeared more closely related to each other than to genus *Hypophthalmus*, which is somewhat surprising given the results of de Pinna (1993). But not too much so if one takes into account the difficulty of establishing a solid, robust hypothesis concerning the phylogenetic position of this latter genus within subfamily Pimelodinae, as recognised by authors such as Howes (1983b) or Lundberg et al. (1991b), and by de Pinna (1998) himself. Also, within family Aspredinidae genus *Bunocephalus* of subfamily Bunocephalinae and *Xyliphius* of subfamily Hoplomyzontinae appeared more closely related to each other than to genus *Aspredo* of subfamily Aspredininae, contrary to that expected given the results of Friel (1994). Lastly, it is noteworthy that within the phylogenetically poorly studied family Schilbidae, the African genus *Siluranodon* appeared more closely related to genera *Ailia* and *Laides* from Asia than to the also African *Schilbe* and *Pseudeutropius*, thus implying that the African members of the family possibly do not constitute a monophyletic clade.

It can be said that the phylogenetic results of the present work pay, to some extent, a tribute to those catfish taxonomists and their classic studies that paved the way, long before the emergence of cladistics, to the studies now made under this paradigm. In fact, these results strongly support the monophyly of certain groups long recognised but questioned in more recent years by, precisely, studies made under the cladistic paradigm. This is the case, for example, of family Pimelodidae, including the pseudopimelodins, pimelodins and heptapterins, family Ariidae, including *Ancharius*, family Amphiliidae, and family Schilbidae. Also, even in what refers to the general interfamilial relationships within the order, some of the results of the present work also approximate to some ideas proposed long ago and outside the cladistic paradigm. Here, I would like to pay tribute to the work of Chardon (1968) who, more than 35 years ago and with no explicit phylogenetic purpose, proved, as shown in Chapter 4, remarkably visionary in view not only of the phylogenetic results of the present work, but also of those results of authors such as de Pinna and of some ideas only now becoming generally accepted (e.g., basal position of the Cetopsidae within the Siluriformes, placement of *Heteropneustes* at the very core of Clariidae, grouping of doradids + mochokids + auchenipterids, etc.). Of course, as in everything, there are always some exceptions, and the results of this work also contradict the monophyly of some other groups that were traditionally accepted in most of the twentieth century, but called into question by recent cladistic studies, e.g. the 'Bagridae' (the present study supports the split of the former 'Bagridae' in Claroteidae, Bagridae and Austroglanididae sensu Mo, 1991) and the 'Sisoridae' (it also supports the split of the former 'Sisoridae' in Sisoridae and Erethistidae sensu de Pinna, 1996).

So, I must thank here ALL those authors that have since years long gone until this very moment, continually helped in attaining our present state of knowledge of the systematics and phylogeny of the various catfish families and the Siluriformes in general, without whose contributions to science this volume would not really have been possible.

It is hoped that the phylogenetic results presented in this chapter will also promote further discussions and pave the way for future studies. For instance, the strict consensus cladogram resulting from the phylogenetic analysis presented here, showing a notably high resolution and consistency index, ought nonetheless to be tested in further works. Concerning this subject, priority should be given, in my opinion, to the inclusion of and/or comparison with other morphological characters pertaining to, for example, the branchial apparatus, urophore complex, vertebral structure, pelvic girdle, or histological structure of the lips. Also, comparison with molecular phylogenetic studies on the higher level phylogeny of Siluriformes, as yet to be published but hopefully appear in the near future, should prove revelatory.

Of course, inclusion of additional taxa on the matrix would also be very useful. This is especially applicable, for example, for particularly 'problematic' catfish genera not included in the present work, such as *Horabagrus* (see de Pinna, 1998), or for key fossil catfishes such as †*Hypsidoris* or †*Andinichthys*. Nevertheless, in my opinion, most catfish families are reasonably well represented in the present phylogenetic analysis. I am perfectly aware that the non-inclusion of more taxa in the analysis will probably be a criticism levelled by some catfish taxonomists, for whom coverage of every possible species and genera of each family is of paramount importance. However, from a strictly phylogenetic point of view, there is no need to include all catfish species and genera in an analysis at the higher level and general interfamilial phylogeny of the order Siluriformes. Of importance is appropriate selection of key taxa. As a practical example, let's take family Pimelodidae. One could argue that only a few species are included in the analysis while the known pimelodid species are numerous. But consider that nine pimelodid genera are included in the analysis, with each of the three pimelodin families represented by at least two genera and, more significantly, at least one of these genera *having a relatively plesiomorphic* position within its respective subfamily. As several studies have strongly supported that each of these three subfamilies is monophyletic (see Chapters 1 and 2), inclusion of these nine pimelodid genera in the analysis thereby, in a purely phylogenetic context, reasonably suffices to test the monophyly or the interfamilial relationships of family Pimelodidae. Of course, a study focused not on the entire order Siluriformes but just family Pimelodidae, ought imperatively to include all pimelodid genera; a study focused on a particular pimelodid genus ought necessarily to embrace every species in this genus, and so on. In fact, it is earnestly hoped that the results provided in this Chapter on the higher level phylogeny of the order Siluriformes will indeed stimulate similar further studies on the phylogeny of particular subgroups of this order, not only different families, but even eventually smaller taxa.

4

Higher-level Phylogeny and Macroevolution of Catfishes: A Discussion

As noted in Chapter 1, this Chapter mainly concerns the macroevolution of the complex catfish systems examined in this study, based on the phylogenetic results obtained in the preceding Chapter. This theme is related to a somewhat controversial old question, namely the 'evolutionary functional school' versus the 'cladistic systematic school'.

This controversial issue has been, and continues to be, discussed, e.g. Cracraft (1981), Szalay (1981), Dywer (1984), Coddington (1988, 1990), Lang (1990), Baum and Larson (1991), de Pinna (1991), Winterbottom and McLennan (1993), Nielsen (1998), Bock (1999), Liem and Summers (2000), Marques and Gnaspini (2001), Swidersky (2001) or Desutter-Grandcolas et al. (2003). Its major arguments can be resumed as follows.

Authors such as Coddington (1988, 1990), Cracraft (1981) or Desutter-Grandcolas et al. (2003), adhering to the major arguments of the cladistic paradigm, declare that evolutionary explanations/discussions should obligatorily be *based on* a comprehensive, explicit cladogram, and not, as done by some 'functional evolutionary morphologists' (sensu Cracraft, 1981), a priori to a phylogenetic comparison. In fact, more than 50 years after the first major work of Hennig (1950: in German, translated into English in 1965 and 1966), and contrary to the theoretical basic lines of cladistics, many authors, e.g. Szalay (1981), Nielsen (1998), Bock (1999), Marques and Gnaspini (2001), continue to claim that certain homoplasic characters ought somehow to be detected and removed *a priori* to phylogenetic analysis. Also, some 'functional evolutionary morphologists' insist on formulating major evolutionary explanations/discussions, such as those concerning the evolution of certain complex systems, or the reconstruction of potential 'ancestral forms', a priori to phylogenetic comparisons. However, in doing so, such evolutionary explanations/discussions would necessarily rely on more or less 'ad hoc' stories, as noted by Cracraft (1981). In fact, as clearly explained recently in Desutter-Grandcolas et al. 2003 up-to-date paper, usually the hypothesis of evolutionary modalities for a character of interest can occur, under the cladistic paradigm, at two different steps in a phylogenetic analysis: 1) they can be formulated apart from phylogenetic inference and tested a posteriori by the independently resultant topology, or 2) they can be suggested by the character state distribution documented *by the cladogram*. An evolutionary hypothesis other than the necessary postulate of 'descent, with modification' should not, however, influence phylogeny reconstruction (Eldredge and Cracraft, 1980). This is particularly true for an 'ad hoc' hypothesis of homoplasy, which is an 'ad libitum explanation, one capable of explaining patterns and non-patterns alike and, in being able to do so..., explains nothing at all' (Kludge, 2001: 202). As underscored by Desutter-Grandcolas et al. (2003), this independence is the guarantee of the explanatory power of the phylogenetic method and should be satisfied to preserve the scientific significance of historical studies.

The above paragraph may seem strange perhaps for some readers in the present context. In fact, the elaboration of evolutionary explanations/discussions a posteriori to, and based on a phylogenetic analysis is one of the basic guidelines of the cladistic paradigm, as is consensually stated in all major recent theoretical and practical phylogenetic textbooks (see, e.g. Kitching et al., 1998, and references therein). However, as shown above, this issue is still, indeed, subject to some controversy, particularly in West European countries in which there is some reluctance to apply an explicit cladistic methodology, especially in countries such as Belgium and the Netherlands with a markedly strong tradition of 'functional evolutionary morphology' (sensu Cracraft, 1981).

My personal aim is to take advantage of both the explicit methodologies of the cladistic school and the 'functional' and philosophical tradition of West European countries. In fact, although I agree with the basic cladistic guidelines in that the major evolutionary explanations/discussions need imperatively to *be based on* an explicit phylogenetic framework, I also denounce the fact that most cladistic studies neglect precisely such evolutionary explanations. Many, if not most, cladistic studies usually end with the description of the obtained cladogram and the respective resultant clades, *without* subsequently discussing the major evolutionary implications of these phylogenetic results.

This traditional dichotomy between classic 'functional evolutionary studies' and classic 'cladistic systematic studies' is evidenced in the major catfish studies published to date. On the one hand, general papers such as Alexander's (1965), Chardon's (1968) or Gosline's (1975), provide excellent morphological descriptions with very interesting functional evolutionary discussions, which, however, were not confronted with, or tested against a comprehensive explicit cladistic cladogram on the higher level phylogeny of Siluriformes. On the other hand, Mo (1991) presented cladograms on the higher level siluriform phylogeny based on an explicit cladistic analysis, but did not provide, subsequently to these cladograms, a tentative discussion on the major evolutionary implications of the phylogenetic results obtained from this analysis.

The present study discusses some major points concerning the macroevolution of certain complex systems within the Siluriformes based on an explicit cladogram on the higher level phylogeny of these fishes. The complex systems discussed here concern those main structural complexes of the cephalic and pectoral girdle regions analysed in the present work, namely: 1) structures associated with movements of the mandibular barbels; 2) pectoral girdle complex; 3) adductor mandibulae complex; 4) palatine-maxillary system; 5) suspensorium; 6) elastic spring apparatus. In addition, a discussion on the origin and biogeographic distribution of the Siluriformes, also based on an explicit phylogenetic framework, is given in the end of this chapter.

4.1 STRUCTURES ASSOCIATED WITH MOVEMENTS OF THE MANDIBULAR BARBELS

Barbels are a general characteristic of Siluriformes (Fig. 1.1), responsible for their popular name 'catfish'. Catfish barbels are clearly involved in near-field chemoreception (while nostrils are concerned with far field: e.g., Herrick, 1903) and mechanoreception including gliding and pressure sensitivity, direct contact and rheotactism (e.g., Hoagland, 1932; Biedenbach, 1971). They are supported by a central rod, comprising a dense network of elastin, with or without 'elastic/cell rich cartilage' (terminology according to Benjamin, 1990). Their skin is covered with mucus cells and taste-buds (e.g., Wright, 1884; Herrick, 1903; Landacre, 1910; Kamrin and Singer, 1953; Desgranges, 1972; Finger, 1976, 1978; Ghiot, 1976; Grover-Johnson and Farbman, 1976; Ghiot and Bouchez, 1980), which are probably not only gustatory, but also mechanosensory (e.g., Rajbanshi, 1966). It is notable that taste-buds are not at all restricted to barbels, but are also present on the whole body surface and particularly close set on the inner side of the lips (e.g., Rajbanshi, 1966).

Catfish barbels are necessary to initiate feeding responses, and contact with food is needed (e.g., Biedenbach, 1971). Their functional value in muddy water dwellers and nocturnal fish is evident and was moreover demonstrated experimentally (Alexander, 1965). Although catfish barbels seem mainly to be associated with search for food and avoiding obstacles, they may be associated with several other functions also, such as locomotion, fright reaction, aggression or sexual attraction (see, e.g., Jayaram, 1978).

Three main types of catfish barbels can be distinguished, namely maxillary barbels, mandibular barbels and nasal barbels (Fig. 1.1).

The maxillary barbels, invariably present in siluriforms, are characterised by their connection to a mobile mechanism, the palatine-maxillary system, comprising the maxilla, autopalatine and more or less specialised ligaments and muscles responsible for their movements (e.g., Alexander, 1965; Gosline, 1975). They are supported by a central rod, the proximal end of which is firmly encrusted in the hollow distal end of the maxillary bone.

The nasal barbels present a rather limited taxonomic distribution within Siluriformes compared with the other two types of barbels (e.g., Burgess, 1989). They are characteristically associated with a cartilaginous basal frame supported by the surrounding bones of the nasal region and are related to no mobile mechanism or any muscular system (e.g., Alexander, 1965; Singh, 1967; Ghiot, 1978; Ghiot et al., 1984; Mo, 1991).

The mandibular barbels are plesiomorphically absent in catfishes, but are present in a great number of siluriforms (see, e.g., Fig. 1.1). They are associated with a basal 'elastic/cell rich cartilage' (see above), which usually consists of an anterior and a posterior part. The anterior part, named the 'supporting part' by Diogo and Chardon (2000b), is usually situated between the base of the barbels and the dentary, to which it is usually firmly attached by connective tissue (see Fig. 4.3). The posterior part, usually longer than the anterior, was termed the 'moving part' by Diogo and Chardon (2000b: 457) since 'it is on this part that the muscles for the movement of the mandibular barbels insert' (see Fig. 4.3).

The structures associated with movements of the mandibular barbels were the subject of some functional studies by, e.g., Munshi (1960), Singh (1967), Singh and Munshi (1968), Ghiot (1978), Howes (1983b), and Ghiot et al. (1984). But these studies were neither based on, nor confronted with, a comprehensive, explicit cladogram on catfish higher level phylogeny. The results of the present phylogenetic analysis provide interesting insights into some major points concerning the macroevolution of these structures.

For example, the results of this analysis indicate that the plesiomorphic configuration concerning this structural complex, i.e., complete absence of mandibular barbels, found in *Diplomystes*, characterises not only diplomystids, but also a significant basal part of the catfish cladogram (see Fig. 4.1). In fact, the transition from 0 to 2 pairs of mandibular barbels seemingly only arrived relatively late in catfish evolution, namely in the node leading to the clade constituted by non-diplomystids and non-loricarioids (Fig. 4.1: CS–0→CS–2). Prior to this node, the only single evolution from CS–0 was the transition to CS–1, i.e., the transition to one pair of mandibular barbels in *Nematogenys*.

Interestingly, Figure 4.1 points out three transitions from CS–2 (two pairs of mandibular barbels) to CS–1 (one pair of mandibular barbels), namely in silurids, pangasiids and *Rita*. Also interesting is the apomorphic (secondary) complete loss of mandibular barbels in *Ageneiosus*. The fact that some catfish taxa have experienced a transition from 2 pairs of mandibular barbels to 1, or even 0 pairs, does not come as a surprise. In fact, some families constituted principally by catfishes with two pairs of mandibular barbels have some members with only one pair of mandibular barbels (*Rita* in Bagridae) or even none (*Ageneiosus* in Auchenipteridae) (see, e.g., Burgess, 1989). However, the relatively late origin of two pairs of mandibular barbels within catfish evolutionary history (Fig. 4.1: CS–0→CS–2) constitutes an interesting new outcome.

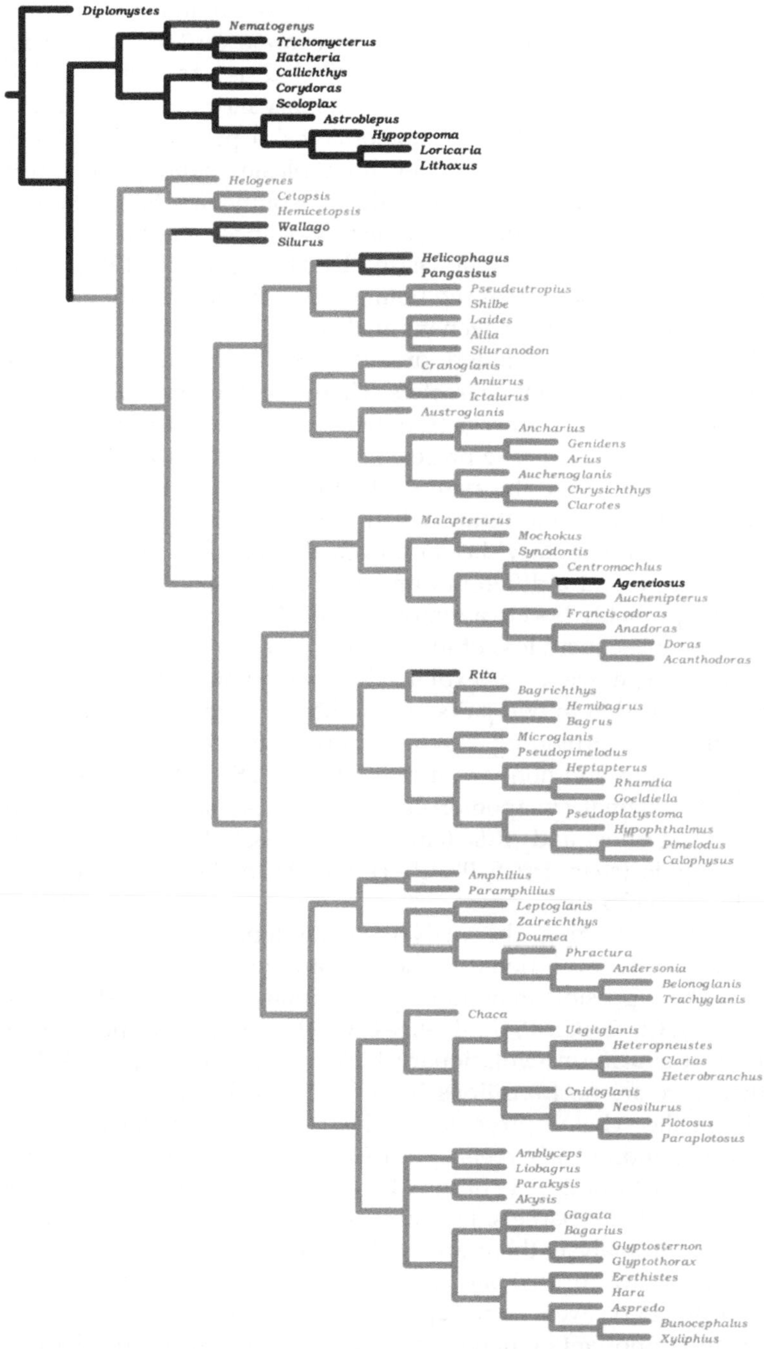

Fig. 4.1 Hypothesis of character state evolution of presence of mandibular barbels (char. 1, unordered): CS-0 (black)= no mandibular barbels; CS-1 (blue)= one pair of mandibular barbels; CS-2 (orange)= two pairs of mandibular barbels [for more details, see text].

The hypothesis of character state evolution present in Fig. 4.1 illustrates an important point concerning the above discussion on the 'functional evolutionary morphologists' *versus* the 'systematic cladists'. As explained above, one of the major lines of the 'functional evolutionary morphology' sensu Cracraft (1981) is to detect 'evolutionary homoplasies' a priori to the phylogenetic analysis and usually remove those homoplasies subsequently from that phylogenetic analysis. This is because these 'evolutionary homoplasies' would constitute a kind of 'interference' in the phylogenetic study.

However, as repeatedly emphasised by Hennig (1950, 1965, 1966, 1981), under the cladistic paradigm the homoplasies should ideally be detected in the *face of* results of the phylogenetic analysis, and not *a priori* to this analysis. In order to do so, Hennig underscored the need for including a great number of somewhat independent characters. Therefore, the homoplasic nature of a particular character will emerge from the overall analysis of all the independent characters examined. The present study clearly illustrates this point. One could *suggest*, a priori to the analysis, that the presence and number of mandibular barbels would constitute a 'bad' character, since it would be 'particularly subject to homoplasy' because even the members of the same catfish family could present a different number of mandibular barbels (see above). This would probably lead to a removal of this 'bad' character from the analysis, with the subsequent loss of information given in the cladogram of Figure 4.1. This Figure, however, as explained above, points out an interesting relatively late transition to two pairs of mandibular barbels in catfish evolutionary history, thus providing a synapomorphy for the clade constituted by non-diplomystids and non-loricarioids (char. 1: CS–0→CS–2). Figure 4.1 also points out potential synapomorphies to characterise the nodes leading to silurids, pangasiids, and *Rita* (char. 1: CS–2→CS–1), the node leading to nematogenyids (char. 1: CS–0→CS–1) and the node leading to *Ageneiosus* (char. 1: CS–2→CS–0).

Also very interesting, the analysis of this figure indicates a much more complex, 'mosaic' evolutionary scheme than would perhaps be imagined a priori: of the 6 possible evolutionary transitions (CS–0→CS–1, CS–0→CS–2, CS–2→CS–1, CS–2→CS–0, CS–1→CS–0, CS–1→CS–2), four indeed seemingly occurred in siluriform evolutionary history. Particularly remarkable is the fact that there are no transitions from CS–1 to CS–0 or from CS–1 to CS–2. That is, in the scheme presented in Fig. 4.1 the character concerning the number of mandibular barbels appears as a flexible feature, with transitions from 0 to either 1 or 2 pairs of mandibular barbels, or from 2 to either 1 or 0 pairs of mandibular barbels. However, within the groups and taxa sampled in the present analysis there are no transitions from 1 pair of mandibular barbels neither to 2 or to 0 pairs of barbels. This could eventually be the case if more catfish taxa were introduced in the phylogenetic analysis.

The most important point here, however, is that inclusion of a large number of features in the phylogenetic analysis did allow discrimination, *a posteriori*, of the homoplasic nature of the character concerning the presence and number of mandibular barbels. Moreover, this character revealed useful phylogenetic synapomorphies and interesting evolutionary aspects concerning catfish macroevolution. These latter interesting insights would be completely lost had the character simply been removed *a priori* to the present phylogenetic analysis.

Returning to the particular aspects concerning the macroevolution of the structures associated with catfish mandibular barbels, another interesting general feature is character 2 of the present cladistic analysis. This character concerns the attachment between the basal cartilages of the mandibular barbels and the mandible. As shown in Figure 4.2, the firm attachment between the cartilages and the mandible only occurred consistently on the catfish clade excluding diplomystids, loricarioids, cetopsids and silurids. There was only a minor independent homoplasic occurrence of CS–1 outside this clade in *Helogenes* and two secondary reversions to CS–0 within it, namely in chacids and aspredinids.

The vast majority of the other characters pertaining to the structures associated with the mandibular barbels concern less general features usually characterising certain catfish subgroups, for example: the large separation between the basal cartilages of the inner and outer mandibular barbels in chacids (char. 6); extreme anteroposterior extension of the basal cartilages of the mandibular barbels in schilbids (char. 8); cartilaginous plates carrying the mandibular barbels in pimelodids (char. 12); mesial cartilaginous complex present in clariids including *Heteropneustes* (char. 17); presence of muscle 1 of the mandibular barbels in pimelodins (char. 20); or presence of muscle 6 of these barbels in cetopsids (char. 24).

In view of the major general points mentioned above concerning the phylogenetic results obtained in the present phylogenetic analysis, the macroevolution of the structures associated with catfish mandibular barbels can be summarised as follows. During the basal phases of catfish evolutionary history, these fishes lacked mandibular barbels. These barbels only occurred consistently in the node leading to the clade uniting non-diplomystid and non-loricarioid catfishes (outside this clade, the only occurrence was that in *Nematogenys*, with transition to one pair and not two pairs of mandibular barbels, as explained above) (Fig. 4.1). Since the first occurrence of the mandibular barbels, these barbels seem to have been associated with more or less developed basal cartilages (see Fig. 3.44), with no catfish discovered to date with mandibular barbels lacking the typical associated basal cartilages. However, the subdivision of these cartilages into supporting and moving parts (see Fig. 4.3) probably occurred only in a subsequent phase since, within the major clade mentioned above uniting the non-diplomystid and non-loricarioid catfishes, the most basal group, i.e. the Cetopsidae (see Fig. 4.1), does not present a differentiation of the cartilages into these two parts.

In those catfishes in which the supporting part of the basal cartilages of the mandibular barbels is firmly attached anteriorly on the dentary (see Fig. 4.2),

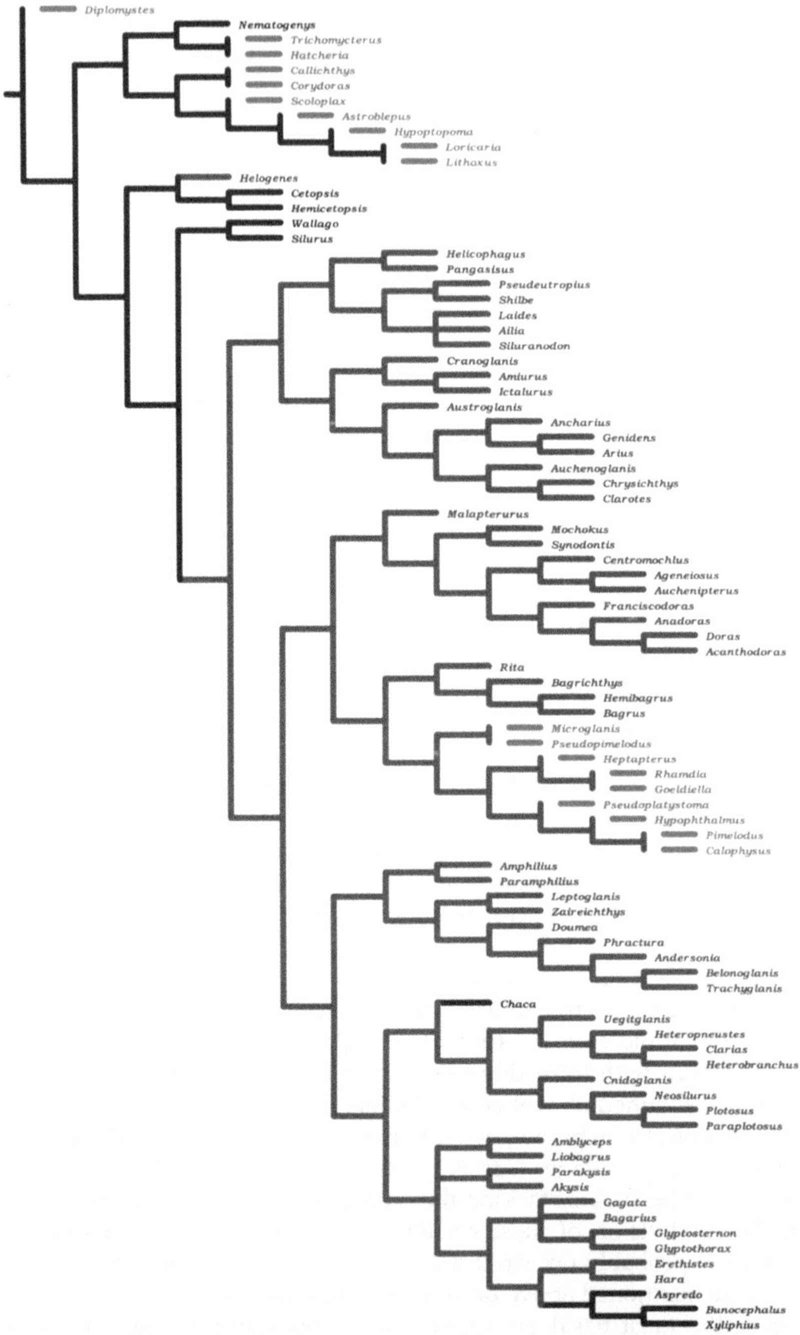

Fig. 4.2 Hypothesis of character state evolution of attachment of cartilages of mandibular barbels (char. 2): CS-0 (black)= no firm attachment to mandible; CS-1 (blue)= firm attachment to mandible; Innaplicable (pink)= since there are no mandibular barbels [for more details, see text].

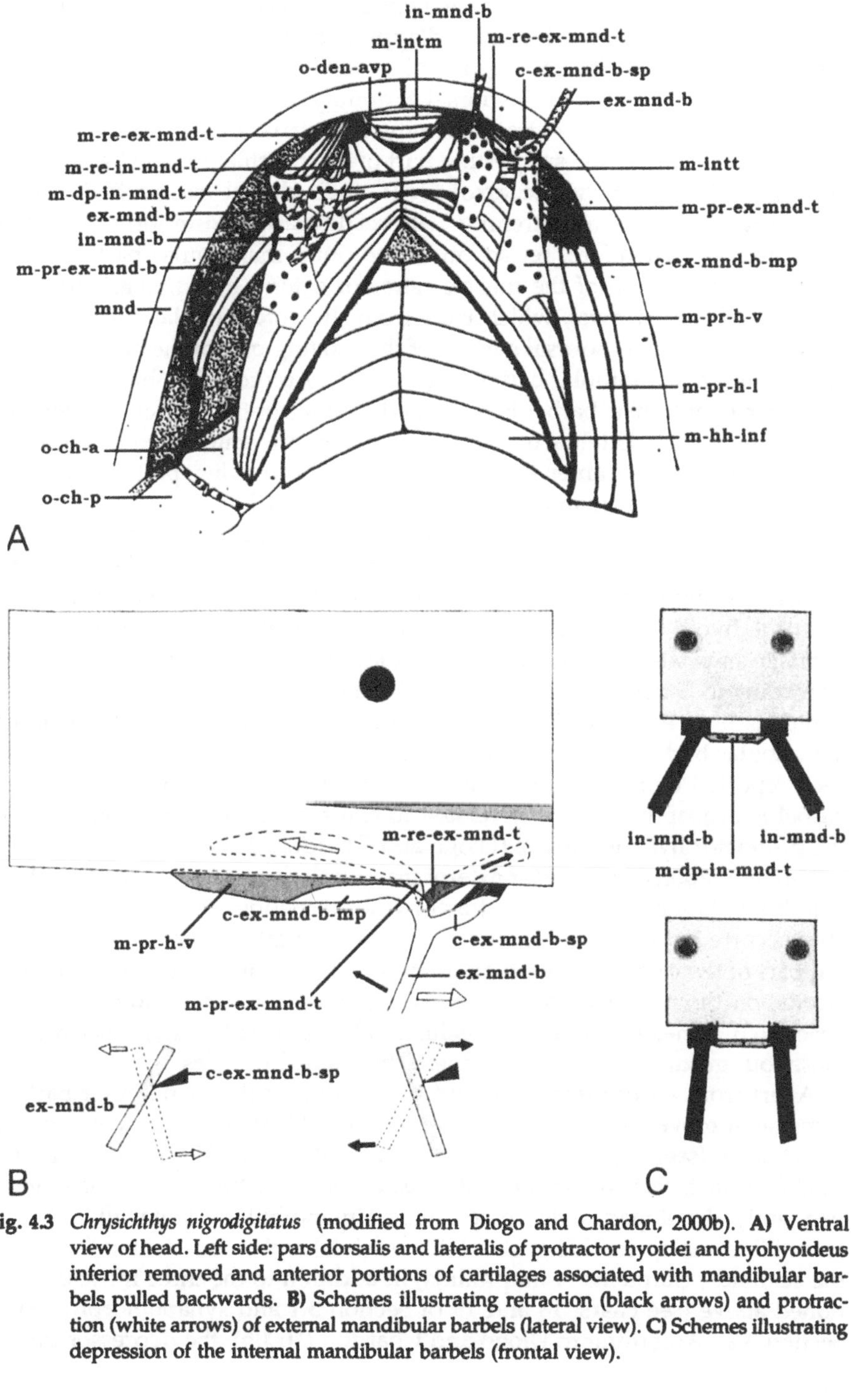

Fig. 4.3 *Chrysichthys nigrodigitatus* (modified from Diogo and Chardon, 2000b). **A)** Ventral view of head. Left side: pars dorsalis and lateralis of protractor hyoidei and hyohyoideus inferior removed and anterior portions of cartilages associated with mandibular barbels pulled backwards. **B)** Schemes illustrating retraction (black arrows) and protraction (white arrows) of external mandibular barbels (lateral view). **C)** Schemes illustrating depression of the internal mandibular barbels (frontal view).

this firm attachment confers to these barbels, as stated by Diogo and Chardon (2000b: 464), 'a solid exterior point d'appui, creating an articulatory system somewhat similar to the rocking palatine-maxillary system present in some catfishes' (Fig. 4.3B). So, 'if the dorsal extremity of the mandibular barbels is pulled posteriorly, their ventral extremity, by means of the solid central point d'appui conferred by the supporting part of their cartilages, will be displaced anteriorly' (Fig. 4.3B: white arrows). If 'their dorsal extremity is pulled anteriorly, their ventral extremity will be displaced posteriorly (Fig. 4.3B: black arrows). Therefore, retraction of the outer mandibular barbel is provoked by contraction of the muscle retractor externi mandibularis tentaculi (Fig. 4.3B: black arrows), which runs from the anteromedial surface of the mandible to the anterodorsal surface of the moving part of the basal cartilage of the external mandibular barbel (see Fig. 4.3). Accordingly, the retraction of the inner mandibular barbel is provoked by the contraction of the retractor interni mandibularis tentaculi, which runs from the anteromedial surface of the mandible to the anterodorsal surface of the moving part of the basal cartilage of the inner mandibular barbel (e.g., Fig. 4.3A).

Protraction of the external mandibular barbel is provoked by the antagonist of the retractor externi mandibularis tentaculi, that is, by the protractor externi mandibularis tentaculi (Fig. 4.3B: white arrows). This muscle runs from the hyoid arch to the anterodorsal margin of the moving part of the cartilage associated with the external mandibular tentaculi (see, e.g., Fig. 4.3). According to Diogo and Chardon (2000b), the pars ventralis of the protractor hyoideus (see, e.g., Fig. 4.3A) is also related to protraction of the external mandibular barbel. As a protractor interni mandibularis tentaculi has never been reported thus far in siluriforms, protraction of the internal mandibular barbel is probably exclusively related to contraction of the pars ventralis of the protractor hyoidei muscle (Diogo and Chardon, 2000b).

Differentiation of a retractor externi mandibularis tentaculi, retractor interni mandibularis tentaculi and protractor externi mandibularis tentaculi probably occurred at about the same time as the firm attachment of the supporting part of the cartilages of the mandibular barbels on the mandible. In other words, on the node leading to the clade uniting the non-diplomystid, non-loricarioid, non-cetopsid and non-silurid catfishes (see Fig. 4.2), since none of these four groups exhibits any of these three small muscles.

Apart from a pure retraction and protraction of the mandibular barbels, some other movements of these barbels are possible in some particular groups of catfishes (see, e.g., Ghiot, 1978; Ghiot et al., 1984; Diogo and Chardon, 2000b). In fact, certain specific, not directly related catfish groups have independently developed a depressor interni mandibularis tentaculi (e.g., the amphiliids, malapterurids or claroteids + ariids) and an intertentacularis (e.g., the schilbids, claroteins or amphiliids) in addition to the three muscles mentioned above (see chars. 18 and 19 of Section 3.1 and synapomorphy list of Section 3.2). According to Diogo and Chardon (2000b), the depressor interni mandibularis tentaculi promotes depression of the internal mandibular barbels (Fig. 4.3C), while the intertentacularis promotes approximation between the internal and external mandibular barbels of the same side.

4.2 PECTORAL GIRDLE COMPLEX

As enumerated in Chapter 1, catfishes exhibit a large number of peculiar morphological specialisations, which make them easily recognisable, even in fossils in which only some small, disarticulated fragments are available (Regan, 1911b; Alexander, 1965; Fink and Fink, 1981; Arratia, 1987; Gayet and Meunier, 2003; and others). One of the more remarkable anatomical specialisations of catfishes, of inestimable value for palaeontologists, is surely the peculiar transformation of all the pectoral girdle, and especially of the first pectoral ray (Reed, 1924; Hubbs and Hibbard, 1951; Alexander, 1965; Lundberg, 1975b; Gosline, 1977; Brosseau, 1978; Grande, 1987; Grande and Eastman, 1986; a.o.).

Contrary to most basal teleosts, in which the pectoral girdle is constituted by a large and variable number of bones and is highly mobile in relation to the neurocranium, in catfishes this girdle is only composed of three skeletal elements and is usually deeply attached to the neurocranium (see Gosline, 1977). The homologies between the three components of catfish pectoral girdle and those of most basal teleosts have been a subject of discussion and uncertainties in the past (Regan, 1911b; Alexander, 1965; Chardon, 1968; Lundberg, 1975b; Gosline, 1977; Brosseau, 1978; Fink and Fink, 1981; Howes, 1985a; Jollie, 1986; etc.). It is commonly accepted nowadays, however, that these bones correspond to the cleithrum, scapulo-coracoid (scapula + coracoid) and posttemporo supracleithrum (posttemporal + supracleithrum) of other teleosts (Schaefer, 1990, 1998; Howes and Fumihito, 1991; Arratia and Gayet, 1995; Gayet and Meunier, 1998; Cabuy et al., 1999; He et al., 1999; etc.).

The particular configuration of these bones is probably related to other peculiar catfish features, such as the ankylosis between the posterior region of the neurocranium, pectoral girdle and anterior vertebra, but also the friction-locking mechanism of the thickened first pectoral ray (see, e.g., Tilak, 1963; Alexander, 1965; Gainer, 1967; Chardon, 1968; Gosline, 1977; Schaefer, 1984; Fine et al., 1997). This friction-locking mechanism, well described by Alexander (1965), seems to be principally related to a protective function (Alexander, 1965). It might, sometimes, also be associated with terrestrial locomotion (Vaillant, 1895; Donnelly, 1973; Gougnard and Vandewalle, 1980; a.o.), sound production (Gainer, 1967; Fine et al., 1997; Kaatz, 1997; Ladich, 1997; Pruzsinszky and Ladich, 1998; a.o.), feeding habits (see, e.g., Tilney and Hecht, 1990) or reproductive behaviour (see, e.g., Winemiler, 1987; Pruzsinszky and Ladich, 1998).

The results of the present cladistic analysis revealed some interesting points concerning the macroevolution of the pectoral girdle complex within Siluriformes. According to Gayet and Meunier (2003), one of the most remarkable and puzzling issues in catfish macroevolution is the fact that the highly modified pectoral girdle and rays characteristic of catfishes seem to have been present since the very beginning of the evolutionary history of this group of fishes.

The highly modified first pectoral ray, as well as the pectoral girdle constituted by the bones posttemporo-supracleithrum, scapulo-coracoid and cleithrum, seem indeed to have been present since the first stages of catfish macroevolution, being inclusively present in diplomystids (see Figs. 3.63, 3.70, 3.71, 3.72). However, one should be very careful in stating that the 'typical' catfish pectoral girdle has been present since the very beginning of siluriform evolutionary history. As underscored by Gosline (1977), Mo (1991), Bornbusch (1995) and Grande and De Pinna (1998), in what concerns the overall aspect of pectoral girdle, in the vast majority of catfishes this girdle is typically formed by a scapulo-coracoid consisting of a broad bony plate visible in ventral and dorsal views. This bone is sutured with the cleithrum along its anterolateral edge and meets its counterpart in an interdigitation of several strong serrations (see Fig. 3.56). However, as emphasised by the aforesaid authors, in diplomystids (see, e.g., Fig. 3.70) as well as some other catfishes such as trichomycterids, nematogenyids, cetopsids or silurids, the scapulo-coracoid is a slender bone not visible in dorsal view and not interdigitating with its counterpart mesially (see Figs. 3.70, 3.71, 3.72).

Mo (1991) and Bornbusch (1995) considered a slender scapulo-coracoid with no medial suture with its counterpart representative of the plesiomorphic condition for catfishes. Grande and De Pinna (1998) suggested that a scapulo-coracoid with no mesial interdigitation with its counterpart represents a catfish plesiomorphy. According to them a scapulo-coracoid medially sutured with its counterpart could either constitute a synapomorphy of the Siluroidea sensu these authors (all catfishes except diplomystids) or a synapomorphy of non-diplomystid and non-cetopsid catfishes (if both Diplomystidae and Cetopsidae were basal to Hypsidoridae and other siluroids; see Grande and De Pinna, 1998: 471). Also according to these authors, the slender scapulo-coracoid with no mesial suture with its counterpart in catfishes such as trichomycterids and nematogenyids would likely be a homoplasic character. This is also the opinion of Bornbusch (1995), who considers that the slender scapulo-coracoid of most silurids is probably due to a homoplasic reversion.

The observations of the present work fully support Mo (1991), Bornbusch (1995) and Grande and De Pinna (1998) in that the overall configuration of the scapulo-coracoid of diplomystids probably represents the plesiomorphic condition for siluriforms. In fact, the compound constituted by both the scapula + coracoid + mesocoracoid of Gonorynchiformes, Cypriniformes, Characiformes, and in particular of Gymnotiformes, clearly resembles the scapulo-coracoid of diplomystids, being a slender structure with a very thin median process that does not suture with its counterpart mesially (see, e.g., Regan, 1911a; Monod, 1963; Alexander, 1964, 1965; Roberts, 1969; de la Hoz, 1974; Gijsen and Chardon, 1976; Gosline, 1977; Howes, 1978; Mago-Leccia, 1978; de la Hoz and Chardon, 1984; Mago-Leccia et al., 1985). However, the results of the present analysis contradict the view that the well-developed scapulo-coracoid, visible in dorsal view, meeting its counterpart in a strong median interdigitation probably would represent a synapomorphy of all non-diplomystid or of all non-diplomystid and non-cetopsid catfishes. These results suggest a somewhat different chronology regarding these main morphological transformations.

According to the present analysis, the three main morphological transitions between the plesiomorphic siluriform configuration of the scapulo-coracoid and cleithrum found in *Diplomystes* and that present in the vast majority of catfishes were respectively: a *broad posterodorsal expansion* (Fig. 4.4A→C) of the mesial surface of the scapulo-coracoid, followed by a *more pronounced ankylosis* (Fig. 4.4A→C) between the anterior margin of the scapulo-coracoid and the posterior margin of the cleithrum plus *interdigitation* (Fig. 4.4A→C) between the scapulo-coracoids.

The first of these three major morphological transitions, the posterodorsal, large laminar projection of the scapulo-coracoid subdividing the arrector dorsalis into two well-developed, well-distinguished divisions (see Fig. 4.4A→C, B→D), constitutes a major event in catfish evolutionary history. According to the results of the present analysis, it seems to have occurred only once within the Siluriformes, namely in the node leading to the clade including all recent groups excluding diplomystids, loricarioids and cetopsids, as shown in Figure 4.5.

Therefore, the situation found in loricarioids such as nematogenyids or trichomycterids appears as a truly plesiomorphic configuration and not, as

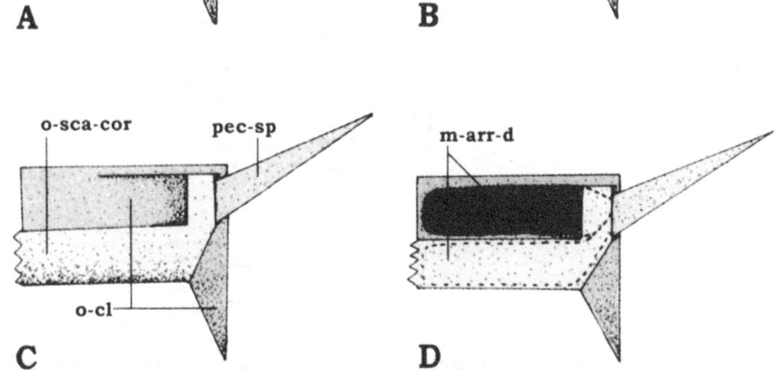

Fig. 4.4 Scheme illustrating main morphological differences between the pectoral girdle **(A)** and arrector dorsalis muscle **(B)** of diplomystids and the pectoral girdle **(C)** and arrector dorsalis muscle **(D)** of most other catfishes [for more details, see text].

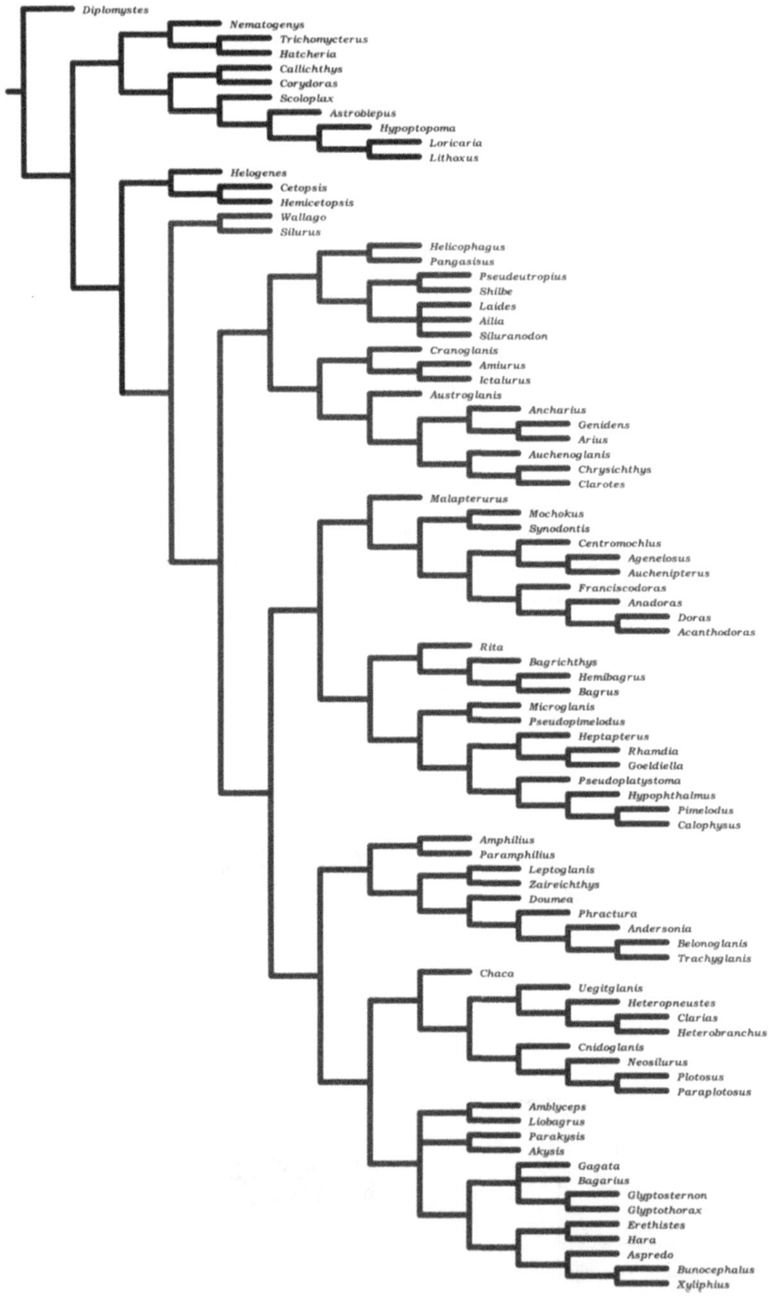

Fig. 4.5 Hypothesis of character state evolution of differentiation of arrector dorsalis (char. 51): CS-0 (black)= arrector dorsalis not differentiated into well-developed dorsal and ventral divisions separated by broad horizontal lamina of the scapulo-coracoid; CS-1 (blue)= arrector dorsalis differentiated into well developed dorsal and ventral divisions separated by broad horizontal lamina of the scapulo-coracoid [for more details, see text].

suggested by some authors (see above), as a secondary reversion. In fact, the observations of this work reveal that not only in these loricarioids, but also in other Loricarioidea, i.e., callichthyids, astroblepids, loricariids and scoloplacids, the overall configuration of the scapulo-coracoid and arrector dorsalis differs morphologically from that found in the great majority of catfishes (see Figs. 3.30, 3.31, 3.32). In these latter four groups the ankylosis between the scapulo-coracoid and the cleithrum is realised in a different manner than in most catfishes. There is no true posterior horizontal lamina of the scapulo-coracoid, nor does the anterolateral margin of this bone form, together with the posterolateral margin of the cleithrum, the typical foramen in which passes the dorsal division of the arrector dorsalis in most catfishes. Consequently, this latter muscle is not subdivided into a dorsal and ventral divisions, as in most catfishes. In no loricarioid examined is the arrector dorsalis truly well differentiated in a dorsal and a ventral divisions separated by a well-developed horizontal lamina of the scapulo-coracoid, as occurs in the great majority of the other catfishes (see Figs. 4.4, 4.5).

The configuration of the arrector dorsalis thus constitutes a very nice example of one of the main phylogenetic outlines of the present work concerning catfish higher level relationships: the markedly basal position of the Loricarioidea within the order Siluriformes. In fact, in terms of the major morphological transitions within catfish evolutionary history, the loricarioids often share with diplomystids a rather plesiomorphic configuration, and hence present a markedly primitive morphological situation in contrast to that found in great majority of other catfishes. This is distinctly evident in Figure 4.5 for the configuration of the arrector dorsalis, as discussed above, but also in several other cases concerning other major morphological key features, e.g., those concerning the presence of mandibular barbels (see Fig. 4.1 and discussion above), shape of the arrector ventralis (see Fig. 4.6 and discussion below), development of the abductor profundus (see Fig. 4.9 and discussion below) and differentiation of the protractor hyoidei (see char. 30 and discussion above), as the reader will be able to verify in the text below.

Interestingly the broad posterodorsal expansion of the mesial surface of the scapulo-coracoid (Fig. 4.4A→C) seemingly arrived at about the same level as another important feature, namely the transition from a mostly well-developed and transversally oriented to mostly thin and anteroposteriorly oriented muscle arrector ventralis (see Fig. 4.6). As shown in Figure 4.6, it was not possible to discern this feature in the cetopsids examined due to the highly peculiar configuration of their pectoral girdle (see char. 48). Thus this transition was preventively assigned to the node leading to the non-diplomystid, non-loricarioid and non-cetopsid catfishes. Already within this clade, the arrector ventralis had become markedly hypertrophied in *Glyptosternon*, and reverted homoplasically to a more transversal orientation in amphiliids (Fig. 4.6). This latter homoplasic event could be related to the occurrence of a peculiar, homoplasy-free feature only found in the Amphiliidae: differentiation of the arrector ventralis into two well-developed

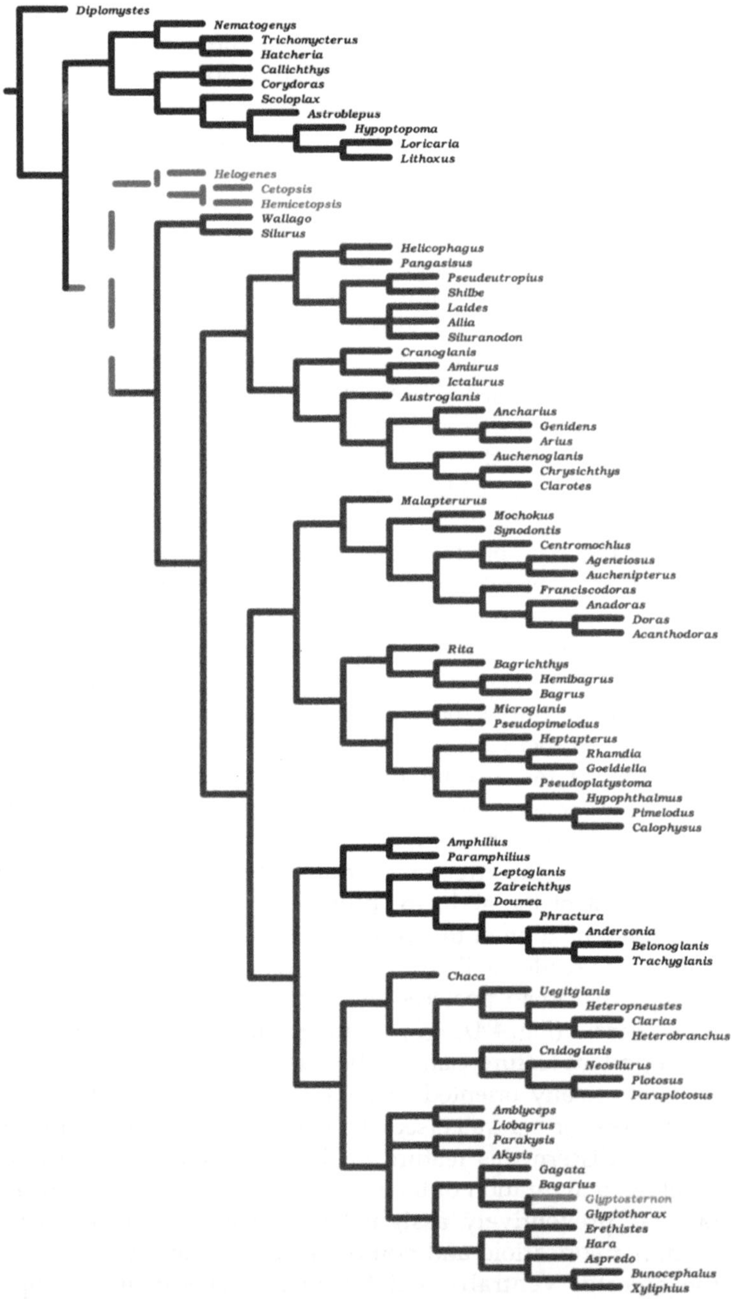

Fig. 4.6 Hypothesis of character state evolution of shape of arrector ventralis (char. 48, unordered): CS-0 (black)= arrector ventralis well-developed and mainly transversally oriented; CS-1 (blue)= arrector ventralis thin and mainly anteroposteriorly oriented; CS-2 (orange)= arrector ventralis hypertrophied, reaching mesial symphisis; Interrogation (pink) [for more details, see text].

bundles in the members of this family, and inclusively in a well-distinguished new muscle in the members of its subfamily Doumeinae (see description of char. 49 above). Interestingly, the amphiliid catfishes also share another rather peculiar synapomorphy related with the configuration of the pectoral girdle complex: the marked interdigitation between the cleithra, a feature found elsewhere only in *Scoloplax* (see description of char. 173 above).

With respect to the other two main general morphological transitions noted above between the plesiomorphic siluriform configuration of the pectoral girdle and that found in the vast majority of catfish, i.e., the *more pronounced ankylosis* between the anterior margin of the scapulo-coracoid and the posterior margin of the cleithrum and *the interdigitation* between the scapulo-coracoids (Fig. 4.4A→C), these seem to constitute synapomorphies of the node leading to all non-diplomystid, non-loricarioid, non-cetopsid and non-silurid catfishes. This is a less inclusive clade than that characterised by the first main synapomorphic morphological transition described above, the *posterior horizontal laminar expansion* (Fig. 4.4A→C) of the scapulo-coracoid subdividing the arrector dorsalis (Fig. 4.4B→D) into two well-developed, well-distinguished divisions, which is present in silurid catfishes (see Fig. 4.5). However, let it be noted that, contrary to this main morphological transition, the pronounced ankylosis between the scapulo-coracoid and the cleithrum and the interdigitation between the scapulo-coracoids are not homoplasy free. They are homoplasically present in callichthyids + scoploplacids + astroblepids + loricariids (in *Astroblepus* the interdigitation between the scapulo-coracoids was secondarily lost), as can be seen in Figures 4.7 and 4.8.

Of interest here is another general feature that also constitutes a synapomorphy of the node leading to the non-diplomystid, non-loricarioid, non-cetopsid and non-silurid catfishes. This concerns character 47, i.e., the transition for a markedly developed abductor profundus almost reaching, or reaching, the midline (see Fig. 4.9). Analysis of the tree shown in Figure 4.9 concerning this character revealed two significant aspects that will be discussed in detail in Chapter 5:1) macroevolution is rather 'complex' and 'mosaic', with homoplasic events seemingly not the exception but rather the rule; 2) these homoplasic events can, however, and *should be*, discriminated *a posteriori* to a phylogenetic analysis since they can, and often do, reveal synapomorphic features for less inclusive clades and since their analysis vis-à-vis the cladogram obtained allows a more objective and detailed discussion of the evolution of the major group under examination.

Such is also the case for uniquely homoplasic features such as seen in the trees of Figures 4.10 and 4.11, respectively: development of the mesial limb of the posttemporo-supracleithrum and suture between this bone and the neurocranium. For example, in the case of the suture between the mesial limb of the posttemporo-supracleithrum and the neurocranium, the present analysis enabled discrimination of this feature as independently acquired in ariids and in the clade mochokids + doradids + auchenipterids (Fig. 4.11), groups in which there is a particularly developed ossification of the skull, as observed earlier by Chardon (1968).

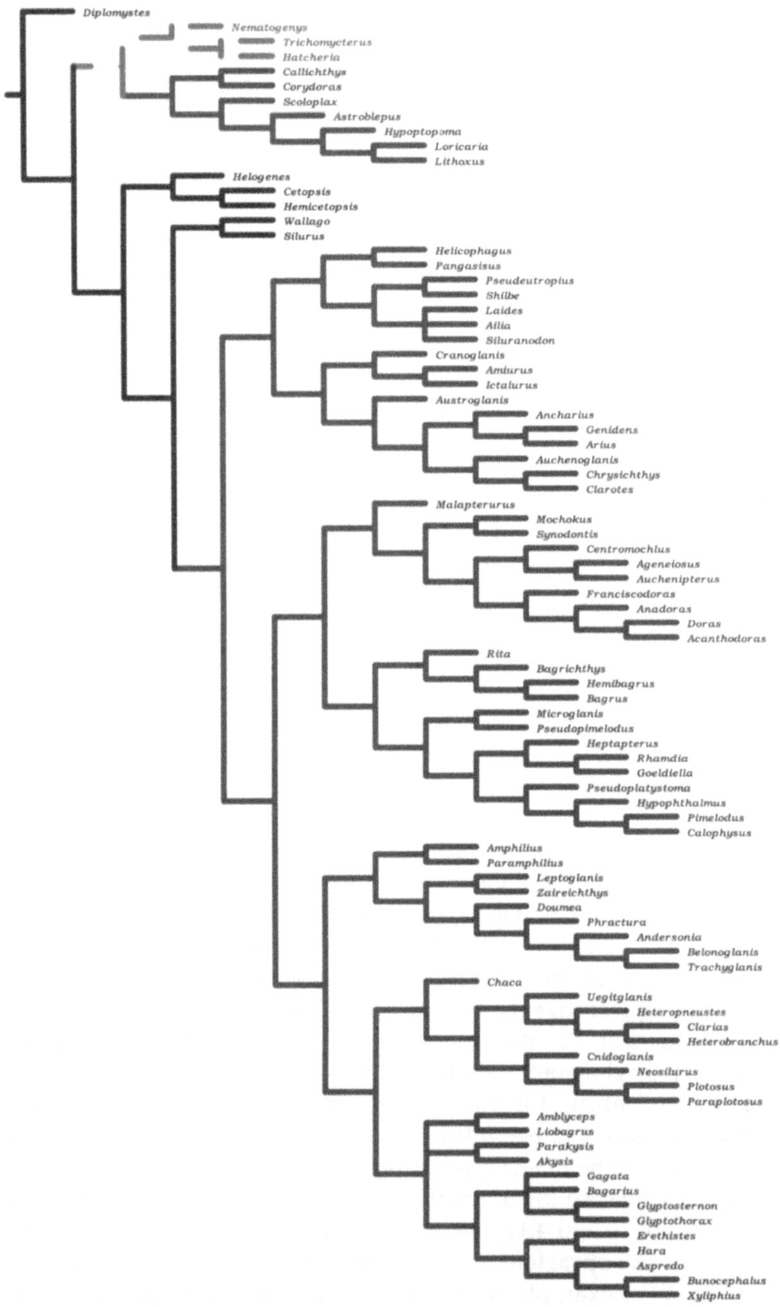

Fig. 4.7 Hypothesis of character state evolution of ankylosis between cleithrum and scapulo-coracoid (char. 174): CS-0 (black)= absence of marked pronounced ankylosis between these bones; CS-1 (blue)= presence of marked pronounced ankylosis between these bones; Interrogation (pink) [for more details, see text].

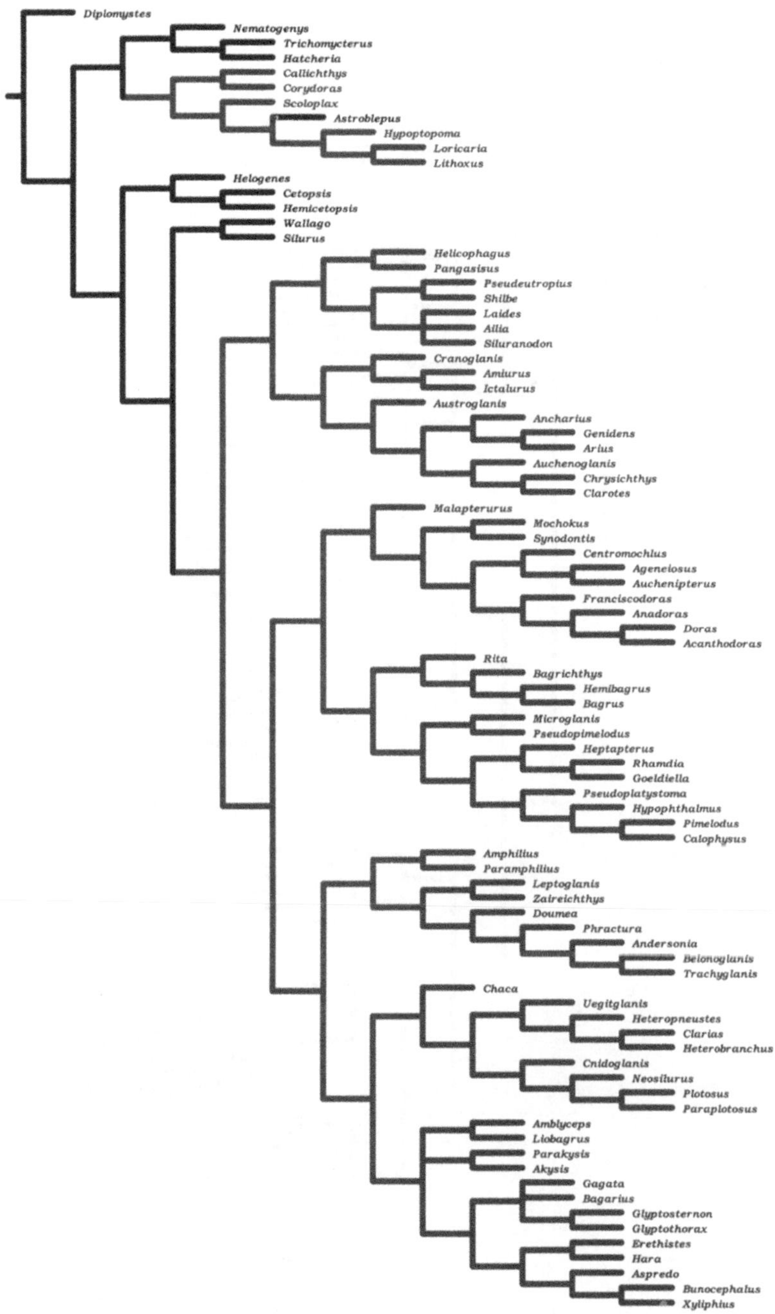

Fig. 4.8 Hypothesis of character state evolution of interdigitation between scapulo-coracoids (char. 180): CS-0 (black)= absence of marked interdigitation between scapulo-coracoids; CS-1 (blue)= presence of marked interdigitation between scapulo-coracoids [for more details, see text].

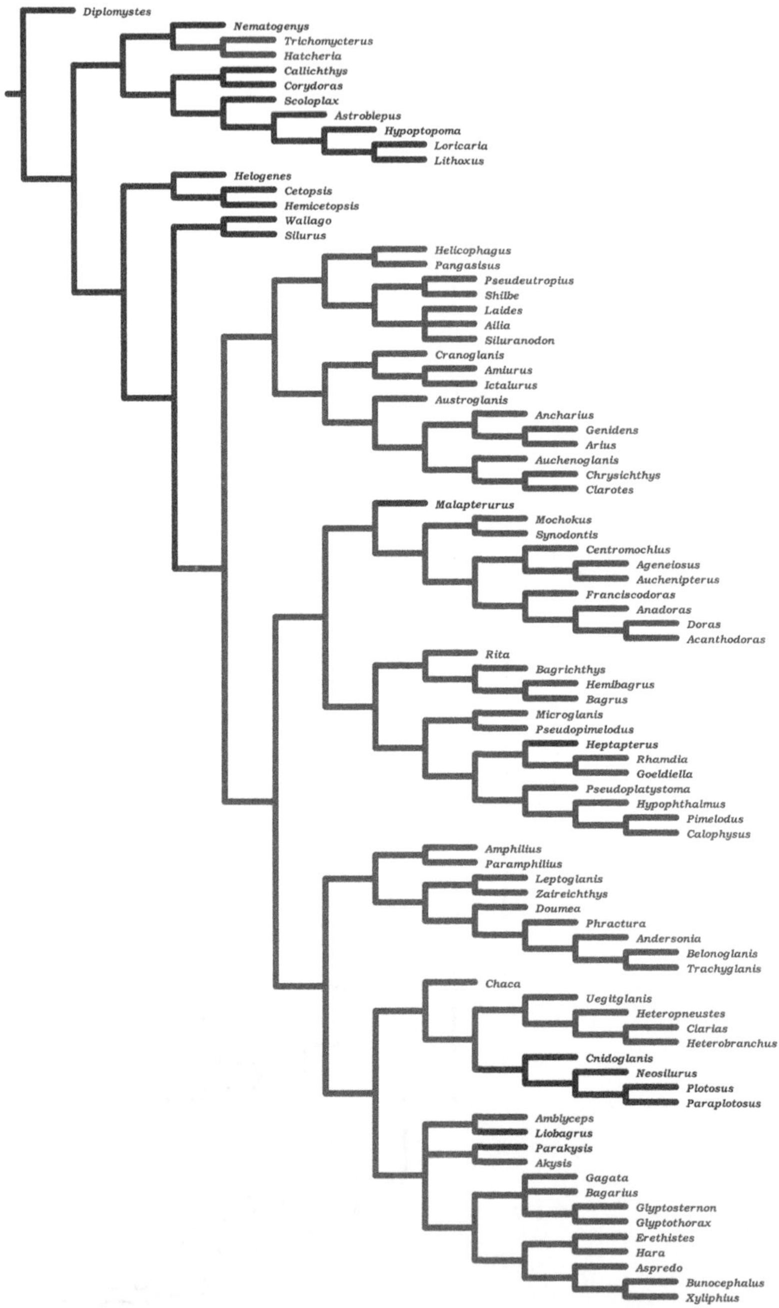

Fig. 4.9 Hypothesis of character state evolution of development of abductor profundus (char. 47): CS-0 (black)= abductor profundus originated far from midline; CS-1 (blue)= abductor profundus almost reaching, or reaching, mesial symphisis of pectoral girdle [for more details, see text].

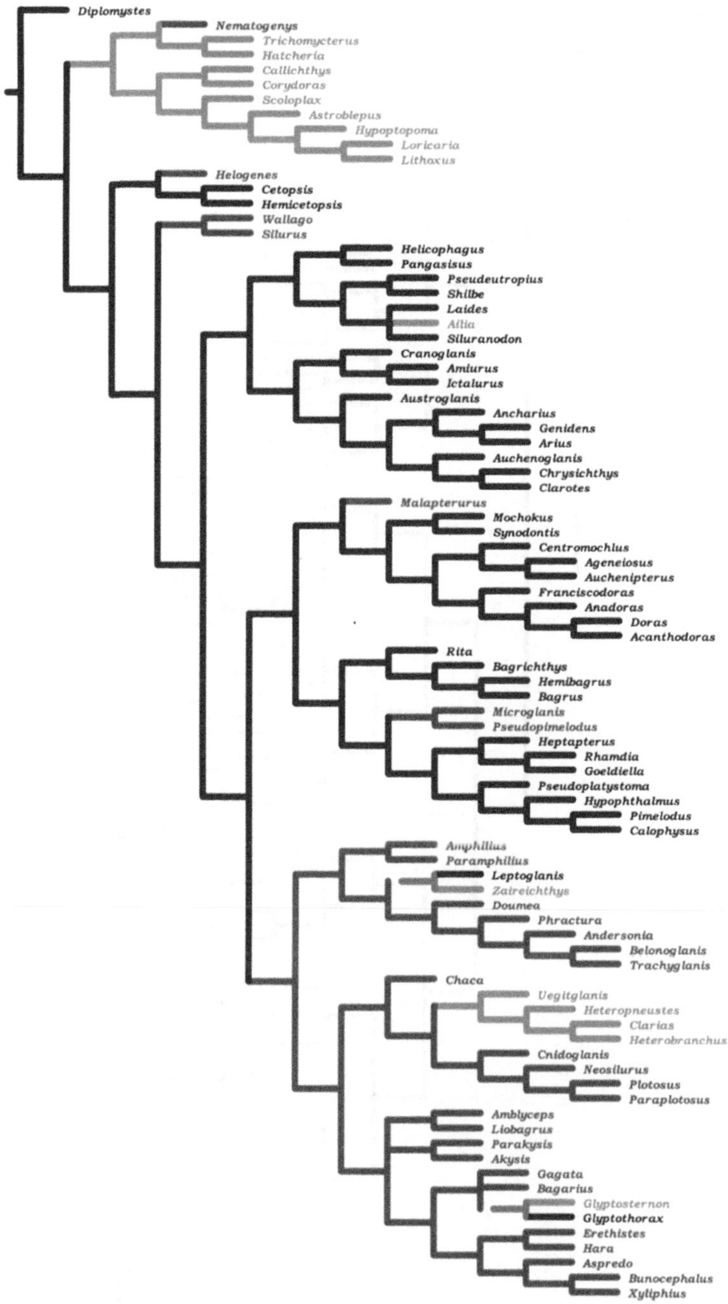

Fig. 4.10 Hypothesis of character state evolution of development of mesial limb of posttemporo-supracleithrum (char. 150, unordered): CS-0 (black)= mesial limb well-developed; CS-1 (blue)= mesial limb markedly compressed anteroposteriorly; CS-2 (orange)= mesial limb undifferentiated; Ambiguity (pink) [for more details, see text].

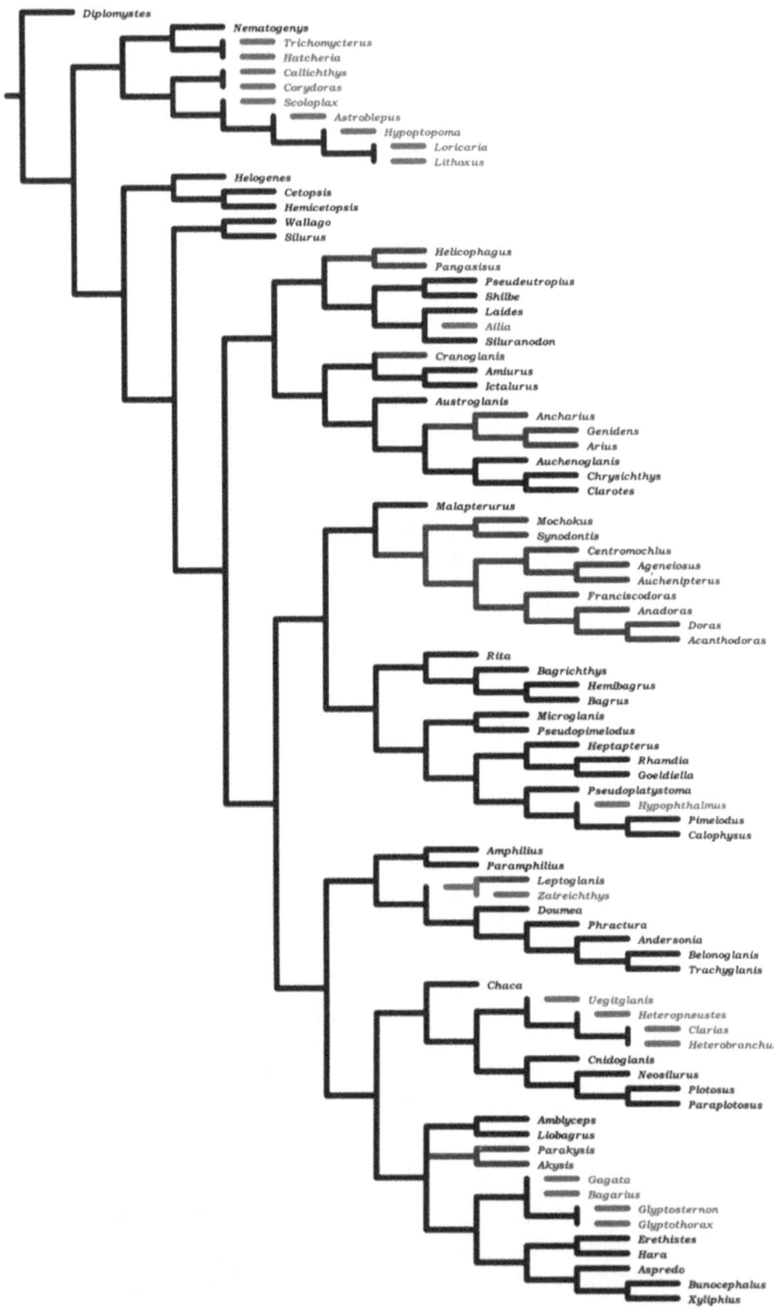

Fig. 4.11 Hypothesis of character state evolution of suture between mesial limb of posttemporo-supracleithrum and neurocranium (char. 155): CS-0 (black)= mesial limb and neurocranium not sutured; CS-1 (blue)= mesial limb and neurocranium firmly sutured; Ambiguity (pink) [for more details, see text].

Let us now discuss the potential evolutionary significance of the main general morphological transformations described above. With respect to the three major osteological events schematised in Figure 4.4, i.e., the posterodorsal expansion of the mesial surface of the scapulo-coracoid, the pronounced ankylosis between the scapulo-coracoid and the cleithrum, and the interdigitation between the scapulo-coracoids, these clearly seem associated to a greater overall solidification of the pectoral girdle complex.

The evolutionary significance of the solidification of this complex has already been discussed by authors such as Alexander (1965), Chardon (1968) or Gosline (1977). This solidification, associated with that of the posterior region of the skull, the anterior vertebrae and the complex of the dorsal spine, as well as with the overall solidification between all these regions, seems to confer a major protection for the particularly delicate and important area of the inner ear and the Weberian apparatus (see Chardon, 1968). The overall ankylosis of these regions is eventually reinforced in some catfish groups by the development of features such as the sutural connection between the mesial limb of the posttemporo-supracleithrum and the neurocranium (see char. 155) or the transcapular process of the neurocranium firmly attaching to the anterior vertebrae (see char. 148).

This overall ankylosis, together with the friction-locking mechanism of the highly modified catfish pectoral spines, would also confer a special defensive protection against predators such as aquatic birds and particularly large predatory fishes (see Alexander, 1965; Gosline, 1977). As stated by Alexander (1965:111), 'a fish with erect spines is more difficult for a predator to swallow, and the predator cannot swallow it without suffering pain or injury'. Hoogland et al. (1957), for example, demonstrated that two predaceous fish, namely *Perca* and *Esox*, seemingly learned not to attack *Gasterosteus*, a genus of gasterosteiform fishes presenting spines similar to those of catfishes, after experiencing its spines a few times. The importance of catfish pectoral spines and their movements could thus explain the functional importance of the differentiation of both the arrector dorsalis (char. 51) and the arrector ventralis (char. 47) mentioned above. Differentiation of these muscles into different bundles and/or separate new muscles could have allowed exploration of a wider range of potential movements of these spines (see, e.g., Alexander, 1965; Diogo et al., 2001c).

Discussion of the major aspects of the macroevolution of the pectoral girdle complex within Siluriformes also raises, however, a particularly puzzling question. If solidification of the pectoral girdle seems to constitute, in major terms, an important event in catfish evolutionary history, with interdigitation between the scapulo-coracoid found in most siluriforms (Fig. 4.8), why is a strong interdigitation between the cleithra only found, among the catfishes examined, in members of two of the 32 extant catfish families, i.e., amphiliids and scoloplacids (see char. 173). Is there a morphological constraint that somehow limits the development of this latter character among catfishes? This puzzling question is of particular interest in the au courant debate on evolutionary constraints, an issue of paramount importance in Gould's (2002) hypothesis on the structure of evolutionary theory that is discussed in Chapter 5.

4.3 ADDUCTOR MANDIBULAE COMPLEX

Let us now discuss the macroevolution of the adductor mandibulae complex within Siluriformes. Given the controversies concerning this subject, as well as the different nomenclatures used to designate the adductor mandibulae sections in catfish, it is necessary to establish first, in a consistent way, the identity and evolutionary track of the different adductor mandibulae bundles within the Teleostei.

Gosline (1989) noted that in teleostean fishes there are two basically different pathways of differentiation in the cheek part of the adductor mandibulae, which are illustrated in Figure 4.12. Figure 4.12A shows the teleostean basal situation of the adductor mandibulae: an undivided cheek muscle attached to the mesial face of the mandible and the Aω already differentiated [the presence of an Aω is an actinopterygian plesiomorphy, with this section probably the first adductor mandibulae bundle to separate as a distinct entity (Edgeworth, 1935; Winterbottom, 1974a; Lauder, 1980)]. This situation is observable, for example, in *Salvelinus* (Lauder and Liem, 1980) and also in *Hiodon*, *Elops* and *Clupea* (Gosline, 1989).

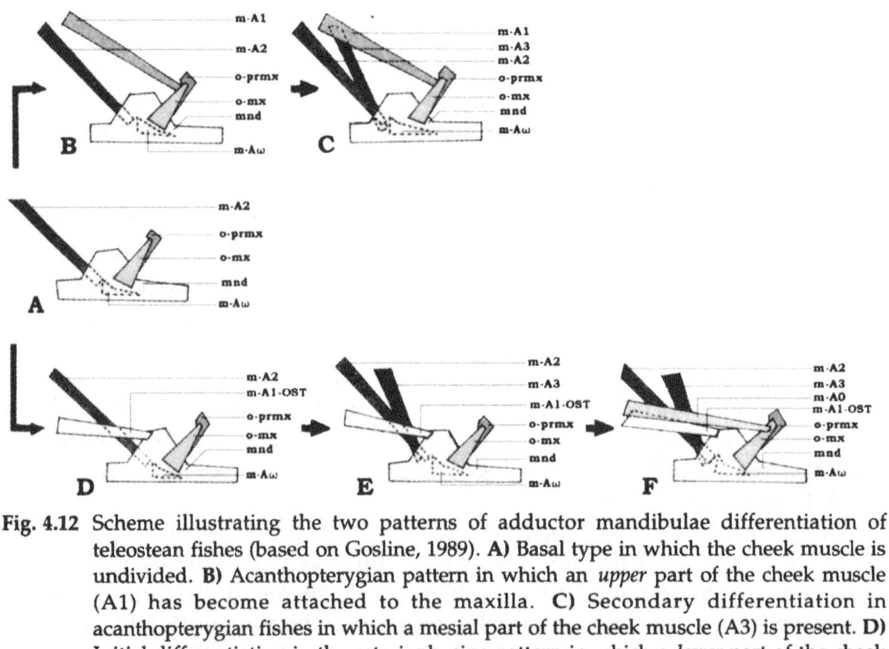

Fig. 4.12 Scheme illustrating the two patterns of adductor mandibulae differentiation of teleostean fishes (based on Gosline, 1989). **A)** Basal type in which the cheek muscle is undivided. **B)** Acanthopterygian pattern in which an *upper* part of the cheek muscle (A1) has become attached to the maxilla. **C)** Secondary differentiation in acanthopterygian fishes in which a mesial part of the cheek muscle (A3) is present. **D)** Initial differentiation in the ostariophysine pattern in which a *lower* part of the cheek muscle (A1-OST) has developed a separate attachment to the back of the mandible. **E)** Secondary differentiation in ostariophysine fishes in which a mesial part of the cheek muscle (A3) is present. **F)** Differentiation in some ostariophysine fishes in which an adductor mandibulae section (A0) has developed, *via* the primordial ligament, an attachment to the maxilla [for more details, see text].

From this configuration, two types of differentiation are possible. In acanthopterygians *an antero-dorsolateral* part of the cheek muscle (A1, see below) appears to have developed an attachment to the maxilla via the primordial ligament (Fig. 4.12B) (Gosline, 1989). This situation seems to be plesiomorphic for acanthopterygians and is present, for example, in *Aulopus* (Lauder and Liem, 1983; Gosline, 1986, 1989) and *Neoscopelus* (Winterbottom, 1974a). A secondary differentiation of the adductor mandibulae (A3, see below), mesial to all the others, is found in most acanthopterygian fishes (Fig. 4.12A).

In ostariophysines an *antero-ventrolateral* (A1-OST, see below) part of the cheek muscle separates and attaches to the postero-dorsolateral face of the mandible (Fig. 4.12A) (Gosline, 1989). This section was effectively present in all the catfishes studied here, inclusively in *Diplomystes* (see Fig. 3.64), which presents the plesiomorphic condition of the adductor mandibulae of these fishes, as is explained below. This situation, where only an A1-OST, an A2 and an Aω are present, e.g., in *Salminus* and *Hepsetus* (Characiformes), seems (*contra* Fink and Fink, 1981) to be plesiomorphic for ostariophysine fishes (Vari, 1979; Howes, 1983b, 1985a, b; Gosline, 1989). The antero-ventrolateral division of the adductor mandibulae present in these fishes seems to have developed as a supplementary system for raising the mandible (Gosline, 1989). Apart from this division, another section of the adductor mandibulae, mesial to all the others (A3, see below), is also present in most ostariophysines (Fig. 4.12E, F) (McMurrich, 1884b; Takahasi, 1925; Edgeworth, 1935; Nawar, 1955; Alexander, 1964, 1965; Osse, 1969; Ballintijn et al., 1972; Winterbottom, 1974a; Vandewalle, 1975, 1977; Gijsen and Chardon, 1976; Howes, 1976, 1983b, 1985a; Vari, 1979; de la Hoz and Chardon, 1984; Vandewalle et al., 1985, 1986, 1993, 1997; Schaefer and Lauder, 1986; Aguilera, 1988; Gosline, 1989; Adriaens and Verraes, 1996; a.o.) and in all catfish studied in the present work (see, e.g., Fig. 3.64).

The situation represented in Figure 4.12E is found in most characiforms (Alexander, 1964, 1965; Gijsen and Chardon, 1976; Howes, 1976, 1983b, 1985ab; Vari, 1979; Gosline, 1989; a.o.) and in a large number of gymnotiforms (Fink and Fink, 1981; Howes, 1983b, 1985b), and seems to be plesiomorphic for siluriforms.

In a large number of ostariophysine fishes, namely in cypriniforms (Winterbottom, 1974a; Fink and Fink, 1981; Howes, 1983b, 1985b; Gosline, 1989; a.o.), some characiforms (Alexander, 1964, 1965; Howes, 1976, 1983a, 1985b; Vari, 1979; Gosline, 1989; a.o.), most gonorynchiforms (Fink and Fink, 1981; Howes, 1985b; Gosline, 1989) and a large number of gymnotiforms (Chardon and de la Hoz, 1973; Howes, 1983b; de la Hoz and Chardon, 1984; Aguilera 1988), a superficial differentiation of the external section (A1-OST) attaches, via the primordial ligament, to the upper jaw as a new section (A0, see below) of the adductor mandibulae (Fig. 4.12F) (Gosline, 1989).

This hypothesis regarding the macroevolution of the configuration of the adductor mandibulae in ostariophysine fishes contradicts Takahasi (1925) and Fink and Fink (1981). These authors considered the presence of a lateral section of this muscle attached to the upper jaw a plesiomorphic situation for these fishes and the conditions in some Characiformes and Siluriformes secondary reductions from a primitive attachment to the maxilla (Fink and Fink, 1981). This opinion is contestable since, as reminded by Howes (1983b), not only in 'some Characiformes', but in the greater majority of these fishes, the most lateral section of the adductor mandibulae attaches to the mandible, and not to the upper jaw. Alexander's (1964), Vari's (1979) and Gosline's (1989) studies support the idea that the attachment of the adductor mandible on the upper jaw is probably a derived character of some characiforms, probably associated with the small-mouth and/or protrusive upper jaw conditions in some specialised species of this group. Moreover, in some Gymnotiformes (*Gymnotus*) the outer part of the adductor mandibulae inserts mainly onto the outer face of the lower jaw, with only a few fibres inserting on the primordial ligament (Howes, 1983b). Moreover, in the morphologically primitive gonorynchiform *Chanos* (Rosen and Greenwood, 1970; Fink and Fink, 1981), the most external bundle of the adductor mandibulae inserts both on the external face of the mandible and on the maxilla (Howes, 1985a; Gosline, 1989). Most other Gonorynchiformes (more specialised than *Chanos*), exhibit a superficial, completely independent bundle attached to the upper jaw, probably associated to their small-mouth condition, as in characiforms (Gosline, 1989). Lastly, in almost all catfish the most external section of the adductor mandibulae attaches to the mandible (see, e.g., McMurrich, 1884b; Takahasi, 1925; Edgeworth, 1935; Eaton, 1948; Nawar, 1955; Munshi, 1960; Alexander, 1965; Winterbottom, 1974a; Howes, 1983ab, 1985a; Schaefer and Lauder, 1986, 1996; Gosline, 1989; Schaefer, 1990; Mo, 1991; Adriaens and Verraes, 1996; a.o.). Thus, in agreement with Alexander (1964, 1965), Vari (1979), Howes (1983b, 1985a) and Gosline (1989), the insertion of the most superficial section of the adductor mandibulae on the mandible seems to be the plesiomorphic condition for ostariophysines. Differentiation of a section attached, via the primordial ligament, to the upper jaw therefore seems to be a derived condition for these fishes (Fig. 4.12A), probably associated with derived characters such as a small mouth and/or a protrusive upper jaw.

There was some controversy in the past concerning the reliability of the path of the ramus mandibularis for identifying subdivisions of the adductor mandibulae. According to Edgeworth (1935) and Winterbottom (1974a), the path of this nerve tract is an 'unreliable character' for recognising the different adductor mandibulae sections, since 'it may pass external to the adductor mandibulae (e. g., *Salmo, Clupea*), external to A2 (e. g., *Pleuronectes*), between A2 and A3 (e. g., *Scomber, Cyprinus, Esox*), external to A1 and internal to A2 and A2A3 (e. g., *Ictalurus, Galeichthys*) and internal to A3 (*Zoaces*)' (Winterbottom, 1974a). However, recent studies (e. g., Howes, 1985b, 1988; Gosline, 1986, 1989) indicate that the course of the ramus mandibularis seems to be a better indicator of cheek sections in the adductor mandibulae than has generally been acknowledged (Gosline, 1989). According to Gosline (1989) 'much of the previously assumed variability in the course of the ramus mandibularis disappears once it is realised that the cheek sections of the adductor mandibulae in acanthopterygians and ostariophysines are not homologous'. My personal observations and comparisons strongly support this last idea. Apart from some minor shifts, the path of the ramus mandibularis of the catfish examined is essentially similar, usually separating the anterior part of the external section of the adductor mandibulae, A1-OST, from the anterior part of the A2 and/or posterior part of the Aω. This type of course of the ramus resembles that of the characiforms and cypriniforms described by Gosline (1989), in which this nerve tract passes between the anterior parts of the A1-OST and A2. Observation of the path of the ramus mandibularis thus supports Gosline's (1989) hypothesis concerning the homologies of the adductor mandibulae sections among ostariophysines (see above).

Accepting the evolutionary scheme proposed by Gosline (1989) and illustrated in Figure 4.12, it is apparent that neither the external section of the adductor mandibulae plesiomorphically present in ostariophysines (A1-OST) nor the section derived from it present in a large number of these fishes (A0), corresponds to the bundle that attaches to the upper jaw in acanthopterygians (A1). Gosline (1989) considered Vetter's (1878) A1, A2, A3 and Aw should be retained for acanthopterygians. Therefore, the section that attaches to the upper jaw in acanthopterygian fishes should be called A1, since it corresponds to Vetter's A1 in *Perca*. The two sections that are homologous (see below) throughout the teleosts (Aω and A2) (Fig. 4.12) correspond respectively to Vetter's Aω and A2 in *Perca*. Finally, the mesial section of the adductor mandibulae present in a great number of acanthopterygians (A3) (Fig. 4.12A) corresponds to Vetter's A3 in *Perca*.

But, as mentioned by Gosline (1989), the question arises as to how the cheek sections of ostariophysine fishes should be labelled if the designations A1, A2 and A3 are retained for acanthopterygians. He suggested three alternative resolutions: 1) use purely descriptive terms for the adductor mandibulae sections of the Ostariophysi; 2) use diagnostic designations other than A1, A2, A3 and Aω that are homologous, or at least represent parallel developments, within the ostariophysines; 3) try to adapt Vetter's designations, starting from the one cheek section (Fig. 4.12: A2) homologous throughout teleosts.

Gosline chose to adopt the first alternative. However, utilisation of purely descriptive terms can be very complicated. In some catfishes, e.g. *Amphilius* (Fig. 3.6), there are nine different adductor mandibulae sections. It is thus too difficult to relate them with descriptive terms such as 'external part of the external division of the adductor mandibulae', 'external part of the internal division of the adductor mandibulae', etc. Moreover, it is also very confusing since no rule was previously established for the use of these descriptive names. Hence an endless number of names could conceivably be given to the same bundle, which would vastly complicate further morphological comparisons. The same problem would result were previously unestablished diagnostic designations other than A1, A2, A3 and Aω used to describe the adductor mandibulae sections present in ostariophysines. Furthermore, two of the adductor mandibulae sections present in ostariophysines clearly correspond to the A2 and Aω of acanthopterygian fishes (Fig. 4.12). Thus, there is no reason (and it would not be correct) to give them names other than A2 and Aω.

Therefore, one should adopt Gosline's third alternative and label the ostariophysine adductor mandibulae Sections A2 and Aω, corresponding to the A2 and Aω of acanthopterygians (Fig. 4.12). As for A3, the situation is more delicate. A3 is seemingly lacking in both the teleostean (Fig. 4.12A), acanthopterygian (Fig. 4.12B) and ostariophysine (Fig. 4.12D) basal configuration of the adductor mandibulae, yet nevertheless is present in a large number of acanthopterygians (Fig. 4.12C) and ostariophysines (Fig. 4.12A) (see above). The A3 of these two groups are very much alike. They are the most mesial adductor mandibulae sections and normally attach posteriorly to the lateral face of the suspensorium and anteriorly to the mesial face of the mandible (Winterbottom, 1974a); they present a quite similar configuration and fibre orientation (Fig. 4.12C, E). Therefore, the A3 of acanthopterygians and ostariophysines are the result of parallel independent specialisations occurring in these two groups and hence are homoplasic structures (Fig. 4.12). From a practical nomenclatural point of view, the name A3, retained for acanthopterygian fishes (Fig. 4.12C) (see above), can also be used to describe the most mesial section of the adductor mandibulae present in the great majority of ostariophysine fishes (Fig. 4.12E, F). In fact, this is the usual procedure for similar structures developed in parallel. It seems logical and facilitates comparisons: the 'retractor tentaculi' and 'elastic string apparatus' of catfish, as well as the 'maxillary barbels' present in various groups of fishes such as Siluriformes and Cypriniformes, are good examples of this procedure.

As for the A1 and A1-OST, the situation is quite different and more complicated, due to the definition of 'homologous', which still remains fuzzy (see, e.g., Hall, 1994; Beaumont, 1998; Gould, 2002). The most widespread definition of the term homologous in vertebrate biology is 'descended from the same ancestral structure'. But the definition of 'structure' poses a problem. For example, if we consider the adductor mandibulae section (A2) present in basal teleostean fishes (Fig. 4.12A) as a single structure, we can consider the A1 and the A1-OST resulting from two different patterns of differentiation of this section (see above) (Fig. 4.12B, D) as homologous. However, if we consider the fibres of this section as individual structures, the A1 and A1-OST are not homologous, since they are derived respectively from its dorsolateral (Fig. 4.12B) and ventrolateral (Fig. 4.12D) fibres. However, for purely nomenclatural purposes, both considerations lead to the same result. In fact, even if these sections are considered homologous, they should not be given the same name, since they are the result of divergent evolutionary steps, which have led to quite different structures (Fig. 4.12B,D). Thus, to distinguish the dorsolateral adductor mandibulae section present in basal acanthopterygians (Fig. 4.12B) termed A1 (see above) from the ventrolateral section of basal ostariophysines, and following Vetter's (1878) nomenclature, in which as a more lateral section it should be attributed an inferior number (the ventrolateral adductor mandibulae section present in basal ostariophysine fishes is lateral to the A2), it was decided to call the ventrolateral section of ostariophysines as A1-OST (Fig. 4.12D).

The lateral adductor mandibulae section that attaches to the upper jaw in some ostariophysines (Fig. 4.12F: A0), e.g., all cypriniforms, some characiforms, most gonorynchiforms and a large number of gymnotiforms (see above), can be considered to some extent homoplasic with the A1 of acanthopterygians (Fig. 4.12B), since both are attached to the maxilla. However, the similarities between these two structures are confined solely to this attachment because their posterior attachment and configuration differ somewhat (Gosline, 1989). To be consistent with Vetter's nomenclature (see above) and since it is external to the A1-OST (Fig. 4.12F), this section is termed here A0 (Fig. 4.12F) (the 'OST' is not necessary, since the name A0 has not yet been used to describe any other adductor mandibulae section). The diagnostic designation A0 was never used by Vetter (1878) for the very simple reason that in his designations A1, A2, A3 and Aω the A1 of acanthopterygians corresponded to the most external adductor mandibulae section of Cypriniformes (A0 in the present work). So, in Vetter's descriptions of the adductor mandibulae of cypriniforms, the A1-OST, A2 and A3 of this work were respectively termed A2, A3′ and A3″. Therefore, the section he labelled A2 in the acanthopterygian *Perca*, he termed A3 in the cypriniforms *Barbus* and *Cyprinus* (Gosline, 1989).

This misinterpretation was followed by most authors. Thus, the A0 of my study was usually called A1 (or A1-a) (Takahasi, 1925; Ballintijn et al., 1972; Chardon and de la Hoz, 1973; Winterbottom, 1974a; Vandewalle, 1975; Howes, 1978; Fink and Fink, 1981; de la Hoz and Chardon, 1984; Aguilera, 1988; etc.). Similarly, sections A1-OST, A2, A3 (or A3′) and A3″ (of this work) of ostariophysines were often called respectively A2 (Takahasi, 1925; Ballintijn et al., 1972; Vandewalle, 1975; Aguilera, 1988; Adriaens and Verraes, 1996; a.o.), A3 (or A3′) (Takahasi, 1925; Ballintijn et al., 1972; Vandewalle, 1975; Adriaens and Verraes, 1996; a.o.), A3″ (or A3″-s) (Takahasi, 1925; Adriaens and Verraes, 1996; a.o.) and A3″-p (Adriaens and Verraes, 1996).

To further complicate this situation, many authors who studied mainly ostariophysine fishes without an A0, e.g. siluriforms or most characiforms (McMurrich, 1884b; Stix, 1956; Alexander, 1964, 1965; Gijsen and Chardon, 1976; Howes, 1976, 1983b, 1985a; Schaefer and Lauder, 1986; a.o.) used the name A1 to designate the A1-OST of the present work (because, with the A0 absent, A1-OST is the outermost adductor mandibulae section of these fishes). Therefore, they used the names A2 and A3 to designate respectively the A2 and A3 of this work. Other authors, e.g. de la Hoz and Chardon (1984), who studied *Sternopygus macrurus*, a gymnotiform ostariophysine fish possessing a configuration of this muscle similar to that illustrated in Figure 4.12F, decided to call A1-a and A1-b respectively the A0 and A1-OST of this work and, therefore, to call (correctly, in our opinion) A2 and A3 respectively the A2 and A3 of this work.

All this confusion about the identity of the adductor mandibulae sections is thus almost exclusively due to a sole main defect: discussion of the configuration and evolution of this muscle was limited to just one particular group of fishes, without major comparisons with other groups and major considerations of the homologies and evolutionary track of its different bundles. The procedure followed here was undertaken precisely to avoid the repetition of this defect. In fact, it is now possible to discuss, in a more comprehensive and well-founded manner, the actual evolution of the catfish adductor mandibulae complex.

As explained above, in the basal adductor mandibulae configuration for ostariophysines this muscle is divided into three sections, namely the Aω, the A1-OST and the A2 (Fig. 4.12D). And what about the basal situation for catfish? Howes (1983b, 1985a), Schaefer and Lauder (1986) and Gosline (1989) considered the plesiomorphic adductor mandibulae configuration for these fishes to be that of *Diplomystes*. According to Howes (1983b, 1985a), Schaefer and Lauder (1986), Arratia (1987) and Gosline (1989), the adductor mandibulae of *Diplomystes* is an almost undifferentiated muscle. However, careful study of this muscle in *Diplomystes chilensis* (see Figs. 3.64, 3.65) revealed that it is divided into four well-distinguishable sections, namely the Aω, A1-OST, A2 and A3, the latter subdivisible into a dorsal (A3-d) and ventral part (A3-v). Despite the presence of these sections and subsections, the plesiomorphic adductor mandibulae configuration of siluriforms seems, indeed, to be that present in *Diplomystes*, as proposed by Howes (1983b, 1985a), Schaefer and Lauder (1986) and Gosline (1989). In fact, the adductor mandibulae of diplomystids presents a configuration quite similar to that observed in certain archaic Characiformes, e.g. *Hoplerythrinus* (Gijsen and Chardon, 1976: Figs. 2, 3, 6), and Gymnotiformes, e.g. *Sternopygus*, considered the most primitive gymnotiform by Fink and Fink (1981). The sole exception is that the latter possesses an adductor mandibulae section attached to the lacrimal (de la Hoz and Chardon, 1984: Figs. 14, 15).

According to Alexander (1965), the configuration of the adductor mandibulae is strongly influenced by the shape of the head. So, in fishes with a narrow cranial roof, the levator arcus palatini is situated very close to the mesial line, and the adductor mandibulae cannot extend mesially to this muscle. The inverse occurs in fishes with a large cranial roof: they present thick adductor mandibulae sections mesial to the levator arcus palatini. Thus, following Alexander's hypothesis, one can, starting from the plesiomorphic configuration of the adductor mandibulae found in *Diplomystes*, track the major morphological transformations leading to the configuration found in other catfishes and, thereby, determine the identity of the different adductor mandibulae sections of these fishes. Some major different types of configuration of the adductor mandibulae complex in siluriforms are schematically illustrated in Figure 4.13, such as *Chrysichthys* (Fig. 4.13B), *Amphilius* (Fig. 4.13C), *Clarias* (Fig. 4.13D) or *Phractura* (Fig. 4.13E). The Figure is explained on the next page.

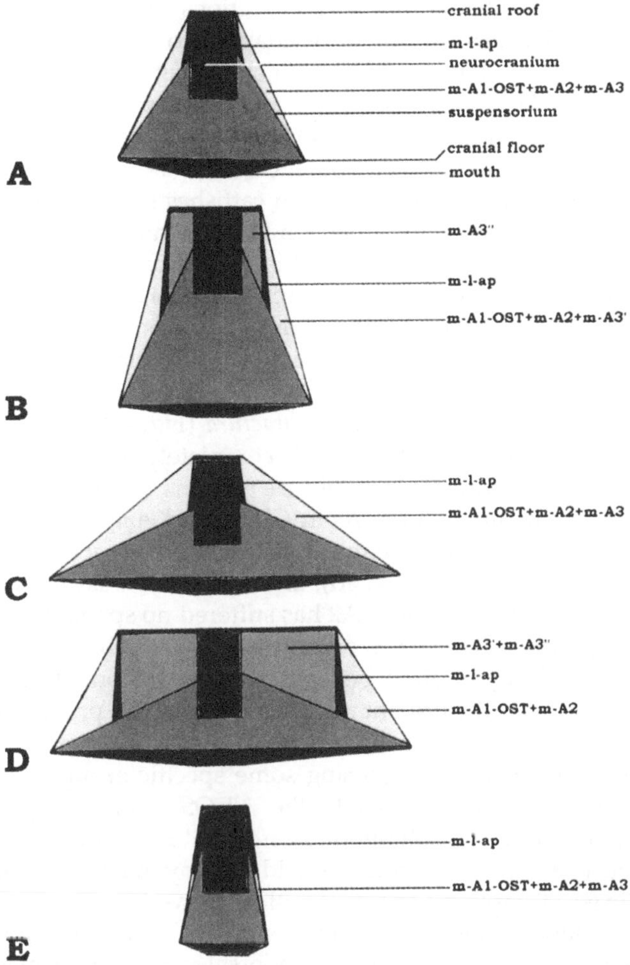

Fig. 4.13 Scheme illustrating the relation between shape of skull, levator arcus palatini and adductor mandibulae in *Diplomystes* **(A)** and some other catfishes, namely *Chrysichthys* **(B)**, *Amphilius* **(C)**, *Clarias* **(D)** and *Phractura* **(E)** [for more details, see text].

As schematised in Figure 4.13A, in *Diplomystes* the cranial roof is narrow and hence the levator arcus palatini lies near the mesial line and A3 is almost entirely lateral to this muscle (see Fig. 3.64B).

In *Chrysichthys* (Fig. 4.13B) the cranial roof is broader than that of *Diplomystes*. Thus, one part of the adductor mandibulae A3, the A3", is situated mesial to the levator arcus palatini. The other adductor mandibulae sections are rather similar to those of *Diplomystes* (see Fig. 3.55).

Amphilius (Fig. 4.12C) presents a very broad cranial floor. As in *Diplomystes*, the adductor mandibulae A3 is external to the levator arcus palatini. However, the area situated lateral to this muscle is much larger than that of *Diplomystes*, which leaves place for a much thicker A1-OST and A2. The A1-OST is strongly developed and differentiated into several sections (see char. 207). *Amphilius* possesses a very flat head, profoundly adapted to a benthic life style. The mandible is depressed dorsoventrally, which probably explains the absence of Aω, a section also lacking in some other catfishes presenting a definite benthic life style such as the remaining amphiliids, clariids, sisorids, or loricariids (see char. 220). In *Amphilius*, as in several other catfishes lacking an Aω, the tendon of A2, typically associated with the Aω in catfishes such as *Diplomystes* and *Chrysichthys*, migrated anteriorly to attach to the dentary (see character 210).

In *Clarias* (Fig. 4.13D) the cranial roof is very broad, but less so than the cranial floor. Consequently, the levator arcus palatini extends quite far to the mesial line, and the A3′ and A3″ situate mesial to this muscle (see char. 212). Aω is missing, probably for the same reason it is missing in *Amphilius* (see above).

With respect to members of genus *Phractura* (Fig. 4.13E), they present a very narrow cranial roof and floor. A3 is completely external to the levator arcus palatini. All the adductor mandibulae sections are very narrow and not one attaches to the neurocranium. As in *Clarias* and *Amphilius*, the mandible of *Phractura* is depressed dorsoventrally (although, contrary to these genera, the head of *Phractura* is not flattened), and the Aω is missing. However, in this genus the anterior insertion of A2 has suffered no spatial modification in relation to the genera where Aω is present (see char. 210).

Generally speaking, it may be said that inclusion of the adductor mandibulae complex in the present phylogenetic analysis proved surprisingly productive, revealing several important synapomorphic, and even autapomorphic features for diagnosing some specific groups. The A2 lying essentially lateral, and not mesial to the A1-OST, for example, seems to constitute an autapomorphic, homoplasy-free feature characterising schilbid catfishes (see char. 209), a group considered by many authors as non-monophyletic (see Chapters 1 and 2). Another derived feature, insertion of all the adductor mandibulae sections in a smooth, well-developed, roughly rectangular surface situated on the posterodorsal surface of the mandible, was found only in malapterurids and in no other catfishes examined (see char. 206). Similarly, differentiation of A3′-v and A3″ into two well-differentiated bundles seems to constitute respectively an autapomorphic feature of amphiliins and cranoglanidids (see chars. 214 and 218). More importantly for the main goal of this work, inclusion of the adductor mandibulae complex in the cladistic analysis revealed some interesting synapomorphic features characterising some important interfamilial clades, such as those concerning character 212, uniting the chacids, plotosids and clariids (see Fig. 4.14) or character 217 supporting the sister-group relationship between ictalurids and cranoglanidids (see Fig. 4.15). The latter character (217), associated with the connection between the adductor mandibulae complex and the primordial ligament, leads us precisely to a discussion of one of the main evolutionary events concerning the macroevolution of this complex, viz. the differentiation of the retractor tentaculi.

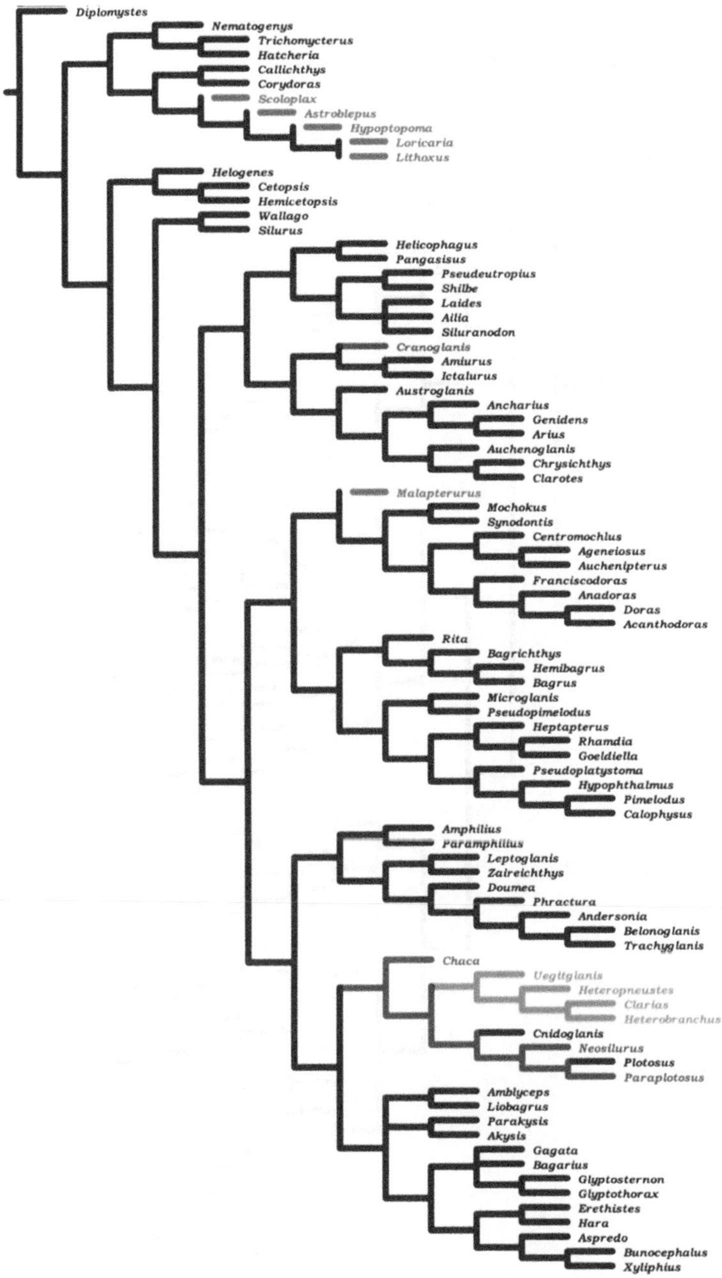

Fig. 4.14 Hypothesis of character state evolution of relation between adductor mandibulae A3′ and levator arcus palatini (char. 212, ordered): CS-0 (black)= A3′ lateral to levator arcus palatini; CS-1 (blue)= A3′ partially mesial to levator arcus palatini; CS-2 (orange)= A3′ completely mesial to levator arcus palatini; Ambiguity or Innaplicable (pink) [for more details, see text].

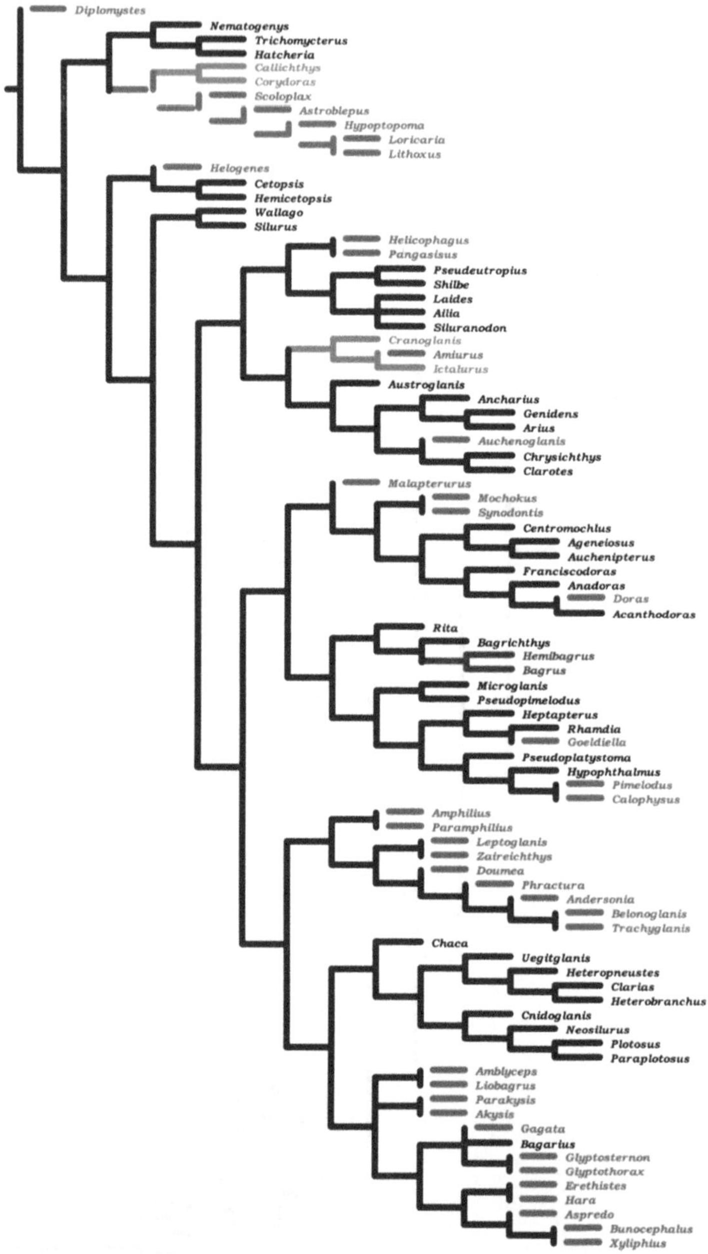

Fig. 4.15 Hypothesis of character state evolution of relation of A3″ with primordial ligament (char. 217, unordered): CS-0 (black)= A3″ not connected with primordial ligament; CS-1 (blue)= A3″ presenting marked anterior bifurcation, with part inserting on mandible and part inserting on anterior portion of this ligament; CS-2 (orange)= A3″ presenting large anterior tendon inserting on mandible and on posterior portion of this ligament; Ambiguity or Innaplicable (pink) [for more details, see text].

As mentioned above, numerous catfish possess a muscle attaching on the maxilla (see Fig. 3.6), which is usually called the retractor tentaculi since its contraction pulls the distal end of this bone backwards, thus retracting the maxillary barbel (Nawar, 1955; Alexander, 1965; Singh, 1967; Gosline, 1975; Fink and Fink, 1981; Howes, 1983ab, 1985a; Schaefer and Lauder, 1986, 1996; Schaefer, 1990; Mo, 1991; Adriaens and Verraes, 1996, 1997a, b; a.o.). The question of the origin of this muscle has been controversial. Some authors, e.g. Takahasi (1925), Edgeworth (1935) and Winterbottom (1974a), considered it to be derived from the most external adductor mandibulae section present in Siluriformes (A1-OST) (Fig. 4.12E). Thus, it would be homologous to the adductor mandibulae A0 (Fig. 4.12F) of other ostariophysine fishes since, like the latter, it inserts on the maxilla. However, other authors, e.g. McMurrich (1884b), Eaton (1948), Alexander (1965) and Howes (1983b), considered that the retractor tentaculi results from differentiation of the adductor mandibulae A3, since it does not occupy the same spatial position as the A0 of other ostariophysines: it is situated mesial to all the other adductor mandibulae sections, whereas A0 is the most lateral bundle of this muscle.

The present study strongly supports the latter hypothesis. In basal loricarioids such as nematogenyids and trichomycterids, as in diplomystids (see Fig. 3.64), there is no association between the adductor mandibulae and the maxillary nor between this muscle and the primordial ligament, and the retractor tentaculi is absent. However, in callichthyids, i.e., in the most plesiomorphic family of the clade including callichthyids + scoloplacids + loricariids + astroblepids, the *innermost* adductor mandibulae section (A3") is associated with the primordial ligament and, through this ligament, to the maxilla, as shown in Figure 4.15. And in the three remaining, more derived groups of this clade, i.e., scoloplacids, loricariids and astroblepids, there is a well-developed, well-differentiated muscle retractor tentaculi situated *mesial* to the adductor mandibulae bundles and attached to the maxilla through a strong anterior tendon. In fact, as shown in Figure 4.15, the deep association between the A3" and the primordial ligament is not exclusively a callichthyid, trait; it is also found, often with different morphological arrangements, in some of the ictalurid, cranoglanidid and bagrid catfishes examined. Such an association has also been reported in other catfishes, such as some pimelodids (e.g. *Megalonema*; Howes, 1983b). Thus, it seems clear that the retractor tentaculi results from the differentiation of the *innermost* adductor mandibulae section, i.e., the A3", attaching via the primordial ligament to the maxillary.

Howes (1983b) suggested that differentiation of the retractor tentaculi probably occurred independently in different catfish lineages. This suggestion can now be corroborated, in an explicit cladistic way, by the confrontation of the distribution of this character in the catfish taxa examined and the cladogram obtained in the phylogenetic analysis of this work. This is shown in Figure 4.16. As can be clearly seen in this Figure, the presence of a retractor tentaculi is a highly homoplasic character, with multiple independent acquisitions and secondary reversions occurring within catfish evolutionary history, as hypothesised by Howes (1983b).

Fig. 4.16 Hypothesis of character state evolution of retractor tentaculi (see char. 230): CS-0 (black)= absence of retractor tentaculi; CS-1 (blue)= presence of well-developed retractor tentaculi; Ambiguity (pink) [for more details, see text].

4.4 PALATINE-MAXILLARY SYSTEM

As explained in Section 4.1, catfish maxillary barbels are characterised by their connection to a mobile mechanism, the palatine-maxillary system, constituted by the maxilla, autopalatine and more or less specialised ligaments and muscles responsible for their movements. Some evolutionary aspects concerning the catfish palatine-maxillary system have been discussed in a few studies, e.g. Alexander (1965), Singh (1967), Gosline (1975), Howes (1983a, b), Ghiot et al. (1984) and Royero and Neville (1997), but these aspects were never confronted with an explicit cladogram on catfish higher level phylogeny.

Siluriform sister-groups, i.e., Gymnotiformes and, in a broader sense, Characiformes, totally lack maxillary barbels and present no structure prefiguring them. The barbels present in some cypriniforms are seemingly not homologous to the maxillary barbels of catfishes (see, e.g., Arratia, 1987, 1992; Arratia and Huaquin, 1995). In fact the catfish palatine-maxillary system is based on three basic siluriform synapomorphies, seemingly present since the very early stages of catfish evolutionary history, being found in all extant catfishes including diplomystids: 1) presence of the characteristic catfish maxillary barbels firmly entrenched on the distal portion of the maxillary bone; 2) uncoupling of the autopalatine from the rest of the suspensorium; 3) differentiation of the adductor arcus palatini in a separate muscle extensor tentaculi, which, in most cases, abducts the maxillary barbel (see, e.g., Regan, 1911a; Takahasi, 1925; Eaton, 1948; Alexander, 1965; Singh, 1967; Gosline, 1975; Howes 1983ab, 1985; Ghiot et al., 1984; Adriaens and Verraes, 1997b; Royero and Neville, 1997). So, unless well-preserved fossils (the maxillary barbels, and the muscles adductor arcus palatini and extensor tentaculi are soft structures of poor conservation) are discovered and described in future, a gap remains between the situation found in basal diplomystids and that found in gymnotiforms and characiforms.

The potential biological significance of the three synapomorphic innovations listed above seems evident. The maxillary barbels are useful, among others, in searching for food, obstacle detection and social behaviour, as explained in Section 4.1. The loose attachment between the autopalatine and the rest of the suspensorium, associated with the differentiation of the extensor tentaculi muscle, permits abduction of the maxilla and its associated barbel to be effected not only by depression of the lower jaw, but directly by contraction of this muscle (see Fig. 4.17A→B, and discussion below).

These three major synapomorphic features are schematised in Figure 4.18, which broadly summarises the main evolutionary transitions pertinent to the evolution of the palatine-maxillary system in catfish. The evolution of this system in siluriforms is often associated with homoplasic events. In fact, Figure 4.18 is mainly based, following an explicit phylogenetic context, on the evolution of this system within the Loricarioidea, which somewhat reflects the major lines of evolution of the palatine-maxillary system in the remaining siluriforms.

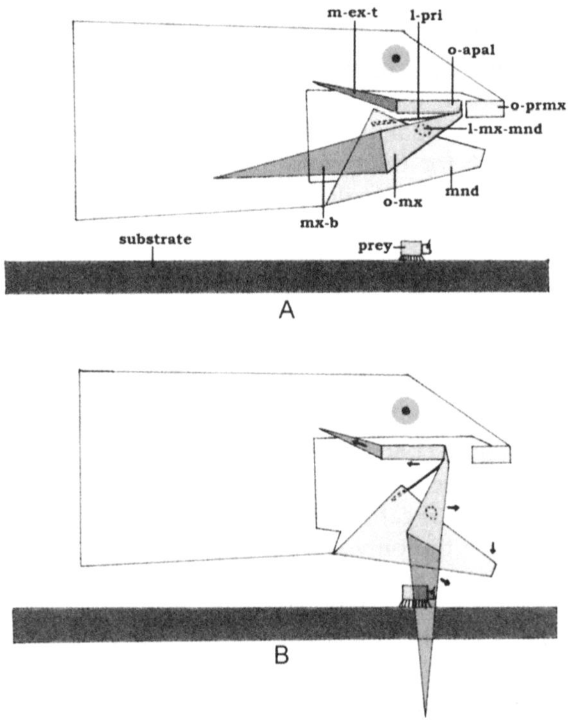

Fig. 4.17 Scheme illustrating the palatine-maxillary system of *Diplomystes chilensis*. **A)** Maxillary barbel adducted. **B)** Maxillary barbel abducted [for more details, see text].

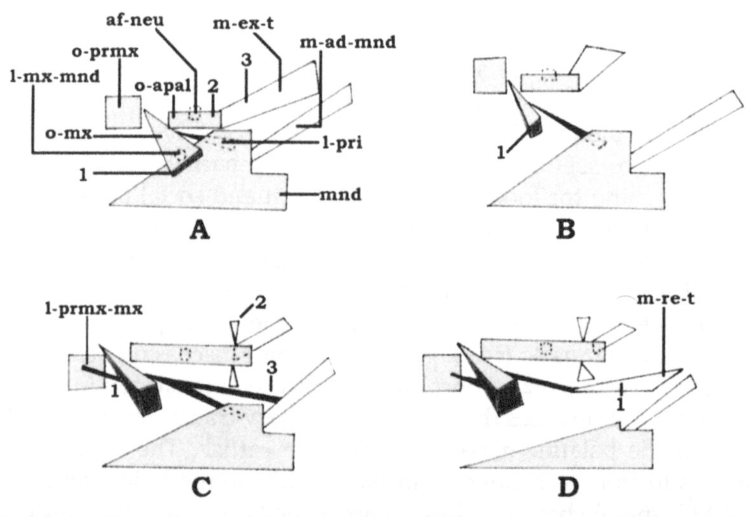

Fig. 4.18 Scheme illustrating some key evolutionary transitions concerning the evolution of the palatine-maxillary system in catfish. **A)** Plesiomorphic configuration found in *Diplomystes*. **B)** Configuration found in catfishes such as *Nematogenys*. **C)** Configuration found in catfishes such as *Callichthys*. **D)** Configuration found in catfishes such as *Scoloplax* [for numbers and further explanations, see text].

Figure 4.18A schematises the plesiomorphic configuration of the catfish palatine-maxillary system found in *Diplomystes*. The numbers represent the three major siluriform synapomorphies described above, i.e., presence of maxillary barbels (1), uncoupling of the autopalatine from the rest of the suspensorium (2) and presence of an extensor tentaculi (3). The mechanism of the palatine-maxillary system of diplomystids differs significantly from the remaining catfishes, being rather simple (Fig. 4.17). The extensor tentaculi muscle pulls the autopalatine backwards, thanks to the mobile articulation between the latter bone and the neurocranium, and thus also pulls the proximal end of the maxilla (Fig. 4.17A→B), which is linked through a cartilaginous joint to the autopalatine (see Fig. 3.64). The backward movement of the proximal part of the maxilla results in an anteroventral displacement of its posterior end because of the thick ligament linking its distal extremity to the lower jaw (Fig. 4.17A→B). The maxillary barbel, whose proximal end is firmly entrenched in the hollow distal extremity of the maxillary bone, follows the maxilla and turns downwards, and forward (Fig. 4.17A→B). Abduction of the maxillary barbel may also result from opening the mouth, thanks to the maxillo-mandibulary ligament linking the distal extremity of the maxilla to the lateral surface of the lower jaw (Fig. 4.17A→B), which rotates the maxilla in the same manner as in many teleosts with a partially freed maxilla (see, e.g., Alexander, 1965; Gosline, 1975). Closure of the mouth thus promotes the opposite movement, that is, adduction of the maxillary barbel (Fig. 4.17B→A). However, adduction of the maxillary barbel is probably not exclusively related to mouth closure. In fact, in an elegant morphofunctional study of the primordial ligament of *Parauchenipterus galeatus*, Royero and Neville (1997: 157) concluded that this ligament 'has rubber properties and because of this can store energy during abduction of the maxilla', 'the stored energy being released when the action of the extensor tentaculi muscle ceases', thus returning the maxilla and its respective barbel to the adducted position. A study of the palatine-maxillary system of some species of genus *Chrysichthys* (Diogo and Chardon, 2000c) and several other catfishes (Diogo et al., 2000a) strongly supported the important role of the primordial ligament in the palatine-maxillary system of catfishes in general.

The main morphological difference between the palatine-maxillary system of diplomystid and non-diplomystid catfishes is that in the latter the distal end of the maxilla is no longer attached to the lateral surface of the mandible by a strong ligament (Fig. 4.18B, innovation 1). This feature confers a greater freedom to the maxilla, and is present in all the non-diplomystids examined (see character 267), including nematogenyids and trichomycterids (Fig. 4.18B).

Another major morphological difference between diplomystids and the great majority of siluriforms is the presence, in the latter, of a de novo strong ligament connecting the premaxilla and the maxilla. This ligament acts, as explained by Alexander (1965) and Gosline (1975), as the fulcrum for maxillary barbel abduction/adduction movements (Fig. 4.18C, innovation 1). However, contrary to the major morphological transition described in the paragraph above, this feature is not present in all the non-diplomystids examined, being absent in nematogenyids and trichomycterids (see Fig. 4.18B). The taxonomic distribution of this character (char. 274) is shown in Figure 4.19. As can be seen, evolution of this character in the basal catfish groups is ambiguous. This is mainly due to the difficulty in discerning the state of this character in cetopsids, as explained in Section 3.1. However, despite this difficulty, the consistent presence of the premaxillo-maxillare ligament in the clade including non-diplomystid, non-loricarioid and non-cetopsid catfishes is not disputable. And, concerning these three basal groups, only two most parsimonious scenarios are possible: A) the ligament was acquired in the node leading to all non-diplomystid catfishes and was subsequently reverted in the node leading to nematogenyids + trichomycterids; or B) absence of the ligament in the nematogenyids + trichomycterids reflects a true plesiomorphic state, with the ligament being independently acquired in the node leading to callichthyids + scoloplacids + astroblepids + loricariids and the remaining catfishes. Both hypotheses are equally parsimonious in theory and thus, as emphasised throughout this text, the author is especially reluctant to choose one or the other based on a somewhat 'ad hoc' evolutionary hypothesis. However, given the major functional significance of this ligament within the evolution of the catfish palatine-maxillary system, the second option would seem, at least at first sight, more reasonable. Nonetheless, this hypothesis must be carefully tested with a more conclusive observation of this character in Cetopsidae.

The two major morphological differences described above between diplomystids and the majority of catfishes (Fig. 4.18B: 1; Fig. 4.18C: 1) confer a rather different, somewhat more complex palatine-maxillary system on the latter, as explained by Gosline (1975).

Two main abduction mechanisms of the maxillary barbels can be recognised in non-diplomystid catfishes, a *'rocking type'* and a *'sliding type'* (Gosline, 1975), with the mechanisms found in those catfishes significantly different to the abduction mechanism described above for diplomystids.

A *rocking palatine-maxillary system* is present in most non-diplomystids (Fig. 4.20). In this type of system, contraction of the extensor tentaculi pulls the posterior end of the autopalatine posteromesially, but the firm articulation between the autopalatine and the neurocranium prevents a longitudinal displacement of the autopalatine. Therefore, the back end of the autopalatine moves essentially medially (Fig. 4.21A→B). The anterior end of the autopalatine and the proximal tip of the maxilla associated with it therefore move essentially laterally. This, by the resistance of the premaxillo-maxillary ligament usually present in most non-diplomystid catfishes (see above), provokes abduction of the maxilla (Fig. 4.21A→B) and associated barbel.

A true *sliding palatine-maxillary system* was only present in the bagrid, pimelodin and heptapterin siluroids examined, as shown in Figure 4.20. In

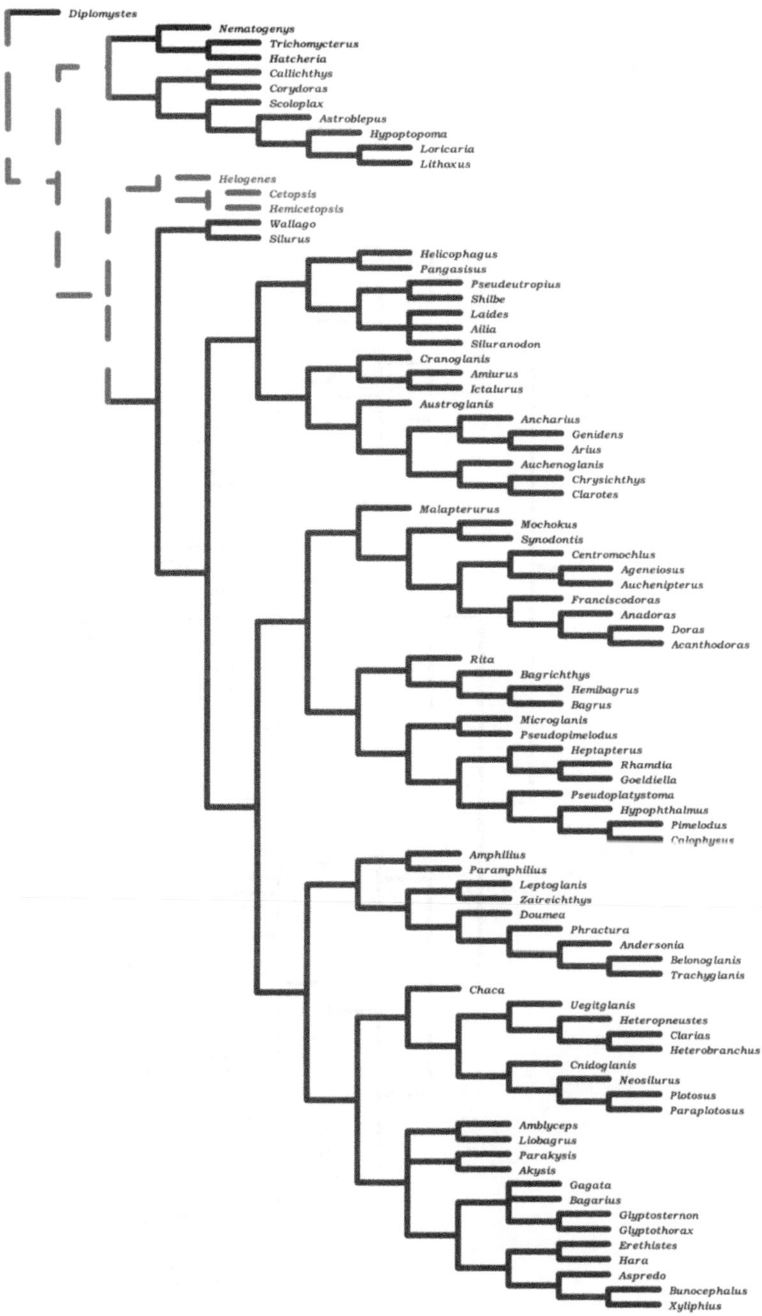

Fig. 4.19 Hypothesis of character state evolution of praemaxillo-maxillary ligament (see char. 274): CS-0 (black)= absence of praemaxillo-maxillary ligament; CS-1 (blue)= presence of well-developed praemaxillo-maxillary ligament; Ambiguity (pink) [for more details, see text].

Fig. 4.20 Hypothesis of character state evolution of rocking/sliding palatine maxillary system (see char. 294): CS-0 (black)= rocking palatine-maxillary system; CS-1 (blue)= sliding palatine-maxillary system; Ambiguity or Innaplicable (pink) [for more details, see text].

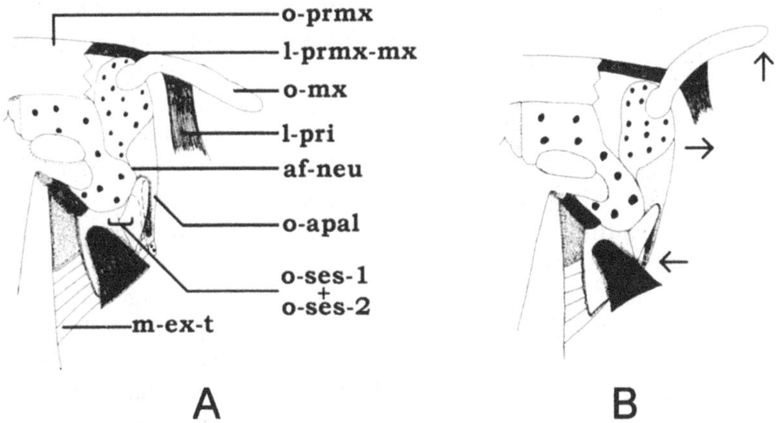

Fig. 4.21 Ventral view of palatine-maxillary system of *Chrysichthys nigrodigitatus*. **A)** Maxillary barbel is adducted. **B)** Maxillary barbel abducted [for more details, see text].

these fishes articulation between the autopalatine is configured in such a way that it allows a wide posterior movement of the autopalatine when the extensor tentaculi contracts (Fig. 4.22A). The proximal tip of the maxilla, associated with the anterior end of the autopalatine, thus moves essentially posteriorly and by the resistance of the premaxillo-maxillary ligament, provokes abduction of the maxilla (Fig. 4.22A) and associated barbel. An interesting outcome of the analysis of this Figure is that it indicates that a true sliding palatine-maxillary system probably evolved only once or twice within recent catfishes. In fact, it could either have evolved in pimelodids + bagrids and subsequently been transformed in the pseudopimelodins or have evolved independently in pimelodins + heptapterins and in the node leading to bagrids (including, or not, genus *Rita*: the situation for *Rita* is not known since it was not possible to discern, in the specimens of this genus, which type of maxillary system prevails, see description of char. 294 in Section 3.1). A homoplasic secondary change to a rocking type in pseudopimelodins would not be too unlikely. The difference between a sliding and a rocking type is relatively

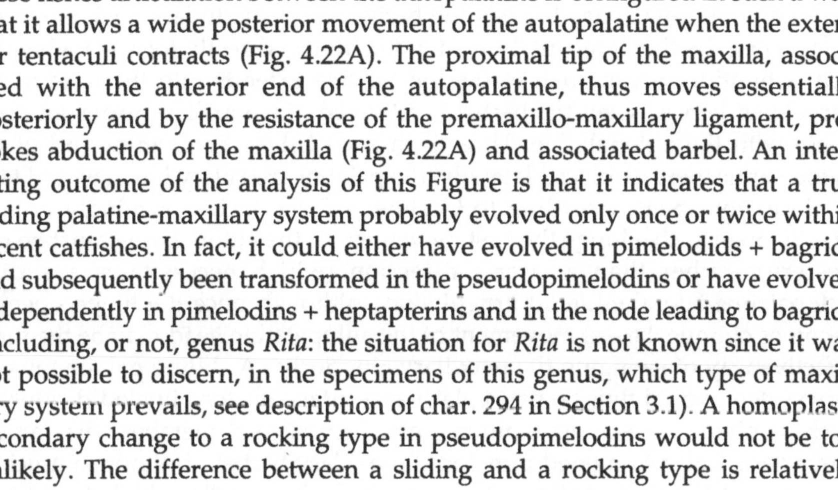

Fig. 4.22 Scheme illustrating palatine-maxillary system of *Bagrus docmak*. **A)** Dorsal view of anterior region of cranium showing abduction of maxillary barbel (on left side of illustration). **B)** Frontal view of cranium showing elevation of maxillary barbel (right side of illustration) [for more details, see text].

small from a morphological point of view, being almost exclusively restricted to a slight change in configuration of the articulatory surface of the neurocranium for the autopalatine.

Moreover, contrary to the example above respecting the development of a *de novo* premaxillo-maxillary system (char. 274), transition from a rocking to a sliding palatine-maxillary mechanism does not seem associated with a major general key innovation, but instead with a rather punctual change related to a particular specific situation. In fact, in a sliding palatine-maxillary system, articulation between the autopalatine and the skull enables a significant translation of the autopalatine in an anteroposterior direction (Fig. 4.22A). So, during the posterior movement of the autopalatine, the proximal tip of the maxilla is retracted and, through the ligamentous connection between the maxilla and the premaxilla, provokes abduction of the maxilla (Fig. 4.22A) and its associated barbel. According to Gosline (1975), this type of palatine-maxillary system would be particularly advantageous for catfishes with markedly long maxillary barbels, as observed in the majority of pimelodids and bagrids (see, e.g., Burgess, 1989). As noted by Gosline (1975), in catfishes with markedly long maxillary barbels the barbels are usually laid back along the sides when the fish is at rest, but extended almost directly forward when the fish is hunting. So, in such forms the maxilla extends into the base of the barbel and hence, when the barbel is retracted, the maxillary forms an acute angle with the autopalatine. Thus, if in such fishes the anterior end of the autopalatine were displaced laterally, the palatine-maxillary hinge would tend to close, not open. According to Gosline (1975: 11), this seemingly implies that 'to open such a hinge the first movement of the autopalatine must be in a posterior rather than a lateral direction (which necessitates a sliding autopalatine-lateral ethmoid articulation), although once the hinge opening becomes oblique, lateral movement of the autopalatine head may be the more effective way of swinging the maxillary anteriorly'. A sliding palatine-maxillary system would thus be particularly advantageous for pimelodid catfishes such as pimelodins or heptapterins, of which several members display the characteristically elongated pimelodid maxillary barbels, but not so much in members of the Pseudopimelodinae in which the maxillary barbels are usually relatively short (see, e.g., Burgess, 1989).

With respect to adduction of the maxillary barbel in non-diplomystid catfishes, this can be realised in three different major ways. The simplest and most direct is through contraction of a retractor tentaculi muscle, derived from an inner bundle of the adductor mandibulae, attaching via the primordial ligament directly on the maxilla (see Fig. 4.18B→D) (see above). However, although differentiation of the retractor tentaculi was, as explained in Section 4.3, a major key-innovation homoplasically acquired in numerous catfish taxa, the plesiomorphic condition for siluroids clearly appears to be that in which this muscle is absent (see Fig. 4.16). Therefore, there should be alternative mechanisms to promote adduction of the maxillary barbel in those siluroids lacking a retractor tentaculi. One of these mechanisms is the second elastic mechanism described above for Diplomystidae, in which adduction of the maxillary barbel is associated with the action of the primordial ligament. Another adducting mechanism is present in those siluroids with a sliding palatine-maxillary system, such as pimelodins, heptapterins and bagrids (including, or not, *Rita*: see above). In these catfishes the anterior portion of the extensor tentaculi muscle is differentiated and configured in such a way that, during its contraction, the palatine is shifted anteriorly and, consequently, the maxilla and maxillary barbel are adducted (e.g., Alexander, 1965; Ghiot, 1978; Ghiot et al., 1984).

Apart from just extension and retraction of the maxillary barbels, differentiation of the muscle extensor tentaculi in several different bundles in some siluroids allows the elevation and depression of these barbels (e.g., Alexander, 1965; Gosline, 1975; Ghiot, 1978; Ghiot et al., 1984). Elevation results from the rotation-elevation of the distal end of the maxilla through a similar rotation-elevation of the autopalatine initiated by contraction of a bundle of the extensor tentaculi inserted on the posterodorsal surface of the autopalatine (see Fig. 4.22: ex-t-4). Contraction of an antagonistic bundle of the extensor tentaculi, inserted on the posteroventral surface of the autopalatine, effects the opposite movement, viz. depression of the maxillary barbel.

Analysis of the evolution of the configuration and differentiation of the muscle extensor tentaculi within Siluriformes is interesting. In fact, tracing out the main different types of extensor tentaculi described in character 226 in the cladogram obtained in the present analysis reveals a somewhat homoplasic and highly complex, but essentially coherent, evolutionary scheme (see Fig. 4.23). Analysis of Figure 4.23 suggests that differentiation of the extensor tentaculi in different bundles only arrived, in a consistent way, in the node leading to malapterurids + mochokids + doradids + auchenipterids + pimelodids + bagrids + amphiliids + chacids + plotosids + clariids + sisoroids. According to the results of the present work, this node is characterised by a differentiation of the extensor tentaculi in various bundles, in which the most posterior ones insert on the posterodorsal and posteroventral extremities of the autopalatine (see Fig. 4.23).

Apart from this clade, the muscle extensor tentaculi seemingly only differentiated into bundles three times, and always in relatively small clades (Fig. 4.23).

Once was in the clade callichthyids + scoloplacids + loricariids + astroblepids, with a transition to an extensor tentaculi differentiated into two elongated ventral bundles attaching on the postero-ventrolateral surface of the autopalatine and a dorsal bundle essentially oriented dorsoventrally and attaching on the posterodorsal surface of this bone (char. 226, CS–5). A subsequent differentiation into two completely separated bundles attached on the lateral and the mesial surfaces of the autopalatine occurred in the loricariids (char. 226, CS–6).

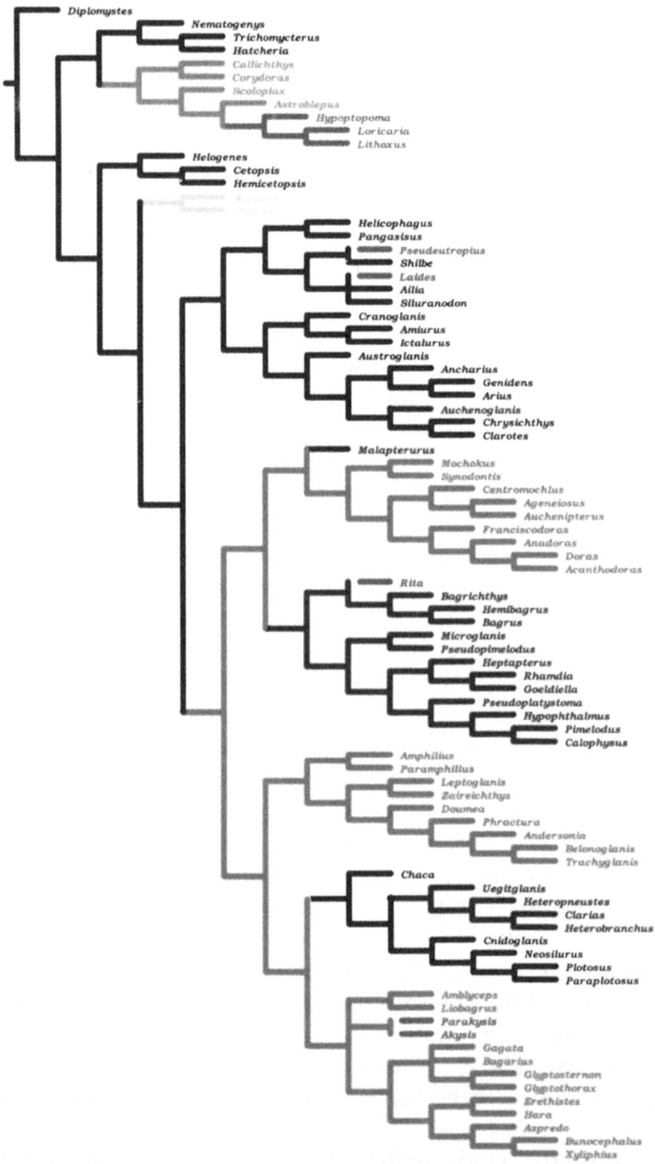

Fig. 4.23 Hypothesis of character state evolution of extensor tentaculi (char. 226, unordered): CS-0 (black)= single bundle; CS-1 (blue)= bundles promoting adduction, abduction, dorsal rotation and ventral rotation of maxillary; CS-2 (orange)= posterior bundles inserted respectively on posterodorsal and posteroventral extremities of autopalatine; CS-3 (brown)= thin bundle inserting on posterior margin of autopalatine and perpendicular bundle on posteromesial surface of this bone; CS-4 (yellow)= bundles attaching on posterior and posterolateral margin of autopalatine; CS-5 (grey)= ventral bundles attaching on posteroventrolateral surface of autopalatine and dorsal bundle on posterodorsal surface of this bone; CS-6 (green)= two completely separated bundles of the muscle extensor tentaculi; Ambiguity (pink) [for details, see text].

The second was in silurids, with a transition to a peculiarly shaped extensor tentaculi that is markedly subdivided into two well-developed subdivisions attaching respectively on the posterior and the posterolateral margins of the autopalatine (char. 226, CS–4).

Lastly, in *Auchenoglanis* the extensor tentaculi differentiated into a thin bundle markedly extended posteriorly and inserted on the posterior margin of the autopalatine, and an anterior bundle almost perpendicular to the first one attaching on the posteromesial surface of the autopalatine (char. 226, CS–3).

The independent occurrence of these three events is not really surprising since they concern the transition from an undifferentiated extensor tentaculi to three rather different configurations of this muscle. Interestingly, one of the three configurations, namely that present in *Auchenoglanis*, is somewhat similar to that found in *Malapterurus* (see Fig. 4.23).

Differentiation of the muscle extensor tentaculi in the catfish groups mentioned in the latter paragraphs seemingly allowed this muscle to perform, in some members of those groups, apart from just abduction of the maxillary barbels, their elevation and depression, as explained above (see Fig. 4.22B). With respect to the differentiation of the extensor tentaculi in an anterior 'adducting' bundle also allowing an adduction of these barbels, this is only found in a very restricted number of catfishes, namely in bagrids, pimelodids and clariids (char. 226, CS–1). The restricted distribution of this feature among catfishes could be explained by the fact that such an adduction of the maxillary barbels by the direct action of the extensor tentaculi is seemingly more efficient in catfishes with a true sliding palatine-maxillary system, like most bagrids and pimelodids, as explained above. Such a mechanism could perhaps also be somewhat effective in catfishes with a rocking palatine-maxillary system, as it is also found in clariids and in the pseudopimelodin pimelodids, groups with an essentially rocking autopalatine. However, as pointed out by Diogo et al. (2000), it should be noted that the 'adducting' extensor tentaculi bundle of clariids is clearly much smaller and poorly developed than that found in most bagrids and pimelodids with a true sliding palatine maxillary-system.

4.5 SUSPENSORIUM AND ASSOCIATED STRUCTURES

Some aspects of the macroevolution of the suspensorium within Siluriformes will now be discussed. But given the controversies concerning the catfish suspensorium, it is fundamental to provide, before passing to an analysis of the evolution of this structure within the order, a fulsome account of the homologies and evolutionary track of these elements within the Teleostei and, in particular, within the Ostariophysi. In fact, the present study provided some new insights into this subject.

The controversy surrounding the catfish suspensorium is mainly due to a major synapomorphy of these fishes, which is precisely the division of the suspensorium into a rostral (autopalatine alone) and caudal (the other skeletal elements) units (Fink and Fink, 1981; Arratia and Schultze, 1991; Arratia, 1992). This seemingly frees the palatine-maxillary system from the more posterior elements, thereby allowing ample movements of the maxillary barbels (see, e.g., Alexander, 1965; Gosline, 1975). The division is ontogenetically present from the first appearance of the splanchnocranium cartilages (Kindred, 1919; Arratia, 1987, 1990, 1992; Howes and Teugels, 1989; Surlemont and Vandewalle, 1990; Kobayakawa, 1992; Vandewalle et al., 1993, 1995a, 1997; Adriaens and Verraes, 1998; a.o.) and is probably required functionally by the early respiratory pattern of the larva (Vandewalle et al., 1985).

The division results in the lack of a solid anterior support for the large posterior portion of the suspensorium and the apparent need for compensatory mechanisms. These latter probably correlate with numerous synapomorphies in the suspensorium of different catfish lineages involving the presence of several ligaments and small bones between the fore-end of the pars quadrata, autopalatine, and ethmoidal region, as discussed below.

There are usually three large bones (not including the preopercular) and some small bones in the pars quadrata, instead of five or six large elements present in that region in other teleosts, and determination of their identity, as mentioned above, has been a matter of major controversy (see, e.g, Regan, 1911; Starks, 1926; Fink and Fink, 1981; Howes, 1983b, 1985a; Arratia, 1987, 1990, 1992; Howes and Teugels, 1989). In several papers, including Arratia's 1992 well-documented review, the three large bones of the pars quadrata were labelled the 'hyomandibula', 'quadrate', and 'metapterygoid', with the symplectic considered to be totally absent and the smaller anterior bones interpreted either as the ectopterygoid and/or as entopterygoid or sesamoid bones. [Note: To render this discussion easier to grasp, Arratia's 1992 interpretation of the evolutionary homologies concerning catfish suspensorial elements will be signalled from here on by quotation marks.] However, many authors have alternative interpretations, of which three are particularly notable. Howes (1985a) suggested that the catfish hyomandibula is the result of the fusion of the hyomandibula sensu stricto and the metapterygoid, with the so-called metapterygoid being the entopterygoid, and the small bones representing sesamoid ossifications. Howes and Teugels (1989) considered that the metapterygoid of other authors is homologous to part of the metapterygoid fused with an ecto- and an entopterygoid, with the smaller anterior bones dubbed sesamoids and/or fragments of the dermal pterygoids. Howes (1983b) hypothesised that, in catfish, the hyomandibula of others authors corresponds to the hyomandibula and metapterygoid of other teleosts, and that the so-called metapterygoid is the result of the fusion of the ecto- and entopterygoid sensu stricto, with the small bones sesamoid ossifications.

Howes' 1983b hypothesis is basically similar to that presented by Diogo et al. (2001a), which is the basis for the nomenclature followed in this work and was established by anatomical dissections, morphological descriptions in the literature, available developmental and palaeontological data, functional morphology and comparisons with other teleosts and specially ostariophysans.

A comparison between siluriforms and gymnotiforms, i.e, catfish sister-group (see Introduction), in particular, proved very illuminating, with interesting evolutionary implications. Components of the suspensorium in gymnotiforms are readily homologised with those of other teleosts (e.g., Chardon and de la Hoz, 1973, 1974, 1977; de la Hoz, 1974; Mago-Leccia, 1978; Fink and Fink, 1981; de la Hoz and Chardon, 1984; Arratia and Schultze, 1991; Arratia, 1992), exception for the so-called 'entopterygoid', which may represent the entopterygoid or the ectopterygoid or both (de la Hoz, 1974) (see below).

The suspensorium of one of the most archaic gymnotiforms (Chardon and de la Hoz, 1974, 1977; de la Hoz, 1974; Mago-Leccia, 1978; Fink and Fink, 1981; de la Hoz and Chardon, 1984; but see Gayet et al., 1994; Albert and Fink, 1996; Albert and Campoz da Paz, 1998, for a different view), *Sternopygus macrurus*, and the South-American trichomycterid catfish *Trichomycterus areolatum*, are very similar (Fig. 4.24), with: 1) a cartilaginous band (A) between two bones; 2) an inverted Y-shaped formation (B) in the middle of the suspensorium; and 3) only one bone (C) situated anterodorsal to this formation and extending up to half the length of the palatine. However, three differences are noteworthy.

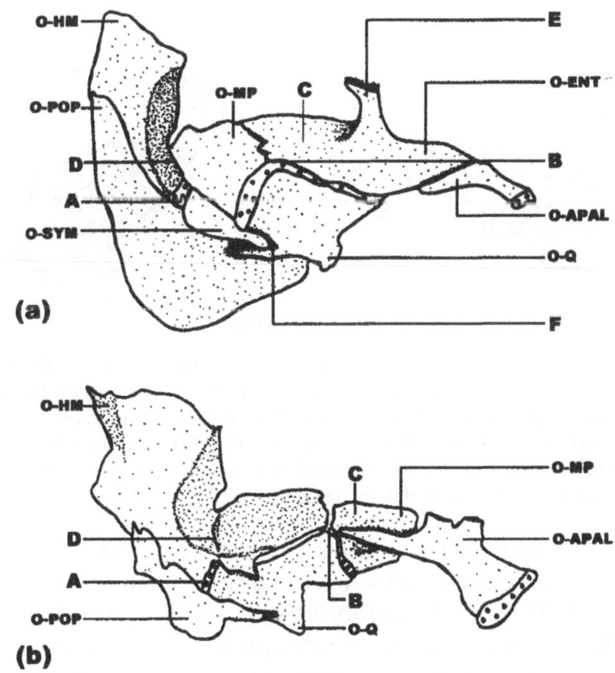

Fig. 4.24 Lateral view of the suspensorium of *Sternopygus macrurus* **(a)** (modified from De la Hoz and Chardon, 1984) and *Trichomycterus areolatum* **(b)** (modified from Arratia, 1990). The nomenclature used here follows that used in the original illustrations [for more details, see text].

Firstly, the A cartilage of *Sternopygus* (Fig. 4.24a) is prolonged by a clear separation (D) between two bones (hyomandibula and entopterygoid) (de la Hoz and Chardon, 1984). In *Trichomycterus*, there is only a partial suture between the bones at the same level, and this is only observed in some species of this genus, such as *Trichomycterus roigi* (see, e.g., Fig. 4.25c) (Arratia and Chang 1975; Arratia et al., 1978; Arratia and MenuMarque, 1981, 1984; Arratia, 1987, 1990, 1992). Complete sutures are found, however, at the same location in malapterurids (see, e.g., Fig. 4.25d) and some diplomystids (see, e.g., Fig. 4.25a, b).

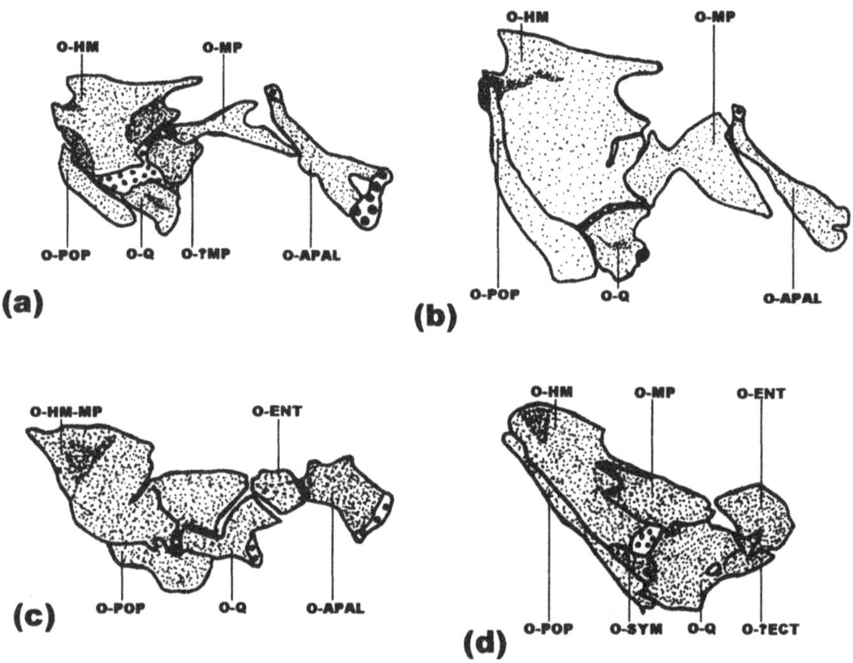

Fig. 4.25 Suspensorium of *Diplomystes camposensis* **(a)**, young specimen of about 28 mm standard length (modified from Arratia, 1987), *Diplomystes camposensis* **(b)**, large specimen (modified from Arratia, 1987), *Trichomycterus roigi* **(c)** (modified from Arratia and Menumarque, 1984) and *Malapterurus electricus* **(d)** (modified from Howes, 1985a, autopalatine not shown). The nomenclature used here follows that used in the original illustrations [for more details, see text].

Secondly, in *Sternopygus* the suspensorium is linked to the neurocranium by an ossified ligament (E), which terminates dorsally in very short fibres (Fig. 4.24a) (de la Hoz and Chardon, 1984), while *Trichomycterus* (Fig. 4.24b), like many other catfishes, has these bones joined by an non-ossified ligament. However, these ligaments seem to be homologous, and the almost total ossification of the ligament in some gymnotiforms is an unusual situation in this group (see de la Hoz, 1974).

Thirdly, *Sternopygus*, like other Gymnotiformes (Chardon and de la Hoz, 1974, 1977; de la Hoz, 1974; Mago-Leccia, 1978; Fink and Fink, 1981; de la Hoz and Chardon, 1984), has a symplectic completely separate from the quadrate (F) (Fig. 4.24a). In *Trichomycterus* (Fig. 4.24b), a single ossification occupies the position of these two bones, with no separation visible at level F. This separation is also lacking in all other catfishes except, according to some authors, e.g. Howes (1985a), some catfishes of genus *Malapterurus* (Fig. 4.25d) (see below).

The above anatomical comparison thus strongly suggests the following evolutionary scenario: 1) the catfish element named 'hyomandibula' by Arratia (see Fig. 4.24b) represents the *hyomandibula* + *metapterygoid* of other teleosts (see Fig. 4.24a); 2) the catfish element named 'quadrate' by Arratia (see Fig. 4.24b) represents the *quadrate* + *symplectic* of other teleosts (see Fig. 4.24a); 3) the catfish element named 'metapterygoid' by Arratia (Fig. 4.24b) corresponds to the 'entopterygoid' of gymnotiforms (Fig. 4.24a). However, as mentioned before, the identity of the 'entopterygoid' of gymnotiforms is somewhat uncertain. Regan (1911) called this bone the 'mesopterygoid' (='entopterygoid') without providing evidence to support its homology with the entopterygoid of other teleosts. This nomenclature was followed by other authors (Chardon and de la Hoz, 1973, 1974, 1977; de la Hoz, 1974; Mago-Leccia, 1978; Fink and Fink, 1981; Arratia and Schultze, 1991; Arratia, 1992; de La Hoz and Chardon, 1984; etc.), but some (Chardon and de la Hoz 1973; de la Hoz, 1974) pointed out that this bone has features typical of the entopterygoid (e.g. ligamentous connection with the neurocranium; relation with the adductor arcus palatini), ectopterygoid (e.g. anterodorsal relation with the palatine) and entopterygoid + ectopterygoid (spatial position) of other ostariophysine fishes. In fact, several arguments seem to suggest, precisely, an evolutionary scenario in which the catfish element named 'metapterygoid' by Arratia (see Fig. 4.24b) corresponds to the *entopterygoid* + *ectopterygoid* of other teleosts and that the small anterior bones present in most catfish are *sesamoid ossifications*, as suggested by Howes (1983b).

The metapterygoid of teleosts results from the ossification of the posterodorsal part of the palatoquadrate, thus dorsally and somewhat posteriorly relative to the quadrate (see, e.g., Starks, 1926; De Beer, 1937; Bertmar, 1959; Daget, 1964; Hunt Von Herbing et al., 1996; Verraes, 1977). It remains in the same position in almost all adults (see, for example, Starks, 1926; Gregory, 1933; De Beer, 1937; Weitzman, 1962; Daget, 1964; Osse, 1969; Roberts, 1969; de la Hoz, 1974; Taverne, 1974; Vandewalle, 1975; Gijsen and Chardon, 1976; Mago-Leccia, 1978; de la Hoz and Aldunate, 1994), except for a very few cases, as, e.g., some clupeids such as *Engraulis encrasicholus* (see Ridewood, 1904: Fig. 135A). In the course of postembryonic development the palatoquadrate fuses with the hyosymplectic, so that the ossifying metapterygoid contacts the future hyomandibula from which it remains separated by a cartilaginous strip, a suture, or a combination of both (Daget, 1964). So, given the position of the metapterygoid in these related groups and primitive teleosts, the true metapterygoid of catfish seems to correspond to the anterior part of the element named 'hyomandibula' by Arratia (see Fig. 4.24b). The fusion between these two bones may be a consequence of the fact that in catfish the pars quadrata and the hyosymplectic are fused from the first appearance of the chondrocranium cartilages (see, e.g., Kindred, 1919; Arratia, 1987, 1990, 1992; Howes and Teugels, 1989; Surlemont and Vandewalle, 1990; Kobayakawa, 1992; Vandewalle et al., 1993, 1995a, 1997; Adriaens and Verraes, 1998; a.o.).

The fact that this true metapterygoid is united by cartilage to the true entopterygoid + ectopterygoid (see below) could lead to the erroneous interpretation that both bones are enchondral and therefore the entopterygoid + ectopterygoid cannot be a dermal compound, as hypothesised above. In fact the cartilage is, very likely, the remnant of the pterygoid process of the pars quadrata. Such a cartilage remains present in some adult ostariophysan fishes, for example *Chanos chanos* (Gonorynchiformes), *Opsariichthys uncirostris* (Cypriniformes), *Xenocharax spilurus* (Characiformes) and *Sternopygus macrurus* (Gymnotiformes) (see, e.g., Fink and Fink, 1981) as well as in some other adult teleosts (see, e.g., Daget, 1964).

One second argument supporting the evolutionary scenario given above is that in some juvenile *Diplomystes camposensis* a broad, completely independent bone lying in the same position and presenting the same configuration as the metapterygoid of other teleosts is present between the catfish elements named 'hyomandibula', 'metapterygoid' and 'quadrate' by Arratia (see Fig. 4.25a: compare with Fig. 4.24a). Arratia (1987) stated that this bone 'does not seem to result from fracture' of one of these three major elements. In adult specimens, it may be completely independent but may also be partially fused with the 'hyomandibula' (Arratia, 1987). In this latter case, only a partial suture is present between these two bones (see Fig. 4.24b). Similar sutures also occur in some trichomycterids (see Fig. 4.25c), plotosids (see Fig. 3.112), and malapterurids (see Fig. 4.25c). The fact that these partial or complete sutures appear in a position very similar to those between the hyomandibula and metapterygoid of other teleosts supports the hypothesis that the catfish element named 'hyomandibula' by Arratia evolutionarily corresponds to the hyomandibula + metapterygoid of other teleosts (Fig. 4.25: compare with Fig. 4.24a).

One third major argument supporting this evolutionary scenario is that in some claroteids (see, e.g., Skelton et al., 1984; Mo 1991), some pimelodids (see, e.g., Arratia, 1992) and some schilbeids (see, e.g., Tilak, 1961) the element named 'metapterygoid' by Arratia in Fig. 4.24b presents a series of ventral teeth (see, e.g., Arratia, 1992: Fig. 36A). The teleost metapterygoid is enchondral and hence does not bear dermal tooth-plates (Jollie, 1986). Arratia (1992) suggested therefore that this toothed catfish element is 'the metapterygoid fused with a dermal tooth-plate'. The phylogenetic results of the present work seriously question this latter suggestion. In fact, the three families mentioned above, including members with a toothed suspensorial element, Claroteidae, Pimelodidae and Schilbidae, do not appear grouped in a monophyletic clade on the cladogram obtained in the present work (see Fig. 3.234). Arratia's suggestion would thus require that the supposed *'metapterygoid' + dermal tooth-plate* compound arose at least three different times among catfishes, hardly likely since the development of such a compound is de facto extremely unusual and rare in teleosts (see, e.g. Taverne, 1974). Such a homoplasic scenario would not be so surprising, however, if the toothed suspensorial element of those siluriforms corresponded to a dermal bone and not to a true enchondral metapterygoid (toothed ectopterygoids and/or entopterygoids are widely distributed and frequent in teleosts: see Taverne, 1974).

The evolutionary homology between the posteriormost suspensorial component of catfishes and the hyomandibula + metapterygoid has been defended, aside from Howes (1983b), by other catfish researchers. Starks (1926), for example, in a study dedicated to the ethmoideal region of several fishes, suggested that in siluriforms 'the metapterygoid, if represented at all, may be incorporated with the pterygoid, but may well be incorporated with the hyomandibula'. In his extensive work on the anatomy and phylogeny of catfish, Mo (1991) pointed out that 'comparing the "hyomandibula" of siluroids with that of non-siluroid fishes, it is very likely that a large portion of the metapterygoid has joined to the hyomandibula at its lower dorsomedial margin in siluroids'. Hoedeman (1960), in a work concerning development of the skull of some callichthyids, suggested that in catfish the 'hyomandibula is ontogenetically fused to the metapterygoid' and that the so-called 'metapterygoid' is a dermal bone. In another developmental study, one concerning the suspensorium of *Clarias gariepinus*, Vandewalle et al. (1993: Fig. 2) interpreted the so-called 'metapterygoid' (Figs. 8, 9) as a dermal bone since 'its ossification seems external to and independent of the processus pterygoquadrato' (Vandewalle et al., 1997). The dermal origin of this suspensorial element in *Clarias gariepinus* was also suggested by other authors, such as Poll (1942).

The discussion above thus indicates that the metapterygoid and hyomandibula of other teleosts are seemingly fused in catfishes, and that the anteriormost of the three major suspensorial siluriform elements is, in fact, a dermal bone. Several arguments (see above) support this latter dermal bone as the evolutionary result of the fusion of the entopterygoid + ectopterygoid of other teleosts. As noted by authors such as Alexander (1965) and Howes (1983b), the catfish element named 'metapterygoid' by Arratia (see Fig. 4.24b) occupies the position of the ectopterygoid and entopterygoid of other teleosts. This similarity is not restricted to the spatial position, but also extends to both the shape of the bone and its relations with other cranial components. This bone presents two more or less developed anterolateral arms in several catfishes examined in this work, such as amphiliids, akysids, cranoglanidids, schilbids, pangasiids, ariids, silurids, pimelodids and auchenipterids (see, e.g., Figs. 3.6, 3.55, 3.96, 3.100). Such a configuration is notably evident in the basal diplomystids, and this since a very early developmental stage (see Fig. 4.25a, b). This configuration is similar to that of the entopterygoid + ectopterygoid of some characiforms (see, e.g., Roberts, 1969: Fig. 18; Gijsen and Chardon, 1976: Fig. 5; Fink and Fink, 1981: Fig. 10A; Miquelarena and Arámburu, 1983: Fig. 6), cypriniforms (see, e.g., Vandewalle, 1975: Figs. 2, 12; Taverne and De Vos, 1997: Fig. 6), gonorynchiforms (see, e.g., Arratia, 1992: Fig. 4D) and some 'fossil Ostariophysi' such as *Lusitanichthys characiformis* (Gayet, 1985: Figs 17, 20) or *Ramallichthys orientalis* (Gayet, 1982: Fig. 10). Moreover, the anteromedial and the anterolateral extremities of the catfish element named 'metapterygoid' by Arratia (see Fig. 4.25a, b) have the same anatomical relations as respectively the entopterygoid and the ectopterygoid of other telosts: the anteromedial margin is typically connected by a ligament to the neurocranium and the anterolateral tip is characteristically situated ventral to the posterior end of the autopalatine (see, e.g., Daget, 1964). Therefore, the anteromedial and anterolateral margins of this catfish element seem to correspond respectively to the anterior tips of the entopterygoid and ectopterygoid of other teleosts and hence this element seems in fact to be an ento-ectopterygoid compound.

Another major argument supporting this evolutionary scenario is the presence, in several catfish groups, either examined in the present work or described in the literature, of an additional tooth-plate (some catfish have more than one) between the catfish element named 'metapterygoid' by Arratia and the ethmoidal region. The identity and evolutionary signification of these tooth-plates have been subject to controversy. Some authors (Jayaram 1966, 1968, 1971; Skelton, 1981; Skelton et al., 1984; Arratia, 1987, 1992; a.o.) suggest that such tooth-plates are associated with the autopalatine. Others (Eigenmann, 1890; Starks, 1926; Gosline, 1975; Fink and Fink, 1981) interpret them as structures associated with the prevomer and to the anterolateral margin of the 'metapterygoid' sensu Arratia. In catfishes such as *Chrysichthys nigrodigitatus* the tooth-plate is embedded in a long ligament between the premaxilla, prevomer, and toothed anterolateral tip of the element named 'metapterygoid' by Arratia (see Fig. 4.27a). The similarity between this configuration and that of some characiforms, for example *Hoplias*, is remarkable, as shown in Figure 4.27. In fact, *Hoplias* has a tooth-plate associated with the toothed anterior portion of the ectopterygoid and the premaxillary (Fig. 4.27b). This resemblance, associated with the presence of such tooth-plates in numerous characiforms (Sagemehl, 1885; Starks, 1926; Weitzman, 1962, 1964; Roberts, 1969; Fink and Fink 1981; a.o.) and several catfishes such as diplomystids, ariids, pimelodids, claroteids, austroglanidids, cranoglanidids, bagrids or schilbeids, as well as in the fossil catfish hypsidorids, support that the anterolateral portion of Arratia's 'metapterygoid' (see Fig. 4.25a, b) is homologous to the anterior tip of the ectopterygoid of other teleosts (see Fig. 4.27b).

This evolutionary hypothesis is supported by other data. Howes (1983b) reported that in catfish such as *Pinirampus pirinampu*, the element named 'metapterygoid' by Arratia has 'a sharply demarcated ventral portion that articulates with the 'quadrate' (sensu Arratia), while the dorsal portion articulates with the 'hyomandibular' (sensu Arratia) process (see Fig. 4.26b). This observation led him to hypothesise that this supposed 'metapterygoid' is, in fact, the evolutionary result of a fusion—which in the particular case of *Pinirampus pirinampu* is incomplete (Fig. 4.26b)—between the ectopterygoid and entopterygoid of other teleosts. Howes and Teugels (1989), in a subsequent embryological study of the pterygoid bones of some catfish, pointed out that the true metapterygoid 'persists in adult siluroids only as a densely ossified area of the palatoquadrate arch and not as a laminar ossification'. Thus, according to these authors, the so-called 'metapterygoid' of catfishes comprises for the most part the entopterygoid and ectopterygoid of other teleosts.

One other major argument supporting the above hypothesis is Grande's 1987 description of *Hypsidoris farsonensis*, the fossil hypsidorid catfish from the Eocene of the Green River formation (see Chapter 1). In his reconstruction of *Hypsidoris farsonensis*, Grande (1987) reported an 'entopterygoid' sutured with a 'metapterygoid' (see Fig. 4.26a). However, as Arratia (1992) argued, this 'condition is unlikely, when you compare it with other primitive siluroids'. Arratia (1992) proposed two alternative hypotheses: 1) that the two bones are not really sutured; 2) that Grande's 'entopterygoid' is, in fact, a fragment of the so-called 'metapterygoid'. However, both of her hypotheses are questionable. Even if we allow that Grande misinterpreted the presence of a suture between the two bones (which is not the case: see below), it seems unlikely that his 'entopterygoid' is homologous to the true teleost entopterygoid, since it does not have the same position and anatomical relations as the latter (see Fig. 4.26a). In reality, this 'entopterygoid' has the typical features of the ectopterygoid of other teleosts (e.g., the spatial relation between its anterior portion and the posterior tip of the palatine) (Fig. 4.26a). Moreover, it clearly corresponds to the anterolateral portion of the element named 'metapterygoid' by Arratia in other catfishes (compare, e.g., Fig. 4.26a to Fig. 4.25a). The second hypothesis above, a post-mortem, incidental fracture of the 'metapterygoid' resulting in the separation of two parts presenting respectively the same spatial position and relations as the ectopterygoid and entopterygoid of other teleosts, also seems very improbable, especially since a similar fracture is also present in some living catfishes (see above). Moreover, a quite similar, complete fracture is equally present in *Astephus antiquus*, a fossil ictalurid catfish that also occurred in the Eocene of the Green River formation (see Grande and Lundberg, 1988: Fig. 10). Therefore, Grande's (1987) 'metapterygoid' and 'entopterygoid' (see Fig. 4.26a) seem to be, in fact, separated by a suture, but correspond respectively to the entopterygoid and ectopterygoid of other teleosts, as explained above.

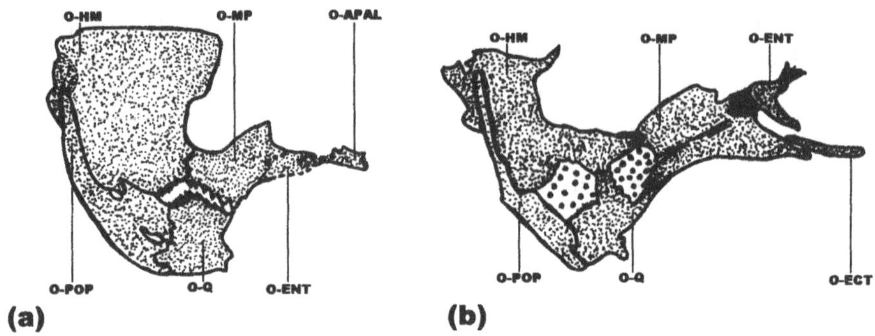

Fig. 4.26 Lateral view of suspensorium of *Hypsidoris farsonensis* **(a)**, fossil catfish from the Eocene (modified from Grande, 1987), *Pinirampus pirinampu* **(b)** (modified from Howes, 1983b, autopalatine not shown). The nomenclature used here follows that used in the original illustrations [for more details, see text].

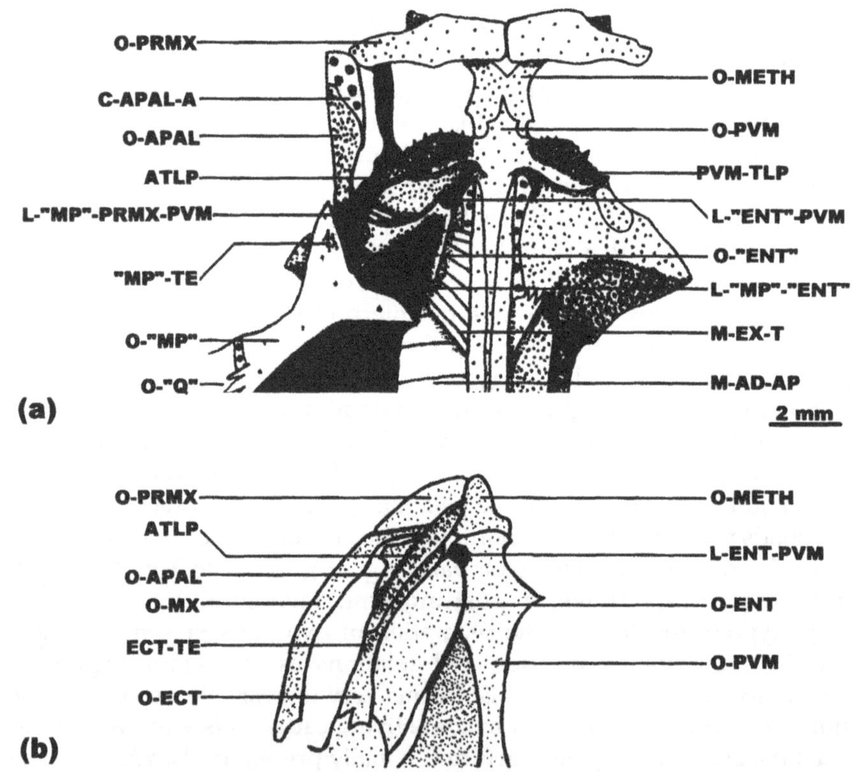

Fig. 4.27 Ventral view of the anterior region of suspensorium of *Chrysichthys nigrodigitatus* (modified from Diogo and Chardon, 2000c) **(a)** and *Hoplias* species **(b)** (modified from Roberts, 1969). The nomenclature used here follows that used in the original illustrations [for more details, see text].

This hypothesis is supported by the fact, largely accepted nowadays, that the bones often named 'entopterygoid' and 'ectopterygoid' in catfish traditional literature (e.g. Regan, 1911a; Alexander, 1965; Gosline, 1975) clearly do not correspond to the entopterygoid and ectopterygoid of other ostariophysans, but rather to sesamoid ossifications (see Arratia, 1992; Diogo et al., 2001a). This important point concerning catfish evolutionary history, and particularly concerning the evolution of the suspensorium within these fishes, is strongly supported by the observations and phylogenetic results of the present study.

In the specimens of *Diplomystes chilensis* examined, the three anterior small bones of the suspensorium clearly appear as sesamoid ossifications related to ligaments and not vestigial bones. This *a priori* morphological observation is supported by the *a posteriori* analysis of this issue in face of the phylogenetic results indicated by the parsimonious distribution of all available characters. For example, even if one codes the existence of the element named 'ectopterygoid' by authors such as Alexander (1965) as a siluriform plesiomorphic feature due to its presence in diplomystids, this suspensorial element clearly appears as the result of several homoplasic, independent events with Siluriformes (see Figure 4.28). This 'ectopterygoid' is thus in no way the remnant of a true homologous ectopterygoid in those catfish groups in which it is present (Fig. 4.28), but clearly a homoplasic element independently acquired in those groups, by independent sesamoid ossifications of the ligaments connecting the anterior part of the suspensorium to the ethmoidal region.

If fact, if the small bones present in the anterior portion of the suspensorium of some catfish were to be interpreted as reduced or vestigial true pterygoids (entopterygoid and/or ectopterygoid), expectedly they would generally be larger in primitive families than in specialised ones. However, in what concerns, for example, the 'entopterygoid' of authors such as Alexander (1965), in plesiomorphic catfishes such as diplomystids, nematogenyids, and the fossil *Hypsidoris farsonensis* (see, e.g., Figs. 3.89, 4.26a) this bone is very small or absent, whereas it is not as reduced in generalised siluriforms such as bagrids, claroteids, pimelodids, schilbeids, silurids, cranoglanidids, malapterurids, austroglanidids or ariids (see, e.g., Fig. 3.34). In certain relatively more apomorphic groups (see Fig. 3.123), such as clariids, amblycipitids, amphiliids, or sisorids, this bone is much larger (see, e.g., Fig. 3.3), being in some cases as broad, or even broader, than the element named 'metapterygoid' by Arratia (see char. 325). It is rather unlikely that basally among catfishes, this 'entopterygoid' had become reduced or even lost, and subsequently reacquired a large size in some more derived catfish. It is rather more plausible that this bone, like the 'ectopterygoid' (see above), is, in fact, a sesamoid ossification that developed homoplasically in certain catfish groups and became, in some cases, progressively larger in some of these groups. These two ossifications are probably functionally related, as explained in the first paragraphs of this Section, to the decoupling of the autopalatine from the rest of the suspensorium, as well as to a certain particular type of specialisation of

Fig. 4.28 Hypothesis of character state evolution of sesamoid bone 2 of suspensorium (see char. 308): CS-0 (black)= presence of sesamoid bone 2 of the suspensorium; CS-1 (blue)= absence of sesamoid bone 2 of suspensorium; Ambiguity (pink) [for more details, see text].

the palatine-maxillary system in some catfish groups. The shape of the sesamoid bones and associated ligaments in the suspensorium of siluriforms seems, in fact, to often be closely associated with the various types of palatine-maxillary system of these fishes. For example, in siluriforms with a 'rocking' palatine-maxillary system (see Section 4.4), these structures are so disposed as to allow a pronounced medial movement of the rear end of the autopalatine (see Fig. 4.21A→B). In catfishes with a 'sliding' palatine-maxillary system (see Section 4.4), contrarily, their configuration allows a large posterior displacement of this bone (see Fig. 4.22A).

This hypothesis is also strongly supported by developmental data. In a detailed embryological study Adriaens and Verraes (1998) show that the so-called 'entopterygoid' of *Clarias gariepinus* is, in reality, a sesamoid ossification of the ligament between the anterior part of the pars quadrata and the prevomer. Also, after studying the development of some silurids, Kobayakawa (1992) stated that the so-called 'entopterygoid' 'has ligaments on both its anterior and posterior sides from the onset of its ossification...; it appears as a small, rod-shaped bone connected by ligaments with the "metapterygoid" and the ventral surface of the lateral ethmoid anteriorly'. According to Arratia (1990) 'the "entopterygoid" in *Nematogenys* arises as an ossification of the ligament extending between the "metapterygoid" and lateral ethmoid, and, late in ontogeny, with the prevomer'. Howes and Teugels (1989) also clearly show that the two anterior bones of the suspensorium of a 72 mm *Pimelodus blochii* result from ossifications in the ligament joining the palatine to the lateral ethmoid, which is already conspicuous in a 40 mm specimen. According to these authors, the catfish bones called 'entopterygoid' and 'ectopterygoid' by authors such as Alexander (1965) are sesamoid bones and/or fragments of the 'metapterygoid'. In fact, it is important to note that the small anterior bones of catfish suspensorium begin to develop *anterior to* the pterygoid process of the pars quadrata (see, e.g., Kobayakawa, 1992: Figs 9, 10; Vandewalle et al., 1993: Fig. 2; 1995a: Figs 9B, 10B; 1997: Figs 7B, 8B). In most teleosts the ectopterygoid and the entopterygoid develop *on the* processus pterygoideus (see, e.g., de Beer, 1937; Daget, 1964; Bertmar, 1959). There is thus a difference between the place of origin of the anterior small bones in catfishes and that of the ectopterygoid and entopterygoid in other teleosts.

This interesting evolutionary scenario, in which the anterior small bones of the catfish suspensorium correspond to de novo, often homoplasic, sesamoid ossifications of the ligaments connecting the pars quadrata to the ethmoidal region, has also been defended in other studies. Already in 1884, for example, McMurrich defended the small bone 'lying behind and within the posterior extremity of the palatine' in *Ictalurus catus* as a sesamoid bone, which he termed 'bone number 4'. Howes (1983b, 1985a) stated, as explained above, that the often called 'entopterygoid' and 'ectopterygoid' of catfish are, very likely, sesamoid bones. Arratia (1987) stated: '...the position of this bone ("ectopterygoid") in diplomystids is not homologous with that of the ectopterygoid in other teleosts; this small "pterygoid" appears as an additional element of the series and it could represent a neomorphic feature'. According to Arratia (1992) the 'pterygoid bones in most catfish are highly specialised sesamoid elements, connected by ligaments to cranial bones or other bones of the suspensorium'.

The last suspensorial elements to be discussed here are the quadrate and the symplectic. A 'typical' teleostean quadrate has a posterior notch in which the symplectic inserts on the lateral side (see, e.g., Taverne, 1974: Fig. 4). The inferior arm of the notch, which probably represents the quadratojugal (Devillers, 1958), is lacking in some teleosts, for example some clupeids (see, e.g., Ridewood, 1904: Figs. 124, 132). As for the symplectic, it remains cartilaginous in some mormyriforms (e.g., Taverne, 1974) and clupeiforms (e.g., Ridewood, 1904). But both the quadrate and the symplectic are present and well ossified in the Gonorynchiformes (e.g., Chardon and de la Hoz, 1974; Mago-Leccia, 1978; Fink and Fink, 1981; Arratia, 1992; Poyato-Ariza, 1996; Grande and Poyato-Ariza, 1999), Cypriniformes (e.g., Vandewalle, 1975; Arratia, 1992; Taverne and De Vos, 1997), Characiformes (e.g., Gijsen and Chardon, 1976; Arratia, 1992), Gymnotiformes (e.g., Arratia, 1992) and 'fossil Ostariophysi' (e.g., Gayet, 1982). Catfish have neither a notch nor a distinct symplectic.

Howes (1985a) described a 'symplectic' in the African catfish *Malapterurus electricus* (see Fig. 4.25d), but this statement was questioned by Arratia (1992). Observations and phylogenetic results of the present work also strongly contradict this statement. In fact, concerning the comparative anatomy of Howes' 'symplectic', this bone presents in no way the typical configuration and spatial relationships of a true symplectic. But perhaps even more elucidating is the fact that *Malapterurus* appears in a relatively apomorphic position within the siluriform cladogram (Fig. 3.123). Therefore, to assume that the *Malapterurus* 'symplectic' is, in fact, the direct remnant of the true symplectic of other teleosts, would imply that this true 'symplectic' would be lost at least 7 times within Siluriformes (see Fig. 3.123). Or, alternatively, that this true 'symplectic' was at a certain time lost or fused in catfish evolutionary history and was subsequently somehow 'regained' in *Malapterurus*, and only in this catfish genus.

A more plausible evolutionary scenario seems to be that in catfishes the symplectic is incorporated into the quadrate, filling the notch usually found in other teleosts. In fact, it is difficult to explain the disappearance of the quadrate notch and the similar shape of the 'quadrate' of Arratia and the quadrate + symplectic of other teleosts (see Fig. 4.24) if we accept that the 'disappearance' of the symplectic in catfish is simply a function of the non-ossification of this element. As in the case of the fusion between the hyomandibula and the metapterygoid, the fusion between the quadrate and symplectic would probably be related to the fact that in catfish the pars quadrata and the hyosymplectic are fused from the first appearance of the chondrocranium cartilages. Most authors (McMurrich, 1884; Harry, 1953; Skelton, 1981; Skelton et al., 1984; Howes, 1983b; Mo, 1991; Arratia, 1992; Adriaens and Verraes, 1998; a.o.) would argue, however, that an ossified symplectic is absent in catfish and that the large suspensorial cartilage usually found in these fishes indicated with an A in Figure 4.24B is precisely the remnant of the symplectic cartilage present early in ontogeny and hence the homologue of the symplectic of other teleosts. But this cartilage differs from the typical symplectic by its position. Even more importantly, *both* the *cartilage* and the *symplectic bone* (which is always situated anterior to the cartilage) are present concomitantly in gymnotiforms (see, e.g., Fig. 4.24; Chardon and de la Hoz, 1974: Figs. 2, 3, 4, 5, 6; Mago-Leccia, 1978: Fig. 12; Fink and Fink, 1981: Fig. 12; Arratia, 1992: Fig. 12A, B, D), characiforms (see, e.g., Weitzman, 1964: Fig. 7; Fink and Fink, 1981: Fig. 10), cypriniforms (see, e.g., Vandewalle, 1975: Figs. 1, 2; Fink and Fink, 1981: Fig. 9; Arratia, 1992: Fig. 4A), gonorynchiforms (see, e.g., Fink and Fink, 1981: Fig. 8; Arratia, 1990: Fig. 2) and a large number of other teleosts (see, e.g., Ridewood, 1904: Figs. 123, 132; Chardon and Vandewalle, 1971: Fig. 2; Vandewalle, 1971: Figs. 6, 11; Vandewalle et al., 1995b: Fig. 2). This implicates that these structures cannot be theoretically homologous (if two structures A and B are present at the same time in a certain species X, it cannot be considered that these two structures are homologous within two different species Y and Z: see, for example, Gould, 1989; Hall, 1994; Beaumont, 1998).

The main evolutionary transitions discussed in the above paragraphs concerning catfish suspensorium are summarised in Figure 4.29. So, as illustrated in this Figure, there are *two* main evolutionary transitions from a typical teleostean suspensorium (Fig. 4.29a) and the configuration found in the most plesiomorphic actual catfishes, the diplomystids (Fig. 4.29c): 1) in the diplomystids, a major decoupling occurs between the autopalatine and the posterior portion of the suspensorium; 2) this posterior portion of the suspensorium is constituted in diplomystids by three, and not six, major osteological elements, which seemingly correspond to the hyomandibulo-metapterygoid, quadrato-symplectic, and ento-ectopterygoid of other teleosts, with the small anterior bones being sesamoid ossifications.

The hypothetical evolutionary signification of these main transitions is of interest here. As explained above, the decoupling between the autopalatine and the posterior portion of the suspensorium is ontogenetically present from the first appearance of the splanchnocranium cartilages. It seemingly frees the palatine-maxillary system from the rest of the suspensorium, thereby allowing ample movements of the maxillary barbels (see, e.g., Alexander, 1965; Gosline, 1975). However, as also explained above, this decoupling probably created, at least in a theoretical morphological point of view, three theoretical functional problems, which are illustrated in Figure 4.29b and respectively indicated by numbers 1, 2 and 3. In fact, contrary to the situation found in most teleosts, in which the suspensorium presents posterior and anterior solid points of articulation with the neurocranium (Fig. 4.29a: Af-I, Af-II), the decoupling resulted in a single major posterior solid articulation between the large posterior portion of the suspensorium and the neurocranium (Fig. 4.29b: 1). This decoupling also resulted in a seemingly problematic lack

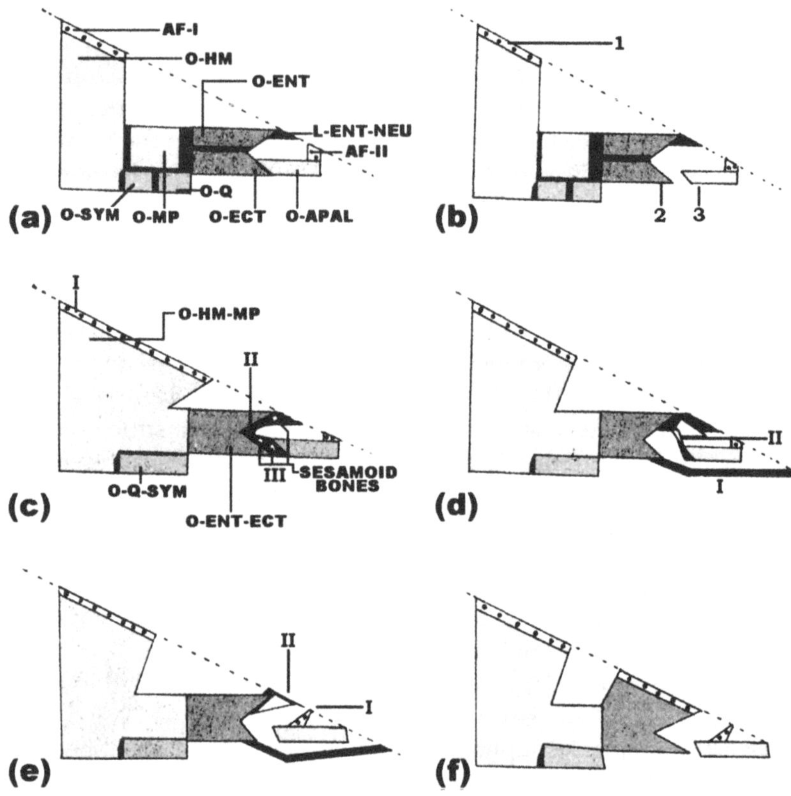

Fig. 4.29 Scheme illustrating the evolutionary hypothesis presented in this book concerning the evolution of catfish suspensorium. Dotted areas represent articulatory facets; black spaces between two suspensorial bones indicate that they are not firmly attached (sutured) to one another. **a)** Typical suspensorium of a primitive teleost. **b)** Hypothetical scheme illustrating functional decoupling between autopalatine and rest of suspensorium. **c)** *Diplomystes chilensis*. **d)** *Chrysichthys nigrodigitatus*. **e)** *Amphilius brevis*. **f)** *Loricaria loricaria* [for numbers and further explanations, see text].

of a point of support for the antero-ventrolateral end of the suspensorium (Fig. 4.29b: 2), since 'the increasing use of mouth opening and mouth closure, as well as orobranchial expansions for feeding and respiratory purposes, will increasingly load the suspensorium ..., [and] powerful biting will exert forces on the suspensorium in an anteroposterior direction' (cited from Adriaens and Verraes' 1998 work on functional aspects of development in *Clarias gariepinus*). In the same way, the decoupling resulted in the lack of a solid point of support for the posterior end of the autopalatine (Fig. 4.29b: 3).

As already mentioned, the three functional points illustrated in Figure 4.29b probably correlate with numerous, often homoplasic, characters concerning the suspensorium of various catfish groups involving mainly the presence of several ligaments and/or small bones between the pars quadrata, autopalatine, and the ethmoidal region.

So, for example, concerning the suspensorium of the most plesiomorphic recent siluriforms, the diplomystids (see Fig. 4.29c), three characters are worthy of mention, which are illustrated in Figure 4.29c and respectively indicated with the numbers I, II and III.

One of these characters, which very likely constitutes a particular apomorphy of diplomystids and not a plesiomorphic situation for siluriforms, as noted by Arratia (1987), is the remarkable widening of the dorsal part of the hyomandibulo-metapterygoid, which increases the length of the articulation on the neurocranium (Fig. 4.29c: I). This remarkable widening, characteristic of diplomystids, together with fusion of the hyomandibula and metapterygoid, quadrate and symplectic, and entopterygoid and ectopterygoid, as well as the better fastening between these elements, characteristic of catfishes (see above) (Fig. 4.29a→c), seemingly reinforces the posterior suspensorium and allows it to work as one single piece, a true functional mechanical unit (sensu Hughes and Ballintijn, 1965). In fact, it is interesting to note that in catfish, from the first appearance of the splanchnocranium cartilages, the pars quadrata and the hyosymplectic, together with the Meckelian cartilage and the hyoid bar, are remarkably grouped in one sole piece, as explained above.

One other interesting feature is the presence, in diplomystids, of a massive ligament between the ento-ectopterygoid, the neurocranium and the autopalatine (Fig. 4.29c: II). This massive ligament seemingly confers a functional point of support to the rear-end of the autopalatine and to the fore-end of the rest of the suspensorium (Fig. 4.29c, compare with Fig. 4.29b). Such ligaments associating the entoectopterygoid, the neurocranium and the autopalatine are also present in a large number of non-diplomystid catfishes, as explained above.

The other interesting feature is the presence, in diplomystids, of small sesamoid bones associated with the ligament referred above. These sesamoid bones seem to be a functional answer to the important forces exerted on the ligaments fixed on the ento-ectopterygoid, due not only to the horizontal tensions on the suspensorium, but also to the increasing mobility of the autopalatine, which is associated with the palatine-maxillary system and thus with the movements of the maxillary barbels. In fact, as mentioned above and shown in Figure 4.28 concerning the sesamoid bone 2, such sesamoid bones were independently acquired homoplasically in various catfish groups. Partial or complete ossification of the ligaments between the neurocranium and the suspensorium is also found in other ostariophysans, for example in gymnotiforms (see, e.g., Chardon and de la Hoz, 1973, 1974; de la Hoz, 1974; de la Hoz and Chardon, 1984).

A somewhat different answer to the three functional points illustrated in Figure 4.29b is found in relatively generalised catfishes such as pangasiids, schilbids, cranoglanidids, austroglanidids, ariids, claroteids, bagrids or pimelodids (see Fig. 3.123), as shown in Figure 4.29d. In such catfishes the dorsal articulatory surface of the hyomandibulo-metapterygoid for the neurocranium is not as developed as in diplomystids (Fig. 4.29d, compare with Fig. 4.29c). However, in these catfish groups the antero-ventrolateral edge of the suspensorium hangs from a strong ligament attached to the prevomer and the premaxillary, seemingly allowing a greater freedom of the posterior portion of the autopalatine (Fig. 4.29d: I). Configuration of the enlarged sesamoid bone/bones and associated ligaments insures concomitantly a support for the posterior end of the autopalatine and a greater mobility of this bone (Fig. 4.29d: II). As explained above, the shape of the sesamoid elements and associated ligaments within these groups seems to be narrowly associated with the different types of palatine-maxillary system of these fishes. Due to the fact that bagrids and pimelodids do not seem to be directly related to pangasiids, schilbids, cranoglanidids, austroglanidids, or claroteids (see Fig. 3.123) and to the high homoplasy concerning catfish suspensorial sesamoid bones and associated ligaments (see above), the suspensorium configuration schematised in Figure 4.29d was, very likely, independently acquired in various siluriform groups.

Another rather different suspensorium configuration is found in specialised catfishes such as amphiliids, as shown in Figure 4.29e. In these catfishes the neurocranial articulatory facet of the autopalatine is situated on the posterior half of this bone (Fig. 4.29e: I). It therefore gives a point of support to the rear-end of the autopalatine (the fore-end of this bone is supported by its narrow connections with the premaxillary and maxillary), which seemingly allows a still greater decoupling between the autopalatine and the rest of the suspensorium (see Fig. 4.29e). The ligament between the ento-ectopterygoid and the ethmoidal region, as well as the sesamoid bone 1 imbedded in it, are no more associated with the palatine-maxillary system (see Fig. 4.29e). The ossification of this ligament is almost complete (sesamoid bone 1 is relatively larger), seemingly fastening the association between the main body of the suspensorium (autopalatine not included) and the neurocranium, and thence creating a somewhat solid 'second articulatory point' between these two mechanical units (see Fig. 4.29e). As noted by He et al. (1999), enlargement of sesamoid bone 1 is particularly notable in the doumein amphiliids, in which this bone is as large as, or even larger than, the ento-ectopterygoid, a configuration also found in *Leptoglanis* and several sisoroids (see Fig. 4.30). As shown in Figure 4.30, the presence of this notable feature in some amphiliids and some sisoroids is, very likely, the result of independent, homoplasic events.

Such a 'second articulatory point' between the main body of the suspensorium and the neurocranium is also found in the loricariids, as a result of a completely different morphological configuration. In fact, in loricariids the dorsal margin of the ento-ectopterygoid directly articulates with the ethmoidal region, as schematised in Figure 4.29f. Some families of specialised catfishes thus independently acquired a suspensorium configuration in which there is an almost complete functional decoupling of the palatine-maxillary system for moving the maxillary barbels while seemingly somehow rebuilding an anterior suspensorial articulation on the skull.

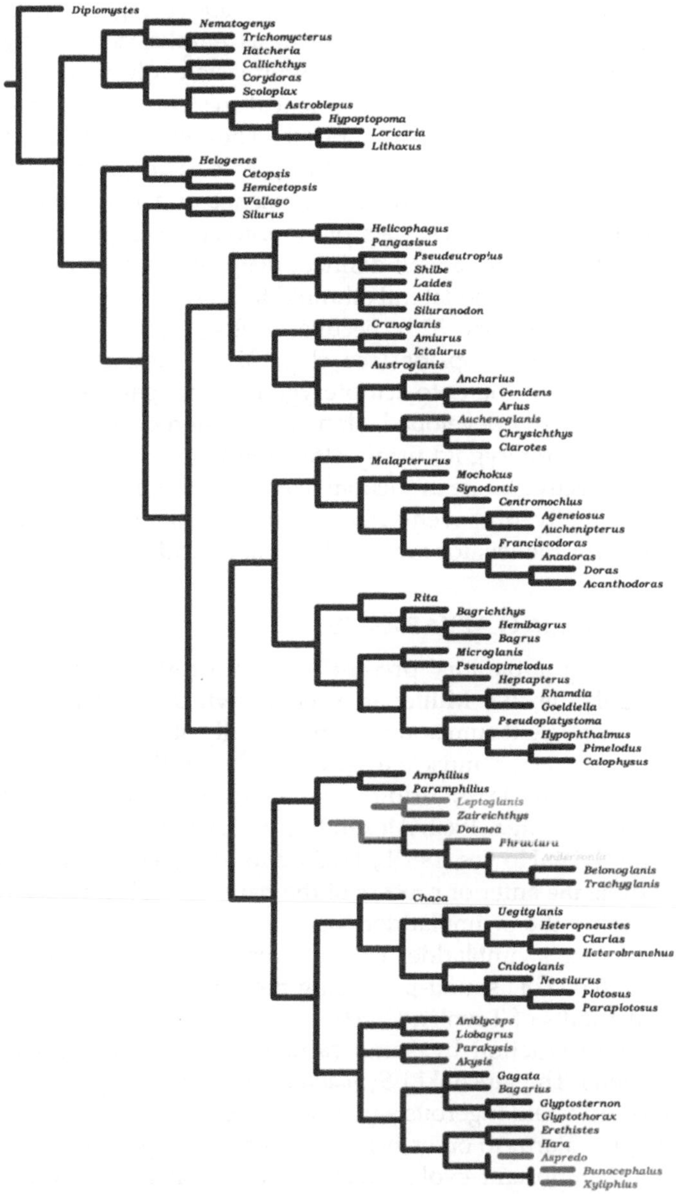

Fig. 4.30 Hypothesis of character state evolution of entoectopterygoid and sesamoid bone 1 of suspensorium (char. 325, ordered): CS-0 (black)= entoectopterygoid well developed, larger than sesamoid bone 1; CS-1 (blue)= entoectopterygoid relatively reduced in size, but larger than sesamoid bone 1; CS-2 (orange)= significant part of entoectopterygoid surrounded by lateral margin of sesamoid bone 1; CS-3 (brown)= entoectopterygoid markedly reduced in size, with mesial surface completely surrounded by sesamoid bone 1; CS-4 (yellow)= entoectopterygoid markedly reduced in size and smaller than sesamoid bone 1; Ambiguity or Inapplicable (pink) [for more details, see text].

In addition to the major points discussed above regarding the general evolution of catfish suspensorium, which, as emphasised throughout this Section, is mainly a history of rather homoplasic evolutionary events, it is important to note here that some specific features concerning this structural complex are relatively poorly homoplasic. They constitute, in fact, strong synapomorphies, or even autapomorphies, of certain particular siluriform subgroups. Some examples of this are: presence of a voluminous, globular structure on the anteroventral surface of the ento-ectopterygoid in astroblepids (see char. 322); presence of a well-defined, deep, anteroposteriorly elongated concavity on the frontal and lateral ethmoid to receive the ento-ectopterygoid in cranoglanidids (see char. 323); peculiarly elongated, roughly rectangular ento-ectopterygoid of cetopsids (see char. 326); presence of a prominent posterodorsal crest of the ento-ectopterygoid in pangasiids (see char. 328); presence of a markedly developed, elongated posterodorsal projection of the hyomandibulo-metapterygoid firmly attaching to the neurocranium by massive, strong connective tissue in trichomycterids and nematogenyids (see char. 338); and essentially dorsoventrally oriented crest of the hyomandibulo-metapterygoid for the levator arcus palatini in silurids (see char. 335).

4.6 ELASTIC SPRING APPARATUS

The 'elastic spring apparatus' present in certain catfishes is formed by the modified, highly flexible Müllerian process, which is separated from the posttemporo-supracleithrum and posteroventrally attached to the swim bladder, and by a 'protractor' muscle inserted on the Müllerian process (see, e.g., Fig. 3.58) (Sörensen, 1894; Bridge and Haddon, 1894; Chranilov, 1929; Schneider, 1961; Tavolga, 1962; Alexander, 1965; Chardon, 1968; Taverne and Aloulou-Triki, 1974; Howes, 1983b, 1985a; Mo, 1991; a.o.). 'When this muscle contracts it pulls the anterior process of the parapophysis forward (the elastic spring), enlarging the swimbladder: when it relaxes, it allows the spring to recoil,..., [and] the swimbladder is thus caused to pulsate, emitting sound' (Alexander, 1965). This sound-producing mechanism was supported by the experimental results of Tavolga (1962).

The sound production by some catfishes may have a social function (Schneider, 1961). The mochokid *Synodontis*, for example, produces a characteristic 'murmur' in a dangerous situation (Taverne and Aloulou-Triki, 1974), probably to give alarm to other fishes. The ability to produce sound is thus, very likely, of important evolutionary signification for the various catfish lineages that acquired it, especially taking into account that these fishes are mainly nocturnal and inhabitants of aquatic habitats with a very reduced visibility, as noted by Alexander (1965).

As described in Section 3.1 (char. 254), such an 'elastic spring apparatus' was found in members of eight catfish families examined, namely doradids, auchenipterids, mochokids, malapterurids, ariids, some pangasiids, cranoglanidids and pimelodin pimelodids. Although some authors (e.g.,

Curran, 1989) hypothesise that catfishes with an 'elastic-spring-apparatus' form a monophyletic group, it is commonly accepted that such an 'elastic spring apparatus' is, in fact, a homoplasic feature (see, e.g., Chardon, 1968; Mo, 1991; de Pinna, 1998).

This latter hypothesis is strongly supported by the present work. As can be seen in Figure 4.31, an 'elastic spring apparatus' seems to have evolved at least 5 times within the Siluriformes, namely: in pseudopimelodins, ariids, cranoglanidids, some pangasiids, and in the node leading to malapterurids + mochokids + doradids + auchenipterids. According to Chardon (1968) the homoplasic occurrence of an 'elastic spring apparatus' in various catfish lineages is not all that surprising. This is because several generalised siluriforms, such as claroteids, bagrids, austroglanidids and many pimelodids, although not presenting a true 'elastic spring apparatus', exhibit a somewhat similar configuration: a great number of the fibres of the muscle epaxialis and/or supracarinalis anterior running in a somewhat oblique direction to attach on a flexible 'anteroventral process' of the parapophysis of the 4th vertebra.

In an excellent, unfortunately still unpublished, Bachelor's thesis on the morphology and evolution of the siluriform dorsal fin and related structures, Royero (1987) provided interesting evidence to support that the 'protractor of the Müllerian process' is likely the result of the differentiation of the supracarinalis anterior (sensu Winterbottom, 1974a). As noted by Royero (1987: 201), the 'protractor of the Müllerian process' usually lies in a position similar to that of the supracarinalis anterior, with the latter muscle not differentiated in those groups presenting a well-developed 'elastic spring apparatus' and its respective 'protractor'. Also, this 'protractor' seems to be innervated by cranial nerves, like the supracarinalis anterior, and not by rachidian nerves, such as the muscle epaxialis. However, the study by Royero (1987) included only four of the eight groups possessing an 'elastic spring apparatus', namely the ariids, mochokids, doradids and auchenipterids. Thus Royero's (1987) observations do not ipso facto apply to the other four catfish families having members with an 'elastic spring apparatus' (see above), since the presence of this apparatus in at least some of those families clearly seems to be the result of homoplasy. For example, according to Royero (1987), the 'protractor of the Müllerian process' of malapterurids described by other authors does not seem to be homologous with that of ariids, doradids, auchenipterids and mochokids directly observed by him: not only does the 'protractor' of *Malapterurus* seem to be rather more complex than that of these four groups, but also, contrary to these groups, in *Malapterurus* both this muscle and the supracarinalis anterior are well developed. In fact, it should be noted that, as described in Section 3.1 of the present work, even within the clade mochokids + doradids + auchenipterids, the 'protractor of the Müllerian process' exhibits rather different configurations. For example, in *Centromochlus* this muscle is markedly subdivided into two well-distinguished divisions, while in *Mochokus* and *Synodontis* this muscle does not originate as usual on the posterior surface of the cephalic region, but instead on the region of the dorsal fin and its support (see characters 255 and 256 in Section 3.1).

Fig. 4.31 Hypothesis of character state evolution of presence of elastic spring apparatus (see char. 254): CS-0 (black)= absence of elastic spring apparatus; CS-1 (blue)= presence of elastic spring apparatus [for more details, see text].

Although not constituting an 'elastic spring apparatus', the situation found in most pimelodins catfishes examined is interesting. In fact, as described by authors such as Bridge and Haddon (1894), Ladich and Fine (1994), Ladich and Bass (1998) and Ladich (2001), most pimelodins analysed present two peculiar muscles seemingly associated with the production of sound by the swim bladder. One of these muscles is the 'drumming muscle' (see char. 258) running from the parapophyses of the fourth vertebra and, eventually, from the posterior surface of the neurocranium, to the anterior and anteroventral surfaces of the swim bladder (see Fig. 3.100). The other muscle is the 'tensor tripodis' (see char. 257), a small muscle running from the posterior surface of the neurocranium to the dorsal surface of the swim bladder near the tripus (see Ladich, 2001: Fig. 2). As noted by Ladich (2001), this second muscle, the 'tensor tripodis', is unique among all sound-generating fishes. According to him, 'because its small size and minute insertion point on the swimbladder', this muscle 'very unlikely serves in sound production'. The close proximity of its insertion to the tripus Weberi would seem to confer a 'protection function of the inner ear similar to the musculus tensor tympani in mammals', reducing 'swimbladder wall vibrations near the Weberian ossicle during and after the action of the drumming muscle to a minimum' (Ladich, 2001: 302-303). Examination of the fine structure of the tensor tripodis by Ladich (2001: 203) supported, according to him, 'the notion that it is able to protect the ears from sensory overload during vocalisation'.

Although the functional signification of the 'drumming' and 'tensor tripodis' seems fairly clear, the evolutionary history concerning the origin of these muscles is far from obvious. Plesiomorphically catfishes lack a 'drumming' and 'tensor tripodis' but, as described in Section 3.1, both these muscles were present in the pimelodins examined in this study, with except for *Hypophthalmus*. The absence of these muscles in *Hypophthalmus* could probably be associated with the fact that in members of this genus, unlike in most pimelodins, the swim bladder is almost completely encapsulated by the well developed parapophyses of the complex vertebra. However, those muscles are apparently not restricted to the pimelodin catfishes, since they were also reported by Ladich in some species of the pimelodid heptapterin genus *Rhamdia* (see Ladich, 2001). Also, in this case there seems to be a relation between the presence of the 'drumming' and 'tensor tripodis' muscles and encapsulation of the swim bladder since, as remarked by Ladich (2001), in some *Rhamdia* species such as *Rhamdia sapo* with a markedly encapsulated swim bladder these muscles are lacking. The situation is even more complex since, as also noted by Ladich (2001), even in heptapterin pimelodids with a free swim bladder, such as those of genus *Pimelodella*, these muscles are absent. In fact, as described in characters 257 and 258 of Section 3.1, none of the heptapterin catfishes examined in the present work exhibited either a 'drumming' or a 'tensor tripodis' muscle.

Therefore, it is difficult to trace the evolutionary history of the origin of the 'drumming' and 'tensor tripodis' muscles within Siluriformes. They could perhaps have evolved in the node leading to heptapterins + pimelodins, thus supporting the sister-group relationship between these two pimelodid subfamilies (see Fig. 3.123) since, aside from pimelodins, they seemingly are present in at least some heptapterin species of genus *Rhamdia*, according to the descriptions of Ladich (2001). However, this would imply a series of secondary losses within each of these two subfamilies, and not only in species with an encapsulated swim bladder since, as already mentioned, those muscles are lacking in some species with an essentially free swim bladder, such as those of genus *Pimelodella*. But the probability of the alternative hypothesis, that is, an independent origin of these muscles in some pimelodin and some heptapterin catfishes appears even more unlikely. As explained above, it would seem somewhat far-fetched that such a peculiar and rare muscle as the 'tensor tripodis', seemingly virtually absent in all other sound-generating fishes, should develop independently in two such closely related catfish groups (see Fig. 3.123).

On the other hand, the first of these two evolutionary scenarios, i.e., an origin of the 'drumming' and/or 'tensor tripodis' muscles in the node leading to heptapterins + pimelodins, might perhaps have interesting evolutionary implications. As a mater of fact, as pointed out earlier, the pseudopimelodin pimelodids present an 'elastic spring apparatus'. Since these catfishes are probably the most plesiomorphic pimelodids, this could indicate that the origin of the 'drumming' and/or 'tensor tripodis' might perhaps be *apomorphic* in relation to the presence of an 'elastic spring apparatus' (see Fig. 4.31). Seemingly this makes more sense, at least from a theoretical point of view, than the scenario proposed by Ladich (2001), according to which the presence of a 'drumming' muscle such as found in pimelodins would be *anterior* to the origin of an 'elastic spring apparatus'. The Ladich scenario implies that the supracarinalis anterior and/or epaxialis (see above) would (1) become directly attached to the swim bladder, as in pimelodins (see Fig. 3.100), but (2) subsequently detach from the swim bladder, running exclusively from the neurocranium to the parapophysis of the 4th vertebra, as in the 'elastic spring apparatus' of pseudopimelodins (see Fig. 3.108).

Another interesting configuration is that found in bagrids. Like those groups presenting a 'protractor' of the 'elastic spring apparatus' (cranoglanidids, ariids, malapterurids, mochokids, auchenipterids, doradids, pseudopimelodins and some pangasiids) or a 'drumming muscle' directly inserted on the swim bladder (some pimelodins and, according to Ladich, 2001, some heptapterins), bagrids seem to present a specialisation that allows production of sound by the swim bladder. As recorded in Chapter 3 (see char. 253), bagrids exhibit a peculiar, apparently autapomorphic, muscle *protractor posttemporalis* extending from the neurocranium to the anterior margin of the posttemporo-supracleithrum (see Fig. 4.32). This muscle was originally described by Mo (1991), who named it the 'retractor posttemporalis'. According to him, the muscle is 'functionally linked with activities of the branchial basket since anterior movement of the posttemporal would inevitably pull the ascending

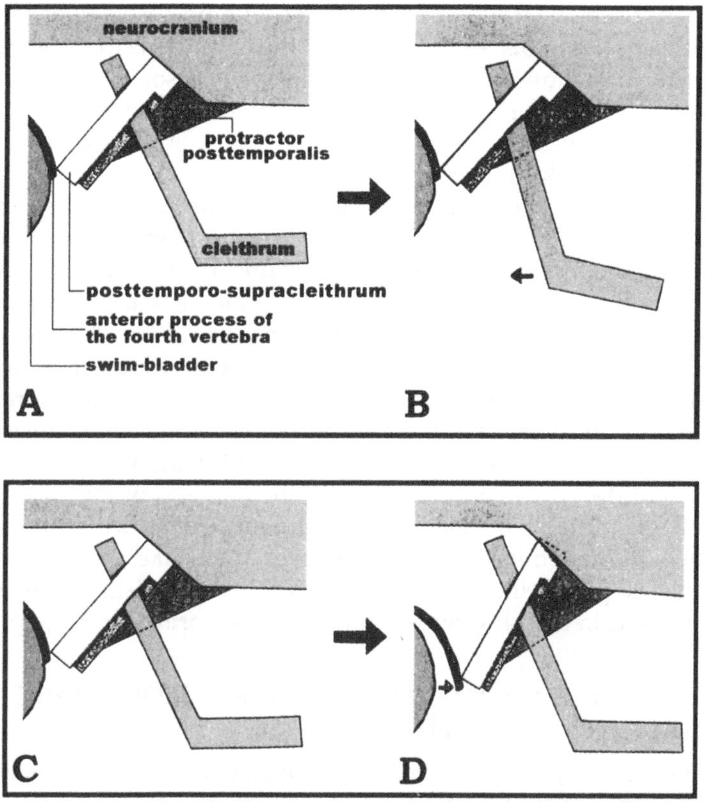

Fig. 4.32 Scheme illustrating the hypothesis on functional signification of presence of protractor posttemporalis in bagrid catfishes, lateral view: **A)** posttemporo-supracleithrum and cleithrum in normal positions; **B)** cleithrum retracted; **C)** posttemporo-supracleithrum and cleithrum in normal positions; **D)** posttemporo-supracleithrum protracted (due to contraction of protractor posttemporalis muscle) [for more details, see text].

portion of the cleithrum with it and consequentially cause the anterior end of the pectoral girdle to turn ventrally ...; this movement would presumably expand the branchial basket'. However, as pointed out by Diogo et al. (1999), Mo's hypothesis seems improbable. Articulation between the posttemporo-supracleithrum and the cleithrum in bagrids, as in most other catfishes (see, e.g., Alexander, 1965; Chardon, 1968; Arratia, 1987), allows each of these bones a great freedom of movement in relation to the other. It is precisely this liberty that permits, with no reciprocal movement of the posttemporo-supracleithrum, retraction of the cleithrum, as shown in Figure 4.32A→B, and hence of the pectoral girdle, when the mouth is opened by the 'hyoid mechanism' (see Adriaens and Verraes, 1997c, 1998). So, as the articulation between the posttemporo-supracleithrum and the cleithrum allows the ventral part of the cleithrum to move posteriorly with no reciprocal movement of the posttemporo-supracleithrum (Fig. 4.32A→B), it also permits an anterior displacement of the ventral part of the latter with no consequent movement of the cleithrum (Fig. 4.32C→D). Therefore, the hypothesis that contraction of the protractor posttemporalis would be principally associated with a ventral displacement of the anterior end of the pectoral girdle and that this displacement per se, via the sternohyoideus muscle, could 'cause expansion of the branchial basket', looks improbable.

Based on anatomical evidence and artificial manipulation, it seems much more plausible that the presence of a protractor posttemporalis is related, as already mentioned, to the production of sound by the swim bladder. So, contraction of the protractor posttemporalis, which is attached on the anterior surface of the posttemporo-supracleithrum, would *protract* this bone and thus the anterior process of the fourth vertebra to which it is associated posteriorly (Fig. 4.32C→D). Relaxation of the protractor posttemporalis muscle would then provoke a rapid and strong posterior movement of the posttemporo-supracleithrum (Fig. 4.32D→C) and consequently of the anterior process of the fourth vertebra, against the fore-end of the swim bladder (Fig. 4.32D→C). The swim bladder and the anterior process of the fourth vertebra would therefore probably work respectively like a drum and associated drumstick, producing sound by means of a mechanism somewhat similar to that found in catfish groups with an 'elastic spring apparatus'.

4.7 A DISCUSSION ON THE ORIGIN AND BIOGEOGRAPHIC DISTRIBUTION OF CATFISHES

As explained in Chapters 1 and 2, the great majority of catfishes, as well as of the other Ostariophysi—Gymnotiformes, Characiformes, Cypriniformes and Gonorynchiformes—are primary division freshwater fishes according to Myers (1938). Only 3 of the 32 extant catfish families, Ariidae, Plotosidae, and to a much less extent, Aspredinidae, possessing marine species. An important aspect that is unfortunately not so often remembered is that the extant marine species of these three families are mainly confined to coastal areas (Burgess, 1989). The particularly wide geographical distribution of catfishes—found in North, Central and South America, Africa, Eurasia, South-East Asia, Japan and Australia, with fossil catfishes having even being reported in Antarctica (Grande and Eastman, 1986)—has long intrigued ichthyologists and biogeographers in general. Researchers such as Briggs (1979: 111) have emphasised that catfishes and other ostariophysans are 'particular useful for studying past continental relationships'.

The biogeographic hypothesis that is nowadays accepted commonly to explain catfish geographical distribution postulates a Gondwanan siluriform origin and a somewhat implicit, but usually not clearly stated, mainly *continental* ulterior passage from the Gondwanan regions to the Northern (Laurasian) areas, i.e, Europe, North America and Asia (see, e.g., Gayet and Meunier, 2003; Teugels, 2003). According to this hypothesis, which I will call 'traditional hypothesis' in this Chapter, the presence of catfishes in these three continents could be mainly due, respectively: to the Eocene connection between Africa and Europe; the Paleocene connection between South America and North America; and the Eocene collision between India and Asia and/or the Eocene passage from Africa to Asia through Europe (see Gayet and Meunier, 2003). Another hypothesis that is put forth by some authors, which I will call marine hypothesis in this Chapter, defends an essentially marine origin/dispersion to explain catfish's wide distribution (see e.g. Chardon, 1967; Gayet and Meunier, 2003).

However, it should be noted that these hypotheses have been mostly formulated outside an explicit phylogenetic framework on catfish intrarelationships. There is, unfortunately, a tendency in several biogeographic works, and particularly in those works concerning major groups of freshwater fishes, to assume certain biogeographic scenarios outside an explicit phylogenetic context as emphasized by authors such as Weitzman and Weitzman (1982), Vari (1988), Lundberg (1998) and Lundberg et al. (1998). As stated by Weitzman and Weitzman (1982), the current distribution of fishes of a given geographical area can be explained best only when our systematic and phylogenetic knowledge of that fish fauna reaches a high level of sophistication. So, a typical problem for biogeographic studies is precisely the lack of available detailed phylogenetic studies on the relationships among the groups to be analysed. As a typical example of a specialised study, many authors have suggested the 'final' (usually meaning Miocene to Pleistocene) uplift of the Andes as the quintessential event in cis/trans Andean vicariance and in the formation of the Amazonas watershed (e.g. Eigenmann, 1909; Géry, 1969; Roberts, 1972). As pointed out by Lundberg et al. (1998: 41), 'such emphasis precedes or overlooks knowledge of the far deeper history of Neotropical fishes, even at fairly fine taxonomic levels'.

In this Section, I will re-examine the biogeographical distribution of catfishes on the basis of results of the cladistic analysis provided above, as well as an up-to-date revision of the data available for different fields such as palaeobiogeography, phylogeny, ecology/physiology, molecular biology, and plate tectonics. I shall argue that the results of the cladistic analysis and the revision of the available data of different fields strongly support an alternative hypothesis on siluriform origin and distribution resulting in a very complex scenario with multiple pre-drift and post-drift continental dispersions, vicariances, and, possibly, some marine migrations. According to this hypothesis: 1) catfishes had a rather old origin in the South-American region than is usually accepted, i.e. the Late Cretaceous, at a time when there were still some remaining Pangean connections between Gondwana and Laurasia; 2) there was a relatively rapid pre-drift continental dispersion of some of the main siluriform groups from the South-American region to Africa and eventually other Gondwanan areas, with some of those groups succeeding to undertake a continental dispersion to the Laurasian continents via the remaining continental Pangean connections existing between the Gondwanan and Laurasian supercontinents; 3) the separation between these two supercontinents, and ulteriorly between the different areas constituting each of them, contributed to important vicariant events; 4) this scenario was further complicated by ulterior events such as the collision of India with Asia, the ulterior re-establishment of certain continental connections that were previously separated (e.g. between the Americas), and eventually also by some marine migrations, thus explaining the highly complex biogeographical distribution of the Siluriformes.

Catfish Higher-level Phylogeny

The phylogenetic results in the present work allow discussion on some interesting points concerning the biogeographical distribution of some major catfish clades. According to Lundberg (1998: 51), due to widespread distribution of catfishes and poorly solved cladograms on the higher-level phylogeny of these fishes available at that time there was no evidence to assume that the ancestral siluriforms 'have been in South America or on the American side of the united southern continents'. The phylogenetic results illustrated in Figure 4.33 change this picture, since all of the eight most basal catfish families of the cladogram of that figure are from South America, with only Neotropical Pimelodidae, Doradidae and Auchenipteridae, appearing together with the Aspredinidae in more apomorphic clades. In the cladograms of Mo (1991) and de Pinna (1998), the most basal catfish were also from South-America, but this exclusively Neotropical basal distribution only applied to two groups (diplomystids and cetopsids).

The phylogenetic results of this work strongly support the existence of four relatively apomorphic catfish groups having members present both in South America and the Old World, shown in Figure 4.33 by 'Y1', 'Y2', 'Y3' and 'Y4' respectively. These are: 1) the Ariidae (Y1), a family restricted to coastal areas, in South America, Africa, Asia, Australia and North America; 2) the clade including the South American auchenipterids + doradids and the African mochokids (Y2); 3) the clade including the South American pimelodids and the Afro-Asiatic bagrids (Y3); 4) and the clade, particularly apomorphic within the Siluriformes, formed by the South American Aspredinidae and the Asian Erethistidae (Y4). The existence of an also relatively apomorphic clade formed by the New World family Ictaluridae, from North America, and the Old World family Cranoglanididae from Asia, should be noted.

One of the clades referred above, namely the one formed by the South American aspredinids and the Asiatic erethistids, was questioned in Gayet and Meunier's (2003: 513-515), an excellent up-to-date overview on catfish palaeontology and palaeobiogeography, since 'the placement of the South American Aspredinidae as the sister-group of the Asian Erethistidae is unlikely outside the Pangean hypothesis', a hypothesis seen as 'unthinkable given the evolutionary level of siluriforms'. In fact, as recognised by Gayet (pers. comm.), a sister group relationship between the Aspredinidae and a group of Asiatic catfishes is not only problematic for the 'traditional hypothesis' commonly accepted nowadays to explain catfish biogeography, but also for the 'marine hypothesis' formulated by authors such as Gayet and Meunier (2003) or

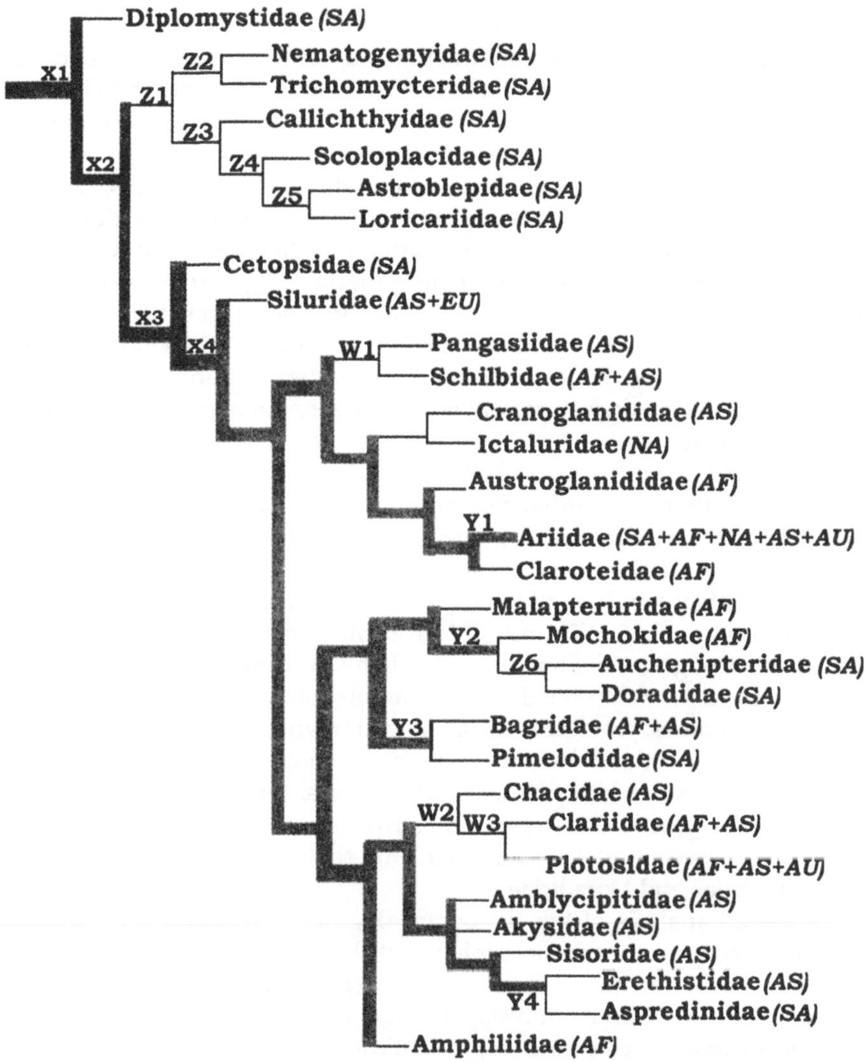

Fig. 4.33 Scheme illustrating the biogeographic hypothesis presented in section 4.7 concerning the distribution of major clades within the Siluriformes in view of the phylogenetic results obtained in the present work: SA= South America; AF= Africa; NA= North America; EU= Europe; AS= Asia; AU= Australia [for more information on numbers and letters, as well as for further explanations, see text].

Chardon (1967). However, it is important to note that the close relationship between aspredinids and Asiatic sisoroids is not only strongly supported in the present cladistic analysis, but also in several other studies such as Ferraris (1989), Mo (1991), Chen (1994), and de Pinna (1993, 1996, 1998).

The clades 'Y1', 'Y2' and 'Y3' are not so problematic as 'Y4' for the "traditional" and "marine" hypothesis, however their presence, with 'Y4', on the cladogram of Figure 4.33 also indicates an origin of Siluriformes significantly older than it is usually accepted. This because it points out that the great majority of the important interfamilial cladogenetic events of catfish evolutionary history have most likely occurred before the separation between the South American and African continents, i.e. at about 112 MY (see Lundberg, 1998): these cladogenetic events are indicated by thick, grey lines in Figure 4.33. This strongly supports the view of Lundberg (1993, 1998), who, based on the distribution of one of these groups, namely mochokids + doradids + auchenipterids, pointed out that considerable diversification within the Siluriformes had already occurred before the final separation of these two continents. It also corroborates the results by Orti and Meyer (1997), based on genetic divergent values derived from mitochondrial DNA sequences, pointing out that *most* of the major groups of the other otophysan order, the Characiformes, were already present before the separation of Africa and South America (see below).

Another indication of a much older origin of catfishes than is usually accepted (i.e. than a Gondwanan origin on the Late Cretaceous) is the biogeographic distribution and phylogenetic relationships of those siluriforms inside clade 'X4' (Figure 4.33). In fact, besides the distribution of the eight most basal catfish families, all South American, indicating that cladogenetic events leading to 'X1', 'X2' and 'X3' probably occurred in South America, there is no strong correlation, after 'X4', between different families of each continent. Indeed, within the 31 catfish families shown in Figure 4.33 (*Note:* genus *Heteropneustes*, previously assigned to his own family Heteropneustidae, is included in Clariidae), there is not even a monophyletic clade constituted by two sister group African families, or by two sister group Asian families. The only cladogenetic event separating two families endemic of the same continent is indicated by 'Z6', leading to the Neotropical families Auchenipteridae and Doradidae.

The overall analysis of the cladogram (Figure 4.33) contradicts the most commonly accepted theories to explain catfish biogeography. If Siluriformes originated in Gondwana at a moment where there were no connections between Gondwana and Laurasia and had evolved almost exclusively in Gondwanan areas much before continental connections between these regions and Laurasian continents came later, a very strong phylogenetic correlation between Gondwanan catfishes, particularly between the South American and African ones, would be expected. However, of all the clades represented in the cladogram (Figure 4.33), only two are exclusively constituted by South American-African families, namely that one uniting the Auchenipteridae + Doradidae + Mochokidae, and that one formed by these three families plus the Malapteruridae.

The cladogram of Figure 4.33 contradicts some of the arguments often used by proponents of the "marine hypothesis". One argument is that many of the oldest catfish fossil records, i.e. of the Late Cretaceous, refer to 'plesiomorphic arioids' (a group in which are included the mostly marine, heavily ossified ariids and often also other groups of heavily ossified siluriforms such as doradids or auchenipterids) were reported from deltaic and marine deposits (see Gayet and Meunier, 2003) . However, the results of the present cladistic analysis did not only contradict a supposedly close relationship between the ariids and the doradids and/or auchenipterids, but also pointed out that Ariidae occupied a rather apomorphic position within the Siluriformes order (Fig. 4.33). This is, by the way, the case of Plotosidae, a family including a large number of marine species, and Aspredinidae, the other catfish family also with some marine members, although in much less proportion than the Ariidae and the Plotosidae (see above). Therefore, under the phylogenetic paradigm, attending to the rather apomorphic position of these three groups, it is far more parsimonious to consider the ability to live in marine environments as an apomorphic, rather rare feature than as a plesiomorphic and rather common character within siluriform evolutionary history.

This calls for attention concerning the 'contemporary importance to historical priority' often discussed in biogeographic and theoretical studies (Gould, 1983). I completely agree with Gould and others that the presence of certain physiological/ecological characteristics in a certain extant group does not *invalidate* the members of this group having presented different physiological/ecological traits in the past. The freshwater extant distribution of most catfish groups does not surely *invalidate* that these groups could have been adapted to marine environments. However, I also think that one should not consider the data from current ecology/physiology as simply irrelevant. Therefore, it should not be considered irrelevant that only three rather apomorphic families among all extant catfish groups present some marine species nowadays, and particularly that, as was referred above and will be discussed with more detail below, these extant 'marine species' seem, indeed, to be confined to coastal areas, and to not be able to live and cross high seas. We should not propose a possible marine origin/dispersion for a group such as catfishes as easily as for a currently mostly marine group such as sharks. One reason for this, under an explicit phylogenetic framework (Fig. 4.33), is that the number of evolutionary steps needed to postulate such a marine origin/dispersion for catfishes is much greater, and, thus, much less parsimonious, than that needed to explain a freshwater origin/dispersion. This stresses the importance of discussing such biogeographic subjects against an explicit phylogenetic scheme on the relationships of the groups been discussed.

I find that the reports of fossil ariids not only from the Maastrichtian but already on the Campanian (see Gayet and Meunier, 2003: Fig. 17.2), far from supporting the "marine hypothesis", in reality support that Siluriformes are very likely much older than commonly accepted, due to the rather apomorphic position of the Ariidae within this order (Fig. 4.33, see also Mo, 1991; de Pinna, 1993). The relatively high proportion of ariid and/or 'arioid' fossils among the older catfishes could, in fact, be due to one or more of the three reasons that follow. Firstly, because ariids and/or 'arioids' are more heavily ossified than most other catfishes, thus facilitating the fossilisation and/or the discovery of these fishes, as written by authors such as Chardon (1968). Secondly, as recognised by authors such as Gayet et al. (1993: 863), 'it is moreover very frequent to find continental vertebrates in coastal marine deposits; such deposits, although formed in a marine environment, are often among the most rich deposits for continental vertebrates' (translation from French, by the author). In respect to this, it is interesting to note that Gayet et al. (1992: 785) did, in fact, explicitly state that the old ariid fossils from the Campanian-Maastrichtian of South-America "lived mainly in freshwaters, their presence in marine waters resulting thus from a post-mortem displacement" (translation from French, by the author). Thirdly, the discovery of relatively few ostariophysan fossils from freshwater deposits of the Cretaceous could precisely be related to the also relatively scarce number of these type of deposits during this Period (1996). In fact, it is striking to note that some of the Cretaceous fossils of otophysan Characiformes, a group currently constituted by freshwater fishes, are also from marine deposits (Gayet et al., 2003; Filleul and Maisey, in press). As for catfishes I have no problem in admitting that *some* Characiformes were, indeed, marine at the Cretaceous, but it is rather difficult to conceive that the *great majority*, or even *all* major characiform clades were marine during that Period, since this would imply a rather impressive number of independent ulterior homoplasious reversions to freshwater environments. And, even more puzzling, if this would have indeed been the case, why would all those characiform groups perfectly adapted to marine environments at that time actually loose, ulteriorly, without seemingly not even a single exception (since there is seemingly not even a single marine characiform group today), an ability that was precisely so important for their own evolution and geographical radiation?

Physiology/Ecology

Apart from the old ariid and/or 'arioid' catfish fossil records, one typical argument used by propounders of the 'marine hypothesis' is the high tolerance to salinity among several extant catfish species, and particularly that species of the families Ariidae, Plotosidae, and, to a much less extent, Aspredinidae, are indeed able to live in marine environments. However, these authors usually seem to forget that a high physiological tolerance to salinity *does not* necessarily confer to a fish the ability to live and/or cross high seas, as stressed by authors such as Rivas (1986): salinity is not even the most important barrier to live and cross open seas, since the latter also differ in the nature of their food supplies, predators, competitors, parasites, physical parameters such as temperature, cover, etc . This seems, by the way, to be precisely supported by the direct empirical observation of the geographical distribution of extant marine catfishes: they are mainly restricted to coastal areas and effectively "never venture to go to the high seas", as noted by Gayet et al. (1992: 785). An important point that should be noted here: the 'marine hypothesis' defended by certain authors to explain catfish distribution assumes that several catfish groups had, in the past, the ability to live and/or cross high seas, that *not even one* group, not even extant ariids, plotosids or aspredinids, seem to display today.

Of course, as stressed by authors such as Unmack (2001: 1065), there is a possibility that fishes that are not adapted to really live in high seas eventually undertake random transoceanic dispersions, such as 'rains and accidental movements by other organisms, whole fish being dropped by birds, or hurricanes' . However, I doubt if such rare phenomena could have contributed in an important way to the biogeographical history of the major Siluriform clades shown in Figure 4.33. Another hypothesis that is sometimes advanced by some authors to explain some transoceanic dispersions of mainly freshwater groups such as catfishes is the existence of 'freshwater oceans'. For example, some authors have hypothesised that some freshwater cypriniforms may have undertook a dispersion 'across the present-day Mediterranean Sea during a 'freshwater phase' that followed the Messinian salinity crisis, during which the basin was transformed into a network of freshwater lakes' (Tsigenopoulos et al., 2003: 208). Some other authors propogate that a similar salinity crisis might have occurred in the arctic ocean during the Late Paleocene/Early Eocene, thus referring to a 'freshwater arctic ocean' (see, e.g., Chang and Maisey, 2003). I can accept, again, that *some* catfishes might have eventually undertook a dispersion across these 'freshwater oceans'. However, one should note that examples of 'freshwater oceans' refer mainly to a time when the major siluriform lineages where already differentiated and already had a rather cosmopolitan distribution (see below, in the discussion concerning 'palaeobio-geography'). To my knowledge, at least, there are no reports of such large, geographically important 'freshwater oceans' in Periods such as the Cretaceous, or the eventual dispersion across these 'freshwater oceans' does not explain the geographical distribution of *most* of the major catfish lineages.

Plate Tectonics

The phylogenetic results discussed above seem to indicate an origin of the Siluriformes much older than usually accepted, i.e. than the Late Cretaceous. Also, contrarily to what would be expected for an origin on an isolated Gondwana, the relationships between the catfish groups of the different Gondwanan regions do not appear to be stronger than those existing between these groups and the taxa occurring in the Laurasian areas, i.e. Europe, North America and Asia. Is it not possible, then, that catfishes originated at a time when there were still some remaining Pangean connections between Gondwana and Laurasia, with the existence of those remaining connections precisely explaining the lack of special correlation between the groups of the different Gondwanan continents and the wide Pangean distribution that these fishes display? In fact, the more I investigated and read about plate tectonics, the more I realised the striking lack of consensus that still dominates several fundamental questions related to this field. This is clearly exemplified by several recent papers, such as Philippe et al.'s 2003 work questioning the timing and place of the latest Pangean connections between Laurasia and Gondwana, Briggs' 2003 article criticising the commonly accepted theories on the biogeographic and tectonic history of India, Dobson's 2003 comments on a supposed Palaeozoic link between eastern Australia and eastern North America, Bosellini's 2002 study re-writing the geodynamics of the eastern Mediterranean, etc.

However, as stressed in Philippe et al.'s 2003 up-to-date paper, there is growing evidence, based on paleobiogeographic data of several dinosaur groups, but also of other organisms such as ferns and lizards, that in Early Cretaceous there were some remaining Pangean connections between Gondwana and Laurasia, and namely between North Africa and North American and/or European continents (see the several examples provided by, e.g., Galton and Taquet, 1982; Buffetaut, 1989; Weishampel, 1990; Le Loeuff, 1991; Buffetaut and Le Loeuff, 1991; Sereno et al., 1994, 1996, 1998; Morell, 1994; Holtz, 1998; Barale et al., 2000; Chure, 2001; Perez-Moreno et al., 1999; Philippe et al., 2003; as well as from the numerous references cited in these papers). Such evidence lead authors such as Le Loeuff (1991) to introduce an explicit Afro-Euro-American palaeobioprovince for the Early Cretaceous. Apart from this strongly supported Pangean connection between Africa and North American and/or European continents, some researchers point out other possible connections in the Early Cretaceous between the Southern continents and some regions currently situated in the North, such as the one between Africa and the Apulia Platform, a sort of Florida peninsula that ulteriorly migrated North, forming the current Italy (see e.g. Bosellini, 2002). But the problem, as emphasized by authors such as Briggs (2003), is that unfortunately many researchers insist on using general geological maps that often date back long and that do not reflect the applicable, recent scientific data available. Thus, they continue not to take into account, in their paleobiogeographic discussions, all the new data available at the moment. Also, as stressed earlier Ebach (2003: 3), many authors, although recognising the importance and the advances of Plate Tectonics, seem to have a tendency to see continental drift 'purely as a geological phenomenon that had virtually no impact on distribution of modern plants and animals'.

According to Sereno et al. (1996: 989), some Pangean connections between North Africa and the North American and/or European continents have 'continued well into the Early Cretaceous', with this explaining, contrarily to what was expected, that the relationships between Early Cretaceous dinosaurs from the Gondwana were not stronger than between these dinosaurs and some others from the Early Cretaceous of Laurasia (see also Morell, 1994). The situation of Early Cretaceous dinosaurs is, in fact, quite similar to that of Siluriformes, in which there is also no special correlation between the different Gondwanan groups (see above). Therefore, it does seem plausible to consider that catfishes, like Early Cretaceous dinosaurs, could have eventually passed through the last Pangean connections between Gondwana and Laurasia. The important thing is, thus, to know if catfishes could have been already there in the Early Cretaceous, or even before that. As explained above, the available data on catfish higher-level phylogeny do provide several different arguments suggesting an origin of the Siluriformes significantly older than is usually acknowledged, that is, than the late Cretaceous. I should argue, in the next paragraphs, that recent paleobiogeographic and molecular data also provide evidence to support this.

Palaeobiogeography of Siluriforms and Other Ostariophysans

The propounders of the "traditional" and "marine" hypotheses usually argue that the direct paleobiogeographic data available for the Siluriformes contradict a Pangean origin of these fishes (see e.g. Gayet and Meunier, 2003). This argument, in fact, is often provided as the 'most important' argument contradicting the diverse remaining evidence, coming from fields such as phylogenetics and molecular biology, supporting a rather old, Pangean origin. However, in my opinion, such an argument should be treated with much caution. In fact, this issue is related with an old and much debated question concerning paleobiogeographic studies: is the absence of evidence, evidence of absence? This subject has long been discussed, and was the subject of Lundberg's (1998: 64) overview on the temporal context of the diversification of Neotropical fishes, in which this author, referring precisely to the otophysan fishes, stated: 'one possible answer, especially for the absence in the Early Cretaceous record of taxa like cichlids placed high up in the tree of acanthomorphs, is yes, they just had not evolved by then; the other possible answer perhaps more applicable to otophysans (which include catfish) for which there is a basis for predicting a ghost-like existence, is no, some had originated and perhaps diversified but they have escaped preservation or detection'. One particularly elucidative example given by this author concerns precisely the catfishes: †*Corydoras revelatus*, a callichthyid record from the late Palaeocene, ca. 58.5 MY, was described more than three-quarters of a century ago by Cockerell (1925), but, despite the heavy bony armour in which callichthyids are encased, there are no subsequent fossil records of these fishes at least until 20 millions of years latter.

As stated Lundberg (1998), one of the problems related with paleobiogeographic hypothesis is that, although palaeontologists do acknowledge, in theory, that 'absence of evidence' does not correspond to 'evidence of absence', they often do refer, in practice, to the absence of evidence as evidence of absence. This is precisely the case concerning the origin of the Siluriformes, in which the defenders of the "traditional" and "marine" hypotheses use the *absence* of catfish fossils older than the Late Cretaceous as an '*evidence*' of the absence of this group before this period. However, as noted in Stiassny et al.'s (in press) up-to-date overview on Gnathostome fishes, within the rayfin group it is not an exception, but rather a rule, that the estimated age of a taxon (indicated by phylogenetic data, molecular clocks, etc.) differs considerably from the age of the older fossil known for that group. One remarkable example, among many and many others, given by these authors to illustrate this is that of bichirs (Polypteridae): although phylogenetically these fishes are placed just one branch above the earliest well-preserved actinopterygian, †*Dialipina*, from the lower Devonian, their fossil record only just extends to the Lower Cretaceous.

I consider that the palaeontological data for catfishes do provide, in reality, strong indirect evidence supporting, as do the phylogenetic and molecular data, a rather older origin of these fishes. As explained above, the propounders of the "traditional hypothesis" usually argue that the presence of these fishes in the Laurasian regions of Europe, North America and Asia can be mainly explained by, respectively: the Eocene connection between Africa and Europe; the Paleocene connection between South America and North America; and the Eocene collision between India and Asia and/or the Eocene passage from Africa to Asia through Europe (see Gayet and Meunier, 2003). But it is important to note that ariid fossil otoliths were reported from North America in the *late Campanian*, that is, *much before* the *Palaeocene* (Gayet and Meunier, 2003: Fig. 17.2). Ariid fossil otoliths were also reported from Europe in the *Palaeocene*, that is, *before* the *Eocene* (Gayet and Meunier, 2003: Fig. 17.2). These records, referring to otoliths, have sometimes been questioned, but the problem is that it is not clear whether they are questioned because they refer to otoliths or because they merely do not fit in the scenario proposed by this "traditional hypothesis" (see Gayet and Meunier, 2003). But the important point is that, besides these sometimes questioned fossils referring to otoliths, there are records referring not to otoliths but to *bone remains* of undetermined siluriform fossils from the *Maastrichtian* of Europe, that is, *much before* the *Eocene*, as shown in Gayet and Meunier (2003: Fig. 17.4). To this, one should add the report on ariid fossils, referring not to otoliths but also to *bone remains*, on the *Maastrichtian* of *Asia*, that is, again *much before* the *Eocene* (Gayet and Meunier, 2003: Fig. 17.2).

Of course, the propounders of the "traditional hypothesis" could say that perhaps what this means is that the land contacts allowing the passage of catfishes from Gondwana to these Laurasian continents occurred previously than they stated. For example, they could explain that the passage to Europe, and thus to Asia, could have been done by means of an eventual land contact between Africa and Europe on the Maastrichtian (Gayet, pers. comm.). However, the presence of catfish fossils on the Campanian of North America would continue to be particularly problematic for the "traditional hypothesis". Some authors do stress that the 'Paleocene' connection between South America and North America might have occurred, in reality, on the Paleocene but also on the Maastrichtian (see e.g. Gayet, 2001); however this would not explain, anyway, the Campanian catfish fossils of North America. The propounders of the "marine hypothesis" (see above) could, nevertheless, always argue that these ariid Campanian fossils might have passed from South America to North America via a direct marine dispersion (see above). This issue regarding possible transoceanic dispersions of catfish taxa has already been discussed.

But what is more important to me here is that, be that as it may, the Late Cretaceous catfishes already had an impressive worldwide distribution, with fossils found in the various continents at that Period being from phylogenetically derived groups such as ariids and presenting a morphology similar to that of extant siluriforms (see above). The paleobiogeographic data available, evidencing such a worldwide catfish distribution already at the Late Cretaceous, does not contradict, in my view, that these fishes originated before this Epoch, as stated above. One other important detail pointing towards a rather older origin of catfishes is precisely that the amazing distribution of recent catfishes, itself indicating a Pangean siluriform origin, appears still more complex at a paleobiogeographic level, with, for example: 1) 'bagrid' fossils being reported from North America (see Gayet and Meunier, 2003), while extant bagrids are only found in Africa and Asia (Fig. 4.33); 2) bagrid, clariid and ariid fossils being reported from Europe (see Gayet and Meunier, 2003), a continent where these three groups are now absent (Fig. 4.33); 3) apparent 'ictalurid' fossil forms being reported from Asia (see Stucky, 1982), a continent lacking extant ictalurids (Fig. 4.33); and 4) undetermined fossil catfishes being found in Antarctica, a continent with no extant Siluriformes (see Grande and Eastman, 1986). The case of bagrids is particularly worthy of mention. These fishes are now found in Asia and Africa, but: bagrid fossils were reported from Miocene and Pliocene of Europe; 'bagrid' fossils, although somewhat dubious, were reported from Eocene of North America; with some researchers even reporting seemingly extant bagrids from the Neotropical region (Lundberg, pers. comm.). If this would be so, bagrids would demonstrate a cosmopolitan distribution perhaps only comparable to that of Ariidae, being reported, thus, from Africa, Asia, and possibly South America (extant forms) and from Europe and North America (fossil forms).

The paleobiogeographic data on other ostariophysan groups also provides strong indirect evidence supporting the view that catfishes might have originated before the Late Cretaceous. In fact, Arratia (1997) described an ostariophysan fossil, †*Tichlingenichthys*, from the Late Jurassic, ca. 150MY. Poyato-Ariza (1996) described a chanid fossil from the Gonorynchiformes, that is, the sister group of the Otophysi, of about 140MY. This means that at about 140MY there were already fishes from an extant ostariophysan family, i.e., the Chanidae. Thus, if members of this extant gonorynchiform family were already there around 140MY, this necessarily means that the first Gonorynchiformes, and, thus, the first Otophysi, were there well before that (if the sister group relationship between the Otophysi and the Gonorynchiformes is accepted: see below). In fact, it is particularly relevant to refer here that, at the time I was writing this text defending a catfish origin older than the Late Cretaceous, Filleul and Maisey (in press) precisely described a characiform fossil, i.e., a fossil from the otophysan order that is seemingly the sister group of the clade Gymnotiformes + Siluriformes (see below), from the Early Cretaceous. This, as stressed Maisey (pers. comm.), does not only make 'us recalibrate the origin of the Otophysi and of the different otophysan orders', but also gives a clear example on how erroneous it is to use the "absence" of records of a certain group in a certain Epoch as "evidence" of the absence of that group at that Epoch, since, in fact, older fossils of existing groups can be, and often do continue to be, discovered.

Catfish Sister-group

As the reader might have noted, when referring to the sister group of the Siluriformes in the paragraph above, I was particularly prudent, referring that this sister group is *seemingly* the order Gymnotiformes. Surely, as explained in Chapter 1, I agree with most authors that the studies of Fink and Fink (1981) providing the first explicit cladistic analysis of ostariophysan relationships, and the paper by the same authors in 1996 corroborating this 1881's study, have indeed largely contributed for our knowledge on this subject. I also agree that the phylogenetic scheme provided by these authors indeed constitutes a solid, well-grounded working hypothesis, and I am completely aware that this scheme is clearly the most commonly accepted nowadays.

However, I also think that one cannot simply neglect the cladistic studies on ostariophysan phylogeny done by other authors after the first cladistic study by Fink and Fink (1981). As regards this, one is obliged to admit that most of these studies have indeed contradicted some of the sister group relationships supported by Fink and Fink. In fact, apart from Orti's 1997 analysis of molecular data from mtDNA and all codon-positions of the ependymin gene, which fully supported Fink and Fink's scheme, the only published works, at least to my knowledge, that theoretically supported this scheme, i.e. Dimmick and Larson (1996) and Arratia (1992), did only support it after the characters of Fink and Fink were introduced in the phylogenetic analysis. The independent analysis promoted by Dimmick and Larson (1996), based on molecular characters without including the morphological characters of Fink and Fink, suggested in reality that the sister group of Siluriformes is a clade formed by both Gymnotiformes and Characiformes, and not by Gymnotiformes alone, as indicated by Fink and Fink. The independent analysis promoted by Arratia (1992), based only on her morphological characters of the suspensorium without including those characters of Fink and Fink (1981), suggested that the characiforms are the sister group of cypriniforms, and not, as suggested by Fink and Fink, of the clade gymnotiforms + siluriforms. It is important to note that the independent results based exclusively on Dimmick and Larson's (1996) molecular data indicating a sister group relationship between the Siluriformes and the clade Characiformes + Gymnotiformes were also supported by Orti's 1997 analysis of the first and second codon positions of the ependymin gene. These results were also supported by Saitoh et al.'s 2003 analysis based on mitochondrial genome. Thus, one cannot say that the relationships among the otophysan orders constitute a completely consensual issue nowadays. In fact, even the sister group relationship between the

Gonorynchiformes and the Otophysi, and, thus, the monophyly of the Ostariophysi as a whole, has been recently put in question by molecular studies such as Saitoh et al. (2003) and Ishiguro et al. (2003).

Anyway, what is important to note here, in my opinion, is that, either accepting the Gymnotiformes, or, instead, the clade Gymnotiformes + Characiformes as the sister group of Siluriformes, this sister-group would be anyway a group almost exclusively constituted by freshwater fishes. Extant Gymnotiformes, as well as the very few known fossils from this group, are freshwater. Among Characiformes, the numerous extant members of this order are freshwater, but some fossils were discovered on marine deposits (see e.g. Gayet et al., 2003), which can either be seen as indicating that at least some ancient members of this group were marine, or, instead, that marine deposits can, indeed, include freshwater organims (see above). Be that as it may, the sister group of Siluriformes, either being Gymnotiformes or the clade constituted by these fishes and Characiformes, is mostly formed by freshwater fishes. And this is something that, in my view, cannot simply be neglected, since, under the cladistic paradigm, the information provided by the sister group should indeed be taken into account when one discusses the evolutionary history of a certain feature within the ingroup being studied. That is, if the 'first' gymnotiform or the 'first' member of the clade Gymnotiformes + Characiformes (in case we consider this latter clade as the sister group of catfishes) were freshwater, this would imply a further extra evolutionary step for the 'marine hypothesis' defended by some authors (i.e., we should now consider a first evolutionary transition conferring catfishes the ability to undertake marine dispersions, and, as explained above, an ulterior series of numerous evolutionary reversions making that nowadays no catfish groups, not even taxa such as ariids or plotosids, can seemingly enter and disperse through open seas: see above).

Other arguments used by some researchers contradicting a Pangean, and sometimes even a Gondwanan origin for catfishes is that gymnotiforms, either extant or fossil, are exclusively known from South-America. There are two points I would like to mention here. One is that, as explained just above, some studies have suggested that perhaps the sister group of catfishes is constituted not only by Gymnotiformes but by these fishes plus Characiformes. Thus, if this would be the case, the sistergroup of catfishes would have a Pangean distribution, as, apart South America and Africa (Gondwana), some characiform fossils were found in Europe (Laurasia) (see, e.g., Gayet et al., 2003; Filleul and Maisey, in press). But the most important point is that this 'sister group distribution' argument seems really flawed to me. Why should one consider that two sister groups have necessarily occupied the exact same geographical areas in the past, if this is not what the most elemental empirical data concerning current examples show us today? Why, if Gymnotiformes are indeed the sister group of Siluriformes, should we consider that both these groups had exactly the same geographical distribution in the Mesozoic? If we take the current geographical distribution of gymnotiforms and siluriforms on the South-American continent, the distribution of siluriforms is *considerably wider* than that of gymnotiforms, with some catfish groups extending, for example, so far South as latitude 47°30′ (Berra, 2001). Thus, as a simple theoretical example to illustrate this point, if there were an eventual separation at about latitude 45° from that small South American area situated South to this latitude and the rest of this continent, catfishes would be present in this small area, while gymnotiforms would not. So, if we can accept this today, why cannot a similar situation have occurred in the Mesozoic? That is, why cannot we conceive that gymnotiforms simply were not so broadly distributed as catfishes at that Period, and, thus, did not succeed to pass from the South-American region to Africa and, thus, to reach the last remaining Pangean connections between Gondwana and Laurasia allowing them to radiate to this last supercontinent (see above)? This seems, in my opinion, clearly the most parsimonious option to consider in this specific case, more parsimonious than to consider necessarily that, if catfishes have done so, Gymnotiformes have necessarily also passed from South America to the other globe regions and ulteriorly were extinct in these latter regions. This hypothesis is, as all hypotheses formulated in the present work, a scientific hypothesis that should be put to test in the future. In reality, in this specific case, this hypothesis can be easily contradicted: it suffices to find either a fossil record or an extant species of Gymnotiformes in Africa or another continent other than South America to prove it wrong.

Molecular Clocks

First of all, I would like to mention that I am completely aware that, as stressed by authors such as Vences et al. (2001: 1095), molecular estimates of the age of divergence between major lineages should be viewed with caution, considering the 'restrictions inherent in molecular clock datings'. However, I do think that one cannot simply neglect all those studies providing such molecular estimates. And, in fact, it is interesting to note that the data provided by these studies does suggest, once again, a rather old origin of the ostariophysan lineages. For example, the recent molecular study by Saitoh et al. (2003), based on an analysis of complete mitochondrial DNA sequences of members of the five ostariophysan orders, as well as of other teleost groups, did precisely suggest a Pangean origin for otophysans. According to Saitoh et al. (2003: 459) 'otophysan basal divergence took place no later than the Jurassic, possibly as early as the end of the Permian Period on the Pangean Continent'. Moreover, it is important to note that, in this same paper, these authors pointed out that, among otophysans, siluriforms probably have a rather old origin, older than that of gymnotiforms and that of characiforms (Saitoh et al., 2003: 470). This strongly supports what is defended in this work, that catfish are a rather old group. Such a rather old origin of the otophysan fishes was also suggested in Orti and Meyer's (1997: 238-239) work using genetic divergence values, according to which '*most* lineages of characiform fishes had originated *before* the vicariant event separating African and Neotropical taxa'.

As concluding remarks of this Chapter, I would thus firstly like to emphasise that a detailed review of all available data, a great part of wish provided in recent studies, from the different fields discussed above do provide evidence to support that catfishes were very likely originated before than it is usually accepted, i.e. than the Late Cretaceous, at a time when there were still some remaining Pangean connections between Gondwana and Laurasia. This origin would have been confined, first, to the South-American side. Then, catfishes would have dispersed from the South-American area to other areas, in a rather complex and geologically rapid 'radiative' pattern, with some groups migrating, via pre-drift dispersion, to Laurasian regions. Therefore, the separation between Laurasia and Gondwana, and ulteriorly between the different areas constituting each of these supercontinents, would have contributed to important vicariant events. This scenario was probably still further complicated by ulterior events such as the collision between India and Asia, the ulterior re-establishment of certain continental bridges such as those between the Americas, and perhaps also by some marine migrations. This would thus explain the highly complex biogeographical distribution of the Siluriformes (Fig. 4.33).

Secondly, I would like to draw attention to the fact that, as stated by authors such as Briggs (2003: 381), although many researchers recognize the important advances on plate tectonics, they continue to use, in the discussion of their biogeographic hypotheses, general maps that do often not provide 'accurate representations of current scientific knowledge'. In my opinion this is very likely the reason why, in reality, although there is growing evidence supporting a rather old origin of modern teleost groups such as the Ostariophysi (e.g. the Late Jurassic/Early Cretaceous: see, for example, Stiassny et al., in press), there seems to be an underlying taboo when it comes to hypothesize an eventual Pangean origin for such modern groups. This precisely because, despite the substantial amount of data supporting the existence of remaining Pangean connections between Laurasia and Gondwana in the Late Jurassic and very likely 'well into the Early Cretaceous' (see above), many ichthyologists continue to use general and sometimes old fashioned maps that do often not take into account such data. If there is growing evidence supporting an origin of ostariophysans in the Late Jurassic and of catfishes at least in the Early Cretaceous, as well as supporting some remaining important Pangean connections between Laurasia and Gondwana at that time, why cannot one hypothesize that such connections could have indeed played a role in the distribution of these groups, and, thus, of 'modern' teleost taxa? Again, this hypothesis is a scientific hypothesis that can, and should, be tested and, hopefully, stimulate further discussions, thus allowing the progress of scientific research. It is therefore hoped that this work could precisely instigate future phylogenetic, palaeontological, tectonic, biogeographical and molecular studies that might bring light to the puzzling, but clearly interesting, questions related to the origin and distribution of catfishes and other ostariophysans, as well as of the Teleostei in general. In fact, concerning this issue, it is interesting to note that, as noted by Lundberg et al. (1998: 43), 'the more we discover in the fossil record, and the more we resolve phylogenetic relationships of fishes, the very much older our estimates of taxic origin get'. So, to close this Section, as well as Chapter 4 devoted to siluriform general macroevolution, I would like to paraphrase the brilliant last paragraph of Lundberg's 1998 overview on the origin and diversification of Neotropical fishes, which, although referring to South American fishes in general, and to a characiform species in particular, applies fittingly to the Siluriformes:

"Probably like most who were taught vertebrate paleontology from the fast-paced, high-turnover perspective of paleo-mammalogy, I am struck by the apparent great antiquity and static history of almost all modern lower taxa of Neotropical fishes that have left fossil traces. Over the course of the latest 13.5 million years many things have changed in the Neotropics – new ranges and heights of the Andes, new drainage basins, an immense Amazonian lake or lakes influenced by a marine transgression, complete right-angle shifts of the main courses of the Amazon and Orinoco, a land bridge to Middle and North America, newly evolved and immigrant mammals plus extinctions of old groups. But as far as we can tell from unmistakable similarities of specialised jaws and teeth, the cambaqui or cachama fish, Colossoma macropomum, *has persisted throughout this period or even longer, apparently unchanging in its diet of fruits and seeds. So also have persisted several other now-ancient lineages of Neotropical fishes. Against this backdrop of ancient age and remarkable conservatorism, one should wonder with concern that if these unusual fishes were to disappear, they would not soon be replaced".*

5

Catfishes, Case Study for General Discussions of Phylogenetic and Macroevolutionary Topics

5.1 PRIMARY HOMOLOGIES, SECONDARY HOMOLOGIES, AND A PRIORI VERSUS A POSTERIORI EXPLANATIONS IN EVOLUTIONARY BIOLOGY

Chapter 4 began with the presentation and a short discussion of an important topic concerning a priori versus a posteriori explanations in evolutionary biology, which has been referenced several times throughout this book. As explained in that Chapter, this controversial issue can be broadly summarised as follows. Some authors, following the basic guidelines of the cladistic paradigm, hold that evolutionary explanations should be *based on* comprehensive, explicit phylogenetic cladograms, and not, as done by some 'functional evolutionary morphologists' sensu Cracraft, 1981, be *a priori* to, or irrespective of, phylogenetic comparisons. The latter hold that homoplasic characters should in some way be detected and removed *a priori* to the phylogenetic analysis, and usually formulate major evolutionary explanations, such as those concerning the evolution of certain complex systems or the reconstruction of potential 'ancestral forms', *a priori* to the phylogenetic comparison. As emphasised in the present work, such *a priori* evolutionary explanations rely necessarily on more or less ad hoc stories, with this being particularly true for those concerning homoplasy or character evolution, which are 'ad libitum explanations, capable of explaining patterns and non-patterns alike and, in being able to do so, ..., explain nothing at all' (Kludge, 2001: 202).

One of the best practical examples, within the present work, to illustrate this point concerns the evolution of a *rocking* versus *sliding* abduction mechanism of the palatine-maxillary system in siluroids, discussed in Section 4.1. When Gosline (1975), in a work that would definitively be ranked by Cracraft (1981) as an example of the 'functional evolutionary morphology' school, provided an insight into the evolution of the catfish palatine-maxillary system, he stated that 'it seems clear' that a rocking system evolved more than once from a generalised sliding one (Gosline, 1975: 23). This is because a sliding mechanism is present in some catfish groups such as pimelodids and bagrids, but also, according to him (and contrary to the observations of the present work: see char. 294), doradids, ariids, schilbids or pangasiids, which are 'at the base of modern siluroid series, whatever point of view considered' (Gosline, 1975: 23). Moreover, according to Gosline (1975: 23), 'at least in some catfishes' the development of a rocking mechanism can readily 'be traced back to forms with sliding autopalatines'.

The quality and excellence of Gosline's 1975 work is beyond dispute. In fact, I have particular personal admiration for the work and life of W. Gosline who, together with McN. Alexander, holds a place in my list of most admired zoologists. Such zoologists with a general interest and a broad philosophical vision are clearly rare in these modern times of essentially narrow and mostly directly focused interests. Gosline's 1975 work, in particular, constitutes one of the main landmarks in catfish bibliography, providing a series of original and interesting insights into the palatine-maxillary system of these fishes, in the good and salutary tradition of functional evolutionary morphology. However, as well as the unquestionable traditional qualities of this school, which were exalted above in Chapter 4, Gosline's 1975 work also illustrates the typical flaws associated with it, as pointed out by authors such as Cracraft (1981), Coddington (1988), de Pinna (1991), Kludge (2001), or Desutter-Grandcolas et al. (2003). Statements such as 'it seems clear' that a rocking palatine-maxillary system evolved more than once from a generalised sliding one, or that the development of a rocking mechanism 'can be easily traced back' to forms with sliding autopalatines, constitute, in fact, somewhat gratuitous conclusions based mainly on ad hoc hypotheses and not tested against any type of explicit phylogenetic scheme. If a rocking mechanism 'can be easily traced back' to forms with sliding autopalatines, why cannot a sliding mechanism 'be traced back' to forms with rocking autopalatines, for example? In which 'strict' known evolutionary rules are such dogmatic statements based?

The problem, in reality, is not so much the enunciation of such hypotheses: scientific hypotheses are always welcome in science. The problem is rather the implicit acceptation of these hypotheses as somewhat dogmatic evolutionary explanations with no *a posteriori* test adduced. In Popper's tradition, an ideal scientific study should start with the enunciation of a certain hypothesis, which *should be tested* subsequently. This test would not serve to 'validate' the hypothesis and confer on it a status of 'scientific truth', but rather to invalidate and eventually discard it if it fails the test or, eventually, to strengthen it if it passes the test. In this latter case, however, the hypothesis, albeit strengthened, would nonetheless continue to constitute just a particular scientific hypothesis. In this respect, I agree with Cracraft (1981) that many 'functional evolutionary morphological' works do not really proceed in this ideal Popperian way. They often *finish* with the enunciation of a certain *evolutionary* hypothesis based mainly on somewhat ad hoc conjunctures without proceeding to the necessary subsequent test, which should imperatively consist of, insofar as evolutionary scenarios are concerned, an *a posteriori* confrontation to an explicit phylogenetic scenario concerning the biological group under analysis.

So, the work of Gosline (1975) is, as mentioned above, excellent in what concerns the functional schemes provided, and has been formulated with remarkable vision and sagacity. But it fails in the implicit acceptation, with more or less dogmatic terms, of those hypotheses without proceeding to any subsequent kind of test. Gosline's 'conclusions' were enunciated more than 25 years ago and underwent no subsequent test, until the present moment, in terms of an explicit phylogenetic scheme and/or other independent data.

In fact, in what concerns the evolution of a rocking versus a sliding palatine-maxillary system in siluroids, the results of the present phylogenetic analysis contradict Gosline's 'conclusions', with the presence of a true sliding mechanism appearing as restricted to a relatively small and apomorphic group within the Siluriformes (Fig. 4.20), as explained in Section 4.4. One could argue that the difference between the phylogenetic scenario proposed in Fig. 4.20 and the 'conclusions' of Gosline (1975) is mainly due to the fact that Gosline considered, *contra* the functional observations made in the present work (see char. 294) as well as in other studies (see, e.g., Alexander, 1965; Royero and Neville, 1997; Oliveira et al., 2002), that ariids, doradids, auchenipterids, schilbids and pangasiids exhibit a true sliding mechanism (see char. 294). One might add that had the author of the present work considered, like Gosline, that these five groups exhibit a true sliding mechanism, he would perhaps have coded this mechanism as the plesiomorphic, general state for non-diplomystid catfishes, with this eventually leading to an evolutionary result similar to Gosline's evolutionary scenario. However, as clearly shown in Fig. 5.1, even if ariids, auchenipterids, doradids, pangasiids and schilbids had been coded (*contra* the observations of this work), altogether with heptapterins, pimelodins and bagrids, in a supposedly plesiomorphic 'sliding' character state, the cladogram obtained would, once again, favour a rather generalised distribution of the rocking mechanism, with the sliding system being only subsequently developed in certain specific, relatively derived clades. This, of course, is not a critique of the excellent work of Gosline, which, as mentioned above, the author particularly admires. It is simply a statement of a typical general flaw often committed in 'functional evolutionary' studies, as noted by many authors, e,g, Cracraft (1981), Coddington (1988, 1990), Kludge (2001) or Desutter-Grandcolas et al. (2003).

The example provided in Fig. 5.1 and discussed above elucidates, in reality, an important point that is also narrowly related to the issue of a priori versus a posteriori evolutionary explanations in biological studies, which is often the subject of misunderstandings, namely the polarity coding of characters in cladistic studies. This misunderstanding is, as explained by Cracraft (1981), often perpetrated by 'functional evolutionary morphologists' who state

Fig. 5.1 Hypothesis of character state evolution of rocking/sliding palatine-maxillary system as shown in Figure 4.20, but with the results of present work modified (compare with Fig. 4.20) in order to discuss Gosline's 1975 evolutionary conclusions and notions of primary and secondary homologies: CS-0 (black)= sliding palatine-maxillary system; CS-1 (blue)= rocking palatine-maxillary system; Ambiguity or Inapplicable (pink) [for more details, see text].

that when coding the polarity of characters, cladists are also making an evolutionary statement a priori to the cladistic analysis. This issue was discussed in detail in de Pinna's 1991 paper, 'Concepts and tests of homology in the cladistic paradigm'. As he explained, when coding the polarity of a certain character, cladists are not making an evolutionary statement or providing an evolutionary explanation, but simply establishing a *primary homology hypothesis* concerning the character being coded. This *a priori primary homology hypothesis* is, however, confronted, *a posteriori,* in the same study, to the results of the cladistic analysis of *all* the data available. Only then can it eventually be passed to a status of *secondary homology hypothesis,* if corroborated vis-à-vis the information based on the distribution of *all* the remaining characters included in the analysis. The eventual evolutionary explanations concerning the character under discussion would only come therefore *a posteriori* to the test of the primary homology vis-à-vis the phylogenetic results based on the taxonomic distribution of *all* the features examined. These phylogenetic results could either *refute the primary homology hypothesis* or, alternatively, *strengthen the hypothesis,* with the hypothesis passing in the latter case, to a status of *secondary homology hypothesis,* but remaining, in the Popperian tradition, just a scientific hypothesis.

Thus, when coding a certain character state as plesiomorphic, cladists are not *making an evolutionary statement,* but simply elaborating a *primary hypothesis* that will always *be tested,* in *the same* cladistic study, by the parsimonious distribution of *all* the available characters. The phylogenetic analysis could, in reality, and often does (see, e.g., Kitching et al., 1998), *refute* the plesiomorphic status initially attributed under the *primary homology hypothesis* to a certain particular feature, if the subsequent parsimonious analysis indicates so. This important point is concisely elucidated in Figure 5.1. It clearly shows that if instead of coding the rocking autopalatine system as plesiomorphic for non-diplomystid catfishes (see char. 294), one had coded the sliding system as the plesiomorphic state, the parsimonious analysis of all the 440 available characters would indicate, anyway, that the rocking system would be the basal, general situation for non-diplomystids, with a sliding system only subsequently acquired independently in some less inclusive, specific clades (Fig. 5.1). That is, the *a posteriori* analysis, based on the phylogenetic distribution of *all* 440 characters available, would *refute* the plesiomorphic status *initially* attributed as a primary homology to the sliding mechanism among non-diplomystids. [Note: diplomystids have a palatine-maxillary system that differs significantly from that of non-diplomystids, as explained in Section 4.4.] This is a very important point often misconstrued by non-cladists, and occasionally even blurred by some cladists.

Of course, one could argue that the coding of a certain specific character influences the whole phylogenetic analysis. But this influence is precisely *minimised* by the use of a *large set* of more or less independent characters. The greater the number of characters used, the less the influence of any particular character in the analysis. This has been constantly highlighted throughout the present study, in the discussion of the evolutionary history of homoplasic features, such as those concerning the width of the mesial limb of the posttemporo-supracleithrum (Fig. 4.10) or the suture between this mesial limb and the neurocranium (Fig. 4.11) (see Chapter 4). The homoplasic status of these features was indeed indicated and discriminated *a posteriori* by the analysis of all the data included in the present phylogenetic study.

In fact, concerning the methodology of removing *a priori* those characters seemingly *susceptible* to homoplasy, one could argue that, in a purely theoretical point of view, all characters are potentially susceptible to homoplasy and hence, following this methodology ad absurdum, would remove all characters from a cladistic analysis. Indeed, until now no evolutionary law has been formulated stating that a certain character *cannot be, a priori,* the subject of homoplasy. Some could eventually argue that some complex systems, such as the Weberian apparatus of ostariophysan teleosts, are seemingly free of homoplasies. However, the picture is definitely not so simple, at least from a theoretical point of view.

Firstly, there are well-documented reports of outstanding homoplasies regarding particularly complex systems, such as the complex electrogenic and electrosensory systems of gymnotiform and mormyriform electric fishes (see, e.g., Alves-Gomes, 1999, 2001). As noted by Alves-Gomes (1999: 1168), the only way to consistently discriminate the homoplasic development of the strikingly similar and remarkably complex electrogenic and electrosensory systems in these two groups of fishes is to confront this character with a cladogram of the higher level phylogeny of the Teleostei (Alves-Gomes, 1999: Fig. 1): 'the two ancestral teleost lineages that later gave rise to the living mormyriforms and gymnotiforms had been separated for at least 140 million years, as inferred from the oldest fossil assigned to the osteoglossomorphs, and there is no evidence that either electric organs or electroreceptors were present in any of the most basal ostariophysans (in the case of gymnotiforms) or osteoglossomorphs (in the case of mormyriforms)'. In fact, some old authors, such as Swainson (1838), had in the nineteenth century placed the gymnotiforms, osteoglossiforms and anguilliforms altogether.

Secondly, one should not be tempted to enunciate a general conclusive dogma such as 'a certain complex character *cannot be, a priori,* the subject of homoplasy' just because the taxonomic diversity documented so far, and the phylogenetic conjectures accepted at the present time, do not contradict it. As emphasised above, a scientific hypothesis should always be seen as one that should imperatively be continually tested vis-à-vis new discoveries and new insights. If, for example, the mormyriform electric fishes had not yet been discovered, one would perhaps be tempted to say that 'such complex systems as the electrogenic and electrosensory systems found in gymnotiforms *are surely free* of homoplasy'. Several species, genera, and even teleostean families are being continually discovered (see the excellent overview 'So many fishes, so little time: an overview of recent ichthyological discovery in continental waters' by Lundberg et al., 2000). Additionally, hundreds of fossils are also being unremittingly reported. So, even systems nowadays referred as *seemingly* homoplasy-free could prove to be homoplasic in the face of new discoveries. The palaeontological findings of Gayet, for example, precisely put in question the rather simple, 'clean' and non-homoplasic linear evolutionary scheme proposed by authors such as Fink and Fink (1981) regarding the origin and evolution of the Weberian complex in the Ostariophysi (see, e.g. Gayet, 1986a,b). It should also be kept in mind that the diversity seen today of either living or fossil forms is not the *final* product of evolution. Evolution is happening and new forms will arise; thus, apparent homoplasic events that have not evolved so far will not necessarily be absent in all future biological evolutionary history. To say that a certain feature *cannot be,* in any possible case or at any possible time, the subject of homoplasy sounds, on the face of the relatively scarce knowledge available on macroevolutionary processes, somewhat dogmatic.

Thirdly, even if this dogma should eventually prove true and, for example, complex systems such as the Weberian apparatus were completely homoplasy-free since forever and ever, where would this lead us? Would one eliminate all the remaining 'bad characters', *a priori susceptible* to homoplasy, and only use these kind of 'privileged' supercomplexes *apparently* immune to homoplasy? Evolutionary events such as development of the Weberian apparatus are surely useful to support the monophyly of a *major clade including four large teleostean orders* (see Chapter I), but are not usually seen *within these orders,* and still less *within their families, genera or species.* Of course, it is preferable to include the maximum possible of those complex structures in a phylogenetic study, which was precisely one of the major methodological concerns of the present work, as explained in Chapters 1 and 2 of this volume (see also the discussion on this subject below). However, one surely cannot expect to use such 'super complex' characters in the search for phylogenetic relationships between, for example, the more than 140 species (Armbruster, pers. comm.) of the loricariid catfish genus *Hypostomus,* which are distinguished mainly by minor, not 'so relevant', morphological differences. So, in defending the removal of 'bad characters' *apparently* (based on their ad hoc *a priori* evolutionary conjectures) susceptible to homoplasy, some evolutionary morphologists are indirectly weakening the power of morphological data in phylogenetic studies. Morphological data would thus be seen as useful for inferring phylogenetic relationships of major clades, but as very limited in the search of relationships among less inclusive groups such as genera and/or species, with the phylogeny of the latter groups being thus obligatorily reserved for the exclusive study of geneticists.

As explained above, the present work, like earlier studies (see the examples provided by Cracraft, 1981; Coddington, 1988, 1990; de Pinna, 1991; Kitching et al., 1998; Kludge, 2001; Desutter-Grandcolas et al., 2003; and references therein), reinforces the point that homoplasic events can, and *should be,* discriminated *a posteriori* against the results of an explicit phylogenetic analysis using *all* the data available, under the basic cladistic paradigm proposed by

Hennig (1950, 1965, 1966, 1981). Homoplasic features within a general clade are not necessarily so for less inclusive groups. These features could thus reveal useful information concerning the phylogeny of some internal groups inside the general clade under analysis. This has been emphasised throughout this book with several elucidating key-examples provided, such as those concerning the shape of the arrector ventralis (Fig. 4.6), interdigitation of the scapulo-coracoids (Fig. 4.8), development of the mesial limb of the post-temporo-supracleithrum (Fig. 4.10), suture between this limb and the neurocranium (Fig. 4.11), relation between the adductor mandibulae and the levator arcus palatini (Fig. 4.14), and differentiation of the extensor tentaculi (Fig. 4.23). Moreover, tracing these features in the cladograms obtained provides inestimable information on the macroevolution of the general group being analysed. More than just enunciating *a priori* that a certain character *seems* to be homoplasic within a certain group and subsequently *removing* it from the analysis, thereby *losing* a whole series of phylogenetic and evolutionary information, it is particularly interesting to precisely *include* it in the analysis and trace out, in the cladogram obtained, *all the eventual evolutionary homoplasic events* regarding this character and thereby its evolutionary history within this group.

5.2 HOMOPLASIES, CONSISTENCY INDEX, AND COMPLEXITY OF MACROEVOLUTION

The discussion above brings us to yet another interesting point: the complexity of macroevolution. In fact, one of the most important aspects of the present work is that, as underscored by several practical examples provided in this study, the often rather 'simple' *a priori* evolutionary scenarios often prove to be considerably simpler than those indicated by a direct confrontation with explicit phylogenetic hypotheses.

This subject is deeply related to my personal scientific development in the last few years. When I first came to Liege, my knowledge of biological evolution was mainly theoretically based. Fortunately, I had the chance to work in Liege with Professors Vandewalle and Chardon, who were the first persons to introduce me to more practical examples of biological evolution, especially those concerning the striking evolutionary history and biological diversity of Siluriformes. As explained in Chapter 1, catfishes constitute a very appropriate, practical case study to discuss general macroevolutionary aspects. I had in particular the chance to learn my first notions in practical anatomical dissection, as well as in functional and evolutionary morphology, which granted me a much broader view concerning the macroevolution of certain complex systems in catfishes. However, I should admit that my general view of macroevolution changed considerably whit the practice of phylogenetics. At the outset of my observations and comparisons on catfishes, my personal feeling was that most evolutionary changes concerning the major structural systems of those fishes would be somewhat 'oriented', in a somewhat 'linear' way. Of course, I felt the incredible complexity and diversity of catfishes, a complexity and diversity surely resulting from several homoplasic events, which was precisely the main reason for choosing this amazing group of fishes as a key group for discussion of theoretical biology. But continual observation of several phylogenetic characters in the various numerous catfishes examined in the present work, and especially the subsequent results obtained from the cladistic analysis of those characters, clearly pointed out a particularly complex, complicated macroevolutionary scenario. True, the decision to include in the present cladistic analysis the maximum possible of complex structural systems, as explained in Chapter 4, succeeded in revealing some robust, relatively homoplasy-free characters, such as the differentiation of the arrector dorsalis into two well-distinguished bundles in the clade including non-diplomystid, non-loricarioid and non-cetopsid siluriforms or the presence of a muscle 6 of the mandibular barbels of cetopsids. However, among the features analysed in this work, including those concerning these complex structural systems, such cases clearly constitute an exception and not the rule. In fact, as indicated by the Consistency Index of the most parsimonious, strict-consensus cladogram obtained in the present work (Fig. 3.123: CI = 0.52), among the characters analysed, for each two evolutionary transitions within catfish evolution, one is, in a rough, approximate average, due to homoplasy.

One could eventually argue that such a level of homoplasy might perhaps be related to an eventual incorrectness of the phylogenetic results obtained in the present work and illustrated in the cladogram of Fig. 3.123. However, again, this is just not that simple. As noted in Section 4.2, the CI value obtained in the present work is significantly higher than that obtained in the two main cladistic studies published so far on catfish higher level phylogeny (Mo, 1991; de Pinna, 1998). As also noted in that Section, the CI of 0.52 is also markedly superior to that expected for a random distribution of 440 morphological characters in 87 different terminal taxa of a so diverse and complex group as the Siluriformes, thus clearly revealing a strong phylogenetic signal (see Sanderson and Donoghue, 1989). In reality, what the present study seems to indicate is that biological macroevolution, at least in this particular case study provided by the order Siluriformes, is a rather complex, 'mosaic' process. The complexity of this process can be illustrated by the practical example given in Section 3.2, concerning one of the best supported clades within catfish higher level phylogeny, namely that formed by the Asian Sisoroidea and the South American Aspredinidae (Ferraris, 1989; Mo, 1991; de Pinna, 1993, 1996, 1998; Chen, 1994; Diogo et al., 2001b, 2002b, in press-b; this work).

As explained in Section 3.2, despite the strong and robust evidence provided by all these studies supporting a close relationship between aspredinids and the remaining Sisoroidea, the Aspredinidae share, as pointed out by de Pinna (1996), some 'striking' morphological similarities with other catfish groups.

For example, as pointed out by authors such as Hamilton (1822) or Chardon (1968) and discussed in the present work (see discussion on clade 64), some remarkable, highly peculiar features are found in both the Neotropical aspredinids and the Asiatic chacids. In particular, the configuration of the posterodorsal region of the skull in the members of these groups is, as noted by Chardon (1968), remarkably similar. For instance, both aspredinids and chacids present a well-developed, deep fossa between the posttemporo-supracleithrum, the parieto-supraoccipital and, eventually, the epioccipital (char. 126). Besides these two groups, such a fossa is found only in the erethistid catfishes examined, that is, in the aspredinid sister-group, according to the phylogenetic results of the present work and the studies of de Pinna (1996, 1998). Also, both aspredinids and chacids present a well-developed, dorsal lamina of the Weberian apparatus coming into contact with the dorsal surface of the body, a feature not found elsewhere in the Siluriformes (char. 141). Another peculiar, and also rather rare feature found in the posterior region of the skull of chacid and aspredinidid siluriforms is the markedly thin and mesially extended dorsomesial limb of the posttemporo-supracleithrum (char. 144). Besides these two groups, it was found only in the specimens of genus *Parakysis* examined. But the morphological similarity between chacids and aspredinids is assuredly not restricted just to the configuration of the structures of the posterodorsal region of the cranium. For example, in both these groups the prevomer is missing (char. 93). The absence of the prevomer is a highly unique and rare feature among Siluriformes, found only, among the catfishes examined, also in the specimens of genera *Microglanis* and *Scoloplax*. According to Chardon (1968), the seemingly convergent evolution of chacids and aspredinids might perhaps be associated with a homoplasic adaptation to a peculiar 'burying' (burrowing) behaviour found in these fishes (see Chardon, 1968: 161).

But, as noted in Section 3.2, the rather 'mosaic', complex morphological combination of characters present in aspredinids means these fishes also share other remarkable, and completely different, peculiar features with groups that in overall shape are morphologically very distinct from chacids, e.g. the auchenipterids and in particular the doradids. Some of these characters concern, for instance, the prominent dorsolateral projections of the laminar bone of the mesethmoid (char. 60), the mesocoracoid arch and the main body of the scapulo-coracoid not distinguished from each other (char. 186), presence of a highly developed anterior process on the dorsal condyle of the pectoral spine (char. 198), and presence of a well-developed anteroventral lamina of the preopercle (char. 366). These characters, and other features, inclusively led Friel to propose an eventual sister-group relationship between the aspredinid and the doradoid catfishes (see Friel, 1994).

However, as noted in Section 3.2, the several synapomorphies supporting the clade chacids + plotosids + clariids and the group doradids + auchenipterids + mochokids, and in particular the 21 synapomorphies, 5 of which completely homoplasy-free among Siluriformes, supporting the grouping of aspredinids + erethistids, clearly seem to indicate that the peculiar features noted above found respectively in aspredinids and chacids and in aspredinids and doradids, are, indeed, seemingly due to homoplasy.

But the *important point here* is that, even if all those phylogenetic studies supporting the aspredinidids + remaining sisoroids were wrong and, for instance, the aspredinidids were eventually more closely related to chacids or, alternatively, doradids, this would still nonetheless imply numerous homoplasic events between Aspredinidae and other catfish groups. The situation may be illustrated as follows. Let's take a hypothetical triangle with the Aspredinidae in the centre, and (1) the Asian Sisoroidea, (2) Doradidae and (3) Chacidae in the vertices respectively. The present cladistic analysis, as well as the works of Ferraris (1989), Mo (1991), de Pinna (1993, 1996, 1998) and Chen (1994), strongly support a close relationship between Aspredinidae and the Asian Sisoroidea. Thus the remarkable, unique similarities found between Aspredinidae and Doradidae on the one hand, and Aspredinidae and Chacidae on the other, are assigned to homoplasic events, revealing a particularly complex, 'mosaic' homoplasic evolution within these groups. However, if alternatively one were to accept the hypothesis Aspredinidae + Doradidae, one would have to accept a great number of homoplasic events to explain the peculiar morphological features shared by aspredinidids and erethistids, as well as a series of homoplasies between aspredinidids and chacids. Lastly, if one were to accept the hypothesis Aspredinidae + Chacidae, one would have to accept that the great number of apomorphic morphological peculiarities shared by aspredinidids and erethistids would be the result of homoplasy and, once more, that there are several homoplasies between aspredinidids and doradids.

Therefore, the choice of any vertex of the triangle would obligatory imply a series of homoplasic events. In reality, as explained above, the phylogenetic scenario proposed in the present work (Aspredinidae + Erethistidae) is the one that required, in this study, the smallest number of homoplasic events. Thus, the high homoplasy illustrated in this example cannot be explained simply by the choice of a particular erroneous phylogenetic scenario, but rather by the markedly complex, 'mosaic' macroevolutionary history of these catfish groups.

This example further illustrates an important point emphasised by authors such as Sanderson and Donoghue (1989), Klassen et al. (1991), Wilkinson (1991), Simonetta et al. (1999), Kitching et al. (1998), and Marques and Gnaspini (2001). As stated by Klassen et al. (1991: 446), the 'amount of homoplasy exhibited by a cladogram is usually considered inversely proportional to the confidence that an investigator will have in both the tree and the dataset from which it was derived'. However, it is important to note that such a 'confidence measurement' only makes full sense when comparing levels of homoplasy exhibited by cladograms concerning a somewhat similar number of characters referring to somewhat similar types of structures in a somewhat similar number of taxa of the *same biological group*. Indeed, there is no reason to believe that all different biological groups, even of a relatively similar size (e.g. total number of species of the group), exhibit exactly the same levels of homoplasy. If one obtains a certain Consistency Index A in the cladogram of a group X and this index A is smaller than that of a cladogram concerning a group Y, there is no reason to consider, a priori, even if the number of taxa and characters analysed is somewhat similar in both cases, that the cladogram of group Y is 'more unlikely' than that of group B. For example, as pointed out by Marques and Gnaspini (2001), some groups of cave animals seem to exhibit a particularly high level of homoplasies. So, a relatively small CI of a cladogram referring to a certain biological group could eventually be simply related to a true high level of homoplasy within that group and not necessarily to a 'bad cladogram'.

Of course, this does not mean that the CI could *never* be used to compare different phylogenetic scenarios proposed by different cladograms. As explained above, if the different cladograms refer to the analysis of a somewhat similar number of characters referring to a somewhat similar kind of structures in a somewhat similar number of taxa of the *same biological group*, such comparisons are probably reasonable. One example in which such comparisons are seemingly reasonable concerns precisely the higher level studies on siluriforms undertaken by Mo (1991) and de Pinna (1998). Those studies refer to somewhat similar structures (anatomical ones, with the great majority related to osteological elements) of the same biological group (the entire order Siluriformes). So, as de Pinna's analysis includes more characters, and in particular more taxa, than Mo's analysis (which, theoretically, should reduce de Pinna's CI: see Sanderson and Donoghue, 1989), and even then presents a considerably higher CI than Mo's, the cladogram of de Pinna seems, at least at first sight, to be somewhat more accurate than that of Mo, as is usually admitted by catfish taxonomists (see Section 1.3). A direct comparison between de Pinna's 1998 and Mo's 1991 phylogenetic hypotheses and that of the present work is, however, considerably more difficult. As mentioned above, a significant part of the cladistic analysis of this work concerns myological characters, which thus constitute a rather different kind of data than the almost exclusively osteological data included in the analyses of Mo (1991) and de Pinna (1998). This point will be discussed below in Sections 5.4 and 5.5.

5.3 FUNCTIONAL UNCOUPLINGS AND MORPHOLOGICAL MACROEVOLUTION

The discussion in Section 5.2 on the complexity of macroevolution does not mean that certain evolutionary generalisations are not eventually allowed. One such generalisation, strongly supported by the examples provided in Chapter 4 concerning the evolution of catfish structural complexes, concerns the important role played by 'functional uncouplings' in morphological macroevolution.

Many contemporaneous authors have emphasised the primordial role of 'functional uncouplings' in the evolution of structural innovations (see, e.g., Schaefer and Rosen, 1961; Vermeij, 1974; Lauder, 1981; Schaefer and Lauder, 1986, 1996; Wainwright and Turinfan, 1993; Galis, 1996; a.o.). According to Vermeij (1974), uncouplings facilitate diversification and speciation by increasing the number of degrees of freedom and allowing more mechanical solutions for functional problems. The major importance of functional uncouplings in catfish macroevolution seems indeed unquestionable. In reality, were this allowed, one could say that siluriform macroevolutionary history is, in a certain way, a history of functional uncouplings, with such uncouplings seemingly having played a markedly important role in the macroevolution of all the six major catfish structural complexes discussed in Chapter 4.

Functional uncouplings are seen, for example, in the evolution of the structures associated with catfish mandibular barbels, discussed in Section 4.1. The small muscles responsible for the movements of these barbels seem to be effectively the result of the differentiation of the cephalic ventral musculature, which originally is essentially functionally associated and anatomically linked with the mechanisms of opening/closure of the mouth and abduction/adduction of the suspensorium. Movements of the mandibular barbels, important for functions such as prey detection or obstacle avoidance, are seemingly relatively independent from the mechanisms of mouth opening/closure or suspensorium abduction/adduction. Also, in order to better assist those movements, the neoformed cartilages of the mandibular barbels are often functionally uncoupled in an anterior 'supporting' part, and a posterior 'moving' part [see Section 4.1].

Functional uncouplings also seem to have played a very important role in the evolution of the catfish pectoral girdle, as remarked in Section 4.2. The typical, stout pectoral spine of catfishes, associated with functions such as producing sound or defence against predators, is the result of an evolutionary decoupling between the first pectoral ray and the remaining ones, which are essentially functionally associated with the movements of the body. In the same way, in order to better assist the motion of this pectoral spine, the arrector dorsalis becomes differentiated within catfish evolutionary history, in an essentially dorsal and an essentially ventral part separated by a large horizontal lamina of the scapulo-coracoid. Additionally, in some cases the muscles arrector ventralis and abductor profundus are also differentiated into two separated bundles (see Section 4.2).

As explained in Section 4.3, the general evolution of the adductor mandibulae complex in teleosts is basically a history of functional uncouplings, with an almost undifferentiated mass of fibres giving a sometimes impressive number of different bundles related to different parts of the mandible, and even to other structures (e.g. maxilla). Within catfishes, functional uncouplings have also played a very important role in adductor mandibulae evolution, as noted by authors such as Alexander (1965), Howes (1983a,b, 1985a) and Schaefer and Lauder (1986, 1996). One of the key innovations concerning the evolution of this muscle in Siluriformes is the homoplasic differentiation, in several different groups, of a muscle retractor tentaculi directly inserting on the maxilla. Contraction of this muscle, contrary to that of the main adductor mandibulae complex, is not mainly associated with the opening/closure of the mouth, but rather with a direct abduction of the maxillary barbels. Again, this is a typical evolutionary decoupling, with a complex A originally associated with a certain mechanism X being uncoupled in a complex B performing a new mechanism Y, with the remaining complex A continuing to perform the original mechanism X. In this way, not only does the organism now have the possibility to perform the new mechanism Y, but this new mechanism Y can eventually also be associated with certain other already available mechanisms (Z, W, etc.). This thus 'facilitates diversification and speciation by increasing the number of degrees of freedom and allowing more mechanical solutions for functional problems', as stated Vermeij (1974: see above). Additional to the retractor tentaculi, in a few catfishes the adductor mandibulae is also differentiated into a retractor palatini and/or a retractor premaxillae, which are directly related to the movements of the premaxilla and/or autopalatine [see Section 4.3].

With respect to the macroevolution of the palatine-maxillary system, functional uncouplings also played a determinant role, as discussed in Section 4.4 and illustrated in Fig. 4.18. The uncoupling of the anterior portion of the adductor arcus palatini in a muscle extensor tentaculi, together with the uncoupling between the suspensorium and the autopalatine, allows that in basal catfishes like diplomystids abduction of the maxilla, and thus of its barbel, could be performed not only by the depression of the lower jaw as in other teleosts, but also by the direct contraction of this new muscle (Fig. 4.17). Loss of the firm ligamentous connection present in diplomystids between the mesial surface of the maxilla and the lateral surface of the mandible subsequently allowed, in most non-diplomystids, the maxilla and its barbel to be relatively free from the movements of the mandible (Fig. 4.18). This independence of the maxilla and its associated barbel were additionally even reinforced with the functional uncoupling in several catfish groups between the retractor tentaculi and the adductor mandibulae, as explained above (Fig. 4.18). Moreover, in order to better assist the movements of the maxillary barbels, the muscle extensor tentaculi became differentiated, in several catfish groups, into different bundles allowing mechanisms of depression and/or elevation of these barbels (Figs. 4.18) [see Section 4.4].

Some major aspects of the evolution of the peculiar suspensorium of catfishes could probably also be explained, as mentioned above, by the important functional uncoupling between the autopalatine and the posterior part of the suspensorium. In fact, as explained at the end of Section 4.5, 'some specialised catfishes independently acquired a suspensorium configuration in which there is an almost complete functional uncoupling of the palatine-maxillary system for moving the maxillary barbels while seemingly somehow rebuilding an anterior suspensorial articulation on the skull'.

Lastly, in what pertains to the evolution of an elastic spring apparatus in catfishes, this, too, supports the important role of functional uncouplings in the macroevolutionary history of structural complexes. As explained in Section 4.6, the protractor of the Müllerian process, responsible for the movements of the elastic spring apparatus and thus for the production of sound by the swim bladder, results very likely from the homoplasic differentiation of muscles that typically perform a rather different function (e.g. supracarinalis anterior and/or epaxialis). This is also the case of the drumming and protractor posttemporalis muscles present respectively in some pimelodids and bagrids, which seemingly also promote production of sound by the swim bladder [see Section 4.6].

5.4 MYOLOGICAL VERSUS OSTEOLOGICAL CHARACTERS IN PHYLOGENETIC RECONSTRUCTIONS

As explained in Section 5.2 it is somewhat difficult to directly compare the consistency indexes of the cladograms referring to the phylogenetic hypotheses of Mo (1991) and de Pinna (1998) to the consistency index of the cladogram obtained in the present work (Fig. 3.123). Contrary to the analyses of Mo and de Pinna, almost exclusively based on osteological information, a significant part of the cladistic analysis of this work concerns myological data. In fact, there is no reason to state, a priori, that the level of homoplasy found in osteological features of a certain biological group is necessarily similar to that found in myological characters of that same group. If, for example, the myological features of a certain biological group were eventually much more homoplasic than the osteological ones, a certain cladogram based on muscular characters of that group could eventually better represent, even if presenting a smaller CI, the true phylogenetic relationships among the group than another cladogram based on osteological characters presenting a greater CI.

This leads us to yet another important point in this chapter: *Are muscular structures more, less, or equally homoplasic than the osteological ones? Are muscular structures as appropriate as osteological ones for the inference of phylogenetic relationships?*

The levels of homoplasy found in different types of data have long intrigued evolutionary scientists. A comparison of the homoplasy levels found in phylogenetic works referring to *animals* versus that found in works concerning *plants* was discussed in the excellent paper of Sanderson and Donoghue (1989). As noted by these authors, it is usually 'widely believed that plants are more homoplasic than most animals' (Sanderson and Donoghue, 1989: 178). For instance, Wagner (1984: 115) stated that 'plants are evidently unusually inclined to have parallelisms', while Cronquist (1987: 24) suggested that 'the relative lack of morphological integration in plants, and poor correlation of evolutionary advances with adaptive zones and ecological niches, combine to permit rampant parallelism, in contrast to the more rigid evolutionary channelling in animals'. As noted by Sanderson and Donoghue (1989), an especially popular argument for increased homoplasy in plants was the reference to their 'relative simplicity' and 'indeterminate growth'. However, as has been continually emphasised throughout the present book, such evolutionary 'arguments' often constitute mainly gratuitous, ad hoc statements, neither based on, nor confronted with explicit phylogenetic scenarios. For example, in what concerns the particular case of homoplasy in *plants* versus *animals*, the comparison promoted by Sanderson and Donoghue between explicit cladistic analyses using these two types of data indicates that 'the levels of homoplasy (between plants and animals) are remarkably similar' (see Sanderson and Donoghue, 1989: 178-179, for more details on this subject).

Another controversial 'versus' issue that has been, and continues to be (see, e.g., Doyle, 1998; Smith, 1998; Benton, 1999a,b; Easteal, 1999; Mallet and Willmott, 2003; Dunn, 2003; a.o.) largely discussed is the use of *molecular* versus *morphological* data in phylogenetic evolutionary studies. This controversial issue was also analysed in some detail in the excellent paper of Sanderson and Donoghue (1989). As noted by these authors, the 'value of morphological data in elucidating phylogeny has frequently been questioned because of the 'susceptibility' of morphology to convergent evolution in natural selection' (Sanderson and Donoghue: 1989). For instance, Sytsma and Gottlieb (1986: 5556) stated that 'primary reliance on morphological data to model phylogenetic relationships may be misleading, no matter how many characters are examined'. On the other hand, as also noted by Sanderson and Donoghue (1989), some researchers have pointed out difficulties in establishing homology in molecular characters, and some others suggested that molecular data may show *more* homoplasy due to the limited number of character states per character (locus) and the difficulty in discriminating genetic homoplasy. In order to analyse this problem, Sanderson and Donoghue (1989) elaborated a comprehensive comparison of 18 molecular and 42 morphological phylogenetic studies, considering, of course, the effects of other factors such as the number of taxa and/or characters (see Section 5.2 above). The result of this comparison was, according to these authors, that 'there is no evidence to date that molecular data are less homoplasic than morphological data' (Sanderson and Donoghue, 1989: 179).

A similar statement was recently made by Benton (1999a), who suggested that probably the more appropriate thing to do is to continue to undertake phylogenetic studies using morphological data, as done in the present work, as well as phylogenetic studies based on molecular data, thereby taking advantage of these two complementary kinds of information. In fact, as elegantly explained by him, more than 'antagonistic enemies', both molecular and morphological studies provide useful, rather complementary kinds of information, each being more appropriate for certain different specific purposes. Thus, for example, molecular data is surely useful in such interesting studies as those concerning population genetics, genetic drift, or chromosomal evolution. Likewise, a morphofunctional account of the macroevolution of certain Vertebrate structural complexes, such as the palatine-maxillary system associated with the movements of catfish maxillary barbels, or the adductor mandibulae complex of these and other teleosts, necessarily rely on anatomical data. Different issues to be analysed need naturally different methodological approaches and hence none of these various approaches should be completely abandoned in favour of another. While this work is being written, some genetic works on the chromosomal evolution in catfishes or on the higher level phylogeny of these fishes are being carried (e.g. M. Hardman, J. Lundberg: pers. comm.) and this should, undoubtedly, be welcomed as a salutary matter of great contentment.

The results of the present study do not provide, of course, information for a direct comparison of the homoplasy levels in animals versus plants, or in morphological versus molecular data. The phylogenetic analysis of this work is wholly restricted to the higher level phylogeny of an animal group, the order Siluriformes, and is exclusively based on morphological data. Moreover, no molecular studies have been published so far on the higher level phylogeny of the Siluriformes to allow a comparison between the phylogenetic results obtained in this work and a general phylogeny of this order based on genetic data.

But the phylogenetic results of the present work do provide, as mentioned before, background material for a discussion on another very interesting, and much less discussed, major issue concerning the use of different kinds of information in phylogenetic reconstructions: the use of *myological* versus *osteological* data. In fact, as far as the author knows, no major studies have been published to date, at least not for Teleostei, providing a direct comparison, *in the same analysis*, between the homoplasy levels and phylogenetic relevance of *a great number of myological characters* (91 in the present work: see below) and *a great number of osteological features* (303 in the present work: see below) in the reconstruction of the relationships within the *same* biological group (in this case, order Siluriformes).

One interesting discussion on the use of myological data in phylogenetic reconstructions was recently presented in an excellent paper by Borden (1999). He described in detail the configuration and variation of 93 muscles in 15 species of genus *Naso*, or unicornfish, of Acanthuridae (Percomorpha), and discussed the phylogenetic implication of the myological results obtained. As noted by him (1999: 191), very few studies in ichthyology focus on myology for a variety of reasons: 'investigators may be reluctant to use myology due, for example, to the plethora of names used to describe the same muscles, to the realisation that osteological proficiency is mandatory in order to identify muscle, leading them to concentrate only on osteology, or to the requirement of potentially finer dissection to preserve muscle bundles and nerves; furthermore, fossil fishes leave few if any myological clues, complicating hypotheses between extinct and extant fishes'. As a consequence, Borden pointed out (1999: 191), 'of those studies using myology as a basis of information, most are functional works often analysing the role of various muscles in feeding or locomotion or comparing a muscle or specific group across a number of taxa systematically and/or ecologically related'. Explicit cladistic analyses based on myological data are thus rather rare. Some of the most relevant examples of such analyses listed by Borden (1999) are those of Winterbottom (1993), using 46 myological characters to reconstruct the relationships among the acanthurid genera, or Borden's own (1998), using about ten myological characters to investigate the phylogeny within the acanthurid genus *Naso*. As these studies are essentially restricted to myology, they do not allow a direct comparison between the homoplasy and phylogenetic relevance of myological versus osteological data. In a study of Cottoidea, Yabe (1985) did include, as noted by Borden (1999), some myological characters together with some osteological ones. But the total number of muscular characters is rather small (only 14, together with an also relatively small (46) number of osteological) and, moreover, Yabe (1985) failed to provide a detailed, direct comparative analysis of the homoplasy and retention levels found in these two different kinds of characters.

As for the Siluriformes, the only relevant, explicit phylogenetic analyses published so far including a relatively significant number of myological characters are those of Howes (1983a) concerning Loricaroidea, with 11 of the 34 characters listed (pp. 337 and 338) referring to muscular features, and Schaefer (1990) concerning the same group, in which 7 of the 72 characters refer to the configuration of muscles (see Section 1.3). But, again, the total number of muscular characters included in these two studies is rather small and not representative, and, further more, neither study has provided a detailed, direct comparative analysis between the homoplasy and retention levels found in the muscular and osteological characters examined.

A direct comparison between the homoplasy levels and phylogenetic relevance of a great number of myological (91) and osteological (303) characters in the reconstruction of the higher level phylogeny of such a diverse and representative group of teleosts as the Siluriformes clearly constitutes one of the most important contributions of the present work.

The list of the 91 myological characters used in this comparison, which refers *exclusively* to the strict configuration of the muscles analysed in the present work, is given in Table 5.1. The 303 osteological characters used in the comparison, which concerns *exclusively* the configuration of the bones and associated cartilages constituting a certain osteological component (e.g. main body of the autopalatine plus the autopalatinum cartilages) but not the ligaments connecting different osteological components, are listed in Table 5.2. For each myological (Table 5.1) or osteological (Table 5.2) character listed, these Tables show its respective CI (Consistency Index) and RI (Retention Index). Those characters that are *autapomorphic for a single terminal taxon*, that is, for a particular catfish genus analysed, and hence not informative of the

Table 5.1 List of 91 myological characters (Char.) included in the cladistic analysis of the present work (see Section 3.1) and their respective CI and RI. Notes: AUT—character autapomorphic for a single genus (thus, not informative for the inference of the phylogenetic relationships between the groups studied); the last cells, indicated TN, MC and MR and marked with thick lines represent respectively, the sum total of the number of myological characters (TN), arithmetical mean of the consistency indexes of these characters (MC), and arithmetical mean of the retention index of these characters (MR) (for further explanations and more details, see text).

Char..	CI	RI	Char..	CI	RI
16	AUT	AUT	214	100	100
18	33	77	215	66	75
19	25	81	216	100	100
20	100	100	217	66	75
21	100	100	218	AUT	AUT
22	AUT	AUT	219	28	68
23	100	100	220	25	40
24	100	100	221	50	50
26	100	100	222	100	100
27	20	76	223	25	40
28	AUT	AUT	224	25	72
29	AUT	AUT	225	50	75
30	100	100	226	66	93
31	100	100	227	16	61
32	33	33	228	33	50
33	33	60	229	100	100
34	100	100	230	100	100
35	22	68	231	50	80
36	100	100	232	100	100
37	33	78	233	100	100
38	100	100	234	AUT	AUT
39	50	0	235	50	66
40	AUT	AUT	236	33	88
41	AUT	AUT	237	33	50
42	10	77	238	AUT	AUT
43	50	0	239	50	75
44	100	100	240	66	85
45	AUT	AUT	241	100	100
46	AUT	AUT	242	50	50
47	14	72	243	AUT	AUT
48	66	95	244	100	100
49	100	100	245	100	100
50	66	85	246	50	60
51	100	100	247	14	82
52	100	100	248	25	88
53	25	77	249	AUT	AUT
54	100	100	250	25	25
55	50	80	251	14	77
206	33	33	252	100	100
207	100	100	253	100	100
208	33	60	254	20	75
209	100	100	255	AUT	AUT
210	20	88	256	100	100
211	50	66	257	100	100
212	40	70	258	100	100
213	66	80	TN=91	MC=0.64	MR=0.80

Table 5.2 List of 303 osteological characters (Char.) included in the cladistic analysis of the present work (see Section 3.1) and their respective CI and RI. Notes: AUT—character autapomorphic for a single genus (thus, not informative for the inference of the phylogenetic relationships between the groups studied); the last cells, indicated TN, MC and MR and marked with thick lines represent respectively, the sum total of the number of myological characters (TN), arithmetical mean of the consistency indexes of these characters (MC), and arithmetical mean of the retention index of these characters (MR) (for further explanations and more details, see text).

Char..	CI	RI	Char.	CI	RI	Char.	CI	RI	Char.	CI	RI	Char..	CI	RI
56	100	100	119	AUT	AUT	181	14	61	309	25	72	378	50	0
57	100	100	120	50	75	182	66	66	310	100	100	379	100	100
58	100	100	121	AUT	AUT	183	66	75	311	100	100	380	AUT	AUT
59	AUT	AUT	122	100	100	184	AUT	AUT	312	66	75	381	16	44
60	33	75	123	40	80	185	20	69	314	AUT	AUT	382	AUT	AUT
61	100	100	124	AUT	AUT	186	AUT	AUT	315	AUT	AUT	383	AUT	AUT
62	100	100	125	100	100	187	100	100	317	AUT	AUT	384	AUT	AUT
64	AUT	AUT	126	50	80	188	50	66	318	50	75	385	100	100
65	100	100	127	AUT	AUT	189	100	100	319	AUT	AUT	386	10	50
66	AUT	AUT	128	100	100	190	50	90	320	AUT	AUT	387	AUT	AUT
67	AUT	AUT	129	AUT	AUT	191	33	60	321	100	100	388	AUT	AUT
68	100	100	130	100	100	192	20	33	322	AUT	AUT	389	66	0
69	25	35	131	16	71	193	50	50	323	AUT	AUT	390	100	100
70	AUT	AUT	132	100	100	194	14	68	324	50	0	391	100	100
71	AUT	AUT	133	100	100	195	AUT	AUT	325	23	65	392	100	100
72	50	88	134	100	100	196	100	100	326	100	100	393	100	100
73	25	60	135	100	100	197	50	0	327	12	46	394	AUT	AUT
74	100	100	136	50	87	198	25	84	328	100	100	395	100	100
75	33	60	137	50	50	199	AUT	AUT	329	AUT	AUT	396	50	75
76	33	66	138	100	100	200	50	80	330	50	0	397	100	100
77	AUT	AUT	139	50	85	201	50	50	331	66	50	398	14	40
78	100	100	140	100	100	202	66	66	332	20	42	399	AUT	AUT
79	100	100	141	50	66	203	100	100	333	AUT	AUT	400	100	100
80	50	0	142	50	85	204	100	100	334	33	63	401	50	83
81	20	20	143	AUT	AUT	205	AUT	AUT	335	100	100	402	50	66
83	AUT	AUT	144	33	50	259	100	100	336	100	100	403	100	100
84	16	68	145	100	100	260	100	100	337	AUT	AUT	404	100	100
85	100	100	146	33	71	261	AUT	AUT	338	100	100	405	100	100
86	20	75	147	100	100	262	AUT	AUT	339	100	100	406	AUT	AUT
87	AUT	AUT	148	30	74	263	50	80	340	100	100	407	33	81
88	66	87	149	100	100	264	50	0	341	33	60	408	25	90
89	100	100	150	15	75	265	14	25	342	100	100	409	AUT	AUT
90	33	60	151	50	0	266	25	76	343	50	80	410	100	100
91	50	50	152	50	66	267	100	100	344	AUT	AUT	411	66	0
92	AUT	AUT	153	50	95	268	100	100	345	100	100	412	66	0
93	25	40	154	AUT	AUT	269	AUT	AUT	346	50	85	413	AUT	AUT
94	AUT	AUT	155	16	70	276	50	0	347	AUT	AUT	414	14	45
95	66	0	156	100	100	277	25	76	348	14	14	415	AUT	AUT
96	33	60	157	33	81	278	50	75	349	AUT	AUT	416	100	100
97	16	66	158	100	100	279	100	100	350	66	80	417	66	66
98	100	100	159	20	55	280	AUT	AUT	351	33	50	418	AUT	AUT
99	50	0	160	50	80	281	100	100	353	AUT	AUT	419	20	82
100	100	100	161	100	100	282	AUT	AUT	354	AUT	AUT	420	100	100
101	AUT	AUT	162	25	50	283	33	88	356	100	100	421	50	85
102	33	66	163	AUT	AUT	284	100	100	357	50	0	422	100	100
103	50	75	165	20	66	285	50	75	358	50	75	423	28	79
104	AUT	AUT	166	50	50	286	50	94	360	75	66	424	50	87
105	100	100	167	50	80	287	100	100	363	AUT	AUT	425	50	80
106	50	0	168	20	73	288	75	85	364	AUT	AUT	426	AUT	AUT
107	100	100	169	100	100	289	100	100	365	100	100	427	66	0
108	100	100	170	100	100	290	28	84	366	66	83	428	AUT	AUT
109	16	41	171	100	100	291	AUT	AUT	367	AUT	AUT	430	100	100
110	AUT	AUT	172	AUT	AUT	292	AUT	AUT	368	AUT	AUT	431	50	75
111	AUT	AUT	173	50	88	293	40	62	369	100	100	432	AUT	AUT
112	50	66	174	50	83	294	50	88	370	AUT	AUT	433	100	100
113	100	100	175	100	100	295	100	100	371	AUT	AUT	434	100	100
114	66	87	176	AUT	AUT	296	100	100	372	AUT	AUT	437	50	87
115	100	100	177	AUT	AUT	297	AUT	AUT	373	AUT	AUT	439	50	75
116	100	100	178	AUT	AUT	298	100	100	375	100	100	440	AUT	AUT
117	33	50	179	AUT	AUT	306	20	20	376	AUT	AUT			
118	100	100	180	33	80	308	20	73	377	AUT	AUT	TN=303	MC=0.66	MR=0.77

phylogenetic relationships *among* the various terminal taxa, are respectively labelled 'AUT' (see Tables 5.1 and 5.2).

Analysis of Tables 5.1 and 5.2 allows a discussion of some interesting points concerning the use of *myological* versus *osteological* characters in phylogenetic studies. All statistical analysis and comparisons were performed using the computer program SPSS, version 8.0.

One point is that there is a much higher proportion of autapomorphic characters characterising the terminal taxa (i.e. the different genera) in the osteological characters examined (83 in a total of 303, i.e., about 27.4%) than in the myological ones (14 in a total of 91, i.e., about 15.4%). This seems to corroborate the opinion of Borden (1999), who pointed out that muscular characters are seemingly less indicative of features to characterise low rank taxa such as species or genera. According to him, muscular characters are somewhat more 'conservative' than osteological ones, with the latter demonstrating a higher variability particularly in small taxa and therefore usually providing more autapomorphic features to characterise such low rank groups. The osteological components included in the cladistic analysis of the present work do, in fact, exhibit a greater morphological variation than the myological ones. The 65 various osteological components (see Section 3.1) provided 303 phylogenetic characters (in an arithmetical mean of about *4.7 phylogenetic characters for each osteological component* examined). The 39 various muscles included in this analysis (see Section 3.1) provided 91 phylogenetic characters (in an arithmetical mean of about *2.3 phylogenetic characters for each muscle* examined).

However, this definitely does not mean that osteological characters are simply more appropriate than myological ones for phylogenetic reconstructions. In fact, in phylogenetic studies concerning major groups, such as this one concerning the Siluriformes, the osteological characters, in providing *a greater percentage of autapomorphies to characterise certain particular, low rank, terminal taxa* (in this specific case, to characterise a particular genus examined) than the muscular ones, reveal thereby *a smaller percentage of informative characters to infer the phylogenetic relationships among these taxa.* Saying that about 27.4% of the osteological characters (versus 15.4% of the myological ones) of the present study provided autapomorphies to characterise terminal taxa, is equivalent to saying that only 72.6% of these characters (versus 84.6% of the myological ones) proved informative in disclosing the relationships among these terminal taxa and thereby disclosing the siluriform higher level phylogeny.

So, to 'counterbalance' the fact that *the osteological structures examined provided more phylogenetic characters* per each structure observed (with an arithmetical mean of about 4.7 phylogenetic characters per each osteological component examined) than myological structures (with an arithmetical mean of about 2.3 phylogenetic characters for each muscle examined), *the myological phylogenetic characters obtained seem to be, on average, at least in the present study, more informative for disclosing the general phylogeny of a major group* such as the

Siluriformes (with 84.6% of the myological characters informative in disclosing the higher level phylogeny of Siluriformes, versus 72.6% of the osteological characters).

Let it be noted, however, that all things taken into account, despite this 'counterbalance', each osteological structure included in the present cladistic analysis did provide, on average, more informative characters for disclosing the general phylogenetic relationships of Siluriformes. The 65 osteological components included in the analysis provided, as already noted, 303 phylogenetic characters, of which 220 are informative for disclosing the higher level phylogeny of this order (with a final arithmetical mean of 3.4 informative characters per osteological component examined). The 39 myological structures included in the analysis provided, as mentioned above, 91 phylogenetic characters, of which 77 are informative in disclosing the higher level phylogeny of this order (with a final arithmetical mean of 2.0 informative characters per myological structure examined).

However, attention must be called to another important point. The 39 myological structures examined in the present analysis provided 91 phylogenetic characters, with a mean of 2.3 phylogenetic characters per muscle examined, against the 4.7 phylogenetic characters provided by each osteological component examined. But, although very likely indicating a true higher osteological variation, as suggested by Borden (1999), this is also clearly importantly related to the historical fact that, as he pointed out, far fewer phylogenetic studies based on myology have been done (see above). In fact, a significant number of the osteological characters included in the present work were inspired in phylogenetic characters pointed out in earlier analyses of other authors (see Section 3.1). This is attributable to the fact that, as explained in Chapter 1, a great number of phylogenetic studies on catfish relationships are available in the literature, particularly on the relationships within certain specific small catfish groups (see Table 1.1), based on osteological data. Those studies caught my attention for several interesting osteological features. A completely different picture was found for myological features, however. As likewise mentioned in Chapter 1, not a single published, explicit catfish phylogeny based on myological data could be found in the literature; just two studies including a few muscular characters in their analyses were discovered (see above). This point should be kept in mind in any discussion on the 'variation' of, and consequent number of informative phylogenetic characters provided by, osteological versus myological structures.

Besides the question discussed above concerning the *number* of informative versus uninformative characters provided by osteological and myological structures, analysis of Tables 5.1 and 5.2 revealed another interesting point: *within the informative characters pointed out by these two types of structures, those referring to the myological structures exhibit, on average, a somewhat higher Retention Index* (= 0.80) *than those referring to the osteological ones (= 0.77)*. However, a statistical analysis reveals that these means are statistically not significantly different at a significance level of 0.50.

Contrarily to the RI, the arithmetical mean of the CI of those informative characters concerning the myological structures (= 0.64) is slightly inferior to that of those informative characters concerning the osteological components examined (= 0.66). However, once again, a statistical analysis reveals that these means are statistically not significantly different at a significance level of 0.05.

The overall interpretation of these indexes, means, averages and examples can be summarised as follows: (1) osteological structures seem to display a greater morphological variation than myological ones; (2) this difference (probably virtually amplified, as explained above, by the fact that the phylogenetic variation of osteological structures has historically been the subject of far more studies and descriptions than that of myological ones) is particularly characters with reference to low rank taxa, i.e., genera or species; (3) myological characters provide a high proportion of informative characters for disclosing the relationships between these low rank taxa, and, thus, for disclosing the higher level phylogeny of the group in which these taxa are included.

Another way to analyse the phylogenetic contribution of the myological versus osteological structures examined in the present work is to simply compare the phylogenetic trees generated directly from these two different kinds of data. These trees are shown in Figures 5.2 and 5.3. Figure 5.2 illustrates the most parsimonious, strict-consensus cladogram obtained by the exclusive cladistic analysis of those 91 myological characters listed in Table 5.1 (see above). Figure 5.3 illustrates the most parsimonious, strict-consensus cladogram resulting from the exclusive phylogenetic analysis of the 303 osteological characters listed in Table 5.2. As can be seen, although the overall Consistency Index of the consensus tree obtained from the muscular characters (CI = 0.48) is a little smaller than that of the consensus tree based on osteological characters (CI = 0.50), the overall Retention Index of the 'myological' tree (RI = 0.81) is relatively higher than that of the 'osteological' tree (RI = 0.73).

The consensus tree obtained from the myological characters appears surprisingly, strikingly resolved, considering that it refers to a cladistic analysis of only 91 characters to infer the phylogeny of a major, markedly complex group encompassing 87 different terminal taxa (Fig. 5.2). Some major points on siluriform higher level phylogeny appear markedly resolved in the 'myological' tree (Fig. 5.2) and, in some cases, even apparently more congruent than in the 'osteological' one (Fig. 5.3). For example, the Loricaroidea, although appearing as non-monophyletic, appear in a somewhat similar, and markedly basal position in the 'myological' tree (Fig. 5.2). In the 'osteological' tree the loricarioid callichthyids + scoloplacids + loricariids + astroblepids appear in a markedly derived position within the Siluriformes, being separated from the loricarioid nematogenyids + trichomycterids by the vast majority of the remaining catfish groups (Fig. 5.3). As Loricaroidea monophyly has been

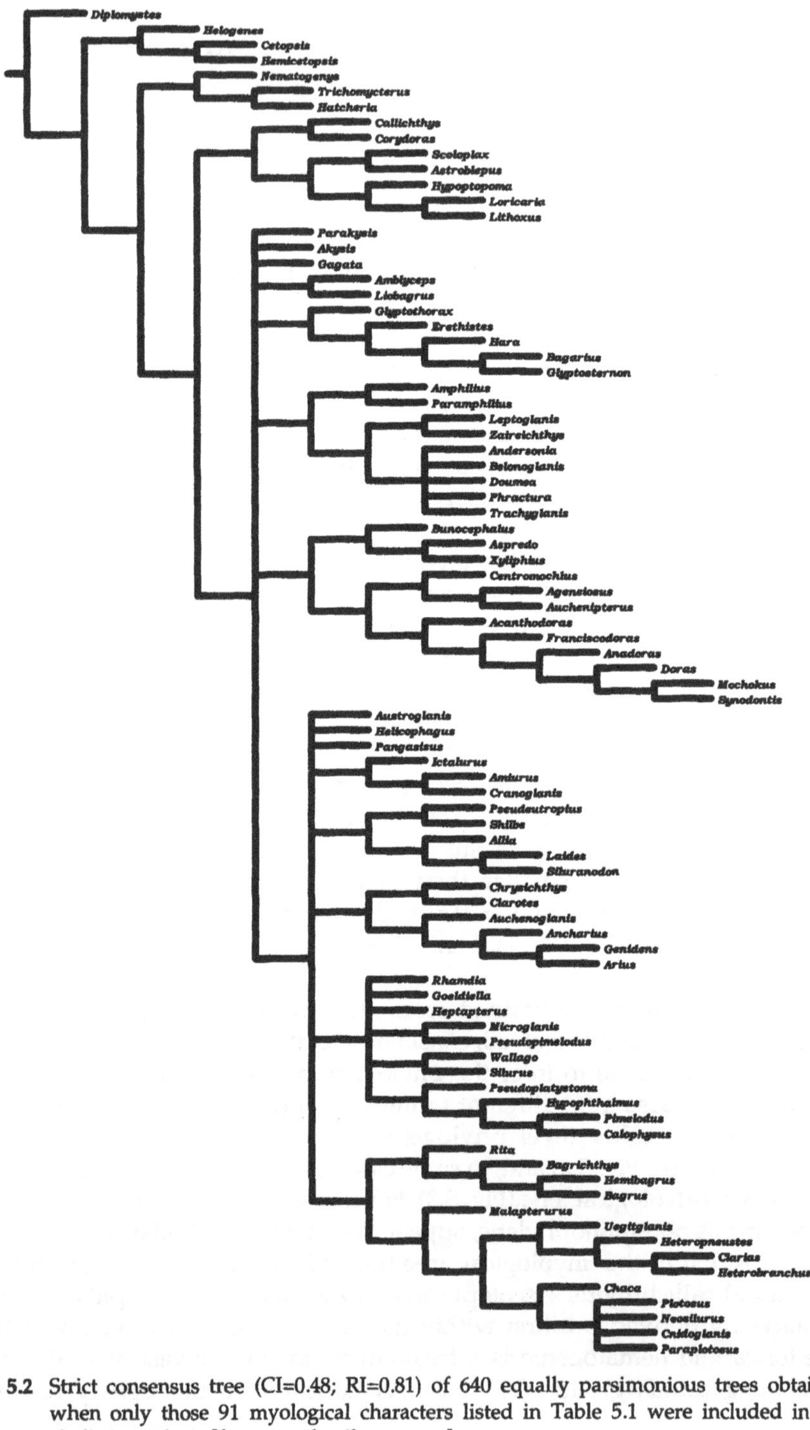

Fig. 5.2 Strict consensus tree (CI=0.48; RI=0.81) of 640 equally parsimonious trees obtained when only those 91 myological characters listed in Table 5.1 were included in the cladistic analysis [for more details, see text].

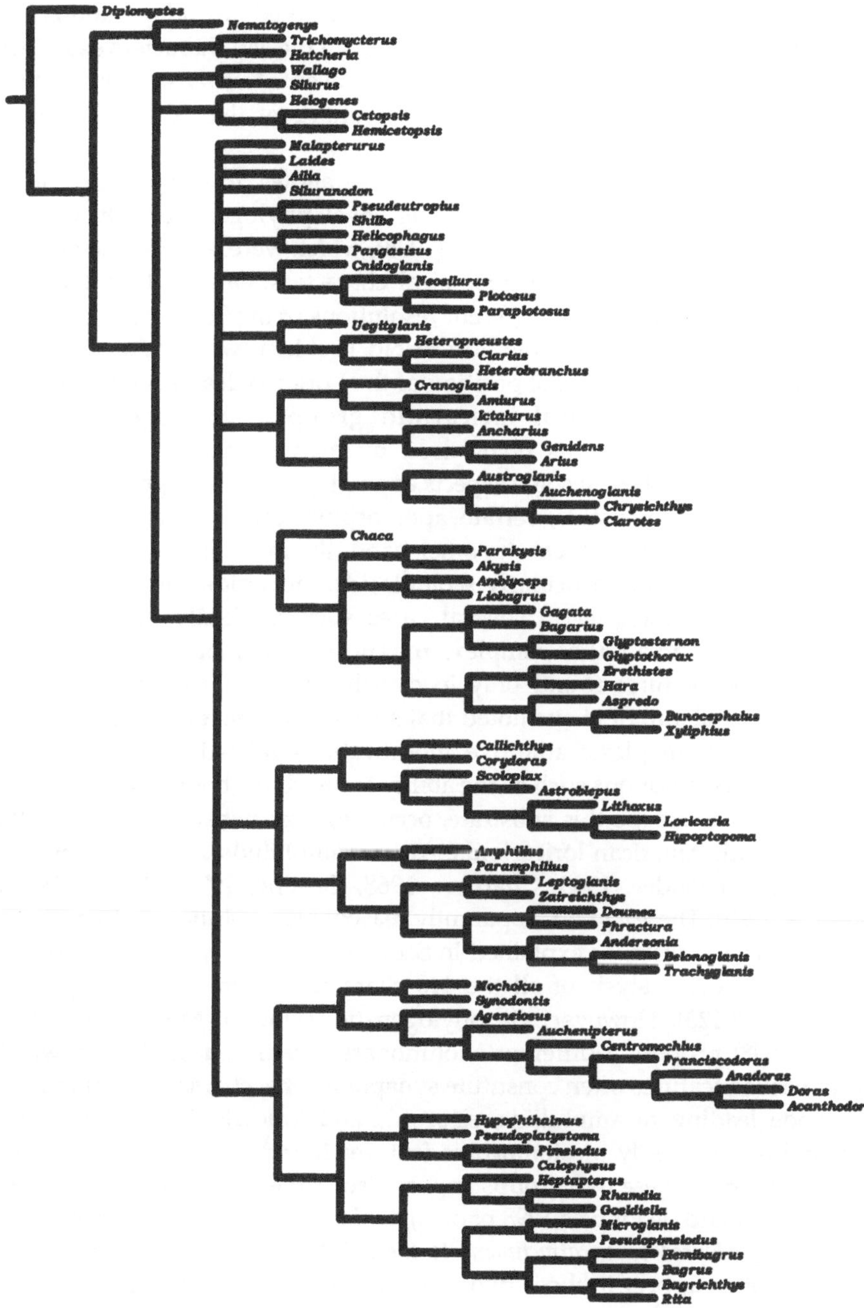

Fig. 5.3 Strict consensus tree (CI=0.50; RI=0.73) of 40 equally parsimonious trees obtained when only those 303 osteological characters listed in Table 5.2 were included in the cladistic analysis [for more details, see text].

supported by several studies, is consensually accepted nowadays, and is also strongly supported by the cladistic analysis of all the 440 characters examined in the present work (see Chapter 3), in what concerns this particular point the 'osteological' tree appears seemingly less congruent than the 'myological' one.

Elaboration of this latter 'osteological' tree allows, in fact, a more direct comparison with the studies on higher level phylogeny of Siluriformes published by Mo (1991) and de Pinna (1998), which were, as explained above, almost exclusively based on osteological characters. One of the interesting points to be compared is the different resolutions of the 'loricarioid problem' proposed in those studies. This 'loricarioid problem' was explained in the discussion of the phylogenetic position of the Loricaroidea in Section 3.2. On the one hand, most basal loricarioid groups, nematogenyids and trichomycterids appear, as recognised by many authors, to be clearly plesiomorphic catfishes, in some aspects appearing as the only groups lacking, together with diplomystids, certain apomorphic features characterising all the remaining siluriforms. On the other hand, the most derived loricarioids, i.e., callichthyids and particularly scoloplacids, loricariids and astroblepids, although also presenting in several cases such markedly plesiomorphic features, exhibit, in a rather complex, 'mosaic' combination (see Section 5.2), some derived features found only in catfishes such as amphiliids and/or sisorids. Several authors have noted that such derived features are problably the result of homoplasic adaptive modifications related to certain major functional specialisations such as the ability to attach the body to the substrate or the ability to scrape this substrate, occurring somewhat in parallel in the derived South American loricarioids, African amphiliids, and Asian sisorids (see, e.g., Alexander, 1965; Chardon, 1968; Gosline, 1975; Howes, 1983a; Burgess, 1989). These *a priori*, apparently reasonable, evolutionary hypotheses (see Section 5.1) were, as explained in Section 3.2, strongly supported by the explicit cladistic analysis of all the 440 characters examined in the present work (Fig. 3.123). However, the phylogenetic results of Mo (1991) and de Pinna (1998) propose a different evolutionary scenario, according to which such derived features often constitute synapomorphic characters occurring in the node leading to amphiliids, sisoroids and loricarioids. Therefore, the 'apparently' markedly plesiomorphic features found in nematogenyids and trichomycterids (these two groups appear also as the most basal Loricaroidea in Mo's 1991 and de Pinna's 1998 papers), which *are not, it should be emphasised, particularly adapted to an attachment to the substrate or to the scraping of this substrate,* often appear in those two studies as secondary reversions.

A different resolution of the 'loricaroid problem' is proposed in the tree based on the 303 osteological characters examined in the present work, wherein the 'plesiomorphic' and the 'apomorphic' loricarioid groups are simply broken apart by most catfish taxa. The 'apomorphic' loricarioids (Callichthyidae + Scoloplacidae + Loricariidae + Astroblepidae) appear in a markedly derived position near the amphiliids, while the 'plesiomorphic' loricarioids (Nematogenyidae + Trichomycteridae) appear in a markedly basal position near the diplomystids (Fig. 5.3). In fact, it is important to note that had muscular characters (Fig. 5.2), but principally the combination of these characters with all the other characters, been included in the cladistic analysis of the present study (Fig. 3.123), evidence would have been provided for the monophyly and the markedly basal position of the Loricaroidea in this analysis (see Section 3.2). This underscores that the inclusion of muscular features, and in particular the conjugation of the complementary information provided by them and osteological features (and by others, whenever possible), allow a much broader and, very likely, better resolved phylogenetic reconstruction of the relationships among the groups under study.

The cladogram illustrated in Figure 3.123 based on all the 440 characters available in the present work is, in fact, significantly more resolved than that illustrated in Figure 5.3 exclusively based on osteological characters. In the 'osteological' tree there are several polytomies, especially concerning the major, most inclusive catfish clades, in a somewhat similar scenario as that seen in the trees of Mo (1991) and de Pinna (1998) (see Figs. 1.6 and 1.11). In this 'osteological' tree 27 extant catfish families (i.e., all extant families except Diplomystidae, Nematogenyidae, Trichomycteridae, Siluridae and Cetopsidae) are included in a largely unresolved polytomy leading to 12 different catfish nodes (Fig. 5.3). In the tree incorporating all the 440 characters available, almost the totality of the nodes appear resolved, with only three trichotomies found, of which only one concerns interfamilial nodes (that leading to amblycipitids, akysids, and the remaining sisoroids) (Fig. 3.123).

Importantly, in what concerns major catfish clades, the 'myological' tree, although based on only 91 muscular characters, appears to be more resolved than the 'osteological' based on 303 osteological characters (Fig. 5.2, compare with Fig. 5.3). In fact, as predicted above in the discussion concerning Tables 5.1 and 5.2, the 'osteological' tree only appears better resolved with reference to less inclusive catfish groups. For example, the 'myological' tree supports the monophyly of Amphiliidae, and even the monophyly, within this family, of the group constituted by leptoglanidins and doumeins, but fails to resolve the relationships *inside* the doumein subfamily (Fig. 5.2, compare with Fig. 5.3).

One other illustration of this concerns, for example, the Plotosidae. The 'osteological' tree is *much less resolved* than the 'myological' one in what concerns the *interfamilial relationships* of the Plotosidae, with this family appearing isolated on the 'osteological' tree, in one of the 12 nodes of a large polytomy (Fig. 5.3). The 'myological' tree supports the close relationship between plotosids, chacids and clariids (Fig. 5.2), as suggested by the 'all-evidence' tree of Fig. 3.123. However, the 'osteological' tree (Fig. 5.3) is *much more resolved* in what concerns the *relationships within family Plotosidae*, with genus *Cnidoglanis* appearing as the sister-group of the clade *Neosilurus* + (*Plotosus* + *Paraplotosus*), as suggested by the 'all-evidence' tree of Fig. 3.123, and contrary to what happens in the 'myological' tree in which the four plotosid genera appear grouped in an unresolved polytomy (Fig. 5.2).

It is interesting to note here nevertheless that the 'myological' tree does succeed, with only 91 characters, in exhibiting some surprising, striking examples of resolution even at the level of some very specific, low rank clades. For example, in some cases the relationships within low rank groups appear in the 'myological tree' as resolved as (e.g. Clariidae), or even inclusively more resolved than (e.g. Pimelodinae or Schilbidae) in the 'osteological' tree (Fig. 5.2, compare with Fig. 5.3).

The overall analysis of the trees based on the myological versus osteological characters examined in the present work thus corroborates the main points underscored in the discussion of Tables 5.1 and 5.2. Osteological structures seem to display a somewhat greater morphological variation than the myological ones (which, let it be noted again, is probably virtually amplified by an historical bias). But this difference refers in particular to low rank taxa such as genera or species. Therefore myological characters provide important information to disclose the relationships between higher clades, and, thus, to infer the higher-level phylogeny of the groups being studied.

As referred above, such direct comparisons of the homoplasy and retention indexes of muscular versus osteological data in phylogenetic reconstructions are unfortunately wanting. Therefore, similar studies concerning other major groups of ostariophysans, as well as teleosts and of vertebrates in general, are clearly needed to determine whether the patterns found here indicate, eventually, a general phylogenetic pattern or, instead, refer to a particular situation found in siluriforms.

Anyway, this study clearly underscores that inclusion of muscular characters, and in particular conjugation of the complementary information provided by myological and osteological features, as well as by other kinds of morphological data (e.g. external anatomy, soft structures etc.), would allow much broader and, very likely, as in the present work, better resolved phylogenetic reconstructions. Moreover, as also emphasised throughout the book, inclusion of myological data in phylogenetic analyses allows a much more complete, integrative posterior discussion on the evolution and potential functional signification of the structures examined, and hence on the general macroevolution of the groups studied. Therefore, despite the intrinsic difficulties in doing so, such as museum availability or the arduousness of muscular dissections, especially in small specimens (see Chapter 1), an effort should be made, in my opinion, by catfish specialists and ichthyologists in general to include myological characters in phylogenetic reconstructions.

5.5 ANALYSIS OF DISTINCT ANATOMICAL REGIONS IN PHYLOGENETIC RECONSTRUCTIONS AND THE CODING OF MULTISTATE CHARACTERS

Another somewhat similar issue often not analysed is the comparison, in the same cladistic study, of the various contributions of the characters concerning the distinct anatomical regions included in the phylogenetic reconstruction of the group under study. Such an analysis is, in fact, from a methodological point of view, relatively simple. For example, in the present work it is interesting to analyse the phylogenetic trees generated by those characters concerning the different anatomical regions described in Section 3.1. Therefore, for purely comparitive proposes, in this analysis the anatomical characters used in the present work were divided into five main categories referring to: 1) *the ventral cephalic musculature and structures associated with the mandibular barbels;* 2) *pectoral girdle and associated musculature;* 3) *neurocranium and anterior vertebrae;* 4) *facial musculature;* and 5) the *splanchnocranium* (see Section 3.1).

The most parsimonious, strict-consensus cladogram (CI = 0.36; RI = 0.63) generated by the cladistic analysis of the 37 characters concerning the ventral cephalic musculature and the structures associated with the mandibular barbels (see Section 3.1: chars. 1-37) is illustrated in Figure 5.4. As expected, the phylogenetic results of this cladistic analysis including only 37 characters to infer the relationships among 87 terminal taxa representing the 32 siluriform families are clearly limited. The cladogram obtained (Fig. 5.4) is largely unresolved and presents some incongruities in relation to the 'all-evidence' cladogram of Figure 3.123. Examples of such incongruities are the sister-group relationship between austroglanidids and cranoglanidids, two groups that appear closely related but not as direct sister-groups in the 'all-evidence' cladogram; sister-group relationship between *Malapterurus* and *Auchenoglanis;* the close relationship between auchenipterid genus *Ageneiosus* (but not of the other auchenipterid genera) and some Loricaroidea; and the non-monophyly of this latter group, as well as of its family Loricariidae (Fig. 5.4, compare with Fig. 3.123). Interestingly, however, these 37 characters do suggest a rather basal position of the nematogenyid and trichomycterid loricarioids (Fig. 5.4, compare with Fig. 3.123). They also support the monophyly of groups such as the mochokids, silurids, cetopsids, aspredinids, clariids, callichthyids, the clade doumeins + leptoglanidins, or the clade callichthyids + scoloplacids + loricariids + astroblepids, as well as of Pimelodidae and its subfamilies Pseudopimelodinae and Pimelodinae (Fig. 5.4, compare with Fig. 3.123). Anyway, the analysis of the largely unresolved tree shown if Figure 5.4 definitely calls attention to an important point: *the distinct limitation of undertaking phylogenetic reconstructions* of such diverse and complex groups as the Siluriformes by *exclusive examination of a single anatomical region,* especially when this region provides a rather limited number of phylogenetic characters.

With respect to the most parsimonious, strict-consensus cladogram (CI = 0.42; RI = 0.75) generated by the cladistic analysis of the 81 characters

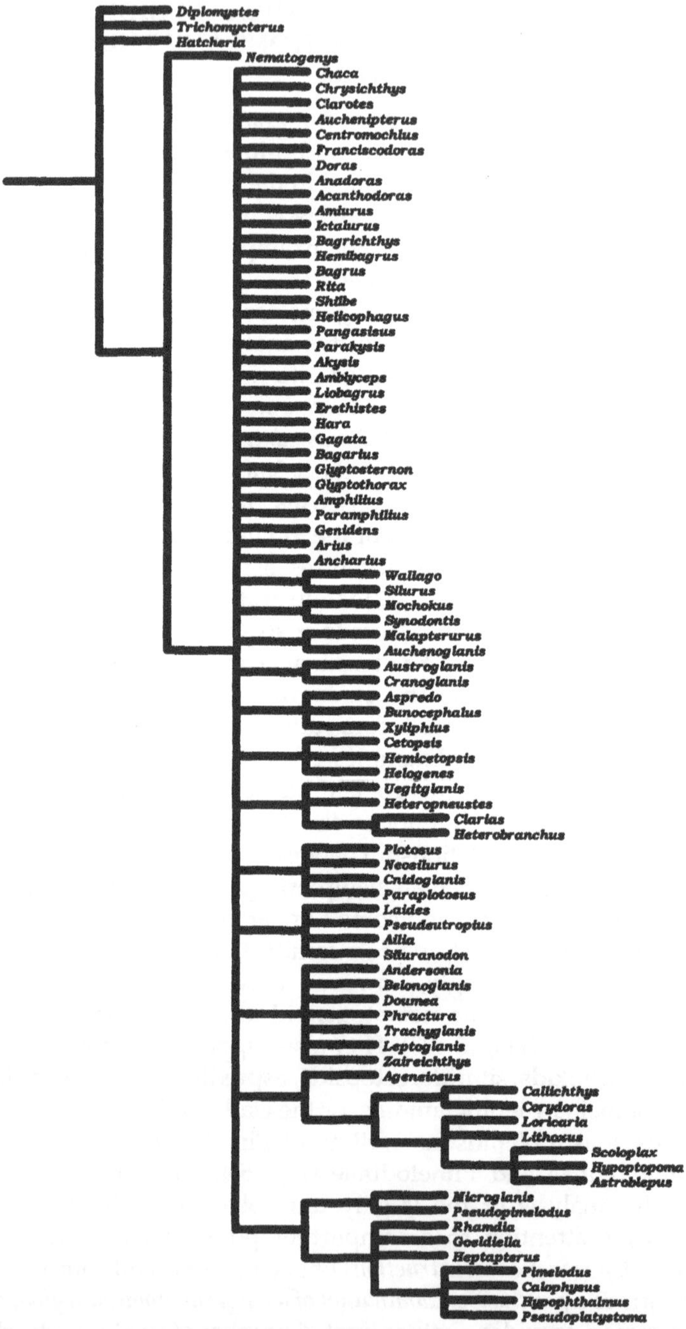

Fig. 5.4 Strict consensus tree (CI=0.36; RI=0.63) of 1000 equally parsimonious trees obtained when only those 37 characters referring to the ventral cephalic musculature and structures associated with the mandibular barbels were included in the cladistic analysis [for more details, see text].

concerning the pectoral girdle and associated musculature (see Section 3.1: chars. 38-55, 143-205), this is illustrated in Figure 5.5. Again, as expected, the phylogenetic results of this cladistic analysis including only 81 characters are clearly limited. The cladogram obtained is largely unresolved and, once again, there are some incongruities in relation to the 'all- evidence' cladogram of Figure 3.123, such as the relationships within loricarioids, the clade malapterurids + plotosids, the non-monophyly of Sisoroidea and family Sisoridae, and the clade clariids + schilbids (Fig. 5.5, compare with Fig. 3.123). However, examination of the cladogram obtained likewise indicates that analysis of the 81 characters referring to the pectoral girdle and its associated musculature did contribute some important points to the overall phylogenetic results of the present work shown in the 'all-evidence' cladogram of Figure 3.123. In fact, these characters further supported the monophyly of groups such as Cetopsidae, Trichomycteridae, Callichthyidae, Loricariidae, Siluridae, Plotosidae, Amphiliidae, Akysidae, Erethistidae, Aspredinidae, Pseudopimelodinae, Claroteidae, Ariidae, Clariidae (including *Heteropneustes*), Mochokidae, or the clade doradids + mochokids + auchenipterids. But the most important contribution of these 81 characters for the overall phylogenetic results of the present work clearly seems to be the indication of a markedly basal position of the whole superorder Loricaroidea, as well as of Cetopsidae, and even Siluridae (Fig. 5.5, compare with Fig. 3.123). Besides these basal nodes, inclusion of the characters of the pectoral girdle and its musculature also contributed some very important points concerning some less inclusive, more restricted nodes. For example, they helped to solve the phylogenetic relationships within Amphiliidae, with the analysis of these characters allowing a striking resolution of the intrarelationships inside this family, which corresponds from point to point to that obtained by the analysis of all the 440 characters examined in this work (Fig. 5.5, compare with Fig. 3.123). This is a paramount finding, principally since the pectoral girdle complex and in particular its associated musculature are not often subject to detailed examination in cladistic analyses undertaken by catfish specialists and ichthyologists in general (see Sections 1.3, 2.1, 2.3 and 3.2).

One region subjected to detailed examination in most cladistic analyses of the phylogeny of catfishes, as well as fishes in general, is the neurocranium. But the most parsimonious, strict-consensus cladogram (CI = 0.50; RI = 0.71) generated by the cladistic analysis of the 87 characters concerning catfish neurocranium and anterior vertebrae (see Section 3.1: chars. 56-142) does not present a resolution higher than that generated by the 81 characters concerning the pectoral girdle and associated musculature (Fig. 5.6, compare with Fig. 5.5). In fact, the cladogram concerning the neurocranium and anterior vertebrae is poorly resolved, not giving, for example, even a single clue to the potential most basal non-diplomystid catfishes; instead, all catfish groups are included in a basal, largely unresolved node (Fig. 5.6). Further, it suggests some points that in view of all the other works available on catfish phylogeny seem, at least at first sight, to be highly incongruous. Examples: genus *Amphilius* appears

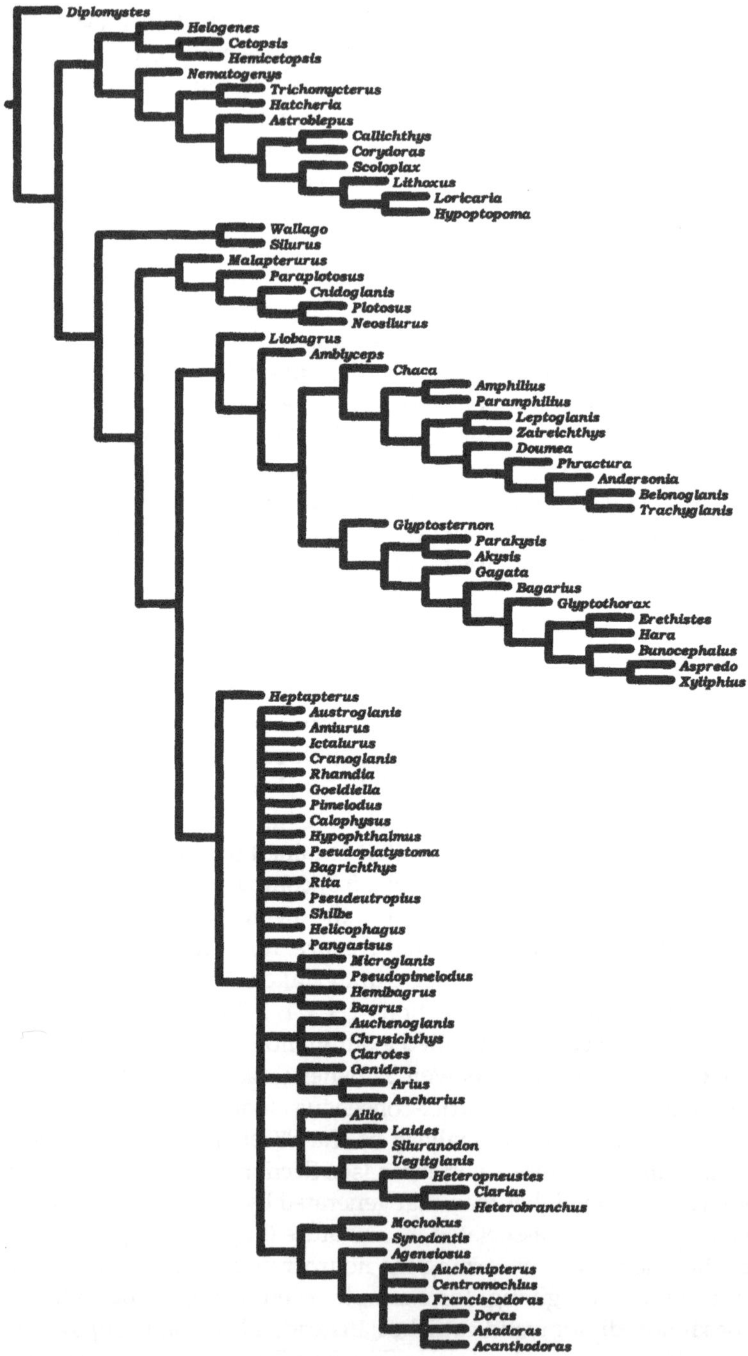

Fig. 5.5 Strict consensus tree (CI=0.42; RI=0.75) of 1000 equally parsimonious trees obtained when only those 81 characters referring to the pectoral girdle and associated musculature were included in the cladistic analysis [for more details, see text].

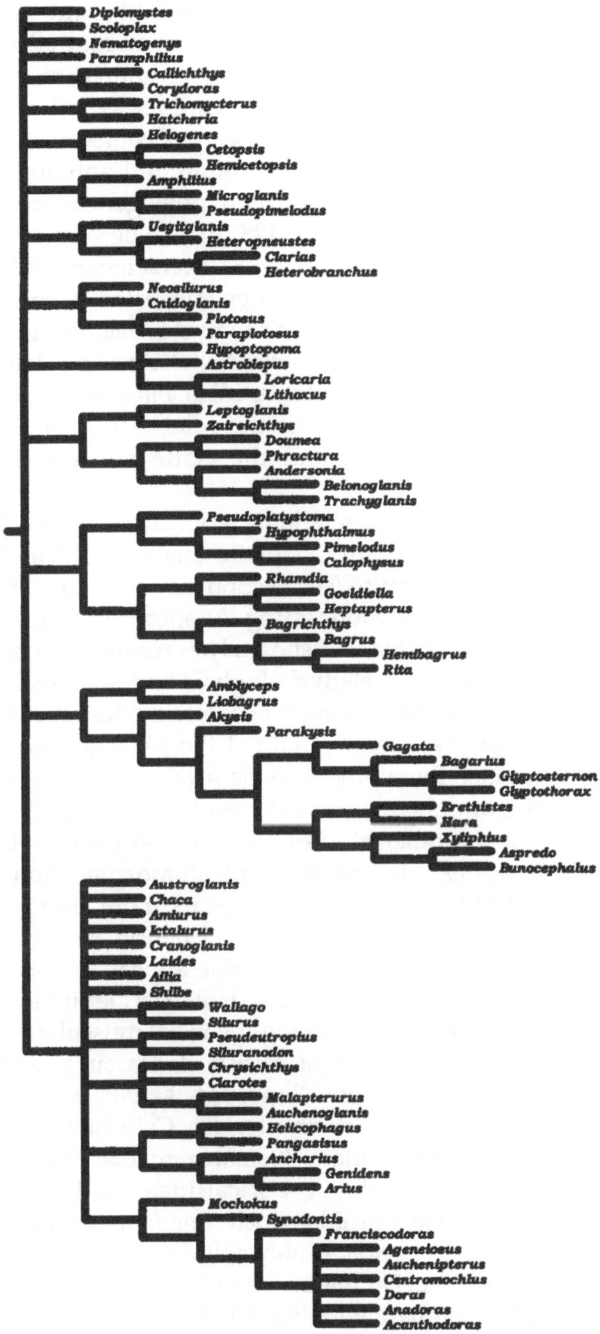

Fig. 5.6 Strict consensus tree (CI=0.50; RI=0.71) of 1000 equally parsimonious trees obtained when only those 87 characters referring to the neurocranium and anterior vertebrae were included in the cladistic analysis [for more details, see text].

closely related to the pseudopimelodin pimelodids, but not to the also amphiliin genus *Paramphilius*; loricarioids appear separated into five non-related groups (scoloplacids, nematogenyids, trichomycterids, callichthyids and astroblepids + loricariids); family Loricariidae is not recognised as a natural group (see Fig. 5.6). This does not mean, of course, that the analysis of the 87 characters concerning the neurocranium and anterior vertebrae did not contribute to the elaboration of the overall phylogenetic scenario obtained in this work and illustrated in Figure 3.123. As can be seen by a comparison of Figures 5.6 and 3.123, analysis of these 87 characters contributed support to the monophyly of several catfish groups suggested in the cladogram of Figure 3.123, such as Callichthyidae, Trichomycteridae, Pseudopimelodinae, Cetopsidae, Clariidae (including *Heteropneustes*), Plotosidae, Doumeinae, Leptoglanidinae, the clade doumeins + leptoglanidins, Pimelodinae, Heptapterinae, Bagridae, Amblycipitidae, Sisoridae, Erethistidae, Aspredinidae, Sisoroidea, Siluridae, Ariidae, Pangasiidae, and the clade mochokids + doradids + auchenipterids.

Examination of the most parsimonious, strict-consensus cladogram (CI = 0.51; RI = 0.82) generated by the cladistic analysis of the 53 characters concerning the facial musculature (see Section 3.1: chars. 206-258), illustrated on Figure 5.7, proved especially interesting. As shown in Figure 5.7, although including only 53 characters, this cladistic analysis resulted in a highly resolved strict-consensus tree, also with relatively high CI and RI. However, when one takes a nearer view of the cladogram, it becomes evident that this obviously scarce number of characters clearly limited the phylogenetic results of the cladistic analysis. In fact, these phylogenetic results are incongruent with the 'all-evidence' cladogram obtained in the present work (Fig. 3.123) and with the vast majority of the other studies available on catfish phylogeny. For example, bagrids appear as the most plesiomorphic non-diplomystid siluriforms, the ictalurid genus *Amiurus* appears more closely related to 29 other non-ictalurid genera than to the also ictalurid genus *Ictalurus*, and families Doradidae, Sisoridae and Erethistidae appear as non-monophyletic groups (Fig. 5.7, compare with Fig. 3.123). However, let it be noted that the analysis of these 53 characters concerning the facial musculature did provide support to an impressive number of catfish clades suggested by the 'all-evidence' cladogram, such as: Bagridae sensu Mo, Pangasiidae, Pseudopimelodinae, Siluridae, Trichomycteridae, Callichthyidae, Loricariidae, Loricaroidea, clade Trichomycteridae + Nematogenyidae, clade Callichthyidae + Scoloplacidae + Astroblepidae + Loricariidae, clade Scoloplacidae + Astroblepidae + Loricariidae, Clariidae (including *Heteropneustes*), Plotosidae, clade plotosids + clariids + chacids, Amblycipitidae, Amphiliinae, Cetopsidae, Cetopsinae, Leptoglanidinae, Doumeinae, Mochokidae, Auchenipteridae, and Aspredinidae (Fig. 5.7, compare with Fig. 3.123). This once again underscores the high phylogenetic potential of myological characters and, in this case, of those characters referring to the facial musculature, for the eventual detection and support of natural groups.

Fig. 5.7 Strict consensus tree (CI=0.51; RI=0.82) of 43 equally parsimonious trees obtained when only those 53 characters referring to the facial musculature were included in the cladistic analysis [for more details, see text].

Lastly, the most parsimonious, strict-consensus cladogram (CI = 0.55; RI = 0.40) generated by the cladistic analysis of the 169 characters concerning the splanchnocranium (see Section 3.1: chars. 259-427) is illustrated in Figure 5.8.

Interestingly, although this cladogram is based on a much higher number of characters than the other cladograms discussed above concerning other anatomical regions, it does not present a particularly high resolution vis-à-vis these cladograms (Fig. 5.8, compare with Figs. 5.5, 5.6 or 5.7). In fact, this cladogram is relatively poorly resolved, with almost the totality of the non-diplomystid siluriforms appearing in a large polytomy including 14 unresolved nodes (Fig. 5.8). Further, it suggests some points that, in view of all the other works available on catfish phylogeny, appear highly incongruous, such as the ictalurid genus *Amiurus* appearing more closely related to *Cranoglanis* than to the also ictalurid genus *Ictalurus* (in the 'all-evidence' cladogram *Cranoglanis* appears indeed near ictalurids, but these latter appear as monophyletic, as consensually accepted nowadays), Bagridae sensu Mo appearing non-monophyletic, and Loricaroidea, Sisoroidea, Doradidae, Pimelodinae, Aspredinidae, Doradidae and Akysidae also appearing as non-monophyletic groups (Fig. 5.8, compare with Fig. 3.123). But, again, this does not mean, of course, that the analysis of the 169 characters concerning the catfish splanchnocranium failed to make a useful contribution to the overall phylogenetic scheme obtained in the present work. Analysis of these 169 characters showed support of the monophyly of several catfish groups suggested in the cladogram of Figure 3.123, such as the Trichomycteridae, the rather basal clade constituted by this family and the Nematogenyidae, Pangasiidae, Claroteidae, Claroteinae, Cetopsidae, Cetopsinae, Siluridae, Clariidae (including *Heteropneustes*), Heptapterinae, Pseudopimelodinae, Mochokidae, Callichthyidae, Loricariidae, Amphiliidae, Doumeinae, Leptoglanidinae, Plotosidae, Amblycipitidae, Erethistidae, Sisoridae and, importantly, inclusively family Schilbidae and the clade constituted by Bagridae plus Pimelodidae (Fig. 5.8, compare with Fig. 3.123).

The overall analysis of the cladograms of Figures 5.4, 5.5, 5.6, 5.7 and 5.8 discussed above thus underscores that examination of a certain anatomical region often provides, per se, some useful, interesting phylogenetic information. It further emphasises that an effort should definitely be made to include several different anatomical regions in the same phylogenetic analysis, in order to obtain the maximum evidence possible and thereby decrease the importance of eventual homoplasies occurring in a certain anatomical region and/or structural complex. This is because, as pointed out some years back by Hennig (1950, 1965, 1966), the homoplasies eventually occurring in two groups of organisms concerning the structures of a particular region (e.g. the presence of one pair of mandibular barbels in two different catfish groups) will not necessarily be paralleled by other homoplasies concerning structures of a completely different region (e.g. the pectoral girdle). This is, in fact, well illustrated by the example discussed in Section 5.2 referring to the striking, and seemingly homoplasic similarities found, on the one hand, in aspredinids and doradids and, on the other, in aspredinids and chacids. These similarities actually concern rather different structures, being probably related to certain

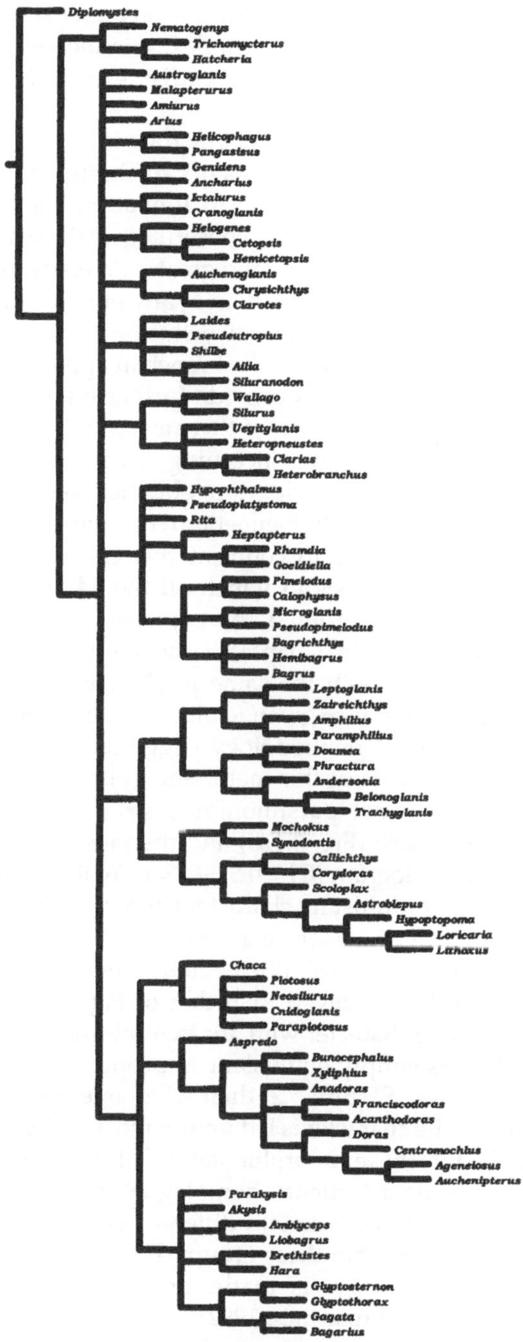

Fig. 5.8 Strict consensus tree (CI=0.55; RI=0.40) of 1000 equally parsimonious trees obtained when only those 169 characters referring to the splanchnocranium were included in the cladistic analysis [for more details, see text].

homoplasic events occurring in a certain anatomical region, but not necessarily in another different region, in a rather complex, 'mosaic' overall distribution of morphological characters in these three catfish groups (see discussion in section 5.2).

So, it would assuredly be interesting to integrate in future other characters of other anatomical regions in this analysis of catfish higher level phylogeny. As explained in Chapters 1 and 2, due to limitations of time it was not possible to include in this book characters concerning all the various anatomical regions. Nevertheless, a greater number and greater diversity of morphological characters have been incorporated here than in any other cladistic analysis on catfish phylogy to date. But an eventual inclusion in future of characters referring to such anatomical regions as the branchial apparatus, pelvic girdle, or caudal skeleton, would doubtless provide further interesting information on the phylogeny and macroevolution of this amazing group of fishes.

Before concluding this section concerning a comparison of the most parsimonious, strict-consensus cladogram obtained in the present work (Fig. 3.123) and other eventual phylogenetic trees, one last, albeit different, additional comparison could be to confront this cladogram with the tree obtained from a cladistic analysis in which all the 440 characters listed in Section 3.1 were coded as unordered. As explained in Section 2.1, in the cladistic analysis undertaken in Chapter 3 the multistate characters were ordered whenever possible, with the same phylogenetic weight and hence the same phylogenetic importance given to all characters used in the analysis, following the 'normalisation' methodology suggested by Wiens (2001) (see Section 2.1 for more details about this methodology).

A comparison of the most parsimonious, strict-consensus cladogram obtained in the present work (Fig. 3.123) and the most parsimonious, strict-consensus 'unordered' cladogram (CI = 0.52; RI = 0.76) illustrated in Figure 5.9 confirms that ordering of multistate characters allows a greater phylogenetic resolution than non-ordering (see, e.g. Mickevich and Lipscomb, 1991; Lipscomb, 1992; Kitching et al., 1998; Wiens, 2001). The 'unordered' cladogram of Fig. 5.9 is markedly less resolved than that of Fig. 3.123. This is natural since, to order a multistate character with, for example, three character states—CS-0, CS-1 and CS-2, is simply equivalent to proposing a further 'primary homology' hypothesis (see Section 5.2) than when one leaves this multistate character unordered. If the character is left unordered, two 'primary homology' hypotheses are posited: 1) the apomorphic state CS-1 found in certain different taxa is hypothesised to be a 'primary homologue' within these taxa; 2) the apomorphic state CS-2 found in certain different groups is hypothesised to be a 'primary homologue' within these groups. If the character is ordered, an additional 'primary homology' is made: state 2 is 'homologous, with modification' of state 1 (see Lipscomb, 1992).

The fact that the ordering of multistate characters results in a less equally parsimonious tree and hence a greater phylogenetic resolution than the non-ordering of these characters is viewed by some authors as a disadvantage of

character ordering (e.g. Hauser and Presch, 1991). However, this seems to be a distortion of the basal, elementary cladistic principles of Hennig (1950, 1965, 1966). Hennig plainly remarked that an effort should be made to include the *maximum 'primary homology' hypotheses as possible,* thus including the *maximum information possible* for investigating the relationships of the group under study (see Section 5.1 for discussion of this subject).

Almost all the differences between the 'unordered' cladogram of Figure 5.9 and the most parsimonious, strict-consensus cladogram presented in Section 3.2 and illustrated in Fig. 3.123 concern the presence, in the 'unordered' cladogram, of a series of unresolved nodes that appear resolved in the latter cladogram. For example, contrary to the latter cladogram, in the 'unordered' one there is a large unresolved pentatomy leading respectively to malapterurids, amphiliids, clade bagrids + pimelodids, clade mochokids + doradids + auchenipterids, and clade chacids + plotosids + clariids + sisoroids, with the relationships between Chacidae, Plotosidae, Clariidae and Sisoroidea appearing also as largely unresolved (Fig. 5.9, compare with Fig. 3.123). Further, the relationships within Doradidae, Amphiliidae, Clariidae, and Heptapterinae are not fully resolved in the 'unordered cladogram' (Fig. 5.9), contrary to what happens in the cladogram of Figure 3.123. It is important to recall that the resolution of the intra-familial relationships within these four groups proposed by the 'ordered' cladogram obtained in the present work is supported by studies of other authors, respectively Higuchi (1992) and de Pinna (1998) [Doradidae], He el al. (1999) [Amphiliidae], Chardon (1968) and de Pinna (1993) [Clariidae], or Lundberg et al., 1991a [Heptapterinae] (see Section 3.2). Interestingly, although in the 'unordered' cladogram the node leading to sisorid, erethistid and aspredinid catfishes appears as unresolved, this cladogram does suggest a better resolved, closer relationship between Akysidae and the clade formed by these catfishes, as suggested by de Pinna's 1996 study (Fig. 5.9, compare with Fig. 3.123).

It is of paramount importance to note here that although the 'unordered' cladogram of Figure 5.9 is less resolved than the cladogram generated in the phylogenetic analysis of Chapter 3, this 'unordered' cladogram is clearly congruent with the phylogenetic results described in that Chapter. In fact, *in what concerns the main subject of the cladistic analysis of the present work, i.e., the higher level phylogeny and interfamilial relationships of Siluriformes, the 'unordered' cladogram contradicts no single point of the cladogram obtained in the cladistic analysis of Chapter 3.* In fact, the only single point in which the 'unordered' cladogram of Figure 5.9 contradicts that of Figure 3.123 is that in the former the pseudopimelodins and heptapterins appear as sister-groups within family Pimelodidae, while the latter suggests a sister-group relationship between the heptapterin and pimelodin pimelodids. Thus, this single contradiction concerns the intrafamilial relationships within a certain catfish family, the Pimelodidae, and not the interfamilial relationships between different siluriform families. Anyway, it is opportune to note here that the close relationship between the heptapterin and pimelodin catfishes, suggested in

Fig. 5.9 Strict consensus tree (CI=0.52; RI=0.76) of 92 equally parsimonious trees obtained when all those 440 characters included in the cladistic analysis of the present work were coded as unordered [for more details, see text].

the cladogram of Figure 3.123, is also strongly supported by recently obtained, but still unpublished molecular sequence data from 12S and 16S mitochondrial rDNA genes, which, according to Lundberg (1998: 59) 'contain a clear phylogenetic signal for Heptapterinae and Pimelodinae, plus monophyly of these two groups together' (see Section 3.2).

With respect to the comparison between the cladogram obtained in the cladistic analysis of Chapter 3 (Fig. 3.123) and the *'ordered, but not normalised'* (see Section 2.1) cladogram illustrated in Figure 5.10 (CI = 0.50; RI = 0.77), these two cladograms are practically equal. Not even a single contradiction is found concerning either the interfamilial or the intrafamilial relationships . In reality, there are only two punctilios (but not contradictions) between the two cladograms: 1) concerning the relationships within family Sisoridae, the 'non-normalised' cladogram of Figure 5.10 is somewhat more resolved, with *Bagarius* and *Gagata* appearing closely related, as suggested in de Pinna's 1996 paper; 2) concerning the relationships within family Amphiliidae, the 'normalised' cladogram of Figure 3.123 is somewhat more resolved, with leptoglanidins and the doumeins appearing closely related, as suggested by He (1997), He et al. (1999), Diogo (2003b), and Roberts (in litt., pers. comm.).

Therefore, to conclude this small comparison of the 'ordered, normalised' cladogram of Figure 3.123, the 'unordered' cladogram of Figure 5.9, and the 'ordered, but non-normalised' cladogram of Figure 5.10, it should be noted that regarding the higher level phylogeny and interfamilial relationships of Siluriformes, the latter two cladograms do not contradict the phylogenetic results obtained in Chapter 3. This provides a further argument for confidence in these phylogenetic results.

5.6 APTATIONS, EXAPTATIONS AND ADAPTATIONS IN MACROEVOLUTIONARY STUDIES

This section discusses a major concept in macroevolution, exaptation. This term was first defined by Gould and Vrba (1982) and recently explained in greater detail in Gould's (2002) last and voluminous book *The Structure of Evolutionary Theory*. However, the controversy leading to the definition of this term began a long ago.

In 1872, Darwin felt the need to compose, for the 6th and final version of the *Origin of Species*, the only chapter ever added to his book, due to publication the year before of a book considered by him a most cogent general critique of natural selection, which was titled, in a parody of Darwin's title, *On the Genesis of Species* (Mivart, 1871). Mivart's pertinent critique of Darwin's theory, centred around 'the problem of the incipient stages of useful structures', or as Gould expressed it, the '5 per cent of a wing principle'. Said Mivart: 'I can understand how wings work for flight once they originate, but how can evolution ever make a wing in Darwin's gradualist and adaptationist mode if five per cent of a wing can't possibly provide any benefit for flight' (Mivart, 1871: 61).

Fig. 5.10 Strict consensus tree (CI=0.50; RI=0.77) of 8 equally parsimonious trees obtained on using ordered characters, as described in Section 3.1, but without the 'normalisation' of their total weights as suggested by Wiens (2001) and explained in Section 2.1 [for more details, see text].

The answer to Mivart's 'incipient stages problem' given in Darwin's 1872 final version of the *Origin of Species* was mainly based on two arguments related to a major issue, the discordance between historical origin and current function. In his first argument Darwin presented an incisive solution which, as remarked by Gould (2002), has unfortunately become enshrined in evolutionary theory, through no fault of Darwin, who never used the term, under the inappropriate name of 'preadaptation': 5 per cent of a wing offers no concevailable aerodynamic benefit, but sequences forged by natural selection only presuppose continuity in differing reproductive success, not continuity in a single function. Thus, the incipient stages may have performed a different function (e.g., thermoregulation), for which their 5 per cent of a wing imparted benefits. Eventually, the enlarging protowing entered the domain of aerodynamic benefit, and the original function changed to the primary utility now exploited by most birds. Darwin's second argument was based in a related but, as noted by Gould (2002), even more generalised, structural principle of redundancy, that is, the inherent capacity of most organs to work in more than one way: an organ need not invent an entirely new function in some mysterious manner, but may evolve by intensifying a previously minor use, or even by recruiting an inherent but unexpressed potential. The modified organ can then eventually abandon its previous major function because other organs can continue their former operation in the service of the same necessary task.

In order to change the use of the 'unfortunate' term 'preadaptation', Gould and Vrba (1982) recommended that features crafted for current use continue to be called *adaptations*, and that features coopted for current use, following an origin for some other reason, be called *exaptations*. According to these authors, one should use the term *aptation*, rather than adaptation, as the general descriptive term for a character now contributing to fitness, with exaptation and adaptation defined as the two subcategories of aptation.

Gould defended this terminology recently explaining that despite some initial critiques, the new terms are now used by many evolutionary biologists (see Gould, 2002: 1234-1246). He called attention to an important point repeatedly noted and addressed in the present work (see Section 5.1), viz. the elaboration of somewhat 'gratuitous' evolutionary statements often made in certain functional/evolutionary works. As pointed out by him, many functional/evolutionary studies try, sometimes rather desperately, to relate the configuration of the anatomical features under study to a necessarily 'adaptive' meaning, since evolutionary statements such as 'structure X is an adaptation for purpose Z' are readily available. However, if one can facilely assume that several anatomical features *changed* their original function at a certain historical moment (see above), why can he not also suppose that in several other features this original function was simply *lost*, with many of these latter features eventually disappearing, but many others, being superfluous but not prejudicial, remaining even nowadays (Gould, 2002). Therefore, there is no reason to suppose theoretically that all, or even the vast majority of those structures now present in extant organisms are necessarily 'adaptive' to the present environment.

Cracraft (1981), de Pinna and Salles (1990), Lang (1990), Baum and Larson (1991), Newman and Muller (2000), as well as Gould (2002) have stated that to classify a given trait as an adaptation for a certain current function, not only must one demonstrate that this trait *enhances organismal performance ('criterion of current utility')*, but also that it *has evolved, following Darwin's definition, via natural selectio*~~ *for its current role ('criterion of historical genesis')*. Alternative hypotheses to natural selection include, for example: '(1) random genetic drift; (2) selection in a different context; (3) developmental or genetic correlation with a different, selected trait; or (4) 'effect sorting' associated with processes analogous to natural selection operating at different levels of biological complexity (e.g., the effects of segregation distortion on the evolution of tail phenotypes in mice)' (Baum and Larson, 1991: 4). But as noted by Cracraft (1981), it is very difficult, if not impossible, to ascertain today whether a certain macroevolutionary trait actually resulted from natural selection for its present function at the moment it evolved many thousands or millions of years ago. Why? Because we have no 'data to relate the intrapopulational phenotypic variability and variation in fitness at that moment, as required by the theory of natural selection' (Cracraft, 1981: 35).

This led to a particularly incommodious but pertinent scenario, as noted by Cracraft (1981): although it appears possible to discriminate, in some cases, adaptations in microevolutionary studies, it becomes much more difficult, if not virtually impossible, to state, in macroevolutionary studies, that a certain feature indeed constitutes a true adaptation as defined by Gould and Vrba (1982). In fact, as explained by de Pinna and Salles (1990: 373), in contrast to other, more general macroevolutionary hypotheses that can to some extent be tested by *a posteriori* confrontations with an explicit phylogenetic scheme 'on the specific relationships and characters shown in the cladogram of a given group of organisms' (see Section 5.1), such *a posteriori* confrontations can only eventually *reject certain potential a priori adaptational hypotheses*, but do *'not provide a method to test adaptations per se'*. This was actually seen in Chapter 4 of the present work concerning the macroevolution of the main siluriform structural complexes examined. Confrontation with the phylogenetic results of this work allowed rejection in certain cases of some a priori adaptational scenarios and enabled hypothesising in other cases that *perhaps* particular features, vis-à-vis these phylogenetic results, *might* be related to *eventual adaptations* to a certain specific condition in catfish evolutionary history.

But what about the second subcategory of *aptations* sensu Gould and Vrba (1982), the *exaptations*? This subcategory differs significantly from macroevolutionary adaptations. Adaptation necessarily refers to 'a character shaped by natural selection for a current use'. Exaptation could refer to 'a character shaped by natural selection for a particular function (an adaptation) and subsequently coopted for a new use', but also to 'a character not necessarily ascribed to the direct action of natural selection (a non-adaptation) and subsequently coopted for a current use' (see above) (Gould, 2002: 1233). The theoretical implication of this is that contrary to what happens with an adaptation, to classify a given trait as an exaptation for a certain current function it is not necessary to demonstrate that this trait evolved originally via natural selection ('criterion of historical genesis') (see above).

As already explained, the concept of exaptation seems to be particularly important with reference to the origin and macroevolution of structural complexes in major groups, for example bird wings, vertebrate jaws, or eye-lens crystallins of both vertebrates and invertebrates (Darwin, 1872; Gould and Vrba, 1982; Mallatt, 1998; Zimmer, 1998; Newman and Muller, 2000; Carrol et al., 2001; Thewissen and Bajpai, 2001; Gould, 2002; etc.). However, as observed by Gould, unfortunately it is often difficult to recognise potential exaptations, since 'anything that currently "works" is often called an "adaptation" by evolutionary researchers', although it might well have originally evolved for some other purpose, thereby actually constituting a potential exaptation by definition (Gould, 2002: 1234).

The case study presented in the present work concerning the macroevolution of the major different catfish structural complexes examined (see Chapter 4) nevertheless provides some good candidates for potential macroevolutionary exaptations. As a matter of fact, as explained in Section 5.3, the macro-evolutionary history of these six complex structures seems to have been associated with several main functional uncouplings. And, as also explained in that section, such decouplings are, as their name indicates, related to historical functional shiftings and thereby to eventual potential evolutionary exaptations.

For example, development of the typical stout 'pectoral spine' of catfishes could constitute a potential example of exaptation. This spine, currently associated with such different functions as sound production or defense against predators, is very likely the result of an evolutionary decoupling of the pectoral fin, essentially functionally associated with body movement [see Sections 4.2 and 5.3].

As also explained in Section 4.3, one of the major innovations concerning the evolution of the adductor mandibulae complex in Siluriformes is the homoplasic differentiation, in several catfish groups, of a muscle retractor tentaculi directly inserting on the maxilla. Contraction of the retractor tentaculi, contrary to that of the main adductor mandibulae complex, is not mainly associated with the opening/closure of the mouth, but rather with the direct abduction of the maxillary barbels. Thus development of the retractor tentaculi in those groups could constitute an example of potential exaptation. [see Sections 4.3 and 5.3].

With respect to the macroevolution of the palatine-maxillary system and the suspensorium, examples of potential evolutionary exaptations can also be given, viz. the configuration of the maxilla and the autopalatine, originally mainly incorporated in the movements respectively of the jaws and the suspensorium, and currently relatively free from those movements and more importantly related to movements of the maxillary barbels [see Sections 4.4, 4.5 and 5.3].

Lastly, the thin, highly mobile Müllerian process associated with sound production in certain catfish groups seems itself to constitute a nice potential exaptation, since several other catfishes not presenting an elastic spring apparatus also exhibit a remarkably thin and mobile anterior process of the fourth vertebra [see Sections 4.6 and 5.3].

The examples given above, as well as many others that could be adduced, seem to indicate therefore that potential exaptations could have indeed played a significant role on the origin and macroevolution of the main catfish structural systems examined in this book. So, despite the difficulties concerning the methodological recognition of potential exaptations (see above), the case study provided here seems to support exaptation as a potentially important concept concerning the origin and macroevolutionary history of major complex structures in higher taxa, as pointed out in the elegant works of such authors as Darwin (1872), Gould and Vrba (1982), Mallatt (1998), Zimmer (1998), Thewissen and Bajpai (2001) or Gould (2002).

5.7 PARALLELISMS, CONVERGENCES AND CONSTRAINTS IN MACROEVOLUTION

This section concerns three other major concepts in macroevolution—evolutionary constraints, parallelisms and convergences. As extensively explained in the up-to-date book of Gould (2002), the importance of parallelism for evolutionary theory has been historically underappreciated. One reason, as described in the historical and conceptual overview on this subject given by him, is the difficulty in properly defining the limits between parallelisms and convergences, and even between parallelisms and homologies. As further noted by him, although nowadays we rather simplistically distinguish between *homologies* and *homoplasies*, with parallelisms and convergences ranked in the category of homoplasies, this was not always the case.

Let us recall that when Lankester proposed the concept of homoplasy in 1870, he defined it as a subcategory of homology. According to him, many similarities not directly due to the inheritance of common ancestral structures have nonetheless arisen as consequences of the inheritance of unique, phylogenetically constrained building patterns and, therefore, deserve inclusion within a broader category of similarity upon descendence (as opposed to similarity derived purely by independent adaptation, with no contribution by constraint from an organism's past history) (see Gould, 2002). These independently evolved, but historically constrained similarities defined by Lankester as 'homoplasy' correspond to what we call now parallelisms. Thus, as noted by Gould, homoplasy has suffered a conceptual, historical movement 'from a subcategory of homology, where it was placed before, to become the diametric opposite of homology, with the domain of homology then shrinking to encompass only Lankester's narrower category of homogeny, and the domain of homoplasy expanding to include all similarities evolved independently and not directly inherited from a common ancestral structure' (Gould, 2002: 1074). But as Gould further noted, 'what looks like an enormous difference—the expulsion of homoplasy as a subcategory of homology (sensu lato), and its establishment as a phenomenon directly contrary to homology (sensu stricto) actually rests upon a small point: the migration of convergence *into* the category of homoplasy as now defined. If we decide that the crucial distinction between homology and homoplasy should rest upon common ancestry versus independent origin, then one important phenomenon, necessarily included within homoplasy by the defining criterion of independent origin for similar structures, shares too much conceptual overlap with homology to permit a clear and comfortable theoretical separation: independent origin channelled by common internal constraints of homologous genes or developmental pathways, in other words, the phenomenon known as parallelism' (Gould, 2002: 1074).

In fact, a detailed reanalysis of the meaning of parallelisms, and in particular the advances in the last two decades in the field of developmental biology, have important evolutionary implications.

One significant implication is the recognition that constraints are fundamental, 'positive' actors, and not merely 'negative' interveners in evolution, since discoveries in the last years have revealed that constraints have been, in fact, directly responsible for homoplasic parallelisms and, consequently, of a great number of evolutionary changes. Of course, constraints are related not just to homoplasic parallelisms. As remarked by Gould, recent discoveries of 'deep homology', i.e., the discovery that major phyla, separated by more than 500 MY of independent evolutionary history, still share substantial, if not predominant, channels of development based on levels of genetic retention, has clearly also contributed to the paramount importance of historical constraints. Evolutionary constraints also, indirectly, modulate homoplasic convergences. In fact, in markedly reducing the total number of possible evolutionary options, constraints augment the probability that the same evolutionary option will be taken by different, distantly related groups of organisms and consequently augment the frequency of homoplasic convergences.

This could help to explain the high levels of homoplasy usually found in most cladistic analyses. As explained in Section 5.2, for a long time evolutionary papers often referred to evolution as a somewhat 'clean', linear process, with homoplasies of course admitted, but somehow considered somewhat peculiar, 'extraordinary' cases. However, the results obtained in direct phylogenetic analyses, like the cladistic analysis of the present work, seem to indicate that evolution is a rather complex, highly homoplasic process, with homoplasies constituting a markedly high fraction of evolutionary changes, especially in major taxa such as the Siluriformes (see Section 5.2).

But one other implication of the reinterpretation of the meaning of parallelisms is precisely reinforcement of the idea that, as discussed on Section 5.1, *all* evidence available should be taken into account in phylogenetic evolutionary studies. Evolutionary parallelisms, although of independent origin, are, by definition, channelled by *certain common* internal constraints of homologous genes or developmental pathways (see above). Therefore, the occurrence of homoplasic parallelisms in various taxa *of a certain major group* could somehow indicate that these different taxa share 'common internal constraints of homologous genes or developmental pathways' and, consequently, are, in fact, phylogenetically associated. This point, concerning the particularly interesting and apparently paradoxical relation between homoplasic parallelisms and phylogenetic reconstructions, led Saether (1983: 343) for example to propose the cladistic concept of 'underlying synapomorphies', defined as 'the capacity to develop synapomorphy', or 'close parallelism as a result of inherited factors within a monophyletic group'.

In order to reinterpret and explain the concepts of convergence and parallelism in the light of the new discoveries of the last decades in developmental biology and genetics, Gould (2002: 1134-1142), known for his analogies between biological concepts and non-biological themes, advanced the interesting concept of *'Pharaonic bricks versus Corinthian columns'*. In 'the light of our burgeoning knowledge of genetic sequences and their actions, homology of some sort or level will always be found in underlying generators of similar end products, if only because all organisms share the same genetic code by common ancestry; but no one would argue that we should redescribe a classic range of convergence as parallelism simply because the markedly different developmental pathways of the two adaptations rest upon the action of genes made of DNA!' (Gould, 2002: 1136). Thus, theoretical criteria need to be developed for ordering and evaluating the highly varied and ever-growing compendium of homoplasic results generated along homologous developmental pathways, since these cases fall along a continuum from narrow and controlling channels of constraint to insignificant sharing of non-specific building blocks. This is where Gould's interesting concept of *'Pharaonic bricks versus Corinthian columns'* (Gould, 2002: 1134) enters the picture:

When Pharaoh 'made the children of Israel serve with rigor' (Exodus 1:13), they fabricated bricks to use in a full range of buildings: 'and they built for Pharaoh treasure cities, Pithom and Raamses' (Exodus 1:11). Now if these bricks built every structure in the city, from great pyramids to public toilets, we might identify a homologous generator of all final products (bricks of the same composition made by the same people in the same way over a continuous stretch of time). But we could scarcely argue that these homologous generators exerted any important constraint over the differing forms of Pharaoh's final products—if only because all realised architectural diversity shared the same building blocks. But if I note a majestic portico of Corinthian columns in front of a building in modern Manhattan, I recognise a strong internal constraint imposed by an architectural module of very different status. The Corinthian column, last and most ornate of the classic orders, consists of a slender fluted shaft, capped by a striking, distinctive, and elaborate capital (the defining 'species' character in a taxonomic analogy) adorned with stylised acanthus leaves ... . Like Pharaonic bricks, Corinthian columns hold clear status as homologous underlying generators for their continuous phyletic history and stable form. But whereas Pharaonic bricks did little to constrain a resultant building by their form or structural character, and would not therefore sustain an interesting interpretation of parallelism for two similar buildings that happened to employ them in construction (if only because many other, very unsimilar, buildings in town also use the same bricks), Corinthian columns do exert a strong structural constraint from an inherited past (a homology) that can help us to identify and distinguish buildings, even after 2500 years after the invention of this unchanged form.

Gould's metaphor is without doubt useful in defining theoretically evolutionary parallelisms as '*Corinthian column* similarities' channelled by a certain particularly strong common internal constraint of homologous genes or developmental pathways. However, as emphasised in the case study of the present work, it is very difficult to discriminate, in practice, in real phylogenetic analyses, such evolutionary parallelisms: how do we differentiate, practically speaking, among characters exhibiting a homoplasic distribution according to the cladistic analysis presented in Chapter 3, those which constitute evolutionary parallelisms and those constituting evolutionary convergences?

One could, for instance, argue that the insertion of the levator operculi in a significant part of the lateral surface of the opercle, a derived feature found in *Nematogenys*, *Austroglanis*, and *Heptapterus*, could constitute a nice, typical illustration of evolutionary convergence. In fact, this feature was independently acquired, within order Siluriformes, in three groups that appear markedly distant on the cladogram obtained in the cladistic analysis of Chapter 3 (see Fig. 5.11).

One could eventually compare this example with the repetitive independent acquisition, four different times, of accessory tooth-plates in the ethmoid region inside that particular, relatively restricted catfish clade including, in this cladogram, pangasiids, schilbids, cranoglanidids, ictalurids, austroglanidids, claroteids and ariids, namely: (1) within Pangasiidae, (2) within Schilbidae, (3) within Claroteidae and (4) within Ariidae (see Fig. 5.12). One might thereby argue that this could perhaps constitute a typical example of evolutionary parallelism eventually channelled by a *certain common* internal constraint of homologous genes, developmental pathways or other inherited factors within this catfish clade.

However, the matter is not so 'black-or-white'. For example, an accessory ethmoidal tooth-plate is also present, aside from this clade including pangasiids + schilbids + cranoglanidids + ictalurids + austroglanidids + claroteids + ariids, in the pimelodin genus *Pseudoplatystoma*, as shown in Figure 5.12. Of course, one could eventually argue that this was perhaps due to a true

Fig. 5.11 Hypothesis of character state evolution of insertion of levator operculi (char. 250): C-S0 (black)= levator operculi not attaching on significant part of lateral surface of opercle; C-S1 (blue)= levator operculi attaching on significant part of lateral surface of opercle; Inapplicable (pink) [for more details, see text].

Fig. 5.12 Hypothesis of character state evolution of presence of accessory tooth-plates on ethmoid region (char. 306): CS-0 (black)= absence of accessory tooth-plates on ethmoid region; CS-1 (blue)= presence of accessory tooth-plates on ethmoid region [for more details, see text].

convergence between the situation found in the node leading to this clade and that found in *Pseudoplatystoma*. However, how can we be sure that the similar tooth-plates found in *Pseudoplatystoma* and the members of that clade are not instead due to an evolutionary parallelism eventually channelled by *the same* common internal constraint of homologous genes, developmental pathways or other inherited factors within the whole order Siluriformes? This constitutes, in my view, one of the major problems in discriminating homoplasic parallelisms and convergences. In fact, if the latter option were accepted, one would have to consider the formation of ethmoidal accessory tooth-plates in 5 different catfish groups as a homoplasic parallelism resulting from the same common internal constraint within the whole order Siluriformes. Why then could we not be just as certain that the attachment of the levator operculi on the lateral surface of the opercle found in 3 different catfish groups is not also due to a parallelism, and not a convergence, as hypothesised above, eventually channelled by another common internal constraint within this order? How many homoplasic events of a certain character are prerequisite within a group such as the order Siluriformes to formally state that the homoplasic development of this character was due to parallelism and not to convergence? To cite a well-known example (see Section 4.6), does the homoplasic development of an 'elastic spring apparatus' in at least 5 different catfish groups (Fig. 4.31) constitute a typical illustration of convergence or, instead, an illustration of evolutionary parallelism? If our focus is only on Siluriformes, we could eventually argue that these five homoplasic events constitute a classical example of convergent evolution within this large and diverse group. However, if we focus instead on a tree incorporating all vertebrate taxa, one could consider somewhat strange that such an 'elastic spring apparatus' did not develop even once in *all* the numerous non-siluriform vertebrate orders, yet was independently acquired at least five different times within the specific order Siluriformes. Such a situation would seem to indicate that catfishes shared a certain common patrimony making the probability of developing this apparatus much greater in these fishes than in other vertebrates.

The foregoing discussion shows that although evolutionary convergences and parallelisms can nowadays be relatively well defined theoretically, in real case studies such as that in the present work it is difficult to envision a single, somewhat unified kind of 'methodological phylogenetic test' that could eventually differentiate convergences from parallelisms. Such discrimination has to be obligatorily done case by case, with a different methodology necessarily applied in each case, which has to include a vast range of information concerning different fields such as comparative anatomy, embryology and/or genetics.

One very important point directly related to this issue, which probably constitutes one of the most studied topics in evolutionary biology but that clearly should be the subject of a much broader, encompassing integrative focus, concerns understanding evolutionary constraints. As already discussed, if the presence of an 'elastic spring apparatus' in at least 5 different catfish groups is in fact the result of a parallel evolution channelled by the same common, strong internal constraint within the entire Siluriformes order, being present in 17 of the 87 catfish genera included in the present cladistic analysis (Fig. 4.31), why didn't this apparatus develop in the other 70 genera examined? Are there some 'negative' constraints eventually limiting the expression of this 'positive' internal constraint? And another example: if the mesial interdigitation of the pectoral girdle seems important for strengthening this girdle in several catfish groups (see Section 4.2), with the scapulo-coracoids interdigitated in the great majority of catfish (char. 180), why are the cleithra only interdigitated mesially in two of the 32 extant catfish families, Amphiliidae and Scoloplacidae (char. 173)? Is this difference due only to a casual distribution of the two homoplasic occurrences of each of these characters? As a matter of fact, both these characters were acquired independently twice in the Siluriformes, with the difference being the inclusiveness of the nodes in which these characters have evolved. On the one hand, the suture between the scapulo-coracoids was acquired in that node leading to callichthyids + scoloplacids + astroblepids + loricariids and in that node leading to the non-diplomystid, non-loricarioid, non-cetopsid and non-silurid catfishes (see Section 4.2). On the other hand, the suture between the cleithra was acquired in those nodes leading respectively to scoloplacids and amphiliids (see Section 4.2). Or is the different taxonomic distribution of these two characters due to a certain specific kind of evolutionary constraint related, for example, to the chondral origin of the scapulo-coracoids versus the dermal origin of the cleithra? These issues, concerning evolutionary constraints, parallelisms and convergences, are definitely interesting, even fundamental central issues in the study of macroevolution that will, very likely, continue to be the subject of special attention in both the short and long term, and an increased understanding of these issues will indisputably contribute significantly to our future overall comprehension of evolutionary biology.

5.8 CORDELIA'S DILEMMA, HISTORICAL BIAS AND GENERAL EVOLUTIONARY TRENDS

Given the discussion above concerning evolutionary constraints and homoplasies, it is interesting to take up a related subject, also well covered in Gould's last book: the occurrence of 'general evolutionary trends'.

Gould (2002) remarked that discourses on the paramount importance and frequency of evolutionary trends have pre-empted much of the research on the history of clades. Concomitantly he added that the importance given to evolutionary trends—a psychological comment about our focus of attention—bears no necessary relationship to the relative frequency or causal weight of this phenomenon in the natural history of these clades. It seems more related to a tradition: trends tell stories and evolution is a narrative science in Western tradition, if not universally, that has always favoured directional tales of conquest and valour while experiencing great discomfort with evolution, aimless and undirected (Gould, 2002: 936-937). So, the great importance usually given to evolutionary trends is most likely explained by a historical bias of not reporting undirected evolution under the false belief that it does not represent 'interesting data' for evolution.

Gould noted that such historical bias is also seen, quite frequently, in palaeontological publications, in which examples of stasis are likewise not reported because of the conviction that such stability represents 'no data'. These biases were cleverly compared by him in another analogy between biological concepts and non-biological themes, with '*Cordelia's dilemma*', memorialising the plight of King Lear's honest but rejected daughter. 'When asked by Lear for a fulsome protestation of love in order to secure her inheritance, Cordelia, disgusted by the false and exaggerated speeches of her sisters Goneril and Regan, chose to say nothing, for she knew "my love's more ponderous than my tongue"; but Lear mistook her silence for hatred or indifference, and cut her tongue off entirely (with tragic consequences later manifested in his own madness, blindness, and death), in proclaiming that "nothing will come of nothing" (Gould, 2002: 764-765). Cordelia's dilemma arises in science when an important signal from nature is either not seen or not reported because scientists read the existing pattern as 'no data', literally as nothing at all. Therefore, 'most clades, while waxing and waning in species diversity through time, show no outstanding overall directionality; but we do not know because the literature has never recognised, or attempted to tabulate, the frequency of such Ecclesiastical clades that change all the time, but "go" nowhere in particular during their evolutionary peregrinations' (Gould, 2002: 937).

There are indeed two major fundamental theoretical contradictions between the supposedly high frequency and fundamental importance of evolutionary trends in the history of natural clades and Darwin's theory. On the one hand, a non-directional vector of environmental change can only elicit a set of meandering responses in the adaptive adjustments of organisms. On the other hand, the more serious challenge of catastrophe and mass extinction raises the spectre of randomness and death for reasons unrelated to the adaptive struggles or normal times—the 'wheel of fortune versus the wedge of progressive trends' (Gould, 1989).

Still, banishing our own scientific bias or desire to see a certain 'progress' in natural history, it nonetheless remains very difficult to defend the existence of some major 'general trend' in evolution. The biological diversity surrounding us (no need to go to Amazonia or some other nice tropical area to experience such) is completely overwhelming. Organisms of all sizes, all colours, with large and small relative cephalic volumes, completely different kinds of communication or locomotion, or no locomotion at all, continue to live here with us on this small and wonderful planet Earth in their own way. And more intriguing is that the most numerous, diverse and widely distributed organisms are those 'minuscule', 'simple', 'old', and oh my! 'acephalous' organisms known as bacteria. So what is this 'general biological evolutionary trend' on the planet Earth?

As with other evolutionary issues, the case study presented in this work allows us to discuss some aspects of this issue on 'general evolutionary trends'. In fact, catfishes are particularly reputed by their 'characteristic', 'general evolutionary trends'. Various characters included in the cladistic analysis of the present work have often been said to constitute 'general trends' within the Siluriformes (see, e.g., Regan, 1911b; Jayaram, 1956, 1966, 1968, 1971, 1978; Tilak, 1961, 1963b,c,d, 1964, 1965, 1967a,b, 1971; Jayaram and Majumdar, 1964; Alexander, 1965; Gosline, 1975; Howes, 1983a,b, 1985; Jayaram and Singh, 1982; Mo, 1991): *strong ankylosis between a posterodorsomesial 'transcapular' process of the posttemporo-supracleithrum and the anterior vertebrae* (char. 148), *strong ankylosis between the mesial limb of the posttemporo-supracleithrum and the neurocranium* (char. 155), *differentiation of the extensor tentaculi into different bundles* (char. 226), *loss of the posterior cartilage of the autopalatine* (char. 283), *loss of sesamoid bone 2 of the suspensorium* (char. 308), *loss of sesamoid bone 1 of the suspensorium* (char. 309), and *loss of the ascending portion of Meckel's cartilage* (char. 419). Of course, the authors defending the existence of these 'trends', like those defending the existence of evolutionary trends in general, are completely aware that natural selection theoretically only yields adaptation to immediate environments, an idea clearly not conducive to sustained general directional trends through geological time given the effectively random fluctuation of most environmental configurations through substantial geological intervals. However, as noted by Gould, such authors defend that such trends refer to 'generalised biomechanical improvements somehow conferring advantages across most or all experienced environments' (Gould, 2002: 889).

So, it is particularly interesting to examine here in detail each of the supposed catfish 'general evolutionary trends' mentioned above, confronting the taxonomic distribution of these features with the cladogram on siluriform higher level phylogeny obtained in the present work and illustrated in Figure 3.123.

With respect to the strong ankylosis between the posterodorsomesial 'transcapular' process of the posttemporo-supracleithrum and the anterior vertebrae, according to the phylogenetic results of the present work this feature evolved only twice within the Siluriformes: in Scoloplacidae and in the clade chacids + clariids + plotosids + sisoroids, with the ankylosis eventually reinforced in some groups within the latter clade, but also lost in some of its taxa (see Fig. 5.13). Therefore, this feature is clearly far from constituting any classic 'general evolutionary trend' evolved several times within the Siluriformes and somehow providing 'generalised biomechanical improvements' and conferring 'advantages across most or all experienced environments' inhabited by these fishes.

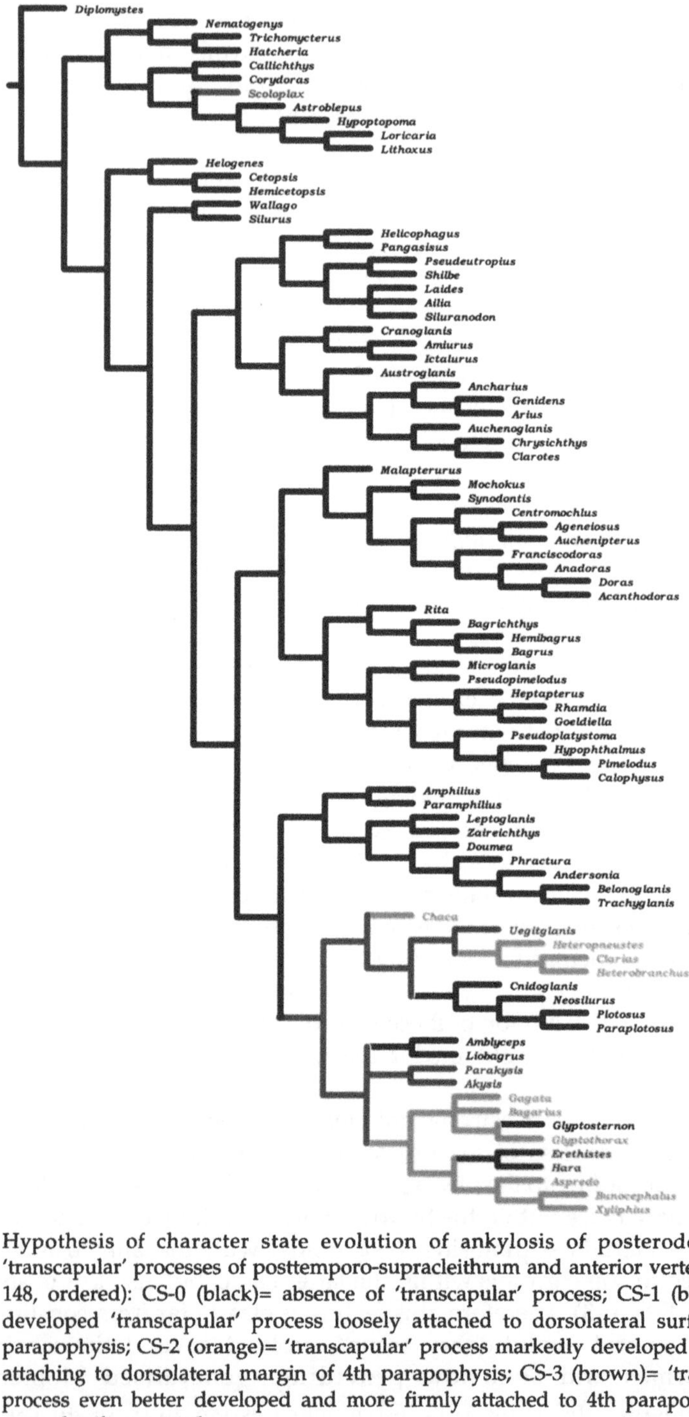

Fig. 5.13 Hypothesis of character state evolution of ankylosis of posterodorsomesial 'transcapular' processes of posttemporo-supracleithrum and anterior vertebrae (char. 148, ordered): CS-0 (black)= absence of 'transcapular' process; CS-1 (blue)= well-developed 'transcapular' process loosely attached to dorsolateral surface of 4th parapophysis; CS-2 (orange)= 'transcapular' process markedly developed and firmly attaching to dorsolateral margin of 4th parapophysis; CS-3 (brown)= 'transcapular' process even better developed and more firmly attached to 4th parapophysis [for more details, see text].

This also applies to the loss of the posterior cartilage of the autopalatine. As illustrated in Figure 5.14, this feature seemingly only occurred two or three times in catfish evolutionary history, being found only in amphiliids and non-nematogenyid loricarioids and hence not found in the vast majority of the catfish families (see Fig. 5.14).

The supposed 'catfish general evolutionary trend' to lose sesamoid bone 2 of the suspensorium ('ectopterygoid' of most authors: see Section 4.5) also seems to constitute an unfounded myth. As explained in Section 4.5, this bone was very likely absent in the node leading to all non-diplomystid catfishes. Its presence in austroglanidids, claroteids, most ariids, some bagrids, pseudopimelodins and some pimelodins is seemingly due, therefore, to its homoplasic development in four relatively small clades, and not to numerous independent losses of this bone within the Siluriformes (see Fig. 4.28).

Contrary to sesamoid bone 2, sesamoid bone 1 of the suspensorium ('entopterygoid' of most authors) was, indeed, seemingly lost at least three different times within the non-diplomystid siluriforms (see Fig. 5.15). However, it should be noted that two of the homoplasic losses of this bone occurred in particularly small clades within the cladogram illustrated on Figure 3.123, namely the mochokids and genus *Rita* (see Fig. 5.15). Consequently, within order Siluriformes, this bone is missing only in the non-nematogenyid loricarioids, mochokids, and one bagrid genus, being present in the vast majority of the catfish families. Therefore, this feature, too, hardly constitutes an unequivocal, typical 'generalised trend' within this order.

According to the phylogenetic results of the present work, the ascending portion of Meckel's cartilage was independently lost five times within the Siluriformes, namely in the clade callicthyids + scoloplacids + loricariids + astroblepids, in silurids, mochokids, amphiliids, and clariids (see Fig. 5.16). However, rather than a 'catfish general evolutionary trend' referring to a 'generalised biomechanical improvement' conferring 'advantages across most or all experienced environments' (see above), this feature seems to correlate strongly with a markedly benthic type of life, as exhibited by callichthyids, scoloplacids, loricariids, astroblepids, mochokids, amphiliids and clariids. This strong correlation is probably explained by the fact that, as pointed out by Alexander, in order to provide further stability against the substratum, those markedly benthic catfishes have a particularly depressed head and, consequently, a particularly depressed mandible (see Alexander, 1965). In fact, the only catfishes not exhibiting such a markedly benthic type of life in which the ascending portion of Meckel's cartilage is also missing are, as mentioned above, the silurids, with this structure thus present, overall, in the vast majority of catfish families (see Fig. 5.16). Therefore, the loss of the ascending portion of Meckel's cartilage is, again, also far from constituting any typical classic example of 'general evolutionary trend' providing 'unequivocal advantages across most or all experienced environments' inhabited by Siluriformes.

Fig. 5.14 Hypothesis of character state evolution of presence of posterior cartilage of autopalatine (char. 283): CS-0 (black)= posterior cartilage of autopalatine present; CS-1 (blue)= posterior cartilage of autopalatine absent; Ambiguity or Inapplicable (pink) [for more details, see text].

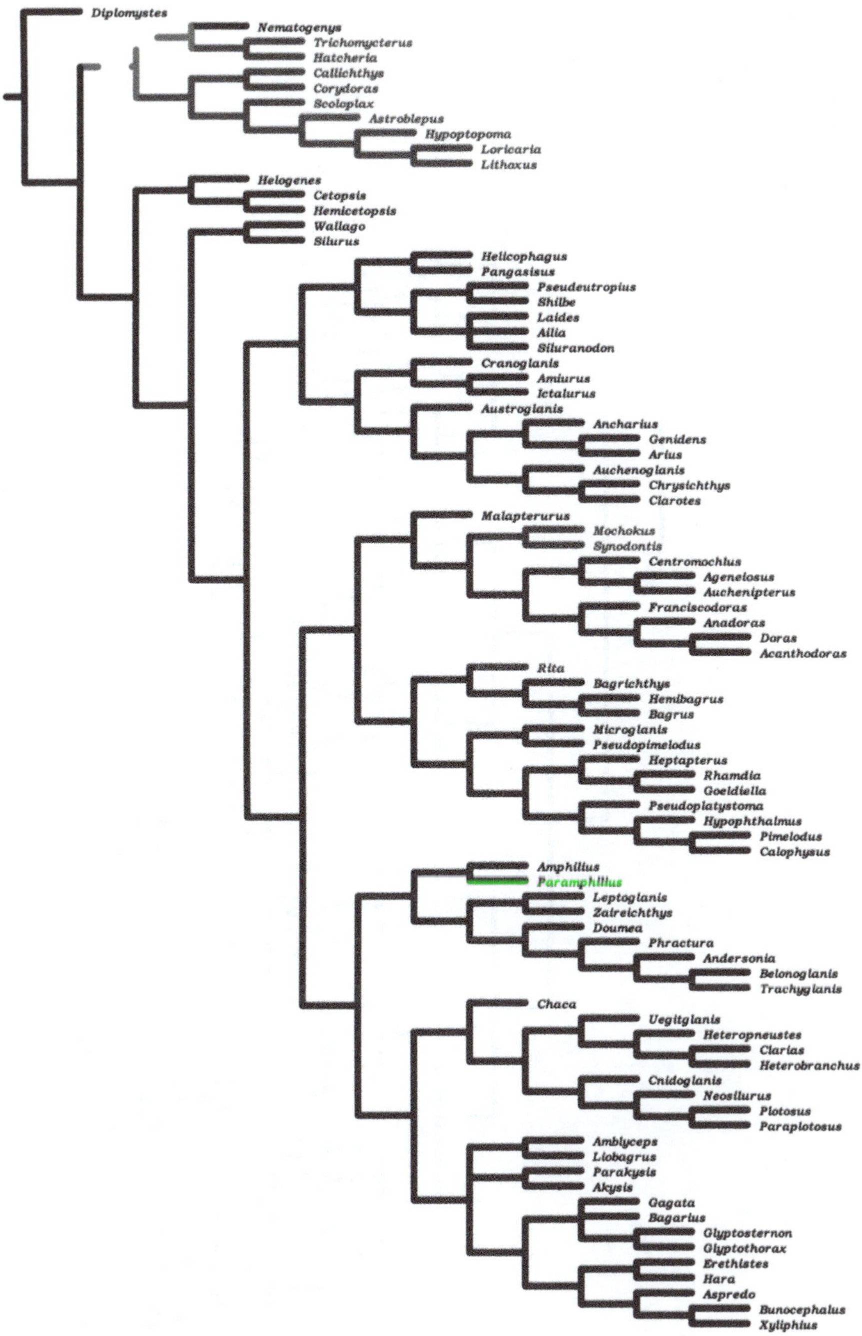

Fig. 5.15 Hypothesis of character state evolution of presence of sesamoid bone 1 of suspensorium (char. 309): CS-0 (black)= presence of sesamoid bone 1; CS-1 (blue)= absence of sesamoid bone 1; Ambiguity (pink) [for more details, see text].

Fig. 5.16 Hypothesis of character state evolution of presence of ascending portion of Meckel's cartilage (char. 419): CS-0 (black)= ascending portion of Meckel's cartilage present; CS-1 (blue)= ascending portion of Meckel's cartilage absent [for more details, see text].

The strong ankylosis between the mesial limb of the posttemporo-supracleithrum and the neurocranium also does not seem to correspond to the classic definition of a 'general evolutionary trend' within order Siluriformes. Although, as explained in section 4.2, this feature apparently developed independently in six different nodes of the cladogram obtained in the present work (see Fig. 4.11), those six nodes concern relatively small restricted groups with, again, a clear correlation, as wisely noted by Chardon (1968), between the presence of this feature and a marked ossification of the cephalic region, as found in groups such as cranoglanidids, ariids, mochokids, doradids, auchenipterids, or genus *Leptoglanis* (see, e.g. Alexander, 1965; Chardon, 1968; Mo, 1991; de Pinna, 1993). Moreover, it is difficult to talk about a 'general evolutionary tendency', or a certain 'progress' towards a strong, firm suture between the mesial limb of the posttemporo-supracleithrum and the neurocranium, when in several taxa commonly considered as 'specialised' catfish groups, such as scoloplacids, astroblepids, loricariids, glyptosternoids or clariids, the mesial limb of the posttemporo-supracleithrum simply became completely undifferentiated (char. 150 and Section 4.2).

With respect to the differentiation of the extensor tentaculi in different bundles, this seemingly occurred, as explained in Section 4.4, in at least five different occasions in the siluriform cladogram obtained from the cladistic analysis of the present work. These are: the clade including non-nematogenyid and non-trichomycterid loricarioids, the silurids, genus *Auchenoglanis*, clade malapterurids + doradids + auchenipterids + mochokids + bagrids + pimelodids + amphiliids + chacids + plotosids + clariids + sisoroids, and, subsequently, within the latter clade, the clariids (see Fig. 4.23).

Contrary to the other features discussed above, differentiation of the extensor tentaculi into different bundles can, in fact, correspond, at least in theory, to the definition of a 'catfish general evolutionary trend'. As already explained, loss of the ascending portion of Meckel's cartilage may correlate strongly with the marked depression of the head in notably benthic fishes. Therefore, the loss of this ascending portion of Meckel's cartilage does not confer, at least in theory, a major theoretical advantage, for instance, to less benthic catfishes presenting a well-developed, high coronoid process, e.g. nematogenyids, cetopsids or claroteids. In reality, as noted above, among the numerous non-markedly benthic catfishes, the only ones in which this ascending portion of Meckel's cartilage is missing are the silurids; otherwise this structure is present in the great majority of catfishes. [Note: in the silurids the coronoid process is only moderately developed, being clearly less developed than in catfishes such as claroteids, nematogenyids or cetopsids.] This is also the case for the strong firm ankylosis between the mesial limb of the posttemporo-supracleithrum and the neurocranium, which, as noted by Chardon (1968), does not seem to be particularly advantageous, quite the contrary, to those catfishes presenting some 'flexibility in the head', such as *Silurus* or *Paramphilius* (see Gosline, 1977: 335). However, as explained in

Section 5.4, the subdivision of the extensor tentaculi into different bundles allows a wider range of movements to the palatine-maxillary system and consequently to the maxillary barbels. Thus, at least theoretically speaking, it is not unreasonable to think that within an order in which the palatine-maxillary system is *invariably* present, the subdivision of the extensor tentaculi could eventually somehow constitute a 'generalised biomechanical improvement' across a large number of different environments. In fact, unlike the loss of the ascending portion of Meckel's cartilage or the strong ankylosis between the mesial limb of the posttemporo-supracleithrum and the neurocranium, subdivision of the extensor tentaculi has occurred in several completely different types of catfish groups: it is not, at least apparently, strongly correlated with a certain type of life or with a certain relatively small catfish group: instead, it is actually present in the vast majority of siluriforms (see Fig. 4.23).

Thus, among the seven characters discussed above which supposedly constitute 'general trends' within Siluriformes, only one, the differentiation of the extensor tentaculi into different bundles, could eventually correspond to the typical definition of a 'general evolutionary trend' within a certain biological group. Therefore, the present work clearly supports the view that the importance usually attributed to 'general evolutionary trends' is overvalued as the result of an historical bias leading us to invariably consider 'non-trends' as 'non-interesting' data and thus to give particular emphasis to purported 'general trends'. In fact, were the 7 characters discussed above eventually to fit the definition of 'general trend', let it be remembered that even so they would constitute only 1.6% of the 440 characters included in the present cladistic analysis.

In this particular respect, the case study of this work concerning the order Siluriformes thus seems to corroborate Gould's premonitory words: 'I suspect that most clades, while waxing and waning in species diversity through time, show no outstanding overall directionality, but we do not know because the literature has never recognised, or attempted to tabulate, the frequency of such Ecclesiastical clades that change all the time, but "go" nowhere in particular during their evolutionary peregrinations'. However, it is very important to recollect here that to say that a certain group 'shows no outstanding overall directionality' is not the same as saying that we can provide no picture of the evolution of that group. So, for example, within vertebrates there is clearly no 'general trend' to develop mammary glands: mammals evolved only once, and we clearly do not see a 'general tendency' in teleosts, birds, crocodiles or frogs to develop mammary glands. But this does not mean that in classical scholarly books, we cannot provide a general picture of vertebrate evolution mentioning, for example, that at a certain moment in time mammary glands were developed in a certain group of vertebrates, the mammals, and that subsequently, within mammals, a certain clade, the giant pandas, in spite of being part of the order Carnivora, developed a particular ability to feed on bamboo, although, again, this is clearly not a 'general trend' either within class Mammalia or within order Carnivora. This is applicable precisely to the case study provided in the present work, concerning order Siluriformes. In fact, although a clear, general 'outstanding overall directionality' clearly seems to not exist within this order, it is possible, as explained in Chapter 4, to provide a picture of certain interesting aspects of the evolution of this group.

5.9 PUNCTUATED EQUILIBRIUM, SPECIATION, LIVING FOSSILS AND MACROEVOLUTION

Figure 5.17 illustrates the phylogenetic results obtained in the cladistic analysis of the present work, as illustrated in the cladogram of Figure 3.123. However, the length of the branches is proportional to the number of evolutionary changes that occurred in those branches, with the total number of unambiguous evolutionary changes listed in Section 3.2 leading to each of the 87 genera examined being indicated after the name of these genera (see the "Synapomorphy list" of Section 3.2 for a detailed list of all these evolutionary changes). So, for instance, within the 440 characters included in the cladistic analysis of the present work, and according to the phylogenetic results obtained from that analysis, from the starting point of the catfish cladogram the total number of those evolutionary changes described in Section 3.2 leading, for example, to *Silurus* and *Doumea* is respectively 24 and 50. The two vertical grey lines of Figure 5.17, introduced in the Figure to facilitate comparisons, indicate respectively a total number of 25 and 50 evolutionary changes on the cladogram.

The phylogenetic information given in Figure 5.17 provides a further example that an explicit phylogenetic analysis of a certain biological group often make us reconsider our previous conceptions about the evolution of that group. For example, in catfish bibliography the loricariids are often seen, as explained in Chapter 1, as one of the most, if not the most, 'specialised' group of catfishes. However, it is interesting to note that within the characters included in the present cladistic analysis, and according to the phylogenetic results of this analysis, the total number of evolutionary changes leading to the loricariid genera *Hypoptopoma* (45), *Lithoxus* (46), and even *Loricaria* (49), are relatively modest compared to those leading to other catfish genera, as shown in Figure 5.17. In fact, *Loricaria*, with its 49 evolutionary changes, holds a humble 18th position, together with amphiliid *Phractura* and plotosid *Neosilurus*, with these genera excluded from the 'G-17' group, restricted as its name indicates to those 17 genera exhibiting a total of at least 50 evolutionary changes:

1) *Bunocephalus* (Aspredinidae) → 65 evolutionary changes
2) *Aspredo* (Aspredinidae) → 64 evolutionary changes
3) *Xyliphius* (Aspredinidae) → 62 evolutionary changes

Fig. 5.17 Phylogenetic tree illustrating the number of total unambiguous evolutionary changes for each genus examined in accordance with the phylogenetic results described in Section 3.2. The horizontal position of these terminal taxa is proportional to the respective total number of evolutionary changes, as indicated by the ruler at the bottom of the Figure. Vertical grey lines indicate respectively a total number of 25 and 50 evolutionary changes [for more details, see text].

4) *Chaca* (Chacidae) → 61 evolutionary changes
5/6) *Clarias* (Clariidae)/*Heterobranchus* (Clariidae) → 60 evolutionary changes
7) *Belonoglanis* (Amphiliidae) → 59 evolutionary changes
8) *Trachyglanis* (Amphiliidae) → 58 evolutionary changes
9) *Heteropneustes* (Clariidae) → 57 evolutionary changes
10) *Plotosus* (Plotosidae) → 55 evolutionary changes
11) *Andersonia* (Amphiliidae) → 54 evolutionary changes
12/13) *Uegitglanis* (Clariidae)/*Glyptosternon* (Sisoridae) → 53 evolutionary changes
14) *Ageneiosus* (Auchenipteridae) → 52 evolutionary changes
15/16) *Paraplotosus* (Amphiliidae)/*Glyptothorax* (Sisoridae) → 51 evolutionary changes
17) *Doumea* (Amphiliidae) → 50 evolutionary changes
18/19/20) *Loricaria* (Loricariidae)/*Phractura* (Amphiliidae)/*Neosilurus* (Plotosidae) → 49 evolutionary changes

The modest ranking of the loricariid catfishes impels us to conscientiously reconsider some aspects of our general picture of catfish evolution. In fact, more than the ranking of loricariids after catfishes such as aspredinids, chacids, clariids, sisorids or amphiliids, which are also usually considered to be 'highly specialised' siluriforms, it is the ranking after some 'generalised' genera such as *Plotosus* or *Paraplotosus* that constitutes an interesting 'surprise'.

It should be noted that this is the first study on catfish higher level phylogeny in which the total number of evolutionary changes of various catfish groups is compared on the basis of an explicit cladistic analysis. Thus, references such as 'highly modifed', or 'specialised' catfish groups in siluriform literature mostly rely on somewhat vague, subjective, indirect comparisons with, for example, particular emphasis given to certain groups and not to others. In fact, one of the possible explanations for the 'surprise' mentioned above could be precisely an issue mentioned in the Section 5.8, namely that concerning the influence of historical bias in biological studies. Loricariidae has been, without doubt, one of the most studied catfish groups, in a morphological, evolutionary, and phylogenetic point of view, with several studies definitely contributing to a consequent progressive growing emphasis on their 'remarkable specialisations'. The difference between the number of available cladistic works dedicated on the one hand to Loricariidae, and, on the other, to catfishes such as chacids or plotosids, for example, is striking (see Table 1.1).

However, it is also important to note here that the ranking of Figure 5.17, based on the total number of evolutionary changes on the cladogram, does not necessarily correspond to what some researchers would call 'levels of specialisation', since this total number of evolutionary changes eventually includes some morphological reversions to a plesiomorphic anatomical configuration. In fact, from the 835 unambiguous evolutionary changes noted in the 'Synapomorphy list' presented in Section 3.2, 70, i.e., about 8.4% of the total, refer to homoplasic reversions. Now, if, for example, during the evolution of a certain taxon there were, let's say, 10 evolutionary changes, of which one corresponds to a reversion, then this taxon would present not 9 apomorphic morphological characters, but only 8 since, de facto, one of the 9 morphological changes to an apomorphic state was reverted subsequently. Thus, reversions should be seriously taken into account when one addresses the issue of the total number of morphological changes within the evolutionary history of a certain group *versus* the number of those anatomical apomorphic features that are, in fact, actually 'visible' in that group. Of course, not all reversions necessarily imply a change to state 0 of a certain character. For example, of the 70 unambiguous reversions in the 'Synapomorphy list' of Section 3.2, three concern evolutionary changes from state 2 to state 1 of a certain character, namely: one in the node leading to bagrids + pimelodids (char. 226: 2→1), one in the node leading to heptapterins + pimelodins (char. 198: 2→1), and one in the node leading to *Rita* (char. 1: 2 →1) (see Section 3.2 for more details).

For just a rough comparison between the 'top 20' ranking given above concerning the total number of historical evolutionary changes and one referring to the actual number of 'visible' apomorphic, 'specialised' anatomical features, taking into account the 70 reversions listed in the 'Synapomorphy list' of Section 3.2, the ranking below was elaborated:

1) *Bunocephalus* (Aspredinidae) → 59 'visible' apomorphic features
2) *Aspredo* (Aspredinidae) → 58 'visible' apomorphic features
3) *Xyliphius* (Aspredinidae) → 56 'visible' apomorphic features
4) *Chaca* (Chacidae) → 55 'visible' apomorphic features
5) *Belonoglanis* (Amphiliidae) → 57 'visible' apomorphic features
6) *Trachyglanis* (Amphiliidae) → 54 'visible' apomorphic features
7) *Andersonia* (Amphiliidae) → 52 'visible' apomorphic features

8/9) *Clarias* (Clariidae)/*Heterobranchus* (Clariidae) → 50 'visible' apomorphic features

10/11) *Heteropneustes* (Clariidae)/*Glyptothorax* (Sisoridae) → 49 'visible' apomorphic features

12/13) *Doumea* (Amphiliidae)/*Synodontis* (Mochokidae) → 48 'visible' apomorphic features

14/15/16) *Loricaria* (Loricariidae)/*Phractura* (Amphiliidae)/*Glyptosternon* (Sisoridae) → 47 'visible' apomorphic features

17) *Uegitglanis* (Clariidae) → 45 'visible' apomorphic features

18) *Zaireichthys* (Amphiliidae) → 45 'visible' apomorphic features

19/20/21/22) *Lithoxus* (Loricaria)/*Cetopsis* (Cetopsidae)/*Amphilius* (Amphiliidae)/*Paramphilius* (Amphiliidae) → 44 'visible' apomorphic features

As can be seen, there are some differences between this ranking and that provided earlier. For example, the plotosids are completely absent from this list concerning the actual, 'visible' apomorphic features. In fact, there are, according to the phylogenetic results obtained in the present work, some morphological reversions in the evolutionary history of plotosid catfishes (see 'Synapomorphy list' of Section 3.2). However, the loricariids still stand in a relatively modest position in the ranking, with *Loricaria* placing 14th together with amphiliid *Phractura* and sisorid *Glyptosternon*. *Lithoxus* appears at the end of the list with the same number of 'visible' apomorphies as three other taxa, of which one is the cetopsid *Cetopsis* (see below). *Hypoptopoma* is missing in the list.

The above discussion emphasises the point that there is an important theoretical difference between the overall 'shape' of a certain taxon, the total number of evolutionary changes that actually occurred in its evolutionary history and its phylogenetic position. It further underscores the highly complex, 'mosaic' taxonomic distribution of morphological characters, as discussed in Section 4.2. For example, the Cetopsinae are found in a rather basal position within the higher level cladogram obtained in the present work, as well as in the cladograms of Mo (1991) and de Pinna (1998), exhibiting some plesiomorphic features found only in *Diplomystes* and the Loricaroidea (see Section 3.2). However, the cetopsins exhibit both a greater number of total historical evolutionary changes and a greater number of total, actually 'visible' anatomical apomorphic features than a series of catfishes situating in a much more derived position within this cladogram, for example *Amblyceps, Liobagrus, Parakysis, Akysis, Gagata, Bagarius, Erethistes* or *Hara* (see Fig. 5.17 and discussion above). The possibility of distinguishing and discriminating the important theoretical differences mentioned above in practical, real, explicit phylogenetic works clearly illustrates one of the main strengths provided by the cladistic tools developed in the last decades.

Analysis of Figure 5.17 also permits us to bring up another subject, which merited special attention in Gould's (2002) general overview of the structure of evolutionary structure: the *'living fossils'*. As noted in several other studies, Figure 5.17 clearly illustrates the markedly plesiomorphic anatomical configuration of the diplomystids within order Siluriformes, with *Diplomystes* plainly deserving its reputation of a 'relict' or 'living fossil' (see Chapters 1 and 2). The amazing anatomical difference between the plesiomorphic configuration of *Diplomystes* and that found in the other extant Siluriformes is remarkably seen in not some but, let it be noted, in *all* complex structures analysed in the present work (see Chapter 4). In fact, this amazing difference within the Siluriformes constitutes an interesting issue not merely for catfish specialists, but is actually an extraordinary case within the Ostariophysi, and even the Teleostei in general, not paralleled in any other ostariophysan order. Interestingly, it is also evident in Figure 5.17 that the silurids *Wallago* and

Silurus appear as the extant non-diplomystid catfishes showing, among the structures included in the present cladistic analysis, and according to the phylogenetic results of this analysis, a fewer number of evolutionary changes. This seems to support Chardon's idea of a markedly plesiomorphic overall configuration of the silurids compared to most other extant non-diplomystid catfishes (see Chardon, 1968).

As stated in Gould's (2002: 816) overview, 'palaeontologists had been truly stymied in their thinking about the important and contentious topic of "living fossils"'. One of the usual conventional explanations for the occurrence of 'living fossils', advanced in most student textbooks, is the old saw that living fossils had probably achieved optimal adaptation to their environments and therefore no alternative construction could selectively replace an ideal form achieved so long ago. However, as Gould noted, no one has ever presented even a vaguely plausible evidence for such a confident assertion: 'Why should horseshoe crabs lie closer to optimality than any other arthropod? What works so well in the design of lingulid versus other brachiopods? What superiority can a lungfish assert over a marlin or a tuna? In fact, since living fossils also (by traditional depiction) present such a "primitive" or "archaic" look, the claim for optimality seemed especially puzzling' (Gould, 2002: 816).

The other usual explanation, in a gradualistic and anagenetic world ruled by conventional selection, held that living fossils had stagnated because they lacked genetic change. However this seemingly 'more plausible' idea was strongly contradicted from the very early days of electrophoresis as a novel method for measuring overall genetic variation, with the well-known work of Selander et al. (1970) showing, for example, that *Limulus*, the horseshoe crab, does exhibit no lowering of genetic variability relative to known levels for other arthropods. This negative pattern has subsequently persisted with 'no standard lineage of living fossils exhibiting depauperate levels of genetic variability' (Gould, 2002: 816).

Gould thus proposed a different and relatively simple explanation for the occurrence of 'living fossils', which is deeply ingrained with some fundamental concepts of the general macroevolutionary theory defended in his voluminous last book. Based on Futuyma's (1986, 1987, 1988) analogy between macroevolution and the conventional Darwinism of natural selection in populations, Gould begins by assuming, like Darwinian gradualists, that morphological change may, indeed, accumulate anywhere along the trajectory of a species, and not exclusively during the geological moment of its origin. Thus, substantial evolution can occur in any local population at any time during the geological trajectory of a species, as documented in a large and developing literature, much beloved by popular sources, illustrating these rapid and apparently adaptive changes in isolated local populations, e.g. substantial evolution of body size in guppies (Reznick et al., 1997) or leg length in anolid lizards (Losos et al., 1997).

But, and this is the crux of the matter, 'these local populations can gain no substantial macroevolutionary expression unless they become "locked up" in a Darwinian individual with sufficient stability to become a unit of selection in geological time, and local populations—as a primary feature of their definition—do not maintain such coherence; they can, in principle—and do, in the fulness of geological time, almost invariably in practice—interbreed with other local populations of their species' (Gould, 2002: 800). In consequence, 'the distinctively evolved adaptations of local populations must therefore be ephemeral in geological terms, unless these features can be stabilised by individuation', with only 'speciation—as the core of its macroevolutionary meaning—providing such individuation by "locking up" evolved changes in reproductively isolated populations that can, thereafter, no longer amalgamate with others' (Gould, 2002: 800).

This helps, in fact, to explain Gould's concept of punctuated equilibrium. Rapid evolution in local populations of guppies and anoles can teach us many important lessons about general evolution, but these local populations usually strut and fret their short hour on the geological stage, and then disappear by death or amalgamation. This therefore produces the ubiquitous and geologically momentary fluctuations that characterise and embellish the long-term macroevolutionary stasis between speciation events (Gould, 2002).

The fundamental implication of this is that morphological change at a macroevolutionary level is not necessarily directly related to *time*, but rather correlates primarily with *frequency of speciation*. Speciation does not necessarily promote macroevolutionary change, but rather 'gathers in' and guards such change by locking and stabilisation for sufficient geological time within a certain unit of evolution. Thus, 'living fossils may simply represent those groups at the left tail of the distribution for numbers of speciation events through time' (Gould, 2002: 817). As Gould noted, such groups cannot theoretically be common, for consistently low diversity makes a taxon maximally subject to extinction in our contingent world of unpredictable fortune, where spread and number represent hedges against disappearance, especially in episodes of mass extinction. But 'every bell curve has a left tail' and, de facto, the occurrence of 'living fossils' among extant organisms is indeed very rare.

Gould's explanation of 'living fossils' seems indeed to be the most plausible one for explaining the occurrence of cases such as the genus *Diplomystes* within the order Siluriformes. The asymmetry found between the clade including the 'relict' *Diplomystes* and that including other extant catfishes is, in fact, astounding, with the striking morphological difference between diplomystid and non-diplomystid catfishes notably accompanied by a marked disparity between the total number of species in the two clades. Family Diplomystidae contributes only 6 species to the more than 2700 extant species of Siluriformes (see Chapter 1). And it seems difficult to accept that this could be due to the fact that diplomystids 'lie closer to optimality' than any other siluriforms or that 'no alternative construction could selectively replace the supposed ideal configuration' achieved so long ago by these catfishes (see above), particularly since diplomystids are, as a matter of fact, presently at serious risk of extinction (see Arratia, 1987).

Order Siluriformes thus provides, once more, a further example permitting the discussion of a particularly interesting macroevolutionary topic. In fact, as stressed throughout this Chapter 5, the study of the phylogeny and macroevolution of this order has allowed to introduce, and to discuss, several general evolutionary topics, such as the role of functional uncouplings in macroevolution, the contribution of myological versus osteological characters for phylogenetic reconstructions, the difference between exaptations and adaptations and between evolutionary parallelisms and convergences, the notion of 'general evolutionary trends' and the concept of 'living fossils'. I would like, in closing this Chapter, to sincerely acknowledge the vision and experienced sagacity of Professors Pierre Vandewalle and Michel Chardon for having chosen this amazingly diverse, interesting, and representative teleost group as the case study for my research, and especially for stimulating and always supporting me throughout its fruition.

Due to the amazing diversity and complexity of catfishes, several aspects of this group remain to be studied. Future research can comprise, for example, inclusion of further terminal taxa and further morphological regions in subsequent phylogenetic analyses, such as the branchial apparatus, pelvic girdle, caudal skeleton and/or structures associated with the dorsal fin. It is hoped that the discussions provided in this volume on the higher level phylogeny and macroevolution of the whole order Siluriformes will also pave the way for further insights into the phylogeny and evolution of some particular subgroups of the order, which could well refer to taxa such as Loricarioidea or Sisoroidea, but also to smaller clades. Each of the characters supporting these subgroups should be the subject of a careful analysis. What is the biological significance of each character? Which characters are most likely associated with the particular success of certain siluriform subgroups? Also, the distribution of various catfish subgroups should be compared with their ecological situation. Thus, a deeper comparison of clades linked to an identical milieu would allow further insights into such interesting issues as evolutionary convergences, parallelisms, or evolutionary trends. Particularly engaging would be, for example, a comparison of marine ariids and plotosids, torrent-dwelling species of amphiliids, sisorids and loricariids, and pelagic schilbids and pangasiids. A large and time-consuming inquiry into the biogeography, ecology and ethology of the members of the various catfish subgroups is without doubt needed to answer and/or discuss such issues. Such an undertaking lay outside the main purpose of the present work, concerned mainly, as explained earlier, with the general phylogeny and macroevolution of order Siluriformes as a whole. However, further insight into the phylogeny, evolution and biogeography of each of the different catfish subgroups, based on the general results of the present work and taking into account a large set of data concerning the biogeography, ecology and ethology of the members of these subgroups, will surely reveal very interesting.

Alternatively, the work presented here for the Siluriformes ought hopefully to also serve as a basis for further expanded research into the biogeography, phylogeny and macroevolution of the entire superorder Ostariophysi, as well as the Teleostei in general. This would provide an even broader background for further discussions on general theoretical phylogenetic and macroevolutionary themes, which clearly continue to be the issues that fascinate me most within the fascinating field of Biological Sciences.

References

Adriaens D, Verraes W. 1996. Ontogeny of cranial musculature in *Clarias gariepinus* (Siluroidei: Clariidae): the adductor mandibulae complex. J Morphol 229: 255-269.

Adriaens D, Verraes W. 1997a. Ontogeny of the suspensorial and opercular muscles in *Clarias gariepinus* (Siluroidei: Clariidae), and the consequences for respiratory movements. Neth J Zool 47: 1-29.

Adriaens D, Verraes W. 1997b. Ontogeny of the maxillary barbel muscles in *Clarias gariepinus* (Siluroidei: Clariidae), with some notes on the palatine-maxillary mechanism. J Zool (Lond) 241: 117-133.

Adriaens D, Verraes W. 1997c. Ontogeny of the hyoid and intermandibular musculature in *Clarias gariepinus*, an African catfish (Burchell, 1822). Zool J Linn Soc 121: 105-128.

Adriaens D, Verraes W. 1998. Ontogeny of the osteocranium in the African Catfish, *Clarias gariepinus* Burchell (1822) (Siluriformes: Clariidae): ossification sequence as a response to functional demands. J Morphol 235: 183-237.

Adriaens D, Verraes W, Taverne L. 1997. The cranial lateral-line system in *Clarias gariepinus* (Siluroidei: Clariidae): morphology and development of canal related bones. Eur J Morphol 35: 181-208.

Aguilera O. 1988. La musculatura estriada en los peces Gymnotiformes (Teleostei-Ostariophysi): musculatura facial. Acta Biol Venez 12: 13-23.

Albert JS, Fink WL. 1996. *Sternopygus xingu*, a new species of electric fish from Brazil (Teleostei: Gymnotoidei), with comments on the phylogenetic position of *Sternopygus*. Copeia 1996: 85-102.

Albert JS, Campoz-da-Paz R. 1998. Phylogenetic systematics of Gymnotiformes with diagnoses of 58 clades: a review of available data. In: Malabarba LR, Reis RE, Vari RP, Lucena ZM, Lucena CAS, (eds.). Phylogeny and Classification of Neotropical fishes. Edipucrs, Porto Alegre, Brazil, pp. 410-446.

Alexander R McN. 1964. The structure of the Weberian apparatus in the Siluri. Proc Zool Soc Lond 142:419-440.

Alexander R McN. 1965. Structure and function in catfish. J Zool (Lond) 148: 88-152.

Alves-Gomes JA. 1999. Systematic biology of gymnotiform and mormyriform electric fishes: phylogenetic relationships, molecular clocks and rates of evolution in the mitochondrial rRNA genes. J Exp Biol 202: 1167-1183.

Alves-Gomes JA. 2001. The evolution of electroreception and bioelectrogenesis in teleost fish: a phylogenetic perspective. J Fish Biol 58: 1489-1511.

Armbruster JW. 1998. Phylogenetic relationships of the suckermouth armoured catfishes of the *Rhinelepis* group (Loricariidae: Hypostominae). Copeia 1998: 620-636.

Arratia G. 1987. Description of the primitive family Diplomystidae (Siluriformes, Teleostei, Pisces): morphology, taxonomy and phylogenetic implications. Bonn zool Monogr 24: 1-120.

Arratia G. 1990. Development and diversity of the suspensorium of trichomycterids and comparison with loricarioids (Teleostei: Siluriformes). J Morph 205: 193-218.

Arratia G. 1992. Development and variation of the suspensorium of primitive catfishes (Teleostei: Ostariophysi) and their phylogenetic relationships. Bonn zool Monogr 32: 1-148.

Arratia G. 1997. Basal teleosts and teleostean phylogeny. Palaeo Ichthyologica 7: 5-168.

Arratia G, Chang A. 1975. Osteocráneo de *Nematogenys inermis* Ghichenot 1848 y consideraciones acerca de la primitividad del género (Peces Siluriformes, Trichomycteridae). Publ Ocasional Mus Nac Hist Nat Chile 19: 1-7.

Arratia G, Menumarque S. 1981. Revision of the freshwater catfishes of the genus *Hatcheria* (Siluriformes, Trichomycteridae) with commentaries on ecology and biogeography. Zool Anz 207: 88-111.

Arratia G, Menumarque S. 1984. New catfishes of the genus *Trichomycterus* from the high Andes of South America (Pisces, Siluriformes) with remarks on distribution and ecology. Zool Jb Syst 111: 493-520.

Arratia G, Schultze H-P. 1990. The urohyal: development and homology within osteichthyans. J Morphol 203: 247-282.

Arratia G, Schultze H-P. 1991. Development and homology of the palatoquadrate within osteichthyans. J Morphol 208: 1-8.

Arratia G, Gayet M. 1995. Sensory canals and related bones of Tertiary siluriform crane from Bolivia and North America and comparison with recent forms. J Vert Pal 15: 482-505.

Arratia G, Huaquin L. 1995. Morphology of the lateral line system and of the skin of diplomystid and certain primitive loricarioid catfishes and systematic and ecological considerations. Bonn zool Monogr 36: 1-110.

Arratia G, Chang A, MenuMarque S, Rojas G. 1978. About *Bullockia* n. gen. and *Trichomycterus mendozensis* n.sp: and revision of the family Trichomycteridae (Pisces, Siluriformes). Stud Neotrop Fauna & Envir 13: 157-194.

Azpelicueta MM, Rubilar A. 1998. A Miocene *Nematogenyis* (Teleostei: Siluriformes: Nematogenyidae) from South-central Chile. J Vert Paleont 18: 475-483.

Bailey RM, Stewart DJ. 1984. Bagrid catfishes from lake Tanganyika, with a key and descriptions of new taxa. Misc Publ Mus Zool Univ Michigan 167: 1-41.

Ballintijn CM, Burg AVD, Egberink BP. 1972. An electromyographic study of the adductor mandibulae complex of a free-swimming carp (*Cyprinus carpio* L.) during feeding. J Exp Biol 57: 261-283.

Barale G, Quaja M, Philippe M. 2000. Le bassin de Tataouine: une référence paléobotanique du Mésozoïque dans le domaine paralique du Gondwana septentrional. Volume des communications du premier colloque du patrimoine géologique, Tunis, Tunisia: 74-91.

Baras E, Laleyé P. 2003. Ecology and behaviour of catfishes. In: Arratia G, Kapoor BG, Chardon M, Diogo R (eds.). Catfishes. Science Publ. Inc., Enfield, NH (USA) pp. 525-579.

Barnard KH. 1942. Note on *Amphilius natalensis* Blgr. (Siluroidea, Amphiliidae). Ann Natal Mus 10: 257-259.

Barnard KH. 1943. Revision of the indigenous freshwater fishes of the S.W. Cape region. Ann S Afr Mus 36: 101-262.

Baskin JN. 1973. Structure and relationships of the Trichomycteridae. Unpubl. PhD thesis. City University, New York, NY.

Baum DA, Larson A. 1991. Adaptation reviewed: a phylogenetic methodology for studying character macroevolution. Syst Zool 40: 1-18.

Beaumont A. 1998. La notion d'homologie. Bull Soc Zool Fr 123: 311-321.

Benjamin M. 1990. The cranial cartilages of teleosts and their classification. J Anat 169: 153-172.

Benton MJ. 1999a. Early origins of modern birds and mammals: molecules vs. morphology. Bioessays 21: 1043-1051.

Benton MJ. 1999b. Reply to Easteal. Bioessays 21: 1059.

Berra TM. 2001. Freshwater Fish Distribution. Academic Press, San Diego, USA.

Bertmar G. 1959. On the ontogeny of the chondral skull in Characidae, with a discussion on the chondrocranial base and the visceral chondrocranium in fishes. Acta Zool 40: 203-364.

Biedenbach MA. 1971. Functional properties of barbel mechanoreceptors in catfish. Brain Res 27: 360-364.

Bleeker MP. 1852. De visschen van den Indischen archipel, Siluri. Act Soc Sci Indo-Needl 4: 1-370.

Bleeker MP. 1862. Atlas Ichthyologique des Indes Orientales Neerlandaises, Tome II, Siluroides, Characoides et Heterobranchoides. Frederic Muller, Amsterdam.

Bleeker MP. 1863. Systema silurorum revisum. Nederl Tydschr Dierk 1: 77-122.

Bock WJ. 1999. Functional and evolutionary explanations in morphology. Neth J Zool 49: 45-65.

Bockmann FA. 1994. Description of *Mastiglanis asopos*, a new pimelodid catfish from northern Brazil, with comments on phylogenetic relationships inside the subfamily Rhamdiinae (Siluriformes: Pimelodidae). Proc Biol Soc Wash 107: 760-777.

Bockmann FA. 1998. Análise filogenética da família Heptapteridae (Teleostei, Ostariophysi, Siluriformes) e redefinição de seus gêneros. Unpubl. PhD thesis. Universidade São Paulo, São Paulo, Brazil.

Borden WC. 1998. Phylogeny of the unicornfishes (*Naso*, Acanthuridae) based on soft anatomy. Copeia 1998: 104-113.

Borden WC. 1999. Comparative myology of the unicornfishes, *Naso* (Acanthuridae, Percomorpha), with implications for phylogenetic analysis. J Morphol 239: 191-224.

Bornbusch AH. 1991a. Redescription and reclassification of the silurid catfish *Apodoglanis furnessi* Fowler (Siluriformes: Siluridae), with diagnoses of three intrafamilial silurid subgroups. Copeia 1991: 1070-1084.

Bornbusch AH. 1991b. Monophyly of the catfish family Siluridae (Teleostei: Siluriformes), with a critique of previous hypotheses of the family's relationships. Zool J Linn Soc 101: 105-120.

Bornbusch AH. 1995. Phylogenetic relationships within the Eurasian catfish family Siluridae (Pisces: Siluriformes), with comments on generic validities and biogeography. Zool J Linn Soc 115: 1-46.

Bornbusch AH, Lundberg JG. 1989. A new species of *Hemisilurus* (Siluriformes, Siluridae) from the Mekong river, with comments on its relationships and historical biogeography. Copeia 1989: 434-444.

Bosellini A. 2002. Dinosaurs "re-write" the geodynamics of the eastern Mediterranean and the paleogeography of the Apulia Platform. Earth-Science Rev 59: 211-234.

Boulenger GA. 1899. Description of a new silurid fish of the genus *Gephyroglanis*, from South Africa. Ann S Afr Mus 2: 227-228.

Boulenger GA. 1904. Teleostei (Systematic Part). In: Harmer SF, Shipley AE (eds.). The Cambridge Natural History (vol. 7). Macmillan & Co., London, UK, pp. 541-727.

Boulenger GA. 1911. Catalogue of the fresh-water fishes of Africa in the British Museum (Natural History) (vol. 2). British Museum (Natural History), London, UK.

Bridge TW, Haddon AC. 1893. Contributions to the anatomy of fishes: II. The air bladder and Weberian ossicles in the siluroid fishes. Phil Trans R Soc London B 84: 65-333.

Bridge TW, Haddon AC. 1894. Notes on the production of sounds by the air bladder of certain siluroid fishes. Proc Roy Soc London 55: 439-441.

Briggs JC. 1979. Ostariophysan zoogeography: an alternative hypothesis. Copeia 1979: 111-118.

Briggs JC. 2003. The biogeographic and tectonic history of India. J Biogeogr 30: 381-388.

Britski HA. 1972. Sistematica e evolução dos Auchenipteridae e Ageneiosidae (Teleostei, Siluriformes). Unpubl. PhD thesis. Universidade São Paulo, São Paulo, Brazil.

Brosseau AR. 1978. The pectoral anatomy of selected Ostariophysi: II. The Cypriniformes and Siluriformes. J Morphol 140: 79-115.

Brown BA, Ferraris CJ. 1988. Comparative osteology of the Asian catfish family Chacidae, with the description of a new species from Burma. Amer Mus Novit 2907:1-16.

Buffetaut E. 1989. Archosaurian reptiles with Gondwanan affinities in the Upper cretaceous of Europe. Terra Nova 1: 69-74.

Buffetaut E, Le Loeuff J. 1991. Late cretaceous dinosaur faunas of Europe: some correlation problems. Cretaceous Res 12: 159-176.

Burgess WE. 1989. An atlas of freshwater and marine catfishes: a preliminary survey of the Siluriformes. H. T.F. Publ., Berkshire, England.

Cabuy E, Adriaens D, Verraes W, Teugels, GG. 1999. Comparative study of the cranial morphology of *Gymnallabes typus* (Siluriformes: Clariidae) and their less anguilliform relatives, *Clariallabes melas* and *Clarias gariepinus*. J Morphol 240: 169-194.

Carroll SB, Grenier JK, Weatherbee SD. 2001. From DNA to Diversity—Molecular Genetics and the Evolution of Animal Design. Blackwell Science, Malden, MA.

Cerny J. 1988. Osteology of the sheatfish (*Silurus glanis* Linnaeus, 1758). Práce Úst Rybár Hydrobiol 6:81-209.

Chang M-M, Maisey JG. 2003. Redescription of *Ellimma branneri* and *Diplomystus shengliensis*, and relationships of some basal clupeomorphs. Amer Mus Novit 3404: 1-35.

Chardon M. 1967. Réflexions sur la dispersion des Ostariophysi a la lumière des recherches morphologiques nouvelles. Ann Soc R Zool Belg 97: 175-186.

Chardon M. 1968. Anatomie comparée de l'appareil de Weber et des structures connexes chez les Siluriformes. Ann Mus R Afr Centr 169: 1-273.

Chardon M, Vandewalle P. 1971. Comparaison de la région céphalique chez cinq espèces du genre *Tilapia*, dont trois incubateurs buccaux. Ann Soc R Zool Belg 101: 3-24.

Chardon M, De La Hoz E. 1973. Notes sur le squelette, les muscles, les tendons et le cerveau des Gymnotoidei. Ann Soc Nat Zool Paris 12 Ser 15: 1-10.

Chardon M, De La Hoz E. 1974. Towards an improved classification of the gymnotoid fishes by the use of splanchnocranium characters. Ichthyologia 6: 15-25.

Chardon M, De La Hoz E. 1977. Remarques anatomiques et fonctionnelles à propos du suspensorium et de la série operculaire chez *Sternopygus macrurus* et *Engenmania virescens* (Teleostei Gymnotoidei). Ann Soc R Zool Belg 106: 177-191.

Chardon M, Parmentier E, Vandewalle P. 2003. Morphology, development and evolution of the Weberian apparatus in catfish. In: Arratia G, Kapoor BG, Chardon M, Diogo R (eds.). Catfishes. Science Publ., Enfield, NH (USA), pp. 71-120.

Chen X. 1994. Phylogenetic studies of the amblycipitid catfishes (Teleostei, Siluriformes) with species accounts. Unpubl. PhD thesis, Duke University, Durham, NC (USA).

Chen X, Lundberg JG. 1994. *Xiurenbagrus*, a new genus of amblycipitid catfishes (Teleostei: Siluriformes), and phylogenetic relationships among the genera of Amblycipitidae. Copeia 1994: 780-800.

Chranilov NS. 1929. Beiträge zur kenntnis der Weber'schen apparates der Ostariophysi, 2. der Weber'schen apparat bei Siluroidea. Zool Jahrb Anat 51: 323-462.

Chure D. 2001. The second record of the African Theropod *Elaphosaurus* (Dinosauria, Ceratosauria) from the Western Hemisphere. Neues Jabr Geol Paleont 9: 565-576.

Cockerell TD. 1925. A fossil fish of the family Callichthyidae. Science 62: 317-322.

Cope ED. 1871. Contributions to the ichthyology of the Lesser Antilles. Trans Amer Phil Soc 14: 445-483.

Costa WJEM. 1994. A new genus and species of Sarcoglanidinae (Siluriformes: Trichomycteridae) from the Araguaia basin, central Brazil, with notes on subfamilial phylogeny. Ichthyol Explor Freshwaters 5: 207-216.

Costa WJEM, Bockmann FA. 1994. A new genus and species of Sarcoglanidinae (Siluriformes: Trichomycteridae) from southeastern Brazil, with a re-examination of subfamilial phylogeny. J Nat Hist 28: 715-730.

Cracraft J. 1981. The use of functional and adaptive criteria in phylogenetic systematics. Amer Zool 21: 21-36.

Cronquist A. 1987. A botanical critique of cladism. Bot Rev 53: 1-52.

Curran DJ. 1989. Phylogenetic relationships among the catfish genera of the family Auchenipteridae (Teleostei: Siluroidea). Copeia 1989: 408-419.

Cuvier G. 1817. Le regne animal, distribue d'apres son organisation. Tome 2. Les reptiles, les poissons, les mollusques et les annelides. Deterville, Paris.

Daget J. 1964. Le crâne des Téléostéens. Mém Mus Natn Hist Nat 31: 163-341.

Darwin C. 1872. The Origin of Species. Murray Publ., London, UK (6th ed.).

Datta NC, Saha AK, Baydya A. 1975. Comparative study of the osteocranium and Weberian apparatus of *Clarias batrachus* (Linn.) and *Heteropneustes fossilis* (Bl.) (Pisces). Zool Anz 195: 374-386.

David L. 1936. *Uegitglanis*, silure aveugle de la Somalie italienne, chaînon entre bagrides et clariides. Rev Zool Bot Afr 28: 369-388.

David L, Poll M. 1937. Contribution à la Faune Ichthyologique du Congo Belge: collections du Dr H. Schouteden (1924-1926) et d'autres récolteurs. Ann Mus Congo Belge (Zool) 2: 19-57.

Day F. 1877. The Fishes of India: being a natural history of the fishes known to inhabit the seas and freshwaters of India, Burma and Ceylon. William Dawson and Co., London, UK.

De Beer GR. 1937. The Development of the Vertebrate Skull. Clarendon Press, Oxford, UK.

De la Hoz E. 1974. Definition et classification des poissons Gymnotoidei sur la base de la morphologie comparée et fonctionnelle du squelette et des muscles. Unpubl. PhD thesis. University of Liège, Liege.

De la Hoz E, Chardon M. 1984. Skeleton, muscles, ligaments and swim-bladder of a gymnotid fish, *Sternopygus macrurus* Bloch & Schneider (Ostariophysi: Gymnotoidei). Bull Soc R Sci Liège 53: 9-53.

De la Hoz E, Aldunate R. 1994. El sistema hioideo-mandibular de *Cheirodon* (Ostariophysi, Characidae): una innovation functional. Ann Mus Hist Nat Valparaíso 22: 83-90.

De Oliveira JC. 1988. Osteologia e revisão sistemática de Cetopsidae (Teleostei: Siluriformes). Unpubl. PhD thesis. Universidade São Paulo, São Paulo, Brazil.

De Pinna MCC. 1988. A new genus of trichomycterid catfish (Siluroidei, Glanapterygidae), with comments on its phylogenetic relationships. Rev Suisse Zool 95: 113-128.

De Pinna MCC. 1989a. A new sarcoglanidine catfish, phylogeny of its subfamily, and an appraisal of the phyletic status of the Trichomycterinae (Teleostei, Trichomycteridae). Amer Mus Novit 2950: 1-39.

De Pinna MCC. 1989b. Redescription of *Glanapteryx anguilla*, with notes on the phylogeny of Glanapteryginae (Siluriformes, Trichomycteridae). Proc Acad Nat Sci (Phil.) 141: 361-374.

De Pinna MCC. 1991. Concepts and tests of homology in the cladistic paradigm. Cladistics 7: 367-394.

De Pinna MCC. 1992. A new subfamily of Trichomycteridae (Teleostei: Siluriformes), lower loricaroid relationships and a discussion on the impact of additional taxa for phylogenetic analysis. Zool J Linn Soc 106: 175-229.

De Pinna MCC. 1993. Higher-level phylogeny of Siluriformes, with a new classification of the order (Teleostei, Ostariophysi). Unpubl. PhD thesis, University New York, NY.

De Pinna MCC. 1996. A phylogenetic analysis of the Asian catfish families Sisoridae, Akysidae and Amblycipitidae, with a hypothesis on the relationships of the Neotropical Asprenidae (Teleostei, Ostariophysi). Fieldiana, Zool 84: 1-82.

De Pinna MCC. 1998. Phylogenetic relationships of Neotropical Siluriformes (Teleostei: Ostariophysi): historical overview and synthesis of hypotheses. In: Malabarba LR, Reis RE, Vari RP, Lucena ZM, Lucena CAS (eds.). Phylogeny and classification of Neotropical fishes. Edipucrs, Porto Alegre, Brazil, pp. 279-330.

De Pinna MCC, Salles LO. 1990. Cladistic tests of adaptational hypotheses: a reply to Coddington. Cladistic 6: 373-377.

De Pinna MCC, Starnes WC. 1990. A new genus and species of Sarcoglanidinae from the Rio Mamoré, Amazon basin, with comments on subfamilial phylogeny (Teleostei, Trichomycteridae). J Zool (Lond) 222: 75-88.

De Pinna MCC, Ferraris CJ. 1992. [Review of] Anatomy, relationships and systematics of the Bagridae (Teleostei: Siluroidei) with a hypothesis of Siluroid phylogeny by T. Mo. Copeia 1992: 1132-1134.

De Pinna MCC, Vari RP. 1995. Monophyly and phylogenetic diagnosis of the family Cetopsidae, with synonymization of the Helogenidae (Teleostei: Siluriformes). Smiths Contrib Zool 571: 1-26.

De Vos L. 1995. A systematic revision of the African Schilbeidae (Teleostei: Siluriformes). Ann Mus R Afr Centr 271: 1-414.

Desgranges JC. 1972. Sur les bourgeons du goût du poisson-chat *Ictalurus melas*: ultrastructure des cellules basales. C R Acad Sci Paris 274: 1814-1817.

Desutter-Grandcolas L, D'Haese C, Robillard T. 2003. The problem of characters susceptible to parallel evolution in phylogenetic analysis: a reply to Marquès and Gnaspini (2001) with emphasis on cave life phenotic evolution. Cladistics 19: 131-137.

Devillers C. 1958. Le crâne des poissons. In: Grassé P (ed.). Agnathes et Poissons, anatomie, éthologie, systématique. Masson et Cle, Paris, pp. 551-687.

Dimmick WW, Larson A. 1996. A molecular and morphological perspective on the phylogenetic relationships of the otophysan fishes. Mol Phyl Evol 6:120-133.

Diogo R. 2003a. Higher-level phylogeny of Siluriformes: an overview. In: Arratia G, Kapoor BG, Chardon M, Diogo R (eds.). Catfishes. Science Publ., Enfield, NH (USA) pp. 353-384.

Diogo R. 2003b. Anatomy, phylogeny and taxonomy of Amphiliidae. In: Arratia G, Kapoor BG, Chardon M, Diogo R (eds.). Catfishes. Science Publ., Enfield, NH (USA) pp. 401-438.

Diogo R, Chardon M. 2000a. Homologies between different adductor mandibulae sections of teleostean fishes, with a special regard to catfishes (Teleostei: Siluriformes). J Morphol 243: 193-208.

Diogo R, Chardon M. 2000b. The structures associated with catfish (Teleostei: Siluriformes) mandibular barbels: origin, anatomy, function, taxonomic distribution, nomenclature and synonymy. Neth J Zool 50: 455-478.

Diogo R, Chardon M. 2000c. Anatomie et fonction des structures céphaliques associées á la prise de nourriture chez le genre *Chrysichthys* (Teleostei: Siluriformes). Belg J Zool 130: 21-37.

Diogo R, Chardon M. 2001. Adaptive transformation of the palatine-maxillary system in catfish: increased mobility of the maxillary barbel. In: Kapoor BG, Hara TJ (eds.). Sensory Biology of jawed fishes—new insights. Science Publ., Enfield, NH (USA) pp. 367-383.

Diogo R, Chardon M. 2003. Homologies and evolutionary transformation of the skeletal elements of catfish (Teleostei: Siluriformes) suspensorium: a morphofunctional hypothesis. In: Val AL, Kapoor BG (eds.). Fish adaptations. Science Publ., Enfield, NH (USA) pp. 275-284.

Diogo R, Vandewalle P. 2003. Review of superficial cranial musculature of catfishes, with comments on plesiomorphic states. In: Arratia G, Kapoor BG, Chardon M, Diogo R (eds.). Catfishes. Science Publ., Enfield, NH (USA) pp.47-69.

Diogo R, Chardon M. In press. Osteology and myology of the cephalic region and pectoral girdle of *Heteropneustes fossilis* (Teleostei: Siluriformes), with comments on the phylogenetic relationships between *Heteropneustes* and the clariid catfishes. Anim Biol.

Diogo R, Vandewalle P, Chardon M. 1999. Morphological description of the cephalic region of *Bagrus docmak*, with a reflection on Bagridae (Teleostei: Siluriformes) autapomorphies. Neth J Zool 49: 207-232.

Diogo R, Oliveira C, Chardon M. 2000a. The origin and transformation of catfish palatine-maxillary system: an example of adaptive macroevolution. Neth J Zool 50: 373-388.

Diogo R, Oliveira C, Chardon M. 2000b. On the anatomy and function of the cephalic structures in *Phractura* (Siluriformes: Amphiliidae: Doumeinae), with comments on some striking homoplasies occurring between the doumeins and some loricaroid catfishes. Belg J Zool 130: 117-130.

Diogo R, Oliveira C, Chardon M. 2001a. On the homologies of the skeletal components of catfish (Teleostei: Siluriformes) suspensorium. Belg J Zool 131: 93-109.

Diogo R, Chardon M, Vandewalle P. 2001b. Osteology and myology of the cephalic region and pectoral girdle of *Bunocephalus knerii*, and a discussion on the phylogenetic relationships of the Aspredinidae (Teleostei: Siluriformes). Neth J Zool 51: 457-481.

Diogo R, Oliveira C, Chardon M. 2001c. On the osteology and myology of catfish pectoral girdle, with a reflection on catfish (Teleostei: Siluriformes) plesiomorphies. J Morphol 249: 100-125.

Diogo R, Chardon M, Vandewalle P. 2002a. Osteology and myology of the cephalic region and pectoral girdle of the Chinese catfish *Cranoglanis bouderius*, with a discussion on the autapomorphies and phylogenetic relationships of the Cranoglanididae (Teleostei: Siluriformes). J Morphol 253: 229-242.

Diogo R, Chardon M, Vandewalle P. 2002b. Osteology and myology of the cephalic region and pectoral girdle of *Glyptothorax fukiensis* (Rendahl, 1925), comparison with other sisorids, and comments on the synapomorphies of the Sisoridae (Teleostei: Siluriformes). Belg J Zool 132: 93-101.

Diogo R, Chardon M, Vandewalle P. 2003a. Functional morphology of catfishes: movements of barbels. In: Arratia G, Kapoor BG, Chardon M, Diogo R (eds.). Catfishes. Science Publ., Enfield, NH (USA) pp. 203-220.

Diogo R, Chardon M, Vandewalle P. 2003b. Osteology and myology of the cephalic region and pectoral girdle of *Liobagrus reini* Hilgendorf 1878, with a discussion on the phylogenetic relationships of the Amblycipitidae (Teleostei: Siluriformes). Belg J Zool 133: 77-84.

Diogo R, Chardon M, Vandewalle P. *In press*-a. Osteology and myology of the cephalic region and pectoral girdle of *Centromochlus heckelii*, comparison with other auchenipterids, and comments on the synapomorphies and phylogenetic relationships of the Auchenipteridae (Teleostei: Siluriformes). Neth J Zool.

Diogo R, Chardon M, Vandewalle P. *In press*-b. Osteology and myology of the cephalic region and pectoral girdle of *Erethistes pusillus*, comparison with other erethistids, and discussion on the autapomorphies and phylogenetic relationships of the Erethistidae (Teleostei: Siluriformes). J Fish Biol.

Diogo R, Chardon M, Vandewalle P. *In press*-c. On the osteology and myology of the cephalic region and pectoral girdle of *Pangasius siamensis*, with a discussion on the synapomorphies and phylogenetic relationships of the Pangasiidae (Teleostei: Siluriformes). Anim Biol.

Dobson JE. 2003. Independent corroboration of a previously proposed Palaeozoic link between eastern Australia and eastern North America. J Biogeogr 30: 473.

Donnelly BG. 1973. Aspect of behaviour in the catfish *Clarias gariepinus*, during periods of habitat desiccation. Arnoldia Rhod 6: 1-8.

Doyle JA. 1998. Molecules, morphology, fossils, and the relationship of angiosperms and gnetales. Mol Phylog Evol 9: 448-462.

Dubale MS, Rao BVS. 1961. Pectoral fin musculature in certain siluroid fishes. J Univ Bombay 29: 89-96.

Dunn CP. 2003. Keeping taxonomy based in morphology. Trends Ecol Evol 18: 270-271.

Easteal S. 1999. Molecular evidence for the early divergence of placental mammals. Bioessays 21: 1056-1058.

Eaton TH. 1948. Form and function in the head of the channel catfish, *Ictalurus lacustris punctatus*. J Morphol 83: 181-194.

Ebach MC. 2003. Area cladistics. Biologist 50: 1-4.

Edgeworth FH. 1935. The Cranial Muscles of Vertebrates. Cambridge University Press, Cambridge, UK.

Eigenmann CH. 1890. The evolution of catfishes. Zoe 1: 10-15.

Eigenmann CH. 1909. The fresh-water fishes of Patagonia and an examination of the Archiplata-Archhelenis theory. In: Reports of the Princeton University expeditions to Patagonia 1896-1899. Princeton University & Stuttgard, Princeton, NJ (USA) pp. 225-374.

Eigenmann CH. 1918. The Pygidiidae, a family of South American catfishes. Mem Carn Mus 7: 259-398.

Eigenmann CH. 1925. A review of the Doradidae, a family of South American Nematognathi, or catfishes. Trans Amer Philos Soc, new series 22: 280-365.

Eigenmann CH, Eigenmann RS. 1890. A revision of the South American Nematognathi or catfishes. Occ Pap Calif Acad Sci 1: 11-508.

Eldredge N, Cracraft J. 1980. Phylogenetic Patterns and the Evolutionary Process, Method and Theory in Comparative Biology. Columbia University Press, New York, NY.

Farris JS. 1988. Hennig 86, version 1.5. Distributed by the author, Port Jefferson Station, New York.

Ferraris CJ. 1988a. The Auchenipteridae: putative monophyly and systematics, with a classification of the Neotropical doradoid catfishes (Ostariophysi, Siluriformes). Unpubl. PhD thesis, University New York, NY.

Ferraris CJ. 1988b. Relationships of the Neotropical catfish genus *Nemuroglanis*, with a description of a new species (Osteichthyes, Siluriformes, Pimelodidae). Proc Biol Soc Wash 101: 509-516.

Ferraris CJ. 1989. On the interrelationships between the Aspredinidae and the Akysidae (Ostariophysi, Siluriformes). Abst 1989 Meet Amer Soc Ichthyol Herpet, San Francisco State Univ: 86, San Francisco, CA.

Ferraris CJ, de Pinna MCC. 1999. Higher-level names for catfishes (Actinopterygii: Ostariophysi: Siluriformes). Proc Calif Acad Sci 51: 1-17.

Filleul A, Maisey JG. In press. Redescription of *Santanichthys diasii* (Otophysi, Characiformes) from the Albian of the Santana formation and comments on its implications for Otophysan relationships. Amer Mus Novit.

Fine M, Ladich F. 2003. Sound production, spine locking, and related adaptations. In: Arratia G, Kapoor BG, Chardon M, Diogo R (eds.). Catfishes. Science Publ., Enfield, NH (USA) pp. 249-290.

Fine ML, Friel JP, McElroy D, King CB, Loesser KE, Newton S. 1997. Pectoral spine locking and sound production in the channel catfish *Ictalurus punctatus*. Copeia 5: 777-790.

Finger TE. 1976. Gustatory pathways in the bullhead catfish. I. Connections of the anterior ganglion. J Comp Neurol 165: 513-526.

Finger TE. 1978. Gustatory pathways in the bullhead catfish. II. Facial lobe connections. J Comp Neurol 180: 691-706.

Fink SV, Fink W. 1981. Interrelationships of the ostariophysan fishes. Zool J Linn Soc (Lond) 72: 297-353.

Fink SV, Fink W. 1996. Interrelationships of ostariophysan fishes (Teleostei). In: Stiassny MLJ, Parenti LR, Johnson GD (eds.). Interrelationship of Fishes. Academic Press, New York, NY, pp. 209-249.

Fitzhugh K. 1989. Cladistics in the fast lane. J New York Entomol Soc 97: 234-241.

Friel JP. 1994. A phylogenetic study of the Neotropical banjo catfishes (Teleostei: Siluriformes: Aspredinidae). Unpubl. PhD thesis, Duke University, Durham, NC (USA).

Friel JP. 1995. *Acanthobunocephalus nicoi*, a new genus and species of miniature banjo-catfish from the upper Orinoco and Casiquiare rivers, Venezuela (Siluriformes, Aspredinidae). Ichthyol Explor Freshwaters 6: 89-95.

Friel JP, Lundberg JG. 1996. *Mycromyzon akamai*, gen. et sp. nov., a small and eyeless banjo catfish (Siluriformes, Aspredinidae) from the river channels of the lower Amazon basin. Copeia 1996: 641-648.

Frost GA. 1925. A comparative study of the otoliths of the Neopterygian fishes, II. Ostariophysi, B Siluroidea. Ann Mag Nat Hist ser 9 16: 433-446.

Futuyma DJ. 1986. Evolutionary Biology. Sinauer, Sunderland, MA (USA).

Futuyma DJ. 1987. On the role of species in anagenesis. Amer Nat 130: 465-473.

Futuyma DJ. 1988. Sturm and Drang and the evolutionary synthesis. Evolution 42: 217-226.

Gainer H. 1967. Neuromuscular mechanisms of sound production and pectoral spine locking in the banjo catfish, *Bunocephalus* species. Physiol Zool 40: 296-306.

Galis F. 1996. The application of functional morphology to evolutionary studies. Tree 11: 124-129.

Galton PM, Taquet P. 1982. *Valdosaurus*, a hypsolophodontid dinosaur from the Lower Cretaceous of Europe and Africa. Geobios 15: 147-159.

Gauba RK. 1962. The endoskeleton of *Bagarius bagarius* (Ham.), Part I. The skull. Agra Univ J Res 11: 75-90.

Gauba RK. 1966. Studies on the osteology of Indian sisorid catfishes, II. The skull of *Glyptothorax cavia*. Copeia 4: 802-810.

Gauba RK. 1968. On the morphology of the skull of catfish *Pseudecheneis sulcatus*. Zool Anz 181: 226-236.

Gauba RK. 1969. The head skeleton of *Glyptosternum reticulatum* McClelland et Grifith. Mon Zool Ital 3: 1-17.

Gauba RK. 1970. The osteology of family Chacidae, 1. The skull of *Chaca chaca* (Hamilton). Mon Zool Ital 4: 21-40.

Gayet M. 1982. Considération sur la phylogénie et la paléobiogéographie des ostariophysaires. Geobios 6: 39-52.

Gayet M. 1985. Contribution à l'étude anatomique et systématique de l'ichthyofaune Cénomanienne du Portugal. Comun Serv Geol Portugal 71: 91-118.

Gayet M. 1986a. *Ramallichthys* Gayet du Cénomanien inférieur marin de Ramallah (Judée), une introduction aux relations phylogénétiques des Ostariophysi. Mém Mus Natl Hist Nat Sér C 51: 1-81.

Gayet M. 1986b. About ostariophysan fishes: a reply to S.V. Fink, P.H. Greenwood and W.L. Fink's criticisms. Bull Mus Natl Hist Sec C 8: 393-409.

Gayet M. 1988. Le plus ancien crâne de Siluriforme: *Andinichthys bolivianensis* nov. gen., nov. sp. (Andinichthyidae nov. fam.) du Maastrichtien de Tiupampa (Bolivie). C R Acad Sci Paris 307: 833-836.

Gayet M. 2001. A review of some problems associated with the occurrences of fossil vertebrates in South America. J South Am Earth Sci 14: 131-145.

Gayet M, Chardon M. 1987. Possible otophysic connections in some fossil and living ostariophysan fishes. Proc Congr Eur Ichthyol Stockholm, 15: 31-42.

Gayet M, Meunier FJ. 1998. Maastrichtian to early late Paleocene freshwater osteichthyes of Bolivia: additions and comments. In: Malabarna LR, Reis RE, Vari RP, Lucena ZM, Lucena CAS (eds.). Phylogeny and Classification of Neotropical fishes. Edipucrs, Porto Alegre, pp. 85-110.

Gayet M, Meunier FJ. 2003. Paleontology and palaeobiogeography of catfishes. In: Arratia G, Kapoor BG, Chardon M, Diogo R (eds.). Catfishes. Science Publ., Enfield, NH (USA) pp. 491-522.

Gayet M, Meunier FJ, Kirschbaum F. 1994. *Ellisella kirschbaumi* Gayet et Meunier, 1991, Gymnotiforme fossile de Bolivie et ses relations phylogénétiques au sein des formes actuelles. Cybium 18: 273-306.

Gayet M, Rage J-C, Sempere T, Gagnier PY. 1992. Modalités des échanges de vertébrés continentaux entre l'Amérique du Nord et l'Amérique du sud au Crétacé supérieur et au Paleocene. Bull Soc Géol France 163: 781-791.

Gayet M, Rage J-C, Sempere T, Gagnier PY. 1993. Réponse des auteurs. Bull Soc Géol France 164: 861-864.

Gayet M, Jegu M, Bocquentin J, Negri FR. 2003. New characoids from the Upper Cretaceous and Paleocene of Bolivia and the Mio-Pliocene of Brazil: phylogenetic position and paleobiogeographic implications. J Vert Paleont 23: 28-46.

Géry J. 1969. The fresh-water fishes of South America. In: Fittkau EJ, Iles J, Klinge H, Schwabe GH, Sioli H (eds.). Biogeography and Ecology in South America. W Junk Publ., The Hague, Netherlands, pp. 828-848.

Ghiot F. 1976. Etude histologique, morphologique et fonctionnelle des barbillons de quelques poissons Siluriformes. Rev Trav Inst Pêches Marit 40: 581-582.

Ghiot F. 1978. The barbel movements of three South American pimelodid catfishes. Zool Anz 200: 395-401.

Ghiot F, Bouchez N. 1980. The central rod of the barbels of a South American catfish, *Pimelodus clarias*. Copeia 1980: 908-909.

Ghiot F, Vandewalle P, Chardon M. 1984. Comparaison anatomique et fonctionnelle des muscles et des ligaments en rapport avec les barbillons chez deux familles apparentées de poissons Siluriformes Bagroidei. Ann Soc R Zool Belg 114: 261-272.

Gianferrari J. 1923. *Uegitglanis zammaronoi*, un nuove Siluride cieco africano. Atti Soc Ital Sci Nat Milano 62: 1-3.

Gijsen L. 1974. Étude comparée du squelette, des muscles et des ligaments de la tête de quatre espéces de poissons téléostéens Characoidei a cáractéres archaiques. Unpubl. Bachelor's thesis, University of Liège, Liege.

Gijsen L, Chardon M. 1976. Muscles et ligaments cèphaliques, splanchnocrâne et quelques possibilités de mouvements dans la tête d'*Hoplerythrinus unitaenatus* (Spix) (Teleostei, Ostariophysi, Characoidei). Ann Soc Nat Zool Paris 12 Ser 18: 251-274.

Gill T. 1872. Arrangement of the familles of fishes, or Classes Pisces, Marsipobranchii, and Leptocardii. Smith Misc Coll 247: 1-49.

Glaw F, Vences M. 1994. A Fieldguide to the Amphibians and Reptiles of Madagascar, Including Mammals and Freshwater Fish. Privately published, Köln (2nd ed.).

Goodrich ES. 1909. Vertebrata Craniata, first fascicle—Cyclostomes and fishes. In: Lankester R (ed.). A Treatise on Zoology, Part 9. Adam and Charles Back, London, UK, pp. 1-518.

Gosline WA. 1975. The palatine-maxillary mechanism in catfishes with comments on the evolution and zoogeography of modern siluroids. Occas Pap Calif Acad Sci 120: 1-31.

Gosline WA. 1977. The structure and function of the dermal pectoral girdle in bony fishes with particular reference to ostariophysines. J Zool (Lond) 183: 329-338.

Gosline WA. 1986. Jaw muscle configuration in some teleostean fishes. Copeia 3: 705-713.

Gosline WA. 1989. Two patterns of differentiation in the jaw musculature of teleostean fishes. J Zool (Lond) 218: 649-661.

Gougnard G, Vandewalle P. 1980. Deplacements terrestres de *Clarias lazera*. Ann Soc R Zool Belg 109: 141-152.

Gould SJ. 1983. What, if anything else, is a Zebra? Nat Hist 90: 6-12.

Gould SJ. 1988. The heart of terminology. Nat Hist 97: 24-31.

Gould SJ. 1989. The wheel of fortune and the wedge of progress. Nat Hist 98: 14-21.

Gould SJ. 2002. The Structure of Evolutionary Theory. Harvard University Press, Cambridge.

Gould SJ, Vrba ES. 1982. Exaptation – a missing term in the science of form. Paleobiology 8: 4-15.

Grande L. 1987. Redescription of*Hypsidoris farsonensis* (Teleostei: Siluriformes), with a reassessment of its phylogenetic relationships. J Vert Paleont 7: 24-54.

Grande L, Eastman JT. 1986. A review of Antartical ichthyofaunas in the light of new fossil discoveries. Palaeont 29: 113-137.

Grande L, Lundberg JG. 1988. Revision and redescription of the genus *Astephus* (Siluriformes: Ictaluridae) with a discussion of its phylogenetic relationships. J Vert Paleont 8: 139-171.

Grande L, de Pinna MCC. 1998. Description of a second species of the catfish †*Hypsidoris* and a revaluation of the genus and the family †Hypsidoridae. J Vert Paleont 18: 451-474.

Grande T, Poyato-Ariza FJ. 1999. Phylogenetic relationships of fossil and recent gonorynchiform fishes (Teleostei: Ostariophysi). Zool J Linn Soc 125: 197-238.

Greenwood PH, Rosen DE, Weitzman SH, Meyers GS. 1966. Phyletic studies of teleostean fishes, with a provisional classification of living forms. Bull Amer Mus Natl Hist 131:339-456.

Gregory WK. 1933. Fish skulls—a study of the evolution of natural mechanisms. Trans Amer Philos Soc 23: 75-481.

Grover-Johson N, Farbman AJ. 1976. Fine structure of taste buds in the barbel of the catfish,*Ictalurus punctatus*. Cell Tissue Res 169: 395-403.

Günther A. 1864. Catalogue of the Fishes in the British Museum (vol. 5). Museum Trustees, London, UK.

Hall BK. 1994. Homology, the Hierarchical Basis of Comparative Biology. Academic Press, New York, NY.

Harry RR. 1953. A contribution to the classification of the African catfishes of the family Amphiliidae, with description of collections from Cameron. Rev Zool Bot Afr 47: 177-232.

Hauser DL, Presch W. 1991. The effect of ordered characters on phylogenetic reconstruction. Cladistics 7: 243-265.

He S. 1996. The phylogeny of the glyptosternoid fishes (Teleostei: Siluriformes, Sisoridae). Cybium 20: 115-159.

He S. 1997. Phylogénie et biogéographie des Sisoridae et des Amphiliidae (Pisces: Siluriformes): deux familles de poissons-chats torrenticoles. Unpubl. PhD thesis. Muséum National d'Histoire Naturelle de Paris, Paris.

He S, Gayet M, Meunier FJ. 1999. Phylogeny of the Amphiliidae (Teleostei: Siluriformes). Ann Sci Nat 20: 117-146.

Hennig, W. 1950. Grundzüge einer theorie der phylogenetischen systematik. Deutscher Zentralverlag, Berlin,

Hennig, W. 1965. Phylogenetic systematics. Ann Rev Entomol 10: 97-116.

Hennig, W. 1966. Phylogenetic Systematics. University of Illinois Press, Urbana, IL (USA).
Hennig, W. 1981. Insect Phylogeny. University of Illinois Press, Urbana, IL (USA).
Herrick CJ. 1903. The organ and sense of taste in fishes. Bull US Fish Comm 1902: 237-272.
Higuchi H. 1992. A phylogeny of the South American thorny catfishes (Osteichthyes; Siluriformes, Doradidae). Unpubl.PhD thesis, Harvard University, Cambridge, MA.
Hoagland H. 1932. Specific nerve impulses from gustatory and tactile receptors in catfish. Proc US Nat Acad Sci 18: 685-693.
Hoedeman JJ. 1960. Studies on callichthyid fishes, 4. Development of the skull in *Callichthys* and *Hoplosternum* (1) (Pisces: Siluriformes). Bull Aquat Biol 1: 73-84.
Holtz TR. 1998. Spinosaurs as crocodile mimics. Science 282: 1276-1277.
Hoogland R, Morris D, Tinbergen H. 1957. The spines of sticklebacks (*Gasterosteus* and *Pygosteus*) as a means of defense against predators (*Perca* and *Exox*). Behaviour 10: 205-236.
Hora SL. 1939. The game fishes of India, *Bagarius bagarius* (Ham). J Bombay Nat Hist Soc 40: 583-593.
Howes GJ. 1976. The cranial musculature and taxonomy of characoid fishes of the tribes Cynodontini and Characini. Bull Br Mus Nat Hist (Zool) 29: 203-248.
Howes GJ. 1978. The anatomy and relationships of the cyprinid fish *Luciobrama macrocephalus* (Lacepède). Bull Br Mus Nat Hist (Zool) 34: 1-64.
Howes GJ. 1983a. The cranial muscles of the loricarioid catfishes, their homologies and value as taxonomic characters. Bull Br Mus Nat Hist (Zool) 45: 309-345.
Howes GJ. 1983b. Problems in catfish anatomy and phylogeny exemplified by the Neotropical Hypophthalmidae (Teleostei: Siluroidei). Bull Br Mus Nat Hist (Zool) 45: 1-39.
Howes GJ. 1985a. The phylogenetic relationships of the electric family Malapteruridae (Teleostei: Siluroidei). J Nat Hist 19: 37-67.
Howes GJ. 1985b. Cranial muscles of gonorynchiform fishes, with comments on generic relationships. Bull Br Mus Nat Hist (Zool) 49: 273-303.
Howes GJ. 1988. The cranial muscles and ligaments of macrouroid fishes (Teleostei: Gadiformes)—functional, ecological and phylogenetic inferences. Bull Br Mus Nat Hist (Zool) 54: 1-62.
Howes GJ, Teugels GG. 1989. Observations and homology of the pterygoid bones in *Corydoras paleatus* and some other catfishes. J Zool (Lond) 219: 441-456.
Howes GJ, Fumihito A. 1991. Cranial anatomy and phylogeny of the South-East Asian catfish genus *Belodontichthys*. Bull Br Mus Nat Hist (Zool) 57: 133-160.
Hubbs CL, Hibbard CW. 1951. *Ictalurus lamda*, a new catfish, based on a pectoral spine from Pliocene of Kansas. Copeia 1951: 8-14.
Hunt von Herbing I, Miyale T, Hall BK, Boutilier RG. 1996. Ontogeny of feeding and respiration in larval Atlantic cod *Gadus morhua* (Teleostei, Gadiformes), I. Morphology. J Morphol 227: 15-35.
Inoue JG, Miya M, Tsukamoto K, Nishida M. 2003. Basal actinopterygian relationships: a mitogenomic perspective on the phylogeny of the "ancient fish". Mol Phyl Evol 26: 110-120.
Ishiguro NB, Miya M, Nishida M. 2003. Basal euteleostean relationships: a mitogenomic perspective on the phylogenetic reality of the 'Protacanthopterygii'. Mol Phylogenet Evol 27: 476-488.
Jayaram KC. 1956. Taxonomic status of the Chinese catfish family Cranoglanididae Myers, 1931. Proc Nat Inst Sci India 21: 256-263.
Jayaram KC. 1966. Contribution to the study of fishes of the family Bagridae, I. A systematic account of the African genera with a new classification of the family. Bull Inst Fond Afrique Noire 28: 1064-1139.
Jayaram KC. 1968. Contributions to the study of the bagrid fishes (Siluroidea: Bagridae), 3. A systematic account of the Japanese, Chinese, Malayan and Indonesian genera. Treubia Mus Zool Bogoriense 27: 287-386.
Jayaram KC. 1971. Contributions to the study of the bagrid fishes, 6. The skeleton of *Rita gogra* (Sykes). J Zool Soc India 22: 117-145.
Jayaram KC. 1978. Functional responses of catfish barbels. Bull Zool Surv India 1: 77-80.
Jayaram KC, Majumdar N. 1964. Siluroid fishes of India, Burma and Ceylon, 15. Fishes of the genus *Chaca* Gray 1831. Proc Zool Soc (Calcutta) 17: 177-181.

Jayaram KC, Singh R. 1982. Contributions to the study of bagrid fishes, 15. A comparative account of the cranial musculature in four bagrid genera with a note on their phylogeny. Rec Zool Survey India 80: 231-250.

Jollie M. 1986. A primer of bone names for understanding of the actinopterygian head and pectoral girdle skeletons. Can J Zool 64: 365-379.

Kaatz I. 1997. The evolutionary origin and functional divergence of sound production in fish: catfish stridulation mechanisms. J Morphol 232: 2675-272.

Kamrin RP, Singer M. 1953. Influence of sensory neurons isolated from central nervous system on maintenance of taste buds and regeneration of barbels in the catfish, *Ameiurus nebulosus*. Amer J Physiol 174: 146-148.

Kindred J. 1919. The skull of *Ameiurus*. Illinois Biol Mono 5: 1-120.

Kitching IJ, Forey PL, Humphries CJ, Williams DM. 1998. Cladistics—the Theory and Practice of Parsimony Analysis. Oxford University Press, New York, NY.

Klassen GJ, Mooi RD, Locke A. 1991. Consistency indices and random data. Syst Zool 40: 446-457.

Kludge AG. 2001. Parsimony with and without scientific justification. Cladistics 17: 199-210.

Kobayakawa M. 1989. Systematic revision of the catfish genus *Silurus*, with description of a new species from Thailand and Burma. Japan J Ichthyol 36: 155-186.

Kobayakawa M. 1992. Comparative morphology and development of bony elements in the head region in three species of Japanese catfishes (*Silurus*: Siluridae; Siluriformes). Japan J Ichthyol 39: 25-36.

Ladich F. 1997. Comparative analysis of swimbladder (drumming) and pectoral (stridulation) sounds in three families of catfishes. Bioacoustics 8: 185-208.

Ladich, F. 2001. Sound-generating and -detecting motor system in catfish: design of swimbladder muscles in doradids and pimelodids. Anat Rec 263: 297-306.

Ladich F, Fine ML. 1994. Localisation of swimbladder and pectoral motoneurons involved in sound production in pimelodid catfish. Brain Behav Evol 44: 86-100.

Ladich F, Bass AH. 1996. Sonic/vocal-acousticolateralis pathways in teleost fishes: a transneuronal biocytin study in mochokid catfish. J Comp Neur 374: 493-505.

Ladich F, Bass AH. 1998. Sonic/vocal motor pathways in catfishes: comparisons with other teleosts. Brain Behav Evol 51: 315-330.

Landacre FL. 1910. On the place of origin and method of distribution of taste buds in*Ameiurus melas*. J Comp Neurol Psychol 17: 1-66.

Lang M. 1990. Cladistics as a tool for morphologists. Neth J Zool 40: 386-402.

Lankester ER. 1870. On the use of the term homology in modern zoology, and the distinction between homogenetic and homoplasic agreements. Annals Mag Nat Hist 6: 34-43.

Lauder GV. 1980. On the evolution of the jaw adductor musculature in primitive gnathostome fishes. Breviora 460: 1-10.

Lauder GV. 1981. Form and function: structural analysis in evolutionary morphology. Paleobiol 7: 430-442.

Lauder GV, Liem KF. 1980. The feeding mechanism and cephalic myology of *Salvelinus fontinalis*: form, function, and evolutionary significance. In: Balon EK (ed.). Charrs, Salmonid Fishes of the Genus *Salvelinus*. Junk Publ., Leiden, Netherlands, pp. 365-390.

Lauder GV, Liem KF. 1983. The evolution and interrelationships of the actinopterygian fishes. Bull Mus Comp Zool 150: 95-197.

Le Loeuff J. 1991. The Campano-Maastrichtian vertebrate faunas from southern Europe and their relationships with other faunas in the world: paleobiogeographical implications. Cretaceous Res 12: 93-114.

Lecointre G. 1995. Molecular and morphological evidence for a Clupeomorpha-Ostariophysi sister-group relationship (Teleostei). Geobios 19: 205-210.

Liem KF, Summers AP. 2000. Integration of versatile functional design, population ecology, ontogeny and phylogeny. Neth J Zool 50: 245-259.

Lipscomb DL. 1992. Parsimony, homology and the analysis of multistate characters. Cladistics 8: 45-65.

Losos JB, Warheit KI, Schoener TW. 1997. Adaptive differentiation following experimental island colonization in *Anolis* lizards. Nature 387: 70-73.

Lundberg JG. 1975a. The fossil catfishes of North America. Univ Mich Mus Paleont Pap Paleont 11: 1-51.

Lundberg JG. 1975b. Homologies of the upper shoulder girdle and temporal region bones in catfishes (Order Siluriformes), with comments on the skull of the Helogeneidae. Copeia 1975: 66-74.

Lundberg JG. 1982. The comparative anatomy of the toothless blindcat, *Trogloglanis pattersoni* Eigenmann, with a phylogenetic analysis of the ictalurid catfishes. Misc Publ Mus Zool, Univ Mi 163: 1-85.

Lundberg JG. 1992. The phylogeny of ictalurid catfishes: a synthesis of recent work. In: Mayden RL (ed.). Systematics, Historical Ecology and North American Freshwater Fishes. Stanford University Press, Palo Alto, CA, pp. 392-420.

Lundberg JG. 1993. African-South American freshwater fish clades and continental drift: problems with a paradigm. In: Goldblatt P (ed.). Biological relationships between Africa and South America. Yale University Press, New Haven, CT, pp. 156-199.

Lundberg JG. 1998. The temporal context for the diversification of Neotropical fishes. In: Malabarba LR, Reis RE, Vari RP, Lucena ZM, Lucena CAS (eds.). Phylogeny and Classification of Neotropical Fishes. Edipucrs, Porto Alegre, Brazil, pp. 49-68.

Lundberg JG, Baskin JN. 1969. The caudal skeleton of the catfishes, order Siluriformes. Amer Mus Novit 2398: 1-49.

Lundberg JG, Case GR. 1970. A new catfish from the Eocene Green River formation, Wyoming. J Paleont 44: 451-457.

Lundberg JG, McDade L. 1986. A redescription of the rare Venezuelan catfish *Brachyrhamdia imitator* Myers (Siluriformes, Pimelodidae), with phylogenetic evidence for a large intrafamilial lineage. Notulae Naturae 463: 1-24.

Lundberg JG, Linares O, Nass P. 1988. *Phractocephalus hemiliopterus* (Pimelodidae, Siluriformes) from the upper Miocene Urumaco formation, Venezuela: a further case of evolutionary stasis and local extinction among South American fishes. J Vert Paleont 8: 131-138.

Lundberg JG, Bornbusch AH, Mago-Leccia F. 1991a. *Gladioglanis conquistador* n. sp. from Ecuador, with diagnoses of the subfamilies Rhamdiinae Bleeker and Pseudopimelodinae n. subf. (Siluriformes: Pimelodidae). Copeia 1991: 190-209.

Lundberg JG, Mago Leccia F, Nass P. 1991b. *Exallodontus aguanai*, a new genus and species of Pimelodidae (Pisces: Siluriformes) from deep river channels of South America, and delimitation of the subfamily Pimelodinae. Proc Biol Soc Wash 104: 840-869.

Lundberg JG, Marshall LG, Guerrero J, Horton B, Malabarba MCSL, Wesselingh F. 1998. The stage for Neotropical fish diversification: a history of tropical South American rivers. In: Malabarba LR, Reis RE, Vari RP, Lucena ZM, Lucena CAS (eds.). Phylogeny and Classification of Neotropical Fishes. Edipucrs, Porto Alegre, Brazil, pp. 13-48.

Lundberg JG, Lottelat M, Smith GR, Stiassny MLJ, Gill AC. 2000. So many fishes, so little time: an overview of recent ichthyological discovery in continental waters. Ann Missouri Bot Gard 87: 26-62.

Maeda H, Watanabe K, Taki Y. 1994. Genetic relationships of Japanese bagrid catfishes. J Tokyo Univ Fish 81: 111–121.

Mago-Leccia F. 1978. Los peces de la familia Sternopygidae de Venezuela. Acta Sci Venez 29 Suppl 1: 1-89.

Mago-Leccia F, Lundberg JG, Baskin JN. 1985. Systematics of the South American freshwater fish genus *Adontosternarchus* (Gymnotiformes, Apteronotidae). Contrib Sci 358: 1-19.

Mahajan CL. 1963. Sound producing apparatus in an Indian catfish *Sisor rhabdophorus* Hamilton. J Linn Soc (Zool) 44: 721-724.

Mahajan CL. 1966a. Sensory canals of the head in *Sisor rhabdophorus* Hamilton. Trans Amer Micr Soc 85: 548-555.

Mahajan CL. 1966b. *Sisor rhabdophorus*, a study in adaptation and natural relationship, I. The head skeleton. J Zool (Lond) 149: 365-393.

Mahajan CL. 1967a. *Sisor rhabdophorus,* a study in adaptation and natural relationship, II. The interrelationships of the gas bladder, Weberian apparatus, and membranous labyrinth. J Zool (Lond) 151: 417-432.
Mahajan CL. 1967b. *Sisor rhabdophorus,* a study in adaptation and natural relationship, III. The vertebral column, median fins and their musculature. J Zool (Lond) 152: 297-318.
Mallatt J. 1998. Crossing a major morphological boundary: the origin of jaws in vertebrates. Zoology 100: 128-140.
Mallet J, Willmott K. 2003. Taxonomy: renaissance or tower of babel? Trends Ecol Evol 18: 57-59.
Marques AC, Gnaspini P. 2001. The problems of characters susceptible to parallel evolution in phylogenetic reconstructions: suggestion of a practical method and its application to cave animals. Cladistics 17: 371-381.
Marshall LG, Sempere T, Butler RF. 1997. Chronostratigraphy of the mammal-bearing Paleocene of South America. J S Amer Earth Sci 10: 49-70.
McMurrich JP. 1884a. On the osteology of *Amiurus catus* (L.) Gill. Zool Anz 168: 296-299.
McMurrich JP. 1884b. The myology of *Amiurus catus* (L.) Gill. Proc Can Inst Toronto N Ser 2: 311-351.
Menon AGK. 1951. On certain features in the anatomy of *Horaglanis* Menon. J Zool Soc India 3: 249-253.
Merriman D. 1940. Morphological and embryological studies on two species of marine catfish, *Bagre marinus* and *Galeichthys felis*. Zoologica 25: 221-248.
Mickevich MF, Lipscomb D. 1991. Parsimony and the choice between different transformations for the same character set. Cladistics 7: 111-139.
Miquelarena AM, Arámburu RH. 1983. Osteologia y lepidologia de *Gymnocharacinus bergi* (Pisces Characidae). Limnobios 2: 491-512.
Mivart StG. 1871. On the Genesis of Species. Macmillan. Press, London, UK.
Mo T. 1991. Anatomy, relationships and systematics of the Bagridae (Teleostei: Siluroidei) with a hypothesis of siluroid phylogeny. Theses Zoologicae 17: 1-216.
Monod T. 1963. Sur quelques points de l'anatomie de *Gonorhynchus gonorhynchus* (Linné 1766). Melanges Ichthyol 68: 255-313.
Montoya-Burgos J-I, Muller S, Weber C, Pawlowski J. 1997. Phylogenetic relationships between Hypostominae and Ancistrinae (Siluroidei: Loricariidae): first results from mitochondrial 12S and 16S rRNA gene sequences. Rev Suisse Zool 194: 185-198.
Montoya-Burgos J-I, Muller S, Weber C, Pawlowski J. 1998. Phylogenetic relationships of the Loricariidae (Siluriformes) based on mitochondrial rRNA gene sequences. In: Malabarba LR, Reis RE, Vari RP, Lucena ZM, Lucena CAS (eds.). Phylogeny and Classification of Neotropical Fishes. Edipucrs, Porto Alegre, Brazil, pp. 363-374.
Morell V. 1994. New African Dinosaurs give an old World novel look. Science 266: 219-220.
Munshi JS. 1960. The cranial muscles of some freshwater teleosts. Indian J Zootomy 1: 59-134.
Myers GS. 1938. Freshwater fishes and West Indian zoogeography. Smithsonian Report for 1937: 339-364.
Nawar G. 1954. On the anatomy of *Clarias lazera,* I. Osteology. J Morphol 94: 551-586.
Nawar G. 1955. On the anatomy of *Clarias lazera,* II. The muscles of the head and the pectoral girdle. J Morphol 97: 23-38.
Nelson JS. 1994. Fishes of the World. John Wiley & Sons, New York, NY (3rd ed.).
Newman SA, Muller GB. 2000. Epigenetic mechanisms of character origination. J Exp Zool 288: 304-317.
Ng HH. 2003. Phylogeny and systematics of Bagridae. In: Arratia G, Kapoor BG, Chardon M, Diogo R (eds.). Catfishes. Science Publ. Inc., Enfield, NH (USA) pp. 439-463.
Ng HH, Kottelat M. 1999. *Hyalobagrus,* a new genus of miniature bagrid catfish from Southeast Asia. Ichthyol Explor Freshwaters 9: 335-346.
Nielsen C. 1998. Morphological approaches to phylogeny. Amer Zool 38: 942-952.
Nixon KC. 2002. Winclada. Published by the author, Ithaca, New York.
Oliveira C, Diogo R, Vandewalle P, Chardon M. 2001. Osteology and myology of the cephalic region and pectoral girdle of *Plotosus lineatus,* with comments on Plotosidae (Teleostei: Siluriformes) autapomorphies. J Fish Biol 59: 243-266.

Oliveira C, Diogo R, Vandewalle P, Chardon M. 2002. On the myology of the cephalic region and pectoral girdle of three ariid species, *Arius heudeloti*, *Genidens genidens* and *Bagre marinus*, with a comparison with other catfishes (Teleostei: Siluriformes). Belg J Zool 59: 243-266.

Orti GS. 1997. Radiation of characiform fishes: evidence from mitochondrial and nuclear DNA sequences. In: Kocher TD, Stepien CA (eds.). Molecular Systematics of Fishes. Academic Press, San Diego, USA, pp. 219-243.

Orti G, Meyer A. 1997. The radiation of characiform fishes and the limits of resolution of mitochondrial ribosomal DNA sequences. Syst Biol 46: 75-100.

Osse JWM. 1969. Functional morphology of the head of the perch (*Perca fluviatilis* L.): an electromyographic study. Neth J Zool 19: 289-392.

Otero O. 1997. Paléoichthyofaune de l'Oligo-Miocène de la Plaque Arabique, approches phylogénétique, paléoenvironnementale et paléobiogéographique. Unpubl. PhD thesis. Université Claude Bernard – Lyon 1, France.

Perez-Moreno BP, Chure DJ, Pires C, Marques da Silva C, Dos Santos V, Dantas P, Povoas L, Cachao M, Sanz JL, Galopim de Carvalho AM. 1999. On the presence of Allosaurus fragilis (Theropoda: Carnosauria) in the Upper Jurassic of Portugal: first evidence of an intercontinental dinosaur species. J Geol Soc Lond 156: 449-452.

Perrière C, Perrière FG. 2003. Poisonous catfishes: venom apparatus, acanthotoxins, crinotoxins and other skin secretions. In: Arratia G, Kapoor BG, Chardon M, Diogo R (eds.). Catfishes. Science Publ. Inc., Enfield, NH (USA) pp. 291-314.

Philippe M, Cuny G, Bamford M, Jaillard E, Barale B, Gomez B, Quaja M, Thevenard F, Thiebaur M, Von Sengbusch P. 2003. The palaeoxylological record of *Metapodocarpoxylon libanoticum* (Edwards) Duperon-Laudoueneix et Pons and the Gondwana Late Jurassic-Early Cretaceous continental biogeography. J Biogeogr 30: 389-400.

Poll M. 1942. Note sur l'ostéologie de *Dolichallabes microphthalmus* Poll et remarques sur l'evolution des Clariidae. Ann Soc R Zool Belg 3-4: 222-235.

Pouyaud L, Teugels GG, Gustiano R, Legendre M. 2000. Contribution to the phylogeny of pangasiid catfishes based on allozymes and mitochondrial DNA. J Fish Biol 56: 1509-1538.

Poyato-Ariza FJ. 1996. A revision of the ostariophysan fish family Chanidae, with special reference to the Mesozoic forms. PaleoIchthyologica 6: 1-52.

Pruzsinszky I, Ladich F. 1998. Sound production and reproductive behaviour of the armoured catfish *Corydoras paleatus*. Envir Biol Fishes 53: 183-191.

Rafinesque CS. 1815. Analyse de la nature, ou tableau de l'univers et des corps organises. Jean Barravecchia, Palermo, Italy.

Rajbanshi VK. 1966. A study on the cutaneous sense-organs of *Wallago attu* (Bleeker). Anat Anz Bd 119: 86-93.

Rastogi M. 1963. The head skeleton of *Silonia silonia* (Ham.). Agra Univ J Res 12: 345-362.

Rastogi M. 1964. The head skeleton of Indian schilbeid, *Clupisoma garua* (Ham.). Agra Univ J Res 13: 205-214.

Reed HD. 1924. The morphology and growth of the spines of siluroid fishes. J Morphol 38: 431-451.

Regan CT. 1911a. The classification of the teleostean fishes of the order Ostariophysi: 1. Cyprinoidea. Ann Mag Nat Hist 8: 13-32.

Regan CT. 1911b. The classification of the teleostean fishes of the order Ostariophysi: 2. Siluroidea. Ann Mag Nat Hist 8: 35-57.

Reichel M. 1927. Etude anatomique du *Phreatobius cisternarum* Goeldi, silure aveugle de Brésil. Rev Suisse Zool 34: 285-403.

Reis RE. 1998a. Anatomy and phylogenetic analysis of the Neotropical callichthyid catfishes (Ostariophysi, Siluriformes). Zool J Linn Soc 124: 105-168.

Reis RE. 1998b. Systematics, biogeography, and the fossil record of the Callichthyidae: a review of the available data. In: Malabarba LR, Reis RE, Vari RP, Lucena ZM, Lucena CAS (eds.). Phylogeny and Classification of Neotropical Fishes. Edipucrs, Porto Alegre, Brazil pp. 351-362.

Reznick DN, Shaw FH, Rodd FH, Shaw RG. 1997. Evolution of the rate of evolution in natural populations of guppies (*Poecilia reticulata*). Science 275: 1934-1937.

Ridewood WG. 1904. On the cranial osteology of the Clupeoid fishes. Proc Zool Soc Lond 29: 448-493.

Risch L. 1981. The systematic status of *Gephyroglanis longipinnis* Boulenger 1899, *Chrysichthys magnus* Pellegrin 1922 and *Gephyroglanis gigas* Pellegrin 1922 (Pisces, Bagridae). Rev Zool Afr 95: 508-524.

Risch L. 1987. Description of four new bagrid catfishes from Africa (Siluriformes: Bagridae). Cybium 11: 21-38.

Risch L, Thys van den Audenaerde DFE. 1981. Note on the systematic status of *Gephyroglanis velifer* Thys 1965. Rev Zool Afr 95: 245-251.

Rivas LR. 1986. Comments on Briggs (1984): Freshwater fishes and biogeography of Central America and the Antilles. Syst Zool 35: 633-639.

Roberts TR. 1967. *Rheoglanis dendrophorus* and *Zaireichthys zonatus*, bagrid catfishes from the lower rapids of the Congo river. Ichthyologica 39: 119-131.

Roberts TR. 1969. Osteology and relationships of characoid fishes, particularly the genera *Hepsetus*, *Salminus*, *Hoplias*, *Ctenoluclus* and *Acestrorhynchus*. Proc Calif Acad Sci 36: 91-500.

Roberts TR. 1972. Ecology of fishes in the Amazon and Congo basins. Bull Mus Comp Zool 143: 117-142.

Roberts TR. 1973. Interrelationships of ostariophysans. In: Greenwood PH, Miles RS, Patterson C (eds.). Interrelationship of Fishes. Academic Press, London, UK, pp. 373-395.

Roberts TR. 1975. Geographical distribution of African freshwater fishes. Zool J Linn Soc 57: 249-319.

Roberts TR, Vidthayanon C. 1991. Systematic revision of the Asian catfish family Pangasiidae, with biological observations and description of three new species. Proc Acad Nat Sci (Phil) 143: 97-144.

Romer AS. 1966. Vertebrate Paleontology. University of Chicago Press, Chicago, USA.

Rosen DE, Greenwood PH. 1970. Origin of the Weberian apparatus and the relationships of the ostariophysan and gonorynchiform fishes. Amer Mus Nov 2428: 1-25.

Roy B. 2003. The great American catfish war. Bus Stand New Delhi 01/24/2003: 11-12.

Royero R. 1987. Morfologia de la aleta dorsal en los bagres (Teleostei: Siluriformes), con especial referencia a las familias Americanas. Unpubl. Bachelor's thesis. Universidad Central de Venezuela, Venezuela.

Royero R, Neville AC. 1997. Polarizing analysis of the crimped collagen ligament of the maxillary barbel in *Parauchenipterus galeatus*, and the functional implications. J Fish Biol 50: 157-168.

Saether OA. 1983. The canalized evolutionary potential: inconsistencies in phylogenetic reasoning. Syst Zool 32: 343-349.

Sagemehl M. 1885. Beiträge zur vergleichenden Anatomie der Fische, III - das cranium der characiniden nebst allgemeinen über die mit einem Weber'schen apparat versehenen Physostomenfamilien. Morphol Jahr 10: 1-119.

Saitoh K, Miya M, Inoue JG, Ishiguro NB, Nishida M. 2003. Mitochondrial genomics of ostariophysan fishes: perspectives on phylogeny and biogeography. J Mol Evol 56: 464-472.

Saxena SC, Chandy M. 1966. Adhesive apparatus in certain Indian hill stream fishes. J Zool 148: 315-340.

Schaefer SA. 1984. Mechanical strength of the pectoral spine/girdle complex in *Pterygoplichthys* (Loricariidae: Siluroidei). Copeia 4: 1005-1006.

Schaefer SA. 1987. Osteology of *Hypostomus plecostomus* (Linnaeus), with a phylogenetic analysis of the loricariid subfamilies (Pisces, Siluroidei). Contrib Sci 394: 1-31.

Schaefer SA. 1988. Homology and evolution of the opercular series in the loricarioid catfishes (Pisces: Siluroidei). J Zool (Lond) 241: 81-93.

Schaefer SA. 1990. Anatomy and relationships of the scoloplacid catfishes. Proc Acad Nat Sci (Phil) 142: 167-210.

Schaefer SA. 1991. Phylogenetic analysis of the loricariid subfamily Hypoptopomatinae (Pisces: Siluroidei: Loricariidae), with comments on generic diagnoses and geographic distribution. Zool J Linn Soc 102: 1-41.

Schaefer SA. 1997. The Neotropical cascudinhos: systematics and biogeography of the *Otocinclus* catfishes (Siluriformes: Loricariidae). Proc Acad Nat Sci Phil 148: 1-120.

Schaefer SA. 1998. Conflict and resolution: impact of new taxa on phylogenetic studies of the Neotropical cascudinhos (Siluroidei: Loricariidae). In: Malabarba LR, Reis RE, Vari RP, Lucena ZM, Lucena CAS (eds.). Phylogeny and Classification of Neotropical Fishes. Edipucrs, Porto Alegre, Brazil, pp. 375-400.

Schaefer SA, Rosen DE. 1961. Major adaptive levels in the evolution of the actinopterygian feeding mechanism. Amer Zool 1:187-204.

Schaefer SA, Lauder GV. 1986. Historical transformation of functional design: evolutionary morphology of feeding mechanisms in loricarioid catfishes. Syst Zool 35: 489-508.

Schaefer SA, Lauder GV. 1996. Testing historical hypotheses of morphological change: biomechanical decoupling in loricarioid catfishes. Evolution 50: 1661-1675.

Schaefer SA, Weitzman HS, Britski HA. 1989. Review of the Neotropical catfish genus *Scoloplax* (Pisces: Loricarioidea: Scoloplacidae) with comments on reductive characters in phylogenetic analysis. Proc Acad Nat Sci (Phil) 141: 181-211.

Schmidt EA. 1994. The evolutionary morphology of the pectoral girdle complex in arioid catfishes (Teleostei: Siluriformes). Unpubl. Bachelor's thesis. Smith College, Northampton, MA (USA).

Schneider H. 1961. Neuere ergebnisse der lautforschung bei Fischen. Naturwissenschaften 15: 513-518.

Selander RK, Yang SY, Lewontin RC, Johnson WE. 1970. Genetic variation in the horseshoe crab (*Limulus polyphemus*), a phylogenetic 'relic'. Evolution 24: 402-419.

Sereno PC, Wilson JA, Larsson HCE, Dutheil DB, Sues H-D. 1994. Early Cretaceous Dinosaurs from the Sahara. Science 266: 267-271.

Sereno PC, Dutheil DB, Larochene M, Larsson HCE, Lyon GH, Magwene PM, Sidor CA, Varricchio DJ, Wilson JA. 1996. Predatory Dinosaurs from the Sahara and Late Cretaceous faunal differentiation. Science 272: 986-991.

Sereno PC, Beck AL, Dutheil DB, Gado B, Larsson HCE, Lyon GH, Marcot JD, Rauhut OWM, Sadleir RW, Sidor CA, Varricchio DJ, Wilson GP, Wilson JA. 1998. A long-snouted predatory dinosaur from Africa and the evolution of Spinosaurids. Science 282: 1298-1302.

Shelden FF. 1937. Osteology, myology, and probable evolution of the nematognath pelvic girdle. Ann NY Acad Sci 37: 1-96.

Simonetta AM. 1999. Is parsimony a justified criterion in the assessment of possible phylogenetic reconstructions? Ital J Zool 66: 159-169.

Singh BR. 1967. Movements of barbels in some siluroid fishes. Zool Anz 178: 402-412.

Singh BR, Munshi JS. 1968. The jaw muscles and their mode of working in two siluroid fishes. Zool Anz 181: 356-370.

Skelton PH. 1981. The description and osteology of a new species of *Gephyroglanis* (Siluriformes, Bagridae) from the Olifants River, South West Cape, South Africa. Ann Cape Prov Mus (Nat Hist) 13: 217-249.

Skelton PH, Risch L, de Vos L. 1984. On the generic identity of the *Gephyroglanis* catfishes from Southern Africa (Pisces, Siluroidei, Bagridae). Rev Zool Afr 98: 337-371.

Smith AB. 1998. What does paleontology contribute to systematics in a molecular world? Mol Phyl Evol 9: 437-447.

Smith AG, Briden JC. 1977. Mesozoic and Cenozoic Paleocontinental Maps. Cambridge University Press, Cambridge, UK.

Smith AG, Smith DG, Funnell BM. 1994. Atlas of the Mesozoic and Cenozoic coastlines. Cambridge University Press, Cambridge, UK.

Soares-Porto LM. 1998. Monophyly and interrelationships of the Centromochlinae (Siluriformes: Auchenipteridae). In: Malabarba LR, Reis RE, Vari RP, Lucena ZM, Lucena CAS (eds.). Phylogeny and Classification of Neotropical Fishes. Edipucrs, Porto Alegre, Brazil, pp. 331-350.

Sörensen WE. 1894. Are the extrinsic muscles of the air-bladder in some Siluridae and the "elastic-string" apparatus of others subordinate to the voluntary production of sound? What is the function of the Weberian ossicles? J Anat Physiol Lond 29: 109-552.
Srinivasa Rao K, Lakshmi K. 1984. Head skeleton of the marine catfish *Arius tenuispinis* Day (Osteichthyes: Siluriformes, Ariidae). J Morphol 118: 221-238.
Srinivasachar HR. 1958. Development of the skull in catfishes, Part V. Development of skull in *Heteropneustes fossilis* (Bloch) (Heteropneustidae). Proc Nat Inst Sci India 24B: 165-190.
Srinivasachar HR. 1959. Development of the skull in catfishes, Part III. Development of the chondrocranium in *Heteropneustes fossilis* (Bloch) (Heteropneustidae) and *Clarias batrachus* (Linn) (Clariidae). Morphol Jahr 101: 373-405.
Starks EC. 1926. Bones of the ethmoid region of the fish skull. Stanford Univ Publications 4: 139-338.
Stevens PF. 1980. Evolutionary polarity of character states. Ann Rev Ecol Syst 11: 333-358.
Stiassny MJL, Wiley EO, Johnson GD, de Carvalho MR. In press. In: Donaghue MJ, Cracraft J (eds.). Assembling the Tree of Life. Oxford University Press, New York, USA.
Stix W. 1956. Vergleichende untersuchungen an der trigeminusmuskulatur der Siluridae (Teleostei). Morphol Jb 97: 45-76.
Stucky RK. 1982. Early fossil catfish from Mongolia. Copeia 1982: 465-467.
Surlemont C, Vandewalle P. 1990. Dévelopment postembrionnaire du squelette et de la musculature de la tête de *Clarias gariepinus* (Pisces, Siluriformes) depuis l'éclosion jusqu'à 6.8 mm. Canadian J Zool 69: 1094-1103.
Swidersky DL. 2001. The role of phylogenies in comparative biology: an introduction to the symposium. Amer Zool 41: 485-487.
Sytsma KJ, Gottlieb. 1986. Chloroplast DNA evidence for the origin of the genus *Heterogaura* from a species of *Clarkia* (Onagraceae). Proc Nat Acad Sci USA 83: 5554-5557.
Szalay FS. 1981. Functional analysis and the practice of the phylogenetic method as reflected by some mammalian studies. Amer Zool 21: 37-45.
Takahasi N. 1925. On the homology of the cranial muscles of the cypriniform fishes. J Morphol 40: 1-109.
Taverne L. 1974. Ostéologie d'*Elops* Linné, C., 1766 (Pisces, Elopiformes) et son intérêt phylogénétique. Acad R Belg 41: 1-96.
Taverne L. 1999. Les poissons cretaces de Nardò, 8°—*Sorbininardus apuliensis*, gen. nov., sp. nov. (Teleostei, Ostariophysi, Anatophysi, Sorbininardiformes, nov. ord.). Studi e Richerche sui giacimenti terziari di Bolca VIII, Spec Vol L Sorbiri, Mus Civ St Nat Verona 23: 77-103.
Taverne L, Aloulou-Triki A. 1974. Étude anatomique, myologique et ostéologique du genre *Synodontis* Cuvier (Pisces: Siluriformes: Mochokidae). Ann Mus R Afr Centr 210: 1-69.
Taverne L, De Vos L. 1997. Ostéologie et morphologie d'un bariliné nouveau du bassin de la Malagarasi (système du Lac Tanganyika): *Opsaridium splendens* sp. n. (Teleostei, Cyprinidae). J Afr Zool 111: 281-300.
Tavolga WN. 1962. Mechanisms of sound production in the ariid catfishes *Galeichthys* and *Bagre*. Bull Amer Mus Nat Hist 124: 5-30.
Taylor WR, Van Dyke GC. 1985. Revised procedures for staining and clearing small fishes and other vertebrates for bone and cartilage study. Cybium 9: 107-119.
Teugels GG. 1996. Taxonomy, phylogeny and biogeography of catfishes (Ostariophysi, Siluroidei): an overview. Aquatic Living Resources 9 (Hors série):9-34.
Teugels GG. 2003. State of the art of recent siluriform systematics. In: Arratia G, Kapoor BG, Chardon M, Diogo R (eds.). Catfishes. Science Publ. Inc., Enfield, NH (USA) pp. 317-352.
Teugels GG, Skelton PH, Levèque C 1987. A new species of *Amphilius* (Pisces, Amphiliidae) from the Konkoure Basin, Guinea, West Africa. Cybium 11: 93-101.
Thewissen JGM, Bajpai S. 2001. Whale origins as a poster child for macroevolution. Bioscience 51: 1037-1049.
Thys van den Audenaerde DFE. 1965. Description d'une nouvelle espèce de *Gephyroglanis* (Pisces, Bagridae) de la rivière Sanaga (Cameroun). Rev Zool Bot Afr 71: 264-273.

Tilak R. 1961. The osteocranium and the Weberian apparatus of *Eutropiichthys vacha* (Ham.) and *Eutropiichthys murius* (Ham.): a study of interrelationships. Zool Anz 167: 413-430.

Tilak R. 1963a. Studies on the nematognathine pectoral girdle in relation to taxonomy. Ann & Mag Nat Hist 13: 145-155.

Tilak R. 1963b. The osteocranium and the Weberian apparatus of a few representatives of the families Siluridae and Plotosidae (Siluroidea): a study of interrelationships. Zool Anz 171: 424-440.

Tilak R. 1963c. The osteocranium and the Weberian apparatus of the fishes of the family Sisoridae (Siluroidea): a study in adaptation and taxonomy. Z Wiss Zool 169: 281-320.

Tilak R. 1963d. Relationships between the osteocranium and the Weberian apparatus in two Indian catfishes of the genus *Clarias* (Clariidae). Copeia 1963: 623-629.

Tilak R. 1964. The osteocranium and Weberian apparatus of the family Schilbeidae. Proc Zool Soc (Lond) 143: 1-36.

Tilak R. 1965. The comparative morphology of the osteocranium and the Weberian apparatus of Tachysuridae (Pisces: Siluroidei). J Zool (Lond) 146: 150-174.

Tilak R. 1967a. Studies on the osteology of the Nematognathine girdle in relation to taxonomy. J Zool Soc India 19: 101-110.

Tilak R. 1967b. Studies on the osteocranium and the Weberian apparatus of Indian siluroids in relation to taxonomy. Zool Anz 178: 61-74.

Tilak R. 1971. A study of the osteocranium, the Weberian apparatus and the girdles of *Chaca chaca* (Hamilton): Family Chacidae, Siluroidei. Zool Anz 186: 417-435.

Tilney RL, Hecht T. 1990. The food habits of two co-occurring marine catfish *Galeichthys feliceps* and *G. ater* along the South-east coast of South Africa. J Zool (Lond) 221: 171-193.

Trajano E. 2003. Ecology and ethology of subterranean catfishes. In: Arratia G, Kapoor BG, Chardon M, Diogo R (eds.). Catfishes. Science Publ., Inc., Enfield, NH (USA) pp. 601-635.

Tsigenopoulos CS, Durand JD, Unlu E, Berreri P. 2003. Rapid radiation of the Mediterranean *Luciobarbus* species (Cyprinidae) after the Messinian salinity crisis of the Mediterranean Sea, inferred from mitochondrial phylogenetic analysis. Biol J Linn Soc 80: 207-222.

Unmack PJ. 2001. Biogeography of Australian freshwater fishes. J Biogeogr 28: 1053-1089.

Vaillant ML. 1895. Sur les habitudes terricoles d'un siluroide African, *Clarias lazera*. Bull Mus Hist Nat (Paris) 1985: 271-272.

Vandewalle P. 1971. Comparaison ostéologique et myologique de cinc cichlidae Africains et Sud-Américains. Ann Soc R Zool Belg 4: 259 292.

Vandewalle P. 1975. Contribution à l'étude anatomique et fonctionnelle de la région céphalique de *Gobio gobio* (Pisces, Cyprinidae), 3: les os, les muscles et les ligaments. Forma et Functio 8:331-360.

Vandewalle P. 1977. Particularités anatomiques de la tête de deux Poissons Cyprinidés *Barbus barbus* (L.) et *Leuciscus leuciscus* (L). Bull Acad R Belg 5: 469-479.

Vandewalle P, Brunin P, Chardon M. 1986. Functional approach to the morphology of the buccal region of *Ctenilоricaria platystoma* (Günther) (Pisces, Ostariophysi, Loricariidae) with respect to a peculiar respiration. Zool Anz 217: 363-373.

Vandewalle P, Surlemont C, Chardon M. 1993. About the early larval development of the anterior suspensorial ossifications of *Clarias gariepinus* (Burchell, 1822). Zool Anz 231: 11-19.

Vandewalle P, Laleyé P, Focant B. 1995a. Early development of cephalic bony elements in *Chrysichthys auratus*. Belg J Zool 125: 329-347.

Vandewalle P, Saintin P, Chardon M. 1995b. Structures and movements of the buccal and pharyngeal jaws in relation to feeding in *Diplodus sargus*. J Fish Biol 46: 623-656.

Vandewalle P, Surlemont C, Sanna P, Chardon M. 1985. Intreprétation fonctionnelle de modifications du splanchnocrâne pendant le développement post-embryonaire de *Clarias gariepinus* (Téléostéens, Siluriformes). Zool Jahr Anat 113: 91-100.

Vandewalle P, Gluckmann I, Baras E, Huriaux F, Focant B. 1997. Postembryonic development of the cephalic region in *Heterobranchus longifilis*. J Fish Biol 50: 227-253.

Vari RP. 1979. Anatomy, relationships and classification of the families Citharinidae and Distichodontidae (Pisces, Characoidea). Bull Br Mus Nat Hist (Zool) 36: 261-344.

Vari RP. 1988. The Curimatidae, a lowland Neotropical fish family (Pisces: Characiformes): distribution, endemism, and phylogenetic biogeography. In: Vanzolini P, Heyer RH (eds.). Proc workshop Neotropical distribution patterns. Academia Brasileira de Ciencias, Rio de Janeiro, pp. 343-377.

Vences M, Freyhot J, Sonnenberg R, Kosuch J, Veith M. 2001. Reconciling fossils and molecules: Cenozoic divergence of cichlid fishes and the biogeography of Madagascar. J Biogeogr 28: 1091-1099.

Vermeij G. 1974. Adaptation, versatility and evolution. Syst Zool 1: 466-477.

Verraes W. 1977. Postembryonic ontogeny and functional anatomy of the ligamentum mandibulo-hyoideu and the ligamentum interoperculo-mandibulare, with notes on the opercular bones and some other cranial elements in *Salmo gairdneri* Richardson, 1836 (Teleostei: Salmonidae). J Morphol 151: 111-119.

Vetter B. 1878. Untersuchungen zur vergleichenden anatomie der kiemen- und kiefer-musculatur der Fische, II. Thiel. Jena Z Naturw 12: 431-550.

Vidthayanon C. 1992. Taxonomic revision of the catfish family Pangasiidae. Unpubl. PhD thesis. Tokyo University of Fisheries, Tokyo.

Wagner WH. 1984. Applications of the concepts of groundplan-divergence. In: Duncan T, Stuessy T (eds.). Cladistics: Perspectives on the Reconstruction of Evolutionary History. Proc workshop Neotropical distribution patterns. Columbia University Press, Columbia, NY, pp. 95-118.

Wainwright PC, Turinfan RG. 1993. Coupled versus uncoupled functional systems: motor plasticity in the queen triggerfish *Balistes vetula*. J Exp Biol 180: 209-227.

Walsh SJ. 1990. A systematic revision of the Neotropical catfish family Ageneiosidae (Teleostei, Ostariophysi, Siluriformes). Unpubl. PhD thesis. University of Florida, Gainesville, FL.

Weishampel DB. 1990. Dinosaur distribution. In: Weishampel DB, Dobson P, Osmolska H (eds.). The Dinosauria. University of California Press, Berkeley, USA, pp. 63-139.

Weitzman SH. 1962. The osteology of *Brycon meeKi*, a generalised characid fish, with an osteological definition of the family. Stanford Ichthyol Bull 8: 1-77.

Weitzman SH. 1964. Osteology of and relationships of South American characid fishes of subfamilies Lebiasininae and Erythrinae with special reference to subtribe Nannostomina. Proc US Natn Mus 116: 127-170.

Weitzman SH, Weitzman M. 1982. Biogeography and evolutionary diversification in Neotropical freshwater fishes, with comments on the refuge theory. In: Prance GT (ed.). Biological Diversification in the Tropics. New York: Columbia University Press, Columbia, NY, pp. 403-422.

Wiens JJ. 2001. Character analysis in morphological phylogenetics: problems and solutions. Syst Biol 50:689-699.

Wilkinson M. 1991. Homoplasy and parsimony analysis. Syst Zool 40: 105-109.

Winemiler KO. 1987. Feeding and reproductive biology of the currito, *Hoplosternum littorale*, in the Venezuelan Llanos with comments on the possible function of the enlarged male pectoral spines. Envir Biol Fishes 20: 219-227.

Winterbottom R. 1974a. A descriptive synonymy of the striated muscles of the Teleostei. Proc Acad Nat Sci (Phil) 125: 225-317.

Winterbottom R. 1974b. The familial phylogeny of the Tetraodontiformes (Acanthopterygii: Pisces) as evidenced by their comparative myology. Smiths Contrib Zool 485: 1-78.

Winterbottom R. 1993. Myological evidence for the phylogeny of recent genera of surgeonfishes (Percomorpha, Acanthuridae), with comments on Acanthuroidei. Copeia 1993: 21-39.

Winterbottom R, McLennan DA. 1993. Cladogram versatility: evolution and biogeography of acanthuroid fishes. Evolution 47: 1557-1571.

Wright RR. 1984. On the skin and cutaneous sense organs of *Amiurus*. Proc Can Inst 2: 251-269.

Yabe M. 1985. Comparative osteology and myology of the superfamily Cottoidea (Pisces: Scorpaeniformes), and its phylogenetic classification. Mem Faculty Fisheries Hokkaido Univ 32: 1-130.

Zimmer C. 1998. At the Water's Edge: Fish with Fingers, Whales with Legs, and How Life Came Ashore but Then Went Back to Sea. Touchstone, New York, NY.

List of Abbreviations

addi-m	additional muscle
af-apal	articulatory facet for os autopalatinum
af-chp	articulatory facet for os ceratohyale posterior
af-cra	articulatory facet for complex radial
af-hmmp	articulatory facet for os hyomandibulo-metapterygoideum
af-hmmp-1	articulatory facet for os hyomandibulo-metapterygoideum 1
af-hmmp-2	articulatory facet for os hyomandibulo-metapterygoideum 2
af-neu	articulatory facet for neurocranium
af-neu-b	additional articulatory facet for neurocranium
af-op	articulatory facet for os operculare
af-pecra1	articulatory facet for pectoral ray 1
af-pecsp	articulatory facet for pectoral spine
af-qsym	articulatory facet for os quadrato-symplecticum
af-scacor	articulatory facet for os scapulo-coracoideum
af-I	articulatory facet between os hyomandibulare and neurocranium
af-II	articulatory facet between os autopalatinum and neurocranium
afo	anterior fontanel
anlp	anterior nuchal plate
apdc	anterior process of dorsal condyle of pectoral spine
atlp	acessory tooth plate
bt	bony tunnel
c-apal-a	cartilago autopalatinus anterior
c-apal-p	cartilago autopalatinus posterior
c-eth	cartilago ethmoideum
c-ex-mnd-b	cartilage of external mandibular barbel
c-ex-mnd-b-mp	cartilage of external mandibular barbel: moving part
c-ex-mnd-b-sp	cartilage of external mandibular barbel: supporting part

c-in-mnd-b	cartilage of internal mandibular barbel
c-l-pri	cartilago ligamentum primordialis
c-Meck	cartilago Meckeli
c-Meck-as	ascending portion of cartilago Meckeli
c-Meck-ho	horizontal portion of cartilago Meckeli
c-mnd-b	cartilage of mandibular barbel
cc	complex centrum
cop	coronoid process of mandible
cor-bri	coracoid bridge
cor-bri-pp	posterior process of coracoid bridge
cor-bri-pvp	posteroventral process of coracoid bridge
cp-mnd-b	cartilaginous plate carrying the mandibular barbels
dc	dorsal crest
dcon	dorsal concavity
drm	drumming muscle
ect-te	ectopterygoid teeth
ex-mnd-b	external mandibular barbel
fo-post	fossa posttemporalis
for-V-VII	foramen trigemino-facialis
in-mnd-b	internal mandibular barbel
isut	imcomplete suture
l-ang-ch	ligamentum angulo-ceratohyale
l-ang-iop	ligamentum angulo-interoperculare
l-ch-ih	ligamentum ceratohyalo-interhyale
l-ch-iop	ligamentum ceratohyalo-interoperculare
l-ent-neu	ligamentum entopterygoideo-neurocranium
l-ent-pvm	ligamentum entopterygoideo-praevomerale
l-entect-apal	ligamentum entoectopterygoideo-autopalatinum
l-entect-mx	ligamentum entoectopterygoideo-maxillare
l-entect-osph-leth	ligamentum entoectopterygoideo-orbitosphenoido-lateralethmoideum
l-entect-prmx	ligamentum entoectopterygoideo-praemaxillare
l-entect-pvm	ligamentum entoectopterygoideo-praevomerale
l-hp-pp5	ligamentum humero-vertebrale
l-meth-apal	ligamentum mesethmoideo-autopalatinum
l-meth-prmx	ligamentum mesethmoideo-praemaxillare
l-mp-ent	ligamentum metapterygoideo-entopterygoideum
l-mp-prmx-pvm	ligamentum metapterygoideo-praemaxillo-praevomerale
l-mx-mnd	ligamentum maxillo-mandibulare
l-pri	ligamentum primordialis
l-pri-1	ligamentum primordialis 1
l-pri-2	ligamentum primordialis 2

l-prmx-apal	ligamentum praemaxillo-autopalatinum
l-prmx-mx	ligamentum praemaxillo-maxillare
l-puh-hh	ligamentum parurohyalo-hypohyale
l-qsym-prmx	ligamentum quadratosymplectico-praemaxillare
l-sca-pecsp	ligamentum between os scapulo-coracoideum and pectoral spine
m-1-mnd-b	musculus 1 of the mandibular barbels
m-2-mnd-b	musculus 2 of the mandibular barbels
m-3-mnd-b	musculus 3 of the mandibular barbels
m-4-mnd-b	musculus 4 of the mandibular barbels
m-6-mnd-b	musculus 6 of the mandibular barbels
m-A0	musculus adductor mandibulae A0
m-A1	musculus adductor mandibulae A1
m-A1-OST	musculus adductor mandibulae A1-OST
m-A1-OST-1	musculus adductor mandibulae A1-OST: part 1
m-A1-OST-2	musculus adductor mandibulae A1-OST: part 2
m-A1-OST-3	musculus adductor mandibulae A1-OST: part 3
m-A1-OST-4	musculus adductor mandibulae A1-OST: part 4
m-A1-OST-5	musculus adductor mandibulae A1-OST: part 5
m-A2	musculus adductor mandibulae A2
m-A3′	musculus adductor mandibulae A3′
m-A3′-d	musculus adductor mandibulae A3′ pars dorsalis
m-A3′-d-1	musculus adductor mandibulae A3′ pars dorsalis: part 1
m-A3′-d-2	musculus adductor mandibulae A3′ pars dorsalis: part 2
m-A3′-v	musculus adductor mandibulae A3′ pars ventralis
m-A3′-v-1	musculus adductor mandibulae A3′ pars ventralis: part 1
m-A3′-v-2	musculus adductor mandibulae A3′ pars ventralis: part 2
m-A3″	musculus adductor mandibulae A3″
m-A3″-1	musculus adductor mandibulae A3″: part 1
m-A3″-2	musculus adductor mandibulae A3″: part 2
m-Aω	musculus adductor mandibulae Aω
m-ab-pro	musculus abductor profundus
m-ab-pro-1	musculus abductor profundus: part 1
m-ab-pro-2	musculus abductor profundus: part 2
m-ab-sup	musculus abductor superficialis
m-ab-sup-1	musculus abductor superficialis: part 1
m-ab-sup-2	musculus abductor superficialis: part 2
m-ad-ap	musculus adductor arcus palatini
m-ad-hm	musculus adductor hyomandibularis
m-ad-mnd	musculus adductor mandibulae
m-ad-op	musculus adductor operculi
m-ad-sup	musculus adductor superficialis
m-ad-sup-1	musculus adductor superficialis: part 1
m-ad-sup-2	musculus adductor superficialis: part 2

m-arr-d	musculus arrector dorsalis
m-arr-d-dd	musculus arrector dorsalis: dorsal division
m-arr-d-vd	musculus arrector dorsalis: ventral division
m-arr-v	musculus arrector ventralis
m-dil-op	musculus dilatator operculi
m-dil-op-1	musculus dilatator operculi: part 1
m-dil-op-2	musculus dilatator operculi: part 2
m-dp-in-mnd-t	musculus depressor internus mandibularis tentaculi
m-ep	musculus epaxialis
m-ex-t	musculus extensor tentaculi
m-ex-t-1	musculus extensor tentaculi: part 1
m-ex-t-2	musculus extensor tentaculi: part 2
m-ex-t-3	musculus extensor tentaculi: part 3
m-ex-t-4	musculus extensor tentaculi: part 4
m-hh-ab	musculus hyohyoideus abductor
m-hh-ad	musculus hyohyoideus adductor
m-hh-inf	musculus hyohyoideus inferior
m-hyp	musculus hypaxialis
m-intm	musculus intermandibularis
m-intt	musculus intertentacularis
m-l-ap	musculus levator arcus palatini
m-l-ap-1	musculus levator arcus palatini: part 1
m-l-ap-2	musculus levator arcus palatini: part 2
m-l-op	musculus levator operculi
m-pr-ex-mnd-t	musculus protractor externus mandibularis tentaculi
m-pr-h	musculus protractor hyoideus
m-pr-h-d	musculus protractor hyoideus pars dorsalis
m-pr-h-l	musculus protractor hyoideus pars lateralis
m-pr-h-v	musculus protractor hyoideus pars ventralis
m-pr-mup	musculus protractor of müllerian process
m-pr-mup-1	musculus protractor of müllerian process: part 1
m-pr-mup-2	musculus protractor of müllerian process: part 2
m-pr-pec	musculus protractor pectoralis
m-pr-post	musculus protractor posttemporalis
m-re-ex-mnd-t	musculus retractor externus mandibularis tentaculi
m-re-ex-mnd-t-mp	musculus retractor externus mandibularis tentaculi: moving part
m-re-ex-mnd-t-sp	musculus retractor externus mandibularis tentaculi: supporting part
m-re-in-mnd-t	musculus retractor internus mandibularis tentaculi
m-re-t	musculus retractor tentaculi
m-sh	musculus sternohyoideus
mcor-ar	mesocoracoid arch
mnd	mandible
mnd-b	mandibulary barbel

mnpl	median nuchal plate
mp-te	metapterygoid teeth
mup	müllerian process
mx-b	maxillary barbel
nlp	nuchal plate
ns-b	nasal barbel
o-ang-art	os angulo-articulare
o-ang-art-mc	mesial crest of os angulo-articulare
o-ang-art-pmp	posteromesial process of os angulo-articulare
o-apal	os autopalatinum
o-apal-dc	dorsal crest of os autopalatinum
o-boc	os basioccipitale
o-ch-a	os ceratohyale anterior
o-ch-a-vlc	ventrolateral crest of os ceratohyale anterior
o-ch-p	os ceratohyale posterior
o-cl	os cleithrum
o-cl-adp	anterodorsal process of os cleithrum
o-cl-alp	anterolateral process of os cleithrum
o-cl-amp	anteromedial process of os cleithrum
o-cl-dp	dorsal process of os cleithrum
o-cl-dp-1	dorsal process 1 of os cleithrum
o-cl-dp-2	dorsal process 2 of os cleithrum
o-cl-hp	humeral process of os cleithrum
o-cl-mg	medial groove of os cleithrum
o-cl-pdsp	posterodorsal spine of os cleithrum
o-com	os coronomeckelium
o-den	os dentale
o-den-avp	anteroventral process of os dentale
o-den-dl	dorsal lamina of os dentale
o-den-pvp	posteroventral process of os dentale
o-ect	os ectopterygoideum
o-ent	os entopterygoideum
o-ent-ect	os entopterygoide-ectopterygoideum
o-epoc	os epioccipitale
o-epoc-pdp	posterodorsal process of os epioccipitale
o-exoc	os exoccipitale
o-exs	os extrascapulare
o-fr	os frontale
o-fr-dmp	dorsomesial process of os frontale
o-hh	os hypohyale
o-hh-d	os hypohyale dorsale
o-hh-v	os hypohyale ventrale
o-hm	os hyomandibulare
o-hm-mp	os hyomandibulo-metapterygoideum

o-hm-mp-sp	hyomandibulo-metapterygoid spine
o-ih	os interhyale
o-io-4	os infraorbitale 4
o-iop	os interoperculare
o-iop-mf	mesial foramen of os interoperculare
o-leth	os lateroethmoideum
o-leth-mf	mesial foramen of os lateroethmoideum
o-meth	os mesethmoideum
o-mp	os metapterygoideum
o-mx	os maxillare
o-mx-pl	posterior lamina of os maxillare
o-ns	os nasale
o-op	os operculare
o-osph	os orbitosphenoideum
o-pa-soc	os parieto-supraoccipitale
o-pa-soc-ll	lateral lamina of os parieto-supraoccipitale
o-pa-soc-pp	posterior process of os parieto-supraoccipitale
o-para	os parasphenoideum
o-pop	os praeoperculare
o-pop-ll	lateral lamina of os praeoperculare
o-post-scl	os posttemporo-supracleithrum
o-post-scl-alp	anterolateral process of os posttemporo-supracleithrum
o-prmx	os praemaxillare
o-prmx-dlp	dorsolateral process of os praemaxillare
o-prot	os prooticum
o-psph	os pterosphenoideum
o-psph-alp	anterolateral process of the os pterosphenoideum
o-pt	os pteroticum
o-pt-plp	posterolateral process of the os pteroticum
o-puh	os parurohyale
o-pvm	os praevomerale
o-q	os quadratum
o-q-sym	os quadrato-symplecticum
o-q-sym-vmf	ventromesial fossa of os quadrato-symplecticum
o-sca-cor	os scapulo-coracoideum
o-sca-cor-plfr	posterolateral foramen of os scapulo-coracoideum
o-sca-cor-pp	posterior process of os scapulo-coracoideum
o-sca-cor-pvmp	posteroventromesial process of os scapulo-coracoideum
o-sca-cor-vlg	ventrolateral groove of os scapulo-coracoideum
o-ses-1	os sesamoid 1 of suspensorium
o-ses-2	os sesamoid 2 of suspensorium
o-ses-3	os sesamoid 3 of suspensorium
o-sph	os sphenoticum
o-sph-adp	anterodorsal process of os sphenoticum
o-spop	os suprapraeoperculare

o-sym	os symplecticum
o-tri	os tripus
pap	papillae
pec-ra	pectoral rays
pec-ra-1	pectoral ray 1
pec-ra-1-ac+dc	anterior condyle + dorsal condyle of pectoral ray 1
pec-ra-1-vc	ventral condyle of pectoral ray 1
pec-sp	pectoral spine
pec-sp-ac	anterior condyle of pectoral spine
pec-sp-amp	anteromesial process of pectoral spine
pec-sp-arp	arch-like process of pectoral spine
pec-sp-dc	dorsal condyle of pectoral spine
pec-sp-dmp	dorsomesial process of pectoral spine
pec-sp-vc	ventral condyle of pectoral spine
pfo	posterior fontanel
pp4	parapophysis 4
pp5	parapophysis 5
pra	proximal radial
pvm-tlp	prevomeral tooth-plate
r-br-II,V,VI,VII,VIII	radius branchiostegus II, V, VI, VII, VIII
rmmnd-lb	lateral branch of ramus mandibularis
rmmnd-mb	mesial branch of ramus mandibularis
sb	swim bladder
stf	supratemporal fossa
t-m-ex-t	tendon of musculus extensor tentaculi
v1	vertebra 1
v5	vertebra 5
vf	ventral fossa
Wa-dl	dorsal lamina of Weberian apparatus

Index